THE SOLAR SPECTRUM

AF860466

ASTROPHYSICS AND SPACE SCIENCE LIBRARY

A SERIES OF BOOKS ON THE RECENT DEVELOPMENTS
OF SPACE SCIENCE AND OF GENERAL GEOPHYSICS AND ASTROPHYSICS
PUBLISHED IN CONNECTION WITH THE JOURNAL
SPACE SCIENCE REVIEWS

Editorial Board

J. E. BLAMONT, *Laboratoire d'Aeronomie, Verrieres, France*
R. L. F. BOYD, *University College, London, England*
L. GOLDBERG, *Harvard College Observatory, Cambridge, Mass., USA*
C. DE JAGER, *University of Utrecht, Holland*
G. H. LUDWIG, *NASA, Goddard Space Flight Center, Greenbelt, Mld., USA*
R. LÜST, *Institut für Extraterrestrische Physik, Garsching-München, Germany*
H. E. NEWELL, *NASA, Washington, D.C., USA*
L. I. SEDOV, *Academy of Sciences of the USSR, Moscow, USSR*

Secretary of the Editorial Board

W. DE GRAAFF, *Sterrewacht 'Sonnenborgh', University of Utrecht, Utrecht, Holland*

PROFESSOR M. G. J. MINNAERT

THE SOLAR SPECTRUM

Proceedings of the Symposium held at the University of Utrecht 26-31 August 1963

edited by

C. DE JAGER

Springer-Science+Business Media, B.V.

ISBN 978-94-017-6370-7 ISBN 978-94-017-6376-9 (eBook)
DOI 10.1007/978-94-017-6376-9

1965

All rights reserved
No part of this book may be reproduced in any form by print, photoprint, microfilm, or any other means without permission from the publisher

Dedicated by the authors to
Professor M. G. J. Minnaert
on the occasion of his seventieth birthday

SOLAR SPECTRUM SYMPOSIUM
AUGUST '63.

PARTICIPANTS

1. H. Bredohl, Liège
2. T. A. Chubb, Washington
3. W. M. Glencross, London
3a. R. Simon, Liège
4. K. A. Pounds, Leicester
5. M. Hotinli, Istanbul
6. Plantevin, Paris
7. L. Neven, Brussel
8. D. J. Faulkner, Canberra
9. E. A. Müller, Genève
10. Y. Cuny, Meudon
11. N. Grevesse, Liège
12. G. Wlérick, Meudon
13. Mrs. Edlén, Lund
14. R. Boyer, Meudon
15. S. Dumont, Meudon
16. A. Unsöld, Kiel
17. M. G. J. Minnaert, Utrecht
18. H. H. Voigt, Göttingen
19. R. Michard, Meudon
20. C. Guillaume, Liège
21. B. Edlén, Lund
22. E. H. Schröter, Freiburg
23.
24. Miss Mulders, Ede
25. T. de Groot, Utrecht
26. L. Deszö, Debrecen
27. K. Hunger, Hannover
28. D. H. Menzel, Cambridge
28a. A. D. Fokker, Utrecht
29. M. de Groot, Utrecht
30. O. Namba, Kiel
31. G. F. W. Mulders, Washington
32. K. H. Böhm, Kiel
33. H. E. Hinteregger, Bedford
34. R. W. Noyes, Cambridge
35. J. C. Brandt, Tucson
36. E. Böhm-Vitense, Kiel
37. J. Waddell, Tucson
38. A. B. Severny, Nauchny
39. H. Nieuwenhuijzen, Utrecht
40. J. D. Purcell, Washington
41. P. Jarzebowski, Wrocław
42. R. N. Thomas, Boulder
43. H. Zirin, Boulder
44. J. N. van Gils, Utrecht
45. Mrs. Crouwel, Utrecht
46. J. B. Zirker, Sunspot
47. J. R. W. Heintze, Utrecht
48. Miss Dorresteyn, Utrecht
49. W. Unno, Tokyo
50. R. Howard, Pasadena
51. E. H. Avrett, Cambridge
52. H. van Regemorter, Meudon
53. H. V. A. M. Maseland, Utrecht
54. C. Zwaan, Utrecht
55. D. W. N. Stibbs, St. Andrews
56. A. Koeckelenberg, Brussel
57. Ch. Moore-Sitterly, Washington
58. W. C. Livingston, Tucson
59. R. H. Garstang, London
60. R. Lüst, München
61. M. Gonze, Brussel,
62. R. G. Athay, Boulder
63. V. Gaizauskas, Ottawa
64. E. van den Heuvel, Utrecht
65. A. Ch. Brinkman, Utrecht
66. M. Bretz, Marseille
67. D. E. Billings, Boulder
68. G. Godoli, Firenze
69. M. R. Kundu, Ithaca
70. J. T. Jefferies, Boulder
71. H. Hubenet, Utrecht
72. F. Q. Orrall, Sunspot
73. G. E. Moreton, Burbank
74. Cl. van 't Veer, Paris
75. S. R. Pottasch, Groningen
76. J. van Nieuwkoop, Utrecht
77. K. O. Kiepenheuer, Freiburg im Breisgau
78. M. Kuperus, Utrecht
79. F. van 't Veer, Paris
80. A. Schadee, Utrecht
81. M. V. Migeotte, Liège
82. L. D. de Feiter, Utrecht
83. J. van Diggelen, Utrecht
84. Su-Shu Huang, Greenbelt
85. L. Houziaux, Liège
86. E. Schatzman, Paris
87. Mrs. Houziaux, Liège
88. R. O. Redman, Cambridge
89. A. K. Pierce, Tucson
90. A. B. Underhill, Utrecht
91. Mrs. de Bruyn, Utrecht
92. J. W. Evans, Sunspot
93. J.-C. Pecker, Meudon
94. H. G. van Bueren, Eindhoven
95. F. Roddier, Paris
96. M. Rigutti, Firenze
97. M. J. Seaton, London
98. J. Houtgast, Utrecht
99. W. de Graaff, Utrecht
100. C. de Jager, Utrecht

INTRODUCTION

A good deal of our information on solar physics and on solar phenomena is derived from the solar spectrum. A quantitative interpretation of this spectrum was only possible after 1920, after the establishment of Bohr's atomic model, the discovery of Saha's law, and the development of spectrophotometry.

The resolving and light gathering powers of our instruments have greatly increased since. We have seen an enormous progress in our theoretical understanding of basic atomic phenomena, and of the intricate problems concerned with the transfer of energy through a complicated structure like the sun's outer layers. In particular the observable part of the solar spectrum tremendously enlarged since the introduction, in the years after 1945, of *radio-astronomy*, enabling us to study the solar spectrum between wavelengths of some mm to about 15 m, of *space research*, giving access to the whole electromagnetic spectrum below 3000 Å, down to about 0.01 Å. Further, the low and high energetic components of the solar particles spectrum have been discovered with space probes (the solar wind), rockets, balloons (the so-called sub-cosmic-ray particles) and cosmic ray monitors (solar cosmic ray bursts). The extreme wealth of this spectrum, much vaster in extent than the earlier investigators could only dream of, is an important source of information.

It looked appropriate to us, after the rapid development of this branch of science, to invite the world's leading solar physicists to Utrecht for a summarizing symposium on the whole solar spectrum.

The time of this symposium has been chosen so as to coincide with the retirement from his University duties of Professor M. Minnaert who has been in the first ranks of solar physicists since the 1920's, and to whom this book is dedicated by the participants in the symposium.

This symposium could not have been possible without the help of many persons and institutes. It is with gratefulness that I acknowledge

– the generous financial assistance obtained from the Netherlands Ministry of Education, Arts and Sciences;

– the assistance obtained from the section "Organization of congresses" of the Ministry of Education, Arts and Sciences, and in particular from Mr. C. W. J. Rosenberg, in organizing the excursion;

– the invitation of the Mayor of Utrecht to receive the participants and their ladies in the City-Hall;

– the willingness of the speakers and participants to submit their communications during and shortly after the symposium, thus enabling a rapid publication;

– the cooperation of D. Reidel Publishing Company in having the Proceedings well printed;

– the helpful cooperation of all members of staff and personnel of the Observatory in making things run smoothly.

C. DE JAGER

TABLE OF CONTENTS

SESSION III

THE ACTIVE PHOTOSPHERE AND CHROMOSPHERE SPOTS, PLAGES, FLARES AND PROMINENCES

SESSION IV

THE CORONA

SESSION V

PARTICLE AND RADIO EMISSION FROM THE SUN

OPENING ADDRESS

Rector Magnificus, dear colleagues and friends, ladies and gentlemen. This symposium on "The Solar Spectrum" has been arranged in Utrecht, the world-center of spectrophotometry, line intensities and solar physics, in order to celebrate the seventieth birthday of Professor M. G. J. MINNAERT. I am glad to use the opportunity of this brief introduction for expressing – as many of us have done already on February 12 – our heartfelt congratulations and our very best wishes for good health, for continued and successful scientific activity and especially for much joy in life and work. Our cordial wishes go just the same with Mrs. Minnaert and the whole family. Their beautiful and original home surely deserves three stars in every astronomers Baedeker!

I think, that among this groups of experts in solar physics I hardly need to say much about Minnaert's scientific merits, about his work on the photometry and the interpretation of solar Fraunhofer lines, about center-limb-variation of lines and continuum, about the flash spectrum and about the moon.

But perhaps not all the younger astrophysicists realize the basis from which Minnaert had to start his scientific work, i.e. the state of solar physics about 1921 when he became "observator" at the Heliophysical Observatory in Utrecht. That was rather modest equipment which Prof. W. H. Julius had just put up in connection with the physics department. And the idea behind it was that many of the solar phenomena known at that time were due to anomalous refraction, i.e. the deflection of rays, near the Fraunhofer lines. Almost the only astrophysicist who before had dared to think quantitatively about the sun in terms of straightforward physics, had been Karl Schwarzschild, and his work – after his untimely death – was almost completely forgotten even in Göttingen and Potsdam, where it had been done.

But let me talk not only about astrophysics! You will agree that few scientists are as manysided and as openminded as our Minnaert. He began – as you know – his studies in biology and obtained his Doctor's degree in it. At international conferences we have admired his versatility in many foreign languages, I hope that his example might encourage even some of our American friends in that direction! But that's not all: From what I have seen, I am fully convinced that Minnaert also understands the language of his charming cats!

The basis of all that and of much more is just his deep love for all the living beings of all kinds. How much has he done for international cooperation in astronomy and science in general and – on a much wider basis – for peace on this earth! The very, very bad experiences which he had made in two world-wars and which might have been sufficient for justifying any degree of mistrust against human reason and moral qualities did not deter him from the belief that it is still useful to work for the case of reason, love and peace again and again.

C. de Jager (ed.), The Solar Spectrum. All rights reserved.

But it would be completely wrong to classify Minnaert as a stern moralist. His basis is just a very fine feeling – I probably should say – for everything. And especially so also in the realm of art. He is an expert musician and even has written an interesting paper on the mechanics of the violin. During his many travels he has made many books full of most charming drawings, mostly done in colored chalk.

May I express our cordial hope, that our friend Minnaert may enjoy in good health for many years to come his scientific, human and art-activities with the old academic wish: "Ad multos annos"!

University of Kiel A. Unsöld

FOURTY YEARS OF SOLAR SPECTROSCOPY

M. MINNAERT
Sterrewacht 'Sonnenborgh', University of Utrecht, Holland

The subject of this lecture is evidently expected to correspond more or less with the subject of this symposium. It seems almost impossible to survey in one hour or so the extremely wide field of solar spectroscopy and to do full justice to its bewildering riches. Fortunately I am speaking for specialists; each name which I may quote evokes directly for you the memory of one of the pioniers of science, with his discoveries, his errors, his contentions, his victories. For me perhaps even more than for you, the story of solar spectroscopy is at the same time the story of my life. I started solar work in the year when Saha established his equation; I married while we made the first photometry of the flash spectrum; our atlas was published at the outbreak of worldwar II. How could I transfer to you the personal associations which for me are connected with the progressive evolution of our science? Still I understand that you wish to hear at least some particular references to my own work, and I shall be happy to include these here and there.

Permit me, therefore, to limit this historical survey to the period of which I myself have been a witness, this means, to about the last 40 years. Of course the story which I intend to tell you will be very incomplete; still worse: it will be only a personal look on the development of solar spectroscopy, which at many places might not be well-balanced. Forgive me all this: it is not easy to summarize in one hour the whole book of Unsöld, plus all the subsequent work! Much of what I will tell you is well known to the elder people among you, but a bird's eye view may have its charm. For the younger colleagues it might be of some interest, to hear how we had to toil and to err, before those simple conclusions could be drawn, which are now accepted as obvious.

1. The Photosphere

I should like to start by discussing the spectrum of the *photosphere*, in the region say between 3000 and 8000 Å, which is still the subject of a classical astrophysical discipline, occupying a central position in solar research. Early in the twenties the Fraunhofer lines had been studied mainly from the point of view of wavelength determination. Their strength had been estimated by ROWLAND on a qualitative scale, most useful for identification, but without theoretical significance. The only intensity profile then known was that of the H and K lines, measured in 1914 by SCHWARZSCHILD, a forerunner in this field as in so many others. But now theoretical physics had progressed just on such points,

C. de Jager (ed.), The Solar Spectrum. *All rights reserved.*

that a measurement of line intensities in the laboratory and in celestial objects had become imperative: PLANCK's law (1900), EINSTEIN's absorption and emission argument (1917), SAHA's formula (1920), LADENBURG's application of quantum theory to dispersion phenomena (1921) had arosen the interest in the *numbers* of quanta emitted or absolved.

At that time I was working at the Physical Laboratory where MOLL had devised his microphotometer (1921), and where this was quickly applied to the measurement of laboratory emission lines. It was obvious that similar measurements had to be made on wide Fraunhofer lines, and I remember our delight, when the first record of the K line was obtained and when all the minute incurvations appeared to correspond to Fraunhofer lines (1923). Almost at the same time VAN CITTERT and I independently devised the method of step weakeners, in order to calibrate our plates. We were still struggling with the theoretical interpretation and the elimination of photographic sources of error of all kinds, when the first similar results were published at Potsdam (1927). This marked the beginning of a long and very stimulating interaction between the work of our German collegues and the Utrecht work leading at the same time to a cordial friendship. The development, from the observational point of view, went quickly ahead. In 1936 MULDERS went to the Mt. Wilson Observatory and took the plates for our Photometric Atlas, while HOUTGAST developed the modified, home made instrument, which could be added to the microphotometer and gave direct intensity recordings. All microphotometer curves were obtained by photographic recording; we worked mostly in the night, because then the microphotometer was free. You were alone in the building, and in the silence of the darkroom, in the dull red light, you were developing your record. There it emerged, slowly emerged, out of nothingness, and as if by magic there appeared on the paper the profile of the cyanogen band, or of the atmospheric oxygen lines, never earlier observed in their true quantitative shape. The Atlas was ready, just before the second world war broke out; and the very last airplane which left Portugal for the United States, the last before the five years long interruption of communications, carried five copies of our Atlas for the collegues over there, our last greeting before the tempest.

The war went over; electronics had considerably developed, there were automatic recorders now, new and better gratings, new photocells, even for the near infrared. Our atlas for the visible spectrum was followed by the infrared atlas of the McMath Hulbert Observatory (1950), extending up to 2.5 μ; and by the Liège atlas (1955) for the subsequent, much broader region, up to 23.7 μ. Between both atlasses there is a gap of 0.3 μ, which has never been recorded up to now, because the water vapour of our atmosphere screens off completely this region. Finally the ultraviolet, which had not yet been adequately studied, was recorded for our sake by the kind help of the Dublin Observatory from 3000 Å on; similar records were recently published by Göttingen (1960) which completed the collection of solar atlasses. Recently the further UV has been explored by means of rockets. The wonderful spectra of TOUSEY and his collaborators (1961) extend from 2200 Å on, with a resolving power entirely equivalent to that of the Göttingen atlas! Even the further region has been gradually conquered by a most remarkable effort of ingenious technique, due to the groups of FRIEDMAN, HINTEREGGER, RENSE, TOUSEY and their collaborators.

These graphs contain an almost inexhaustible wealth of information. In order to put this material to use, the first step was obviously the derivation of a catalogue of Fraunhofer lines, with strengths, expressed in equivalent widths. Such a catalogue had been made by ROWLAND already in 1893–1896 and revised by ST JOHN, Miss MOORE, and others in 1928; but so much more was known about the solar spectrum now, and we were able to determine quantitative intensities. HOUTGAST and other Utrecht astronomers had started the systematic measurements of equivalent widths already during the war. We had the good fortune, that Mrs. MOORE and Dr. BABCOCK were just then preparing a new edition of Rowland's table and that our intensities could be incorporated in this great enterprise. How many letters were interchanged in the course of this collaboration! It took much more time than we expected. At each consecutive meeting of the IAU we had the pleasure to meet and to discuss the progress of the work, but also we felt bitterly the regret that it was not yet finished and the anxiety that we would perhaps never reach the goal. Thanks to Mrs. MOORE, who carried the heaviest burden, this goal is practically attained. The ledgers are ready and the proofsheets are being corrected. What we have felt ever and ever again in the course of these years, that is the deepest admiration for Henry Rowland, who accomplished a similar enterprise 70 years ago, with so much less technical means, and whose work is still now a marvel of perfection.

At the very start of our intensity measurements (1923), I had introduced the equivalent width as a measure for the importance of a Fraunhofer line. The later development has confirmed that this quantity indeed is of the greatest use for theory. But the recent and future development of solar spectroscopy is directed towards a more and more precise determination of each individual profile. For this purpose, the elimination of the instrumental imperfections is imperative, in any case for the medium and faint lines. Already in 1932–1933 BURGER and VAN CITTERT had devised a beautiful procedure for this purpose; later VAN DE HULST developed more refined and effective methods (1941–1947). The resolving power of our spectrographs is now great enough, but the elimination of stray light puts higher and higher requirements. The vacuum spectrograph of the McMath Hulbert observatory reaches almost the limit of the most perfect seeing, but it has still a noticeable amount of stray light; the method of the Sacramento Peak Observatory, using twice the same grating, marked a new progress towards the solution of this difficult problem. Photoelectric recording has increased the precision at least by a factor of 10, eliminating all photographic uncertainties; and there seems a vast field open for investigation, when the slightest depressions in the continuum, of the order of

1 % or even 1 ‰, – why not? – will be determined in a reliable way by repeated recording. Wonderful, entirely new methods, as e.g. the metal vapour monochromatic analysis by BLAMONT and RODDIER (1962) are very promising. Great progress is also to be expected from a systematic exploration of the spectrum at different distances from the limb, the most crucial observations relating to the utmost boundary of the solar disc. HOUTGAST's dissertation (1942) was one of the earliest systematic investigations in that direction; the results clearly showed the high theoretical significance of such work. Later many others have followed this path; but, compared to all that still has to be done, I dare say that the work has just begun.

So much about the observational techniques and investigations.

The significance of solar spectroscopy has been, during a long time, the identification of the elements composing the solar gases. Since KIRCHHOFF and ROWLAND, an immense progress has been reached, partly because the laboratory spectra have been much more completely investigated, but mainly because we *understand* the spectra so much better. The identification of an element is no more a question of counting how many lines are found to coincide, how many not, and estimating the probability that the element is present. Saha's theory made possible to distinguish for each element the lines of atoms and ions; moreover, knowing the term analysis and the band structure, we may predict to a certain extent, even without precise calculation, which lines may be expected to be strong and which should be weak. This considerably increases the reliability of our identifications.

The results in this field were almost entirely obtained by Mrs. MOORE, during a whole life of uncessant, intelligent and responsible work. They are found in their latest shape in the twice Revised Rowland Table, which I already quoted and about which we had the privilege to hear her speech yesterday. (See page 89.)

This brings me to the theoretical research on the solar spectrum. How does a Fraunhofer line originate? From the very start, two different conceptions appeared: Kirchhoff assumed absorption and emission processess, STOKES considered resonance scattering. In elementary courses, KIRCHHOFF's theory was quoted; the other conception was taken up by SCHWARZSCHILD (1914) and at Utrecht by JULIUS. I have had the privilege to work during several years with this noble man, and I think that full right should be done to his memory in the history of solar spectrocopy. Julius applied to solar physics what he called the theory of anomalous dispersion, which is nothing else than the classical electromagnetic dispersion theory. Near a spectral line, the refractive index becomes abnormally great or abnormally small, and Julius considered, what would be the astronomical consequences. Between 1901 and 1910 under the influence of the entirely phantastic theory of Schmidt, he first considered the large scale curvature of the rays in the solar atmosphere. But from 1910 on he understood that wavelengths near a spectral line would be strongly scattered, for Rayleigh's formula is proportional to $(n - 1)^2$. Fraunhofer lines would be due to scattering. Some years later, unfortunately, though not forgetting about the scattering, he emphasized more and more the effects of refraction by small scale inhomogeneities in the solar gases and was lead astray, because he worked only qualitatively.

At the moment when our first profiles were measured, I directly tried the scattering theory and combined the damping formula with Schuster's transfer solution; the theory explained the broad wings of strong resonance lines, but failed in the central parts, and I was disappointed. However UNSÖLD at the same time had made the test in a more clever way: he understood that the theory could only apply to the wings, he noticed that the profiles of doublets were in the correct relation, that center and limb corresponded, and by his optimism found the clue to the quantitative analysis and to the understanding of the Fraunhofer lines (1927). Curiously enough, it was discovered later by Unsöld himself that for most lines the process of scattering is *not* important compared to absorption, but quite some years elapsed before the very special type of absorption was pointed out by him, that plays the main role on the sun, and is entirely different from absorption in the laboratory.

As long as the investigations were restricted to the wings, only a few strong lines could be used for abundance determinations. However, RUSSELL, another pioneer, had already found an entirely different method, based on the old

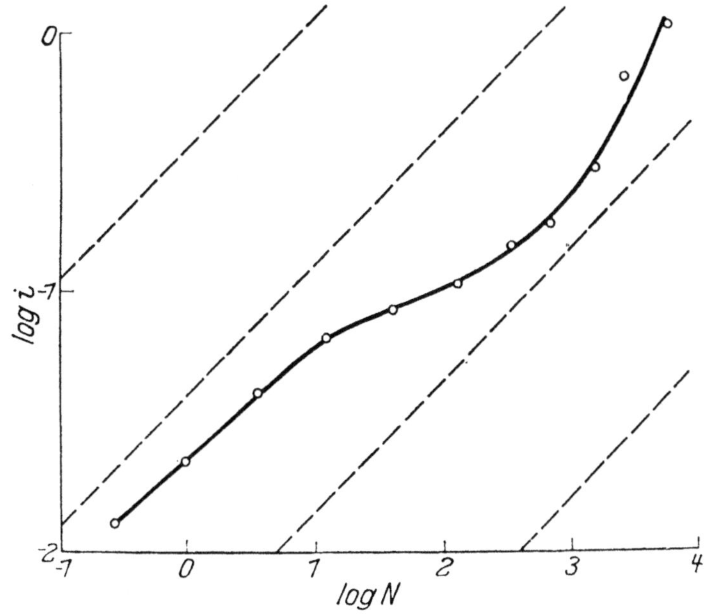

Fig. 1. The first curve of growth for Fraunhofer lines, at 4500 Å; only measurements (MINNAERT and VAN ASSENBERGH: 1929, *Z. Physik* **53**, 248).

visual Rowland estimates, which allowed to use any lines, even of those elements for which transition probabilities are completely unknown! He calibrated the Rowland scale by multiplets, used what we now call the "sum f-rule" and found the number of absorbing atoms of each kind. These admittedly rough determinations, made in 1928, are still now, 35 years later, the only ones available for 17 out of the 64 elements, for which abundances have been determined in the sun! Notwithstanding our admiration for Russell's wonderful work, I thought that it was necessary to get rid of the Rowland scale, to substitute for its estimates the better defined equivalent width, and to find how it increases as a function of the number of atoms. Even with our not too perfect spectrograph this could be achieved. We found a curve for the blue region of the spectrum (1929), then another, more or less similar, for the green region (1930).

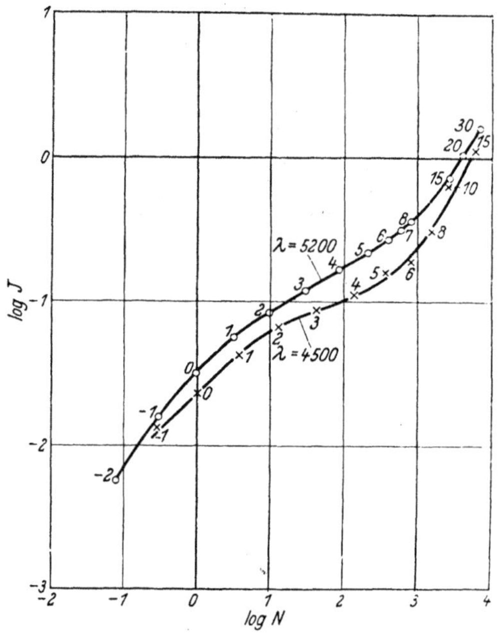

Fig. 2. Curves of growth at 4500 Å and 5200 Å; only measurements, still without theory (MINNAERT, M. and MULDERS, G.: 1930, *Z. Astrophys.* **1**, 192).

These curves had a curious bend and looked very incomprehensible. However, by a piece of good fortune, just at that moment a paper had been published by SCHÜTZ (1930), who had found a similar curve in laboratory spectra. I asked Mulders to study the paper and to see whether it could not be applied in our case; he thought that it looked hopeful. The next days were very exciting; we computed from the morning to the evening and obtained the first theoretical curves of growth for the sun. Our incomprehensible graphs had found an explana-

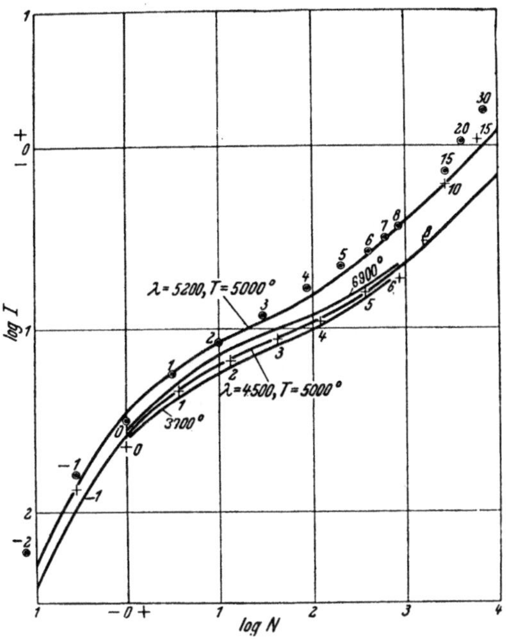

Fig. 3. A first comparison of theoretical curves of growth with observations (MINNAERT, M. and MULDERS, G.: 1931, *Z. Astrophys.* **2**, 165).

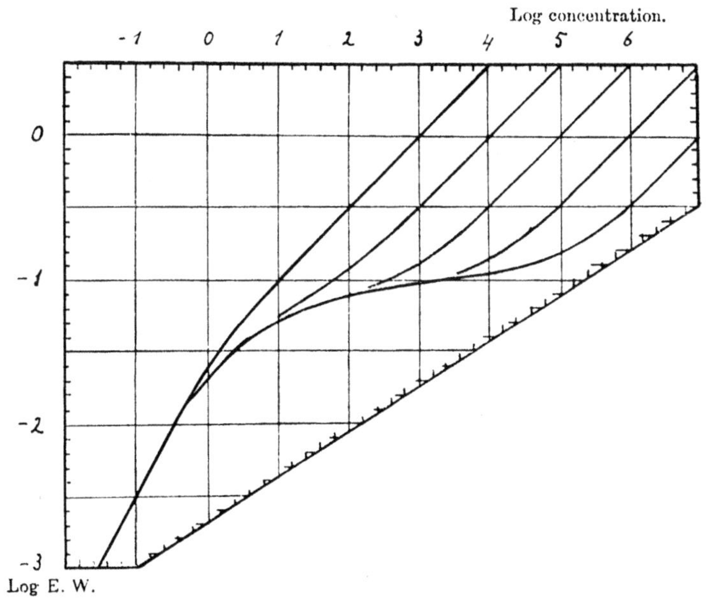

Fig. 4. Theoretical curves of growth for the sun. The first theoretical curves of growth or different values of the ratio between Doppler and damping widening parameters (MINNAERT, M.: 1934, *The Observatory* **57**, 333).

tion: apparently the Fraunhofer lines were broadened as well by Doppler effect as by damping; the damping could be found, and proved to be more than ten times the classical damping. This was considerably more than what radiation damping could explain, this could only be collisional damping. We enthusiastically wrote a paper for the *Zeitschrift für Astrophysik*, added a rather accessory calculation about the radiation damping of the green magnesium lines and sent the paper to Grotrian; but two or three weeks later, we noticed that in that accessory part we had made an obvious and serious error. Fortunately the manuscript had not yet gone to the printer and the error could be corrected – it would have spoilt the paper, which now remained one of our finest remembrances (1931). This is the story of what I called "the curve of growth" (1934), a term where you are free to find a reminiscence from the biological studies of my youth.

The analysis of the solar spectrum by this method leads both to the determination of abundances, as also to some fundamental properties of the solar atmosphere. It has been applied later innumerous times to the analysis of stellar spectra. And I think that this is a good example, showing how solar physics explores the field and opens the way for stellar physics. In applying this analysis, STRUVE and ELVEY (1934) discovered the influence of turbulence, later macro- and microturbulence were distinguished; and this now gave a stimulus to solar physicists, who discovered on their turn both these turbulence effects in the sun, where they are so much weaker. Here on the other hand there was the additional opportunity, to distinguish between the centre and the limb and to find the variation of turbulence with heigth.

Since those years, early in the thirties, an astonishing development has started, in which there was a continual strong interaction between theory and observation. Work in many different directions was taking place at the same time; almost as in a modern theater play, where a complicated stage is built up, actors playing in different rooms, but all the time influencing each other. Our understanding of the solar spectrum progressed, but always by successive approximations. All this makes the history of that period a very complicated one. I would not like to let myself go astray in trying to follow the chronological development, with all personal reminiscences connected to this evolution for one who has followed it with such an intense interest. Let me rather try to obtain a more systematic survey, by distinguishing first, the progress in our understanding of the fundamental atomic processes; secondly our gradually improving models of the solar atmosphere, based on the knowledge of these atomic properties; third, the connection between these models and the observed continuum, the line profiles, the equivalent widths.

There is no doubt, that the discovery of the absorption by the H^- ion by WILD (1939) introduced one of the most important new concepts into solar spectroscopy. I remember how we had felt that something up to then was wrong:

With VAN DEN BRIEL I had shown, that there ought to be well-observable discontinuities at the end of some magnesium series, for example – but they could not be detected. This and many other difficulties were now at once solved. It was now understandable, how a gas could produce such an evenly distributed absorption. The difficult calculation of this absorption as a function of wavelength, carried out by CHANDRASEKHAR and BREEN became of fundamental importance (1945–1946). The other contributions to the continuous absorption, known up to then, proved to be of relatively minor importance, though not negligible: these were the continua beyond the series limits of the hydrogen atoms and of some metals. The graphs of Miss VITENSE (1950), which have been so widely used, have now been reborn in an even more perfect shape by the publication of BODE (1963). No other particles have been found to be of appreciable importance.

The selective absorption by the atoms and ions produces the Fraunhofer lines. Of primary significance of course is the absorbing power of each particle for its individual lines, the so-called *f*-value. Early pioneers in the field of physical optics had already deduced some of these values from experiments, interpreted by the correspondence principle as a number of virtual resonators. The sudden development of quantum-mechanics opened at once the possibility to calculate by theory such *f*-values. At least, the road was open; but there was still a long way to go before the mathematical procedures and approximations were sufficiently explored. Even now, these theoretical methods are almost powerless in the cases of complicated systems of many electrons, as are often the atoms of astrophysically important elements. Fortunately there are many experimental methods, by which the *f*-values may be found. Gradually, thousands of these numbers have been determined, by theory or by experiment, or by both. At the Utrecht Physical Laboratory a great contribution was made to our knowledge in this respect, by the school of ORNSTEIN, VAN CITTERT, BURGER, DORGELO, SMIT and a whole series of young men, earning their dissertation and the doctor's degree in this new field. Gradually however, physicists became less and less interested, precisely at the moment when astrophysics had more and more demands for these fundamental data. There is no doubt that the urge of solar and stellar spectroscopy has considerably contributed to encourage our tired physical collegues to go on with their efforts or to resume them. You may remember how I tried since many years to collect in the *IAU Transactions* piece by piece such information from all sources available in the physical litterature. And you will understand how happy I was to see the first bibliographical catalogue by KOLESNIKOV and LESKOV (1958) followed by that of GLENNON and WIESE (1962), or, still more recently, to admire the preprints of the Bureau of Standards, the first of which relate to oxygen, where we find for the first time systematic, quite reliable tables of *f*-values, drawn from all available sources and critically weighted.

The *f*-values determine the absorption of light by the atoms. Another question is, how the excited electron comes back to its original orbit. This amounts essentially to the distinction between so-called thermal emission and scattering, which I mentioned already and which years ago played an important part in the discussion. For the moment there is somewhat less interest for this question, since it became probable that in most cases thermal emission prevails.

For the cases where scattering occurs, HOUTGAST (1942) and SPITZER (1944) showed that this scattering is mainly non-coherent: theory became more complicated, but the agreement with the observations was improved. We might have to give more attention to the effects of scattering, which are difficult to include in the calculations but which influence many profiles in a degree, easily detectable by modern photometric methods.

Some of the most interesting atomic properties are at the basis of the line broadening, in all its different varieties. For our most important atom, hydrogen, the classical papers of HOLZMARK, published more then 40 years ago, are still fundamental. Many efforts have been made to improve on this beautiful theory; KOLB and GRIEM have reached important progress, though the real goal does not yet seem to have been attained. For the other atoms we are accustomed to apply the collision theory after WEISSKOPF and LINDHOLM; but it remains vexatious that a really adequate treatment, comprising as well statistical as collisional broadening and all intermediate cases, is not yet available. We have learned how damping in the solar atmosphere is mainly due to the VAN DER WAALS forces of neutral hydrogen atoms, and we know also about other effects; but precise quantitative calculations are difficult as long as the quadratic Stark effect and the polarizability of the atoms is not better known.

Finally there is the line broadening due to isotopic shifts, clearly shown to exist, but difficult to take into account with precision. Here again many necessary physical data are lacking.

Reviewing the enumeration of these atomic properties, one wonders how it has been possible to reach so many important results on the solar spectrum, knowing so little! Apparently the main effects are so predominant, that even a rough simplification gives quite satisfactory results. We should however be well aware that a *detailed* agreement with modern, precise observations can only be reached when *all* the effects have been taken into account. Nature does not try for our sake to keep things simple!

We have now discussed all physical constants which are necessary to construct a model of the solar atmosphere. All, except two: the chemical composition and the hydrodynamical state of motion. The chemical composition cannot be known *a priori*; it should be the coping-stone, the final result of our confrontation with the observations. But the hydrodynamical state of motion must be predictable if we were clever enough. As things are now, we can only rely on a few general theorems, and we have to deduce the real situation from a comparison with the observation. This unfortunately increases the number of unknowns.

Originally, our solar models were one-layer models, that one layer being so chosen as to represent as well as possible the representative layer where the Fraunhofer lines are formed. On this basis was done up to fifteen years ago all the early work on quantitative solar analysis, first by Russell, then by Unsöld and others. Think how primitive the model was, and how much was nevertheless reached! Early enough, some hydrodynamic features could be easily introduced: it was only necessary to add all through that representative layer a certain amount of random currents, which was called micro- and macro-turbulence depending on the size of the turbulent elements. Nothing more was necessary, and HUANG could develop his fine method of line-width correlation (1952), which yields directly both types of turbulence, and which has been recently refined at Utrecht by VAN DEN HEUVEL. It is interesting that by this method the effects of turbulence are found, independently of our knowledge of the chemical constitution; we thus disentangle the two unknowns of the theory.

But inevitably there came a moment, when discrepancies with the observations were found, clearly due to the model not being adequate. Perhaps even stronger was the uneasy feeling of the theoretician, who knows that the one layer model used is, after all, nonsensical and could never exist, while he sees so clearly the possibility of doing better. It thus became necessary to take into account how the properties of the solar gases vary with height; we had to describe these layers of gas as a continuum, without any boundaries, while temperature and pressure, ionization, continuous and selective absorption, damping, all vary with depth, each according to a different function, macro- and microturbulence being added in some not easily predictible way; and then line-formation in such an atmosphere had to be studied. The pioneer in this direction was Anton PANNEKOEK, my venerated elder colleague and friend, who already in the thirties, and especially in his fundamental paper of 1936 emphasized the necessity of such an analysis and developed the basis for its application by a wonderful feeling for the technique of numerical work. UNSÖLD at that time thought that this refinement was hardly worth-while, and perhaps it came too early indeed; but still it prepared the field for the moment when times would be ripe.

Since then, the solar models soon became more and more realistic, more and more refined. As I already told, the introduction of the H^- absorption changed entirely the results. The French astrophysicists introduced very precise methods to test the flux constancy; the blanketing effect was taken into consideration better than ever before. First UNSÖLD, then I, between 1932 and 1948, in mutual interaction, showed that the formation of Fraunhofer lines even in such a complicately build atmosphere could be calculated in a relatively simple way if the lines are weak, by the introduction of our weighting functions; and PECKER was able to extend this method to strong lines. MUGGLESTONE recently introduced a variant, the method of the Planckian gradient, which may have several advantages (1958). This refined analysis was applied for the first time by CLAAS (1951). The curves of growth could be so much perfected now, since they were constructed for each individual atomic or ionic spectrum; or, still better, for each spectral line.

However, a new complication had arisen. DE JAGER (1952 and 1954), BÖHM (1954) and others had proved more and more clearly that it is necessary to introduce inhomogeneities in the horizontal direction, corresponding to visible granulation; and so there are now two-stream models, three-stream models and even multi-stream models. For each stream, there is microturbulence and macroturbulence, and each of them varies with height in a not yet completely ascertained manner; and moreover, there is the possibility that this turbulence should be anisotropic. Things have become terribly complicated indeed; after a lecture which I had given at Cambridge on this subject, I remember REDMAN saying that the would never more work on Fraunhofer lines! Please note that the complication had been still further increased by REDMAN himself (1943), when he introduced the "rugosity effect", which is now interpreted as "optical rugosity" by PECKER and LEFÈVRE (1960–1961): in the center of a Fraunhofer line the cold regions are more opaque and partly mask the hot ones near the limb. The introduction of inhomogeneities gives new possibilities, but has to be kept within certain bonds, since SCHWARZSCHILD obtained his wonderful stratosphere photographs of the granulation, which seem to eliminate almost any blurring effect of the atmosphere, and which show that there is only a moderate contrast between granules and intergranules.

The study of the granulation became at once more complete and interesting when RICHARDSON and SCHWARZSCHILD (1950) produced their spectra with "wiggly lines"; the still more beautiful spectra of MCMATH and MOHLER were the sensation of the IAU meeting of 1955. It was now possible to investigate directly the differences between the granules and the intergranular matter. There came soon a number of papers, comparing: the brightness, the radial velocity, the equivalent widths, the central intensities and even the magnetic fields; and the cross-correlations of all these quantities were studied. Every one expected that the rising elements would be found to be hotter and brighter; actually surprisingly little correlation was found; the central intensities in ordinary lines and those in Hα did not correlate, which showed that the lower motions do not extend up to the higher layers. Or perhaps the correlation disappeared because of phase-shifts. As a climax finally came the brilliant discovery of LEIGHTON, EVANS and MICHARD (1960–1962): radial pulsations with a period of about five minutes, disclosed from the corresponding Doppler effects. During the oscillation the standing gravitational waves are gradually transformed into running acoustical waves. So this is the mechanism, visible before our eyes, which transmits the energy of the lower convective motions into the chromosphere!

The situation has not yet completely been cleared up. It seems that the deeper layers show convective circulations on a large schale, while the higher layers pulsate in small scale units. We see now, how little justified we were in using loosely the term "turbulence", when lines were broadened by differential motions. In applying theory to the solar spectrum it will be necessary henceforth to distinguish between the line profile at one point of a sharp solar image, and the profiles corresponding to the averaged solar surface. A Fraunhofer line, photographed now, may show slight local asymmetries; two minutes later the effect may be reversed.

The study of the differential motions between hot and cold elements, combined with the effects of optical rugosity, might play an important role in the explanation of the wavelength shifts and asymmetries of the Fraunhofer lines and of their variation from center to limb. However these effects seem to require considerably greater temperature differences and inhomogeneities of a considerably smaller size than would be derived from the direct granulation photographs. A close discussion will be necessary here.

But let us leave aside these minute details and come back to the model of the photosphere, of which we now better understand the heterogeneity and the so-called turbulence. Frankly speaking: no existing model seems to explain all observations in a really satisfactory way. Many of us had tried: my colleague DE JAGER, who is a master in combining all available data into a consistent picture, has given several gradually improved solutions; and this symposium was opened by the remarkable lecture of Dr. Pecker (p. 28) on the same subject. In the region between $\tau = 0.1$ and $\tau = 2$, where most Fraunhofer lines are formed, the model seems rather well established by the consensus of all. Most interesting however are the higher regions and the transition to the chromosphere, where a temperature minimum occurs; it must be possible to investigate these layers: (1) by a study of the molecules, which are very sensitive to pressure; (2) by the investigation of the far UV spectra, where the transition between absorption and emission lines is observed; (3) by detailed photometry of the lowest flash spectra.

As to the layers below $\tau = 2$, we may hope to draw interesting conclusions from the lines of very high excitation potential, in the near IR region where the continuous absorption is a minimum. But here the limit will be quickly reached, where observational methods will have to be replaced by theoretical work, so brilliantly started by BÖHM and Mrs. BÖHM-VITENSE (1953–1955).

The difficulties of finding a satisfactory model are so great, that some have lost their confidence in our classical methods and have doubted the existence of LTE in the upper photospheric layers. There seems no doubt, that in these layers the ionization does deviate appreciably from LTE. Does this also apply to the *excitation* of the atoms? The courageous and original work of THOMAS, PECKER and their students has not entirely convinced all workers in the field. Though a definitive verdict on their theories does not yet appear possible, we all should recognize that their attempts have called attention to many phenomena not yet sufficiently understood. Moreover they are forcing us to analyze in detail the atomic processes, occurring within each volume element; and this analysis, difficult as it is, will indubitably bear rich fruits. This is, I should think, a program for the future: to calculate for a given atom the complete set of transitions, by radiation, by collisions, by interchanges between all atomic states and with the free electrons. Computing machines may open the possibility of such work.

Finally then, we come to the determination of the chemical composition. As long as our theories were in a preparatory stage, some guess about the composition was necessary before anything could be said on the depth and the thickness of the so-called representative layer. The relative abundances of the metals had been determined with fair certainty by RUSSELL, but the abundance of hydrogen remained a problem. In 1929 Russell guessed that the abundance of hydrogen relative to metals was 60 to 1 and he writes: "The abundance of hydrogen is almost incredibly great". He keeps to this estimate in 1931, but the same year MENZEL from his flash spectrum derived a ratio of fully 6000! In recent years, this number has been increased to about 20000. STRÖMGREN had namely shown (1940), that any spectrophotometric measurement of a spectral line leads to a straightforward determination of the ratio between the metal and the hydrogen abundance, there being no necessity of measuring the hydrogen lines themselves, of which the profiles are not completely understood. The abundances of the individual elements could now be determined with more and more precision, because our knowledge of the solar model and our understanding of line-formation became more satisfactory; of greater importance even were the improved f values. The most recent systematic analysis of the solar atmosphere was carried out, with wonderful photoelectric records and modern theoretical improvements, by GOLDBERG, MÜLLER and ALLER (1960). Just as an example to show the importance in such work of good physical constants, I may remind you about the discussion on the abundance of iron, a so important element, where the results changed at once by a factor of almost 3 at the moment when better absolute f values were obtained by the atomic ray method!

There remains the problem of the helium abundance, not yet solved for the stars, even less solved for the sun. Our colleague, Prof. UNDERHILL, takes a very firm position in this question, for which she has strong arguments; but these are not directly connected to the sun. There also remain as unknowns two inert gases, most rare earths, the halogenes, several other elements, partly already treated by Russell but never again since then.

The importance of the abundance problem has become especially great from the point of view of stellar evolution and element formation. For no star is the analysis so detailed and reliable as for the sun.

From this bird's-eye view on the history of solar spectrum research, with its epic progress, its dramatic peripeties, its moments of despair, its revivals of hope and confidence, I am inclined to draw the following conclusion, bearing on the method of scientific research: new improvements and refinements in the conceptions should be presented neither too soon, not to late. If they come too soon, they will hamper progress by introducing unnecessary complications, at a moment when more important results could still be reached with the existing methods. If they come too late, time may be lost by working with an insufficiently detailed analysis, while increased precision would be needed. It requires a great amount of feeling and wisdom, however, to apply this rather trivial moral in the right way!

2. The Sunspots and the Chromosphere

This symposium is not only devoted to the spectrum of the normal photosphere. Its scope is much more extensive. The solar spectrum is such an inexhausible source of information that the entire outer structure of a stellar globe may be deduced from it.

In the deeper layers, *the riddle of sunspots* is a continuous challenge to solar observers. The problems in the study of their spectrum are almost the same as those of the photosphere, but for a certain number of additional complications. From the very first photometric measurements, it was clear that the stray light of the surrounding, dazzlingly bright photosphere on the tiny dark patches which we observe would be extremely dangerous. And all these years this remained a continual menace, with which everyone struggled; only quite recently ZWAAN showed – rather convincingly, I should say – how surprisingly great this contribution is.

Another difficulty is the lower temperature, owing to which many molecules are produced, which again indirectly influence the continuous absorption and moreover absorb thousands of Fraunhofer lines, covering a great part of the spectrum, blending with most atomic lines, and undoubtedly producing a strong blanketing, perhaps not yet sufficiently taken into account. Finally there are the electromagnetic forces, which we had not yet to consider when discussing the photosphere. We now understand, that they must be intimately interwoven with the hydrostatic and hydrodynamic effects. – Because of these various complications, the model of a sunspot cannot yet be considered as sufficiently ascertained.

Many years ago, WANDERS started measurements of the continuous spectrum and studied the energy-wavelength curve and the limb darkening (1935). At that time, RUSSELL's conception of an ascending column in adiabatic expansion was generally accepted. We were very surprised to find a complete discrepancy, and, after many trials, we finally concluded that only radiative equilibrium could explain the measurements. This result, obtained with modest means, was soon confirmed at the Mt Wilson Observatory and at several other places. Later the facts as well as the conclusions have been doubted; but in any case, this field of study had been opened.

Several authors have tried since to determine the spot model, either from the continuum, or from the lincs. They have variously postulated hydrostatic equilibrium of the spot column, or equality of pressure with the photosphere at each level, or a contribution of the magnetic pressure, or radiative equilibrium. It looks almost impossible that all these conditions should be fulfilled at the same time. And even if a consistent model could be constructed, we are still confronted with the difficulty, to adapt the scale of heights to the scale of the adjacent photosphere. The models of MICHARD and recently that of VAN 'T VEER show interesting possibilities that should be carefully considered.

3. The Chromosphere

Let us now gradually ascend into the higher layers above the photosphere. We first have to do with the chromosphere, observed best of all during solar eclipses. This word evoques a long history of heroic efforts of the last generation, all trying to record by photography the elusive flash spectrum, often in difficult circumstances, during the hectic seconds of a total eclipse, and always with the threatening of clouds. The spectrum, obtained by PANNEKOEK and me in 1927, yielded the first photometric study of the flash, all intensities being given in absolute units. MENZEL in 1932, REDMAN, EVANS and ATHAY in 1952, HOUTGAST at several successive eclipses obtained better and better results, showing especially how the spectrum fades out in the course of a few seconds, so that the distribution of the emission as a function of height may be studied; and even determining more or less reliable profiles. The aim of the observers is, to reach a resolving power sufficient for profile determinations, and at the same time an increasingly high time-resolution, in order to observe even the lowest layers. The considerable energy ratios which have to be bridged require a very reliable photometric technique.

These observations have been the starting point for a real understanding of the chromosphere.

In the first place the integration along the line of sight and the self-absorption were eliminated, by different semi-empirical or theoretical methods. Then it was tried to explain the radiation of each volume element by a judicious choice of T and P. About the years 1950–1955 it became clear that this is not possible, and a model of hot and cold elements, juxtaposed, became more and more generally assumed. In the survey papers of UNSÖLD and VAN DE HULST, 1955 and 1953, this point of view is only tentatively mentioned. In the article of DE JAGER, 1959, it is already the main feature of the model, on which the whole discussion is centered. In the development of our conceptions, there is thus an interesting analogy with the development of photospheric research, where also a two-streams model was introduced; in the chromosphere, however, the differences in temperature of the two components are much greater and the relative importance of these components varies considerably with height. –

Again, as in the case of the photosphere, the question is put, whether local thermodynamic equilibrium applies; here there is no doubt, at least in the higher layers: it doesn't. Here we are really *obliged* to examine in detail the excitations and recombinations and to take into account all atomic processes. The directly observed phenomenon of the spicules, which attracted new attention since 1945 (ROBERTS), was evidently the basic for the introduction of the heterogeneous model of the chromosphere. It gave also the answer to the old question about the support of the chromosphere and it was the starting point for the introduction of pressure waves, shock waves, magnetohydrodynamic waves.

The study of quiescent prominences is closely connected to that of the chromosphere. I remember how SLOB and I, already in 1930, adapted our eclipse spectrograph to a study of their spectra, and how we compared the H and the He lines and constructed a curve of growth. Later BRÜCK and MOSS found different results, which seemed to show that the H/He ratio would vary from one point of a prominence to the other. As so many times, UNSÖLD took the material in hand and was able to reach a quite satisfactory agreement. This was actually the first determination of the He abundance in the sun, which was improved only after 1953.

During a very long time there remained a gap between the chromospheric layer and the coronal layers. I am not aware that astrophysicists felt very strongly the need for such a transition. In any case it was more or less a surprise when these transitional layers were revealed in an exceedingly beautiful way by their far ultraviolet emission spectrum shortward of 1 900 Å, photographed since 1955 in such a clever way by say 20 different rockets, each of them having perhaps 100 seconds available above the level of 120 km. That is to say that all documents on those layers have been obtained in about half an hour (not counting the Orbiting Solar Observatory). There appear most of the successive degrees of ionization of Si up to Si XII; Ne V, VI, VIII, an element not earlier detected on the sun; Fe IX and X, finally Fe XV and Fe XVI. There is no doubt, we have reached the corona, which we now observe in front of the disk. But please note that by this method we observe *only a small part* of the spectrum of each of these layers; we are able to observe it, because in that spectral region the lines are not superposed on a bright continuum of the deeper layers.

A full knowledge of the spectrum of these transitional layers and of the corona itself will be reached only when we shall be able to observe by rockets or from satellites the corona from aside, at different heights above the limb, by preference during a solar eclipse. Thus there is still a great amount of new information to be gathered about the solar spectrum.

4. The Corona

I am now coming at the spectrum of the *corona* itself. The starting point for an understanding of the corona was LUDENDORF's observation about the energy-wavelength curve, which showed that it reproduced entirely the photospheric curve (1925). A few years later (1930), I showed how the brightness distribution in the corona, combined with the polarisation, informs us about the emission by each volume element. The equations could be found in principle, in a very old, but excellent paper, by SCHUSTER (1879); but it was quite a job to disentangle the two interacting polarized components. I was even able to include the anisotropy of the scattering, which by later authors was mostly neglected. The conclusion was, that the continuous spectrum could be explained, either as scattered by free electrons, or as a mixture of bluish Rayleigh scattering and of reddish proper emission at a low temperature. I was at that moment almost convinced that the first assumption was the right one. But I felt obliged to mention also the second hypothesis, which at that moment was the only possibility if one wished to explain at the same time the white colour of the corona and YOUNG's measurements of the polarisation, which gave *very* different values in the different colours. Soon enough, it was proved that these measurements were completely in error; and so my quite improbable escape solution could be rejected. I quote this story only, to show how uncertain we still were, at that moment, about the nature of the corona. The next step was the identification of the coronal lines, achieved by GROTRIAN in 1939 for two iron lines, shortly afterwards by EDLÉN for all the other lines (1941). This problem had been considered as the last, outstanding riddle about spectral lines in the universe; it was solved much later than that of the nebular lines and required a deep insight in atomic structure. Since then, some new lines have been discovered, better intensity measurements were made, recently image converters have even been used for the infrared lines. A tragic moment was the issue of the 1952 eclipse, when LYOT obtained his wonderful corona spectrum, and died two months afterwards, while still in Egypt.

In broad lines our modern conception of the corona and of the coronal emission is safe and beyond any doubt. But the more *details* have been worked out, the more riddles were encountered.

Consider e.g. the composition of the corona. EDLÉN had already been pointed out, that the metal ions must be excited by collisions, the lines being radiated subsequently by spontaneous emission. When SEATON, GARSTANG and VAN REGEMORTER had furnished the transition probabilities and cross-sections, more quantitative calculations became possible and abundances could be determined; and look, such astonishing results as have been found for the iron abundance, which came out to be ten or twenty times higher than in the photosphere! (ŠKLOVSKIJ, WOOLLEY and ALLEN, POTTASCH). Also silicium and other elements deviate. – Then there is the question of temperature, which is found to be about 2 million degrees from the linewidth, and only 1 million by other methods. – In both cases a solution has been tried by assuming an inhomogeneous corona, composed out of juxtaposed cold and hot elements (BILLINGS and others 1962). In that case, for example, it would not be possible to determine coronal temperatures from the ratio between the red and the green line. Other authors have kept to the model of a homogeneous corona, they explain the apparent higher temperatures from linewidths by assuming shock waves (LÜST and Miss TREFFTZ); or by considering the motions of the ions in the weak, variable macroscopic electric fields of the corona, as it has been observed in laboratory experiments (MOGILEVSKIJ, 1962).

The choise between a homogeneous and an inhomogeneous corona must be of very great influence on all theoretical results. It would be most interesting if the inhomogeneous structure already found for the photosphere, the chromosphere, and the prominences, would also be a feature of the corona.

Surveying our overall knowledge of the coronal spectrum, we have first the X-ray region, up to 50 Å, where very little is known, except the total flux over some sections and its great variability (FRIEDMAN). Then the region up to 1 800 Å, where the rocket spectra disclosed a host of lines, still for a great part unindentified. Then comes the gap between 1 800 Å and 2 200 Å, where ordinary rocket spectra are of no use. Then the well-known visual and near infrared. A gap in the further infrared, probably not too interesting. And finally the radio-emission.

5. Radio-emission

The spectral properties of *the radio-emission* are entirely different from all other spectroscopic features. This emission is roughly monochromatic; however, the wavelengths are not determined by the electronic orbits in the atom, but by the collective oscillatory motions of great swarms of free electrons.

Moreover the source of the emission is very *localised* and it *moves* rapidly. One wonders what it would look in the visual region, if we could see objects, resembling blue stars, creeping up through the corona in the course of a quarter of an hour, slowly turning to green, yellow and red; or also a kind of rockets, ascending in a few seconds, showing a similar color transformation. How great are the difficulties of observation, since we are obliged to work with instruments, always tuned to one frequency only at the same time, while our radiotelescopes never give a tolerable image of the disturbances!

The history of solar radio-spectroscopy is mainly the history of the Australian work on this subject. At each meeting of the IAU, at each important Symposium on radio-astronomy, highly competent specialists as WILD, SMERD, CHRISTIANSEN, headed by the dynamic personality of our regretted PAWSEY, were able to announce spectacular progress.

The first radiospectrograph, built on the swept-frequency principle, was already in operation in 1949; it encompassed the range of 70 to 130 MHz. In 1950 WILD and MCCREADY published the first typical examples of bursts, which they classified in types I, II, III. In 1951–1952 PAYNE-SCOTT and LITTLE observed with the interferometer the rapid ascent of disturbances in the corona. In 1952 the first harmonic was discovered and the range of the spectrograph was extended to the interval 40–240 MHz. In 1957 WESTFOLD and later ROBERTS interpreted Type II by the shock front theory.

The Australians missed the discovery of Type IV: it was due to BOISCHOT and DENISSE in 1957, the first spectra being taken in 1958 by HADDOCK. But already the next year the Australian McLEAN had found the close association of this type with the magnetic storms, and in the meantime WILD and his associates had discovered Type V in 1958. Moreover they had combined the spectrograph with an interferometer and could show directly that the disturbance moved away from the sun and emitted longer and longer waves the higher it came. The same year ROBERTS discovered the echo's. KOMESAROFF studied simultaneously the spectrum and the polarization, and thus was able to ascertain the precise polarization for each type of burst. GIOVANELLI showed the association between radiophenomena and optical events on the sun. Finally I might mention the Australian discovery of 1962: the observation by SMERD, WILD and SHERIDAN, demonstrating that the first harmonic comes from lower heights than the fundamental.

As you see, the history of solar radio-astronomy extends only over hardly fifteen years, but how momentuous these years have been! I have emphasized the importance of the Australian work; but let there be no misunderstanding: many others cooperated and gave important contributions to our knowledge. International cooperation is very well organized in this branch of solar physics. It has in a certain sense found its expression in the synthetic dynamic spectra, first constructed by TAKAKURA and by HACHENBERG, later by FOKKER, in which the results of many observatories on different frequencies have been combined into a complete diagram of the radio-emissions on the several frequencies, giving a synthetic survey of all what happened at all heights in the corona during a period of perturbation.

The interpretation of the radio-spectra is not based on the structure of the atom, as it is the case for the other spectral regions. Here we need electromagnetic theory, magnetohydrodynamics and plasma theory, which in recent years have opened new fields, in almost the same way as atomic theory did at one time. Many problems are still unsolved, as e.g. the nature of the different groups of bursts and the precise mechanism by which they originate in the corona.

This brief survey may have called to your mind how much has been achieved in a period, relatively so short. The solar spectrum has been found to be an extremely rich source of information on all that is happening in the outer layers of the sun. It does not give us a photograph of the phenomena, but a kind of transform, which we have to re-interpret and which then becomes a very complete, sometimes wonderfully detailed picture of those realms, for ever inaccessible to man.

I cannot close this summary, without mentioning two events which in my opinion have been of great importance for the development of our science: in 1938 and in 1955 were published the two consecutive editions of UNSÖLD's book. This invaluable treatise has perhaps contributed more than anything else to the formation of a whole generation of solar spectroscopists.

There is always a peculiar fascination in unrevelling the threads of history, in finding out how a branch of science started, developed and reached its present

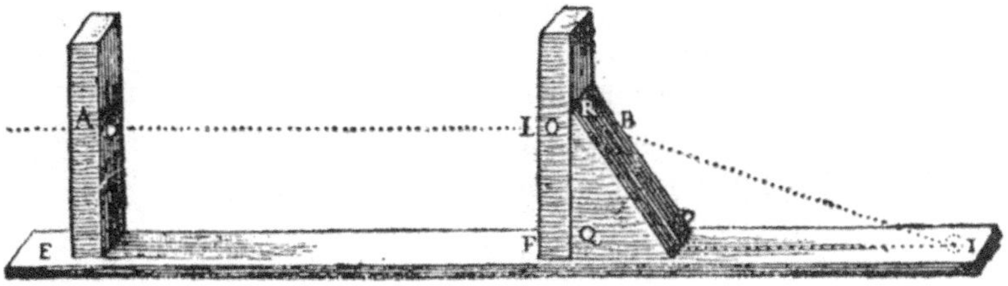

Fig. 5. Refraction through a prism. The position of the point I is measured; this determines the refractive index of the prism (DESCARTES, R.: 1637, "La Dioptrique", *Discours X*, p. 137).

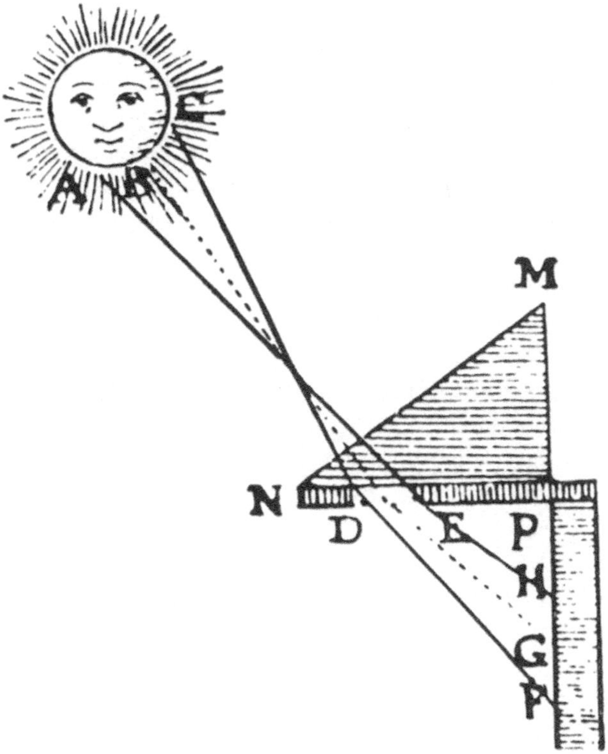

Fig. 6. Dispersion through a prism. The light is blue at H, red at F (R. DESCARTES: 1637, "Les Météores", *Discours VIII*, p. 255; application to the rainbow: p. 341).

exuberant splendor. We have surveyed the history of these last 40 years. But where did solar spectroscopy *start*? We jump back to Fraunhofer, we jump still farther back to Newton, analyzing in 1666 a beam of sunlight with his prism. But was NEWTON the first one who used a prism and studied the dispersion? Certainly not. Thirty years earlier, in 1636, DESCARTES had made experiments with a glass block, which he first calls a "triangle" (Dioptrique, *Discours* X) later "triangle ou prisme" (Météores, *Discours* VIII). In the first treatise he only measures the refraction. But in the second one he uses a narrow beam of sunlight and he studies the dispersion phenomena in order to understand the colours of the rainbow; and from the sign of the dispersion he derives that the red rainbow arc must be the outer one.

So by historical evidence at the very origin of solar spectroscopy we find: man's wonder and delight at the glory of the rainbow!

My dear colleagues,

I appreciate immensily, that you have been willing to come to Utrecht on this occasion, to give us the honour of your presence and the benefit of your recent scientific work.

This symposium, organized by my Utrecht collaborators, is at the same time for me the end of my official career at this observatory. You understand that it will not be the end of my astronomical activity, though the nature of my occupations will evidently change. I am sure that we shall meet many times more. But in any case this symposium is an occasion to thank you from all my heart for the help and inspiration which you have given me during many years, all of you, primarily by your scientific work. At moments of melancholy and doubt, which I have known as anyone else, I retired into the silence of our library and started reading the recent literature. You spoke to me; new suggestions, new possibilities for research came up in my mind, I was seized by the logic and by the beauty of nature, by the logic and beauty of science. I got new confidence in myself, new hope and optimism. – But I am also deeply indebted to you for your personal kindness and friendship, often a lifelong friendship, which added human warmth to the austerity of scientific work. Forgive me my weaknesses, keep giving me your confidence as I am giving you mine. May I be able to work for some more years and to be of some use to you.

Dear Professor Unsöld,

I do not know how to thank you for your kindness in addressing me as you did, yesterday. The praise that you have given me shows the warmth of your feelings, more than my own merits. You have said things that I will not easily forget.

Dear Utrecht friends,

In organizing this symposium, you have spared me any trouble, any worries, just giving me the pleasure to listen to all these most interesting lectures and discussions, all the time rejoicing in the recent progress of solar spectroscopy. You also gave me the opportunity to deliver this lecture, which to prepare was for me a considerable work but at the same time a real pleasure. Be sure that I feel deeply grateful for these wonderful days and for your initiative.

From my survey and by the reading of the Utrecht publications our foreign colleagues will have understood, I am sure, that the research problems at our observatory are strongly interconnected and that there is a close cooperation between our research workers. If personally I have done work of some use in astronomy, it is for a great part thanks to this cooperation. There will come other opportunities to address you. But already now I feel urged to say that I have the greatest confidence in the future development of the Utrecht observatory, headed henceforth by a man, so dynamic, so competent, so humanly sympathetic as is my colleague and friend De Jager with, next to him, Professor Anne Underhill, who very quickly has become dear to all of us. May thorough and successful scientific work make your life bright and happy!

I thank you for all these good years.

SESSION I

THE PHOTOSPHERE

LA STRUCTURE DE LA PHOTOSPHERE

JEAN-CLAUDE PECKER
Observatoires de Paris-Meudon et de Nice

Introduction

En 1953, Marcel MINNAERT publiait dans le volume *The Sun* édité par G. P. KUIPER, la principale étude d'ensemble sur la photosphère du soleil écrite au cours des vingt dernières années. Depuis cet article magistral, la connaissance de la photosphère n'a pas changé considérablement. Mais sans doute, d'une façon presque insensible, les points de vue ont changé. L'article de Minnaert, en effet, sépare nettement l'étude de ce que l'on déduit du spectre continu – le *modèle* de la photosphère – de l'étude des raies, puis de celle de la granulation. Les interactions mutuelles de ces différents domaines, dont l'observation se fait séparément, presque comme s'ils étaient indépendants les uns des autres, sont cependant loin d'être négligeables: La construction d'un modèle, à partir des données de l'observation, doit tenir compte des inhomogénéités; les populations des niveaux, que l'étude des raies prouve être non strictement boltzmanniennes, ne doivent donc pas être supposées telles dans l'interprétation du spectre continu. L'existence de *couplages* entre les champs de rayonnement et les propriétés locales, entre les champs de vitesse et la structure thermodynamique de l'atmosphère solaire, commandent maintenant les recherches, et si ces études n'ont pas encore abouti à une compréhension totale de la photosphère, du moins à ce changement de point de vue correspond-il aujourd'hui une phase critique des conceptions uselles de la photosphère, une destruction provisoire, prélude à une connaissance améliorée. Ce sont certains aspects importants de cette critique que je voudrais examiner aujourd'hui.

Qu'on veuille bien excuser cet exposé de n'être pas, et de loin, complet au point de vue de la bibliographie, et aussi de laisser d'importants points sous silence: notamment la structure de la photosphère active, dont la thèse de F. van 't Veer vient de moderniser l'étude; les différences possibles entre régions polaires et équatoriales de la photosphère: MICHARD (1962), PLASKETT (1962); et bien d'autres. Je laisserai aussi de côté, peut-être pour marquer la façon dont je conçois l'évolution des idées, les recherches sur les modèles *calculés* à partir de l'hypothèse de l'équilibre radiatif. Et l'on comprendra que les tables, modèles, abondances, que je donnerai en exemple ne puissent être considérés que comme des valeurs "provisoires", non comme des valeurs "adoptées".

 All rights reserved.

1. Les limites de la photosphère

A. INTRODUCTION

C'est de moins en moins qu'il est aujourd'hui possible de parler de *photosphère*. Cette notion apparaissait pourtant claire au début du siècle, dans la construction formaliste et sans consistance interne qui décrivait le soleil: la photosphère était considérée comme une région chaude, responsable seulement du spectre continu, et surmontée d'une région, la couche renversante, responsable seulement des raies (Figure 1). Lorsqu'on en vint à décrire l'atmosphère solaire à

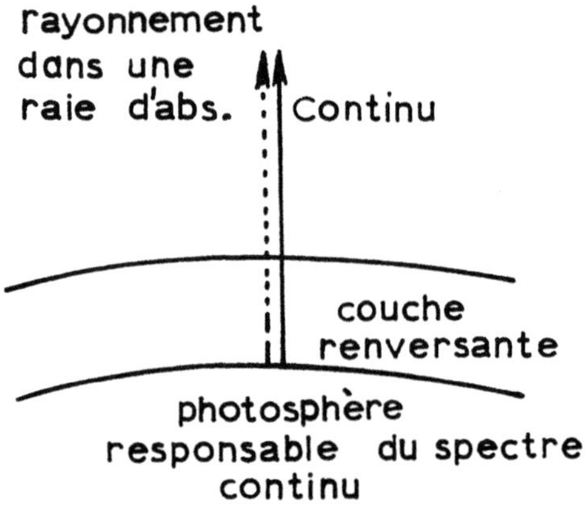

Fig. 1. Photosphère et couche renversante: conception ancienne.

l'aide d'hypothèses physiques cohérentes, la photosphère, région en équilibre radiatif, devint plus accessible à une véritable analyse théorique. Comme une atmosphère en équilibre radiatif représente assez bien les observations du disque solaire, mais qu'il était néanmoins impossible d'ignorer l'extension chromosphérique ou coronale en dehors de ce disque, il fallut bien alors définir la profondeur optique (par exemple à 5000 Å, ce qui sera noté par l'indice 0) au-dessous de laquelle le modèle en équilibre radiatif ne devait plus être considéré comme valable. Cette profondeur optique τ_{l_0} est celle (Figure 2) d'une couche telle que, pour le rayon tangent à cette couche, l'épaisseur optique de la chromosphère, à la longueur d'onde considerée, soit égale à l'unité: c'est en somme celle qui définit le *bord net* du disque solaire. Il est évident que τ_{l_0} dépend de la longueur d'onde d'une part, de l'opacité des couches extérieures d'autre part. On admettait alors que pour $\tau > \tau_{l_0}$, c'est à dire dans la photosphère, l'équilibre radiatif était réalisé, ainsi que l'équilibre thermo-dynamique local. La température y décroissait donc vers l'extérieur, tout comme si l'on considérait la profondeur τ_{l_0} comme une transition ayant une *réalité physique* et marquant un passage abrupt de l'équilibre radiatif au régime de dissipation des ondes mécaniques chauffant la chromosphère, de l'ÉTL au Non-ÉTL, de la photosphère à la chromosphère.

Dans le traité d'Unsöld d'abord, puis dans l'article consacré par MINNAERT (1953) à la Photosphère dans le traité de Kuiper, et qui constitue une somme de nos connaissances sur la photosphère, c'est une telle description qui rend compte de la structure de l'atmosphère solaire.

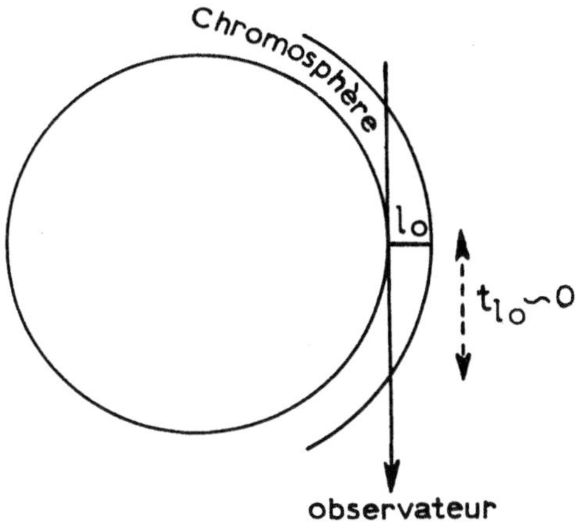

Fig. 2. Définition classique de la profondeur τ_{l_0} de transition entre la photosphère et la chromosphère.

Depuis lors, l'hypothèse de l'équilibre radiatif d'une atmosphère homogène est toujours considérée comme résumant correctement le processus physique essentiel qui maintient la photosphère en son état d'équilibre et qui conditionne le rayonnement solaire dans le domaine visible. Mais des différences (même faibles) entre un modèle théorique utilisant cette hypothèse, et les données (d'une complexité accrue) de l'expérience, ont paru souvent d'un tel intérêt physique que la notion même de photosphère a perdu son caractère original, et que l'on a peu à peu abandonné l'espoir (et pour longtemps!) de pouvoir en construire des *modèles* à partir des seules hypothèses physiques simples, telles que celles de l'équilibre radiatif, de l'ÉTL, de l'équilibre hydrostatique. Les modèles que nous utilisons aujourd'hui sont inspirés des observations, de façon plus ou moins judicieuse; ils essaient de décrire la variation (avec la profondeur), et les fluctuations, à une profondeur déterminée, de tous les paramètres physiques de l'atmosphère solaire, qu'ils soient des paramètres physiquement indépendants, ou non (auquel cas, une certaine cohérence interne doit être respectée, ce qui n'est certainement pas toujours le cas); l'étude de ces "dépendances" n'est d'ailleurs pas très claire, et reste souvent matière à contestation.

Nous continuerons néanmoins à suivre l'usage établi, et à utiliser le mot "photosphère". Avant d'essayer d'en définir les limites, précisons donc bien que la physique de la photosphère n'est plus compréhensible que si l'on tient compte des interactions avec ce qu'on nomme la chromosphère, comme avec les couches les plus internes du soleil.

L'astrophysique d'aujourd'hui est une astrophysique des *couplages*; il n'est plus possible de traiter telle ou telle région d'une étoile en la considérant isolément, sans se préoccuper de ce qui se passe dans d'autres régions, pas plus qu'il n'est possible, en un point quelconque de cette atmosphère, de traiter séparément la physique de la matière et du champ de gravitation, d'une part, et d'autre part, celle du champ de rayonnement et des phénomènes électro-magnétiques.

B. TRANSITION ENTRE PHOTOSPHÈRE ET CHROMOSPHÈRE

La première conséquence de l'attitude qui a été définie ci-dessus, c'est que les limites de la photosphère ne sont pas aussi bien définies qu'on l'admettait jadis – sans d'ailleurs alors le discuter. Certes dans le domaine visible, et dans le continu, τ_{l_0} varie peu (entre les limites 0.1 et 0.001 en gros). Mais, dans les domaines radio ou UV, τ_{l_0} est extrêmement petit; dans le proche IR, au contraire, τ_{l_0} est nettement plus grand que les valeurs usuelles. La profondeur τ_{m_0} correspondant au *minimum* de température (entre photosphère et chromosphère) est sans doute définie de façon plus physique que la profondeur correspondant au *"bord" du soleil* à une certaine longueur d'onde. Mais où exactement situer ce minimum de température? Les résultats des mesures sont encore peu concluants à ce point de vue. J'y reviendrai ci-dessous. Quant à définir la transition par le point où "commence" la dissipation des ondes mécaniques, où "s'amorcent" les écarts à l'ÉTL, causes de remontée de la température vers l'extérieur, il faut aussi y renoncer, car ce point serait sans doute assez profondément situé dans l'atmosphère solaire, plus en tous cas, et de beaucoup, que si la profondeur de la transition était définie à l'aide d'un des deux critères ci-dessus et de toutes façons, bien difficile à définir sans ambiguïté!

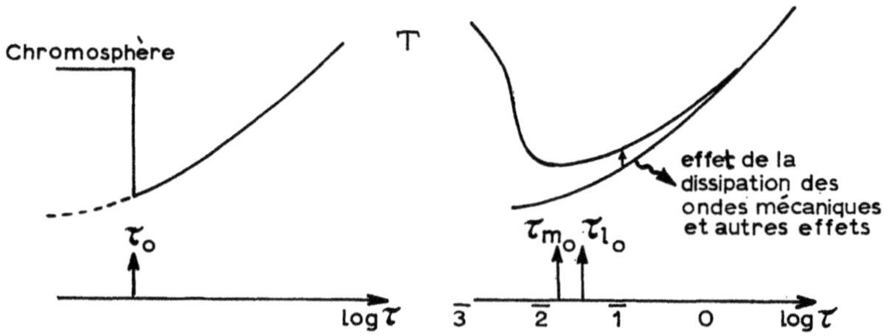

Fig. 3. La transition entre chromosphère et photosphère est continue.

La transition entre photosphère et chromosphère est donc un phénomène progressif et non pas abrupt (Figure 3). Si l'on considère l'opacité moyenne (définie en première approximation) des couches photosphériques, on peut calculer la variation, en fonction de la longueur d'onde, de la profondeur moyenne τ_0 d'ou vient le rayonnement choisi, correspondant à $\tau_\lambda = 1$ à la longueur d'onde choisie. On voit alors (Figure 4, tracée en utilisant les tables d'absorption de

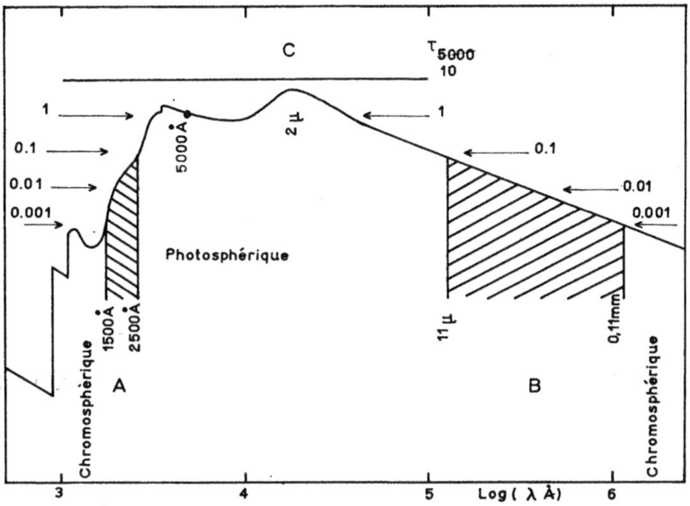

Fig. 4. Origine des différentes parties du spectre continu du soleil. (On notera que DE JAGER (1963) situe à 15 μ la région de l'IR, presque dénuée de raies, qui correspond à la région 1800 Å de l'UV.)

VITENSE (1951) pour $\theta = 1.2$ et $\log p_e = 0$) que les régions hachurées ($0.1 > \tau_0 > 0.001$) correspondent aux régions du spectre où s'opère la transition entre le spectre photosphérique de Fraunhofer, et le spectre chromosphérique. L'étude détaillée de ces deux régions doit donc permettre, en principe, de fixer la profondeur optique du minimum de température, et plus généralement, d'étudier la région de transition.

1. À partir des mesures de l'IR lointain

L'une de ces régions reste encore pratiquement inobservée; elle s'étend de *l'infrarouge aux ondes millimétriques*, et son étude n'est pratiquement possible qu'en ballon. Il est évident qu'une telle recherche est à encourager vivement. La variation centre-bord de l'intensité dans le spectre continu (bien que là, en raison de la forme de la loi de Planck, l'intensité dépende assez peu de la température) y présente certainement la transition continue entre un assombrissement, et un anneau brillant, et doit permettre l'établissement d'un modèle. L'absence de raies d'une part, d'autre part la connaissance assez bonne de l'absorbant (H^-, transitions libre-libre) permettraient de faire ce travail dans les meilleures conditions possibles.

2. À partir des mesures autour de $\lambda = 1800$ Å

L'autre région a déjà été photographiée plusieurs fois par l'équipe du Dr. Tousey. Elle montre *dans la zone de 1800 Å* un passage assez abrupt du spectre photosphérique usuel (raies d'absorption) à un spectre chromosphérique (raies d'émission). La figure 5, déjà ancienne (TOUSEY, 1961), en montre l'aspect. Les résultats montrés récemment par Tousey au IVe *Space Science Symposium* (1963) sont encore beaucoup plus fins, et devraient permettre une analyse quantitative complète.

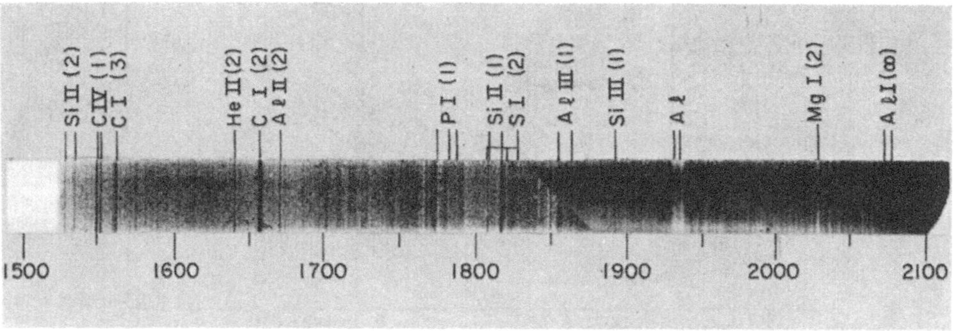

Fig. 5. D'après DETWILER, PURCELL, TOUSEY, 1961. (Permission US Navy.)

L'exploitation de ce phénomène en vue de calculer τ_{m_0} a déjà été suggérée par DE JAGER (1961) et l'auteur (PECKER, 1961). Les calculs publiés sont très grossiers; aussi en discuterons-nous à nouveau les résultats maintenant. Il est en effet possible de refaire ces calculs en tenant compte (aussi bien que possible!) de la stratification des propriétés de l'atmosphère solaire. Il ne faut toutefois pas se leurrer; cette analyse est essentiellement provisoire, pour les raisons que nous énumérerons ci-dessous.

On peut admettre d'abord (mais ce n'est sans doute pas très bon, car cette région spectrale est sûrement riche en raies issues de niveaux assez bas) que l'hypothèse ÉTL (hypothèse a) n'est pas trop mauvaise, et que par suite le caractère pour les raies d'être d'émission est du uniquement à l'inversion du gradient de température. Les premières raies d'émission sont observées à 1800 Å. À cette longueur d'onde-là, la profondeur optique de formation du spectre continu doit donc être de l'ordre de grandeur de l'unité (hypothèse b). La connaissance de la variation de la quantité τ_λ avec τ_0 permet donc en principe de fixer la valeur de τ_0 correspondant à $\tau_\lambda = 1$. Ce calcul suppose que les abondances utilisées dans les tables de VITENSE (1951) sont valides (hypothèse c), et que les opacités pour chaque élément sont correctes, malgré l'approximation hydrogénoide faite dans chaque cas par Vitense (hypothèse d), et la supposition de l'ÉTL (hypothèse a). Enfin, le calcul utilise nécessairement une relation (modèle) entre T, p_e, et τ_0.

Nous avons fait les calculs pour deux modèles: celui de PAGEL (1956), et celui, utilisé à MEUDON, dans une série de publications récentes (voir par exemple LEFÈVRE et PECKER, 1961). Aucun autre modèle comportant une chromosphère n'est en effet disponible. Les modèles en question sont homogènes et ceci constitue, la dernière hypothèse (e) importante de ce calcul. Les modèles (avec quelques autres) sont reproduits sur la figure 6.

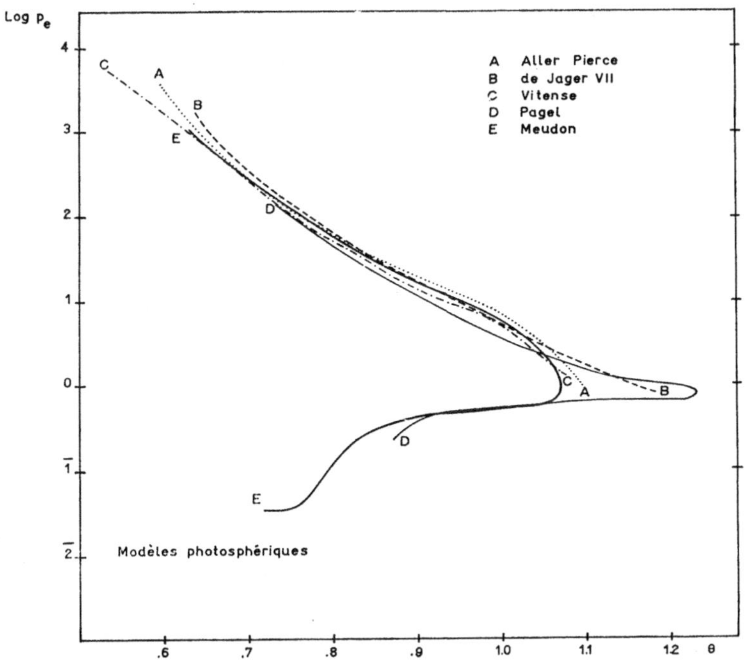

Fig. 6. Modèles de la photosphère. Nous avons reproduit sur cette figure certains des modèles moyens de la photosphère. Nous devons préciser que nous ne considérons pas ces modèles comme meilleurs que d'autres, mais ils sont typiques. Il s'agit uniquement de modèles contruits à partir des mesures, des relations d'équilibre hydrostatique et, pour la plus grande partie, des relations de l'ÉTL. Notations: (A) ALLER, PIERCE (GOLDBERG, MÜLLER, ALLER, 1960), (B) DE JAGER VII (1952), (C) VITENSE (1954), (D) PAGEL (1956), (E) Meudon: LEFÈVRE et PECKER (1961).

La figure 7 reproduit le résultat des calculs et constitue un raffinement de la figure 4. On a simplement calculé la quantié:

$$\tau_0 = \int_0^1 \frac{\kappa_0}{\kappa_\lambda} \, d\tau_\lambda \,. \tag{1}$$

On voit alors que, pour les deux modèles, la profondeur du minimum de température est:

Modèle D, log $\tau_{m0} = \bar{3}.7$ à $\bar{3}.9$ (le minimum du modèle se trouve à log $\tau_0 = \bar{2}.2$)
Modèle E, log $\tau_{m0} = \bar{3}.5$ à $\bar{3}.9$ (le minimum du modèle se trouve à log $\tau_0 = \bar{3}.7$)
Nous serions donc tenté de choisir $\bar{3}.7$, et le modèle E.

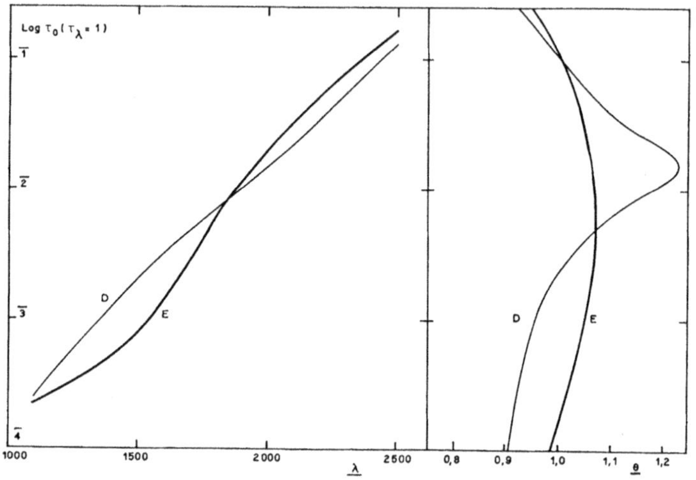

Fig. 7. On voit pour les modèles D et E (voir légende de la figure 6) le résultat du calcul mentionné dans le teste (à gauche). A droite, représentation des deux modèles, avec la même échelle de profondeurs optiques que dans la figure de gauche.

Ce résultat est intermédiaire entre les résultats antérieurs grossiers de l'auteur et ceux de DE JAGER (1961).

Comment les diverses hypothèses pourraient-elles influencer ce résultat?

L'hypothèse (b) ($\tau_\lambda = 1$) n'est vraie en toute rigueur que pour un modèle où la fonction-source est une fonction linéaire de la profondeur. Ici au contraire cette fonction-source est proche d'une fonction parabolique, au voisinage du minimum de température. D'après les recherches antérieures de DUMONT et PECKER (1960, 1961), la valeur de τ_λ doit donc considérablement différer de l'unité. On pourrait trouver 0.3, changeant d'autant les valeurs de τ_{m_0} données ci-dessus, et les abaissant à $\bar{3}.0$ environ au lieu de $\bar{3}.7$.

L'hypothèse (a) (ÉTL pour les raies) est sans doute fausse; nous l'avons dit. Les raies formées le plus souvent (peut-être!) par collisions, ont une fonction source minimum pour une valeur de plus petite que τ_{m_0}.

En choisissant cette valeur comme celle de τ_{m_0}, nous sous-estimons τ_{m_0}. Mais de quelle quantité? Il est bien difficile de le dire. Si l'on se réfère à la figure 8, reproduisant la température de brillance T_b dans le spectre en fonction de la longueur d'onde, l'erreur ici n'est sans doute pas très importante, car le minimum assez plat de température de brillance se produit entre 1600 et 2100

Å. Pourtant la figure 9, reproduite d'après Tousey, semble indiquer que le minimum de T_b correspond à une longueur d'onde inférieure à 1800 Å. Cette constatation est contradictoire avec la précédente; elle correspond à une surestimation de τ_{m_0} si l'on choisit l'apparition du caractère d'émission comme critère de la chromosphère au lieu du signe du gradient de température.

L'hypothèse (d) et l'hypothèse (e) sont inchiffrables. Quant à l'hypothèse

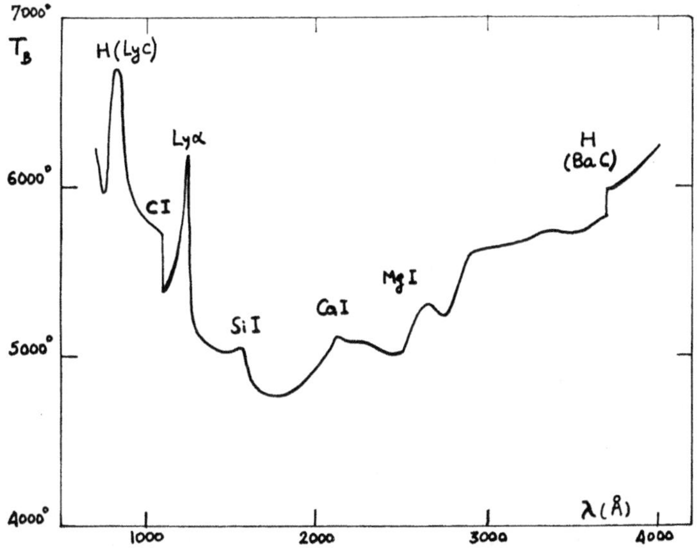

Fig. 8. Températures de brillances d'après PAGEL (1963). Le graphique original de Pagel, compilé d'après les différentes observations, a été modifié légèrement dans la région 1800 Å pour se conformer à l'allure générale des données de Tousey (Figure 9 ci-dessous).

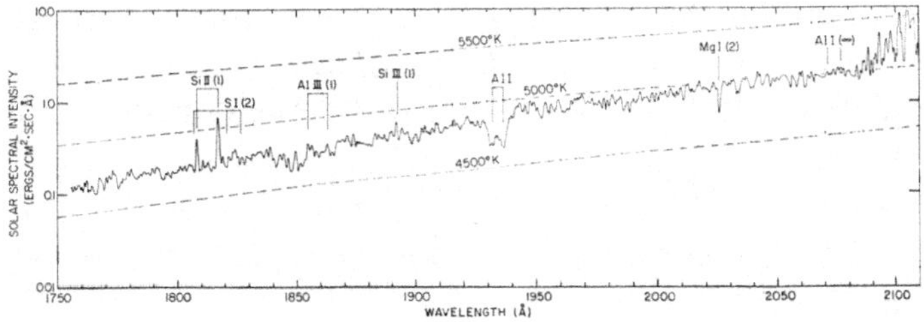

Fig. 9. Enregistrement du spectre solaire UV au voisinage de 1800 Å d'après TOUSEY. (Permission US Navy.)

(c) (abondances), on peut dire que dans la région considérée, les absorbants continus principaux étant le Magnésium et le Calcium, une sous-estimation de l'abondance de ces éléments par Vitense conduirait à une sous-estimation du rapport x_λ/x_0 et donc à une sur-estimation de τ_{m_0}. Là encore, de quel ordre de grandeur est l'erreur? Si on se base sur la comparaison entre les abondances utilisées par VITENSE (1951), et celles de GOLDBERG, MÜLLER et ALLER (1960), on trouve une erreur de l'ordre de 0.10 en logarithme (en sous-estimation si l'on admet que ces derniers auteurs sont plus proches de la vérité).

Il est donc en réalité très prématuré de conclure. Par ailleurs, une information additionnelle doit être utilisée: les intensités des raies (d'émission ou d'absorption) au voisinage de 1800 Å, ainsi que les données des figures 8 et 9. Ces levées informations supplémentaires, si les autres hypothèses pouvaient être par ailleurs, devraient nous permettre de déterminer mieux le modèle au voisinage de τ_{m_0}.

Notons en effet que le modèle D (forts gradients) doit conduire à des raies très nettement plus fortes que le modèle E.

On pourrait au contraire envisager d'utiliser les données IR pour fixer la variation de la température, et de déduire des mesures UV les caractéristiques de l'opacité.

3. *A partir du profil des raies d'absorption*

Naturellement, il ne faut pas oublier qu'une source importante d'information sur la transition entre chromosphère et photosphère est constituée par *certaines portions du profil des raies d'absorption*. D'un intérêt particulier à cet égard sont les raies pour lesquelles la fonction-source est d'origine collisionnelle, car ces raies sont une meilleure image de la température électronique, ainsi que Thomas l'a souvent fait remarquer (voir notamment THOMAS et ATHAY (1961). Citons ainsi les raies de Ca II et celles, dans le proche UV, de Mg II. Mais bien d'autre raies peuvent être utilisées, faibles ou fortes. C'est d'ailleurs d'une telle utilisation que provient le modèle de PAGEL (1955, 1956) qui a utilisé les raies infrarouge du fer neutre.

De même, le modèle de BÖHM-VITENSE (1954) tient compte des raies de l'hydrogène. Toutefois, ces deux modèles sont fort différents, puisque celui de BÖHM-VITENSE ne comporte pas de chromosphère: le raison en est simplement une confusion, alors explicable, entre T_{exc} et T_e. Alors que PAGEL justifie, pour les raies qu'il utilise, l'hypothèse de l'ETL, cette justification n'est pas possible – nous y reviendrons – dans le cas de l'hydrogène. Il est évident, que, dans chaque cas, le passage de T_{exc} à T_e est difficile et laborieux, même si T_{exc} est étroitement lié à T_e (raies dominées par les collisions). De plus, dans les raies, l'opacité varie si vite d'un point à un autre que le manque de pouvoir de résolution spectrale est aussi un manque de résolution en profondeur, et que la structure de la région de transition est donc difficile à extraire des profils de raies. Le cas le plus prometteur serait celui des raies UV de Mg II.

C. RÉGIONS PROFONDES DE LA PHOTOSPHÈRE

Si, dans le spectre continu, les régions de l'IR lointain et de l'UV vers 1800 Å permettent de préciser la structure de la région de transition entre la photosphère et la chromosphère, il est en revanche évident à partir de l'examen de la figure 4, que la région au voisinage de 2 microns doit permettre l'étude des régions les plus profondes de la photosphère. Cette région a été en fait mesurée, et les mesures incorporées dans la détermination des modèles; la discussion en a été faite bien des fois (voir notamment GOLDBERG et PIERCE, 1959; DE JAGER, 1959). C'est H^- qui est l'absorbant principal, et on peut considérer les calculs ÉTL de cette absorption comme corrects (ainsi que l'a montré PAGEL, 1957), surtout dans ces régions profondes.

En revanche, aucune observation de granulation n'a été faite dans cette région; cela serait pourtant de la plus haute importance, car c'est là que l'on peut pénétrer le plus profondément, et que par conséquent l'étude des inhomogénéités fournirait le plus de renseignements précis sur le mécanisme profond de leur origine.

2. Le Modèle thermodynamique de la photosphère

A. PRINCIPES DE L'OBTENTION EMPIRIQUE D'UN MODÈLE $T(\tau)$, À PARTIR DU SPECTRE CONTINU

À chaque longueur d'onde λ, l'intensité observée en un point du disque (défini par la valeur $\mu = \cos\theta$ du cosinus de l'angle des rayons lumineux avec la droite qui joint le centre du soleil au point étudié) dépend de la variation avec la profondeur de la fonction-source S, et de l'opacité, qui elle-même dépend de la population des différents atomes ou particules absorbantes dans leurs différents états d'excitation ou d'ionisation.

Si t est la profondeur optique, à cette longueur d'onde, le long du rayon visuel, alors l'expression rigoureuse de l'intensité est:

$$I = \int_0^\infty S(t) e^{-t} \mathrm{d}t \ . \tag{2}$$

Pour extraire un "modèle" de cette relation, il faut plusieurs hypothèses complémentaires.

Tout d'abord, (hypothèse (a) = sphéricité), on admet qu'entre la profondeur optique t et la profondeur optique τ suivant la normale aux couches de la photosphère, existe la relation:

$$\mu t = \tau$$

Enfin, on admet que la fonction-source dans le continu est identifiable à la fonction de PLANCK de la température électronique (hypothèse (b): ÉTL-Continu):

$$S(\tau) = B_\lambda[T(\tau)] \ . \tag{4}$$

Enfin, on admet que la variation de S (et donc, selon l'hypothèse (b), de T) est régulière, monotone même, et l'on exclut ainsi a priori la détection d'inhomogénéités.

Nous reviendrons sur ces hypothèses. Dans ces conditions, on peut inverser la relation intégrale:

$$I(\mu) = \int_0^\infty B(\tau)\ e^{-\tau/\mu} \frac{\mathrm{d}\tau}{\mu} \tag{5}$$

et en tirer la variation de B avec τ à la longueur d'onde considérée, soit la relation $T(\tau_\lambda)$. Le plus souvent on considère la relation $T(\tau_0)$ comme caractéristique du modèle. Notons ici (nous y reviendrons) que ce n'en est *qu'une* caractéristique: il nous faut aussi la relation entre la profondeur géométrique h et la valeur de τ_0, ainsi que les populations des différents niveaux, pressions, etc., en fonction de τ_0 ou de h.

L'inversion de la relation (5) est d'abord un problème de nature numérique et mathématique. L'hypothèse b lui donne un certain sens. Il est en effet évident que si on trouve une solution $S(\tau)$ à l'équation (5), une autre solution serait obtenue en *ajoutant* à la première solution telle fonction (Figure 10) dont la transformation par l'intégrale du second membre de (5) serait nulle (puisque

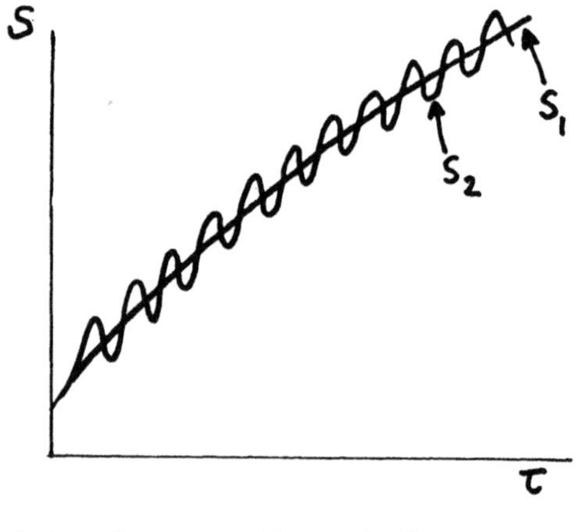

Fig. 10. Non-unicité de la fonction-source. La courbe S_1 représente une solution; la courbe S_2, obtenue en superposant à S_2 une fonction convenablement choisie, est aussi une solution.

en effet la transformation en question est linéaire). Barbier (1951) en effet a déjà remarqué cette circonstance. Il est inutile de revenir longuement sur ce point développé aussi par Böhm (1961). Mais il importe de ne pas le sous-estimer. En effet, même en admettant que la fonction $I(\mu)$ est connue pour $0 \leqslant \mu \leqslant 1$ (intervalle complet), avec une précision donnée (par exemple avec trois chiffres significatifs exacts), il est impossible, sinon de trouver une solution S qui rende bien compte de la fonction I à la précision souhaitée, mais en tous cas d'évaluer la précision de cette solution. Quelques méthodes classiques ou nouvelles de cette inversion de la relation de Laplace sont décrites dans l'appendice.

Toutes les méthodes mises en oeuvre jusqu'à présent ont des inconvénients communs: (a) Un accident local de $S(\tau)$ sera dilué dans $I(\mu)$ et ne pourra pas être mis en évidence de façon nette; (b) même pour des fonctions régulières, aucune des méthodes en question n'a réellement permis de calculer $S(\tau)$ avec une précision suffisante, puisque les résidus $I(\mu)_{\text{obs}} - I(\mu)_{\text{calc}}$ restent plus importants que les erreurs d'observation; de plus, l'utilisation, à partir des mêmes valeurs mesurées de $I(\mu)$, de différentes méthodes, conduit à des fonctions $S(\tau)$ d'allures tout à fait différentes; (c) Enfin, les mesures ne vont pas jusqu'à l'extrême bord solaire. Le calcul n'utilisera donc que les valeurs de l'intensité correspondant à des points assez distants du point $\mu = 0$. La solution $S(\tau)$ obtenue permettra de calculer $I(\mu)$ aux points où on n'a pu le mesurer. Les différentes méthodes utillisées rendent parfois correctement (en gros) compte de la partie mesurée de l'intensité, mais aux points de petits μ, on en déduit des valeurs absurdes, bien plus mauvaises souvent que celles que fourniraient une extrapolation graphique.

En admettant que ces difficultés soient levées, ou en tout cas oubliées momentanément, nous disposons donc d'une relation entre T, et, pour chaque longueur d'onde λ, la profondeur optique τ_λ à cette longueur d'onde.

Si l'on admet qu'un seul modèle $T(\tau_0)$ doit être cohérent avec ces différents modèles monochromatiques, alors on peut déduire des mesures ce modèle et la variation de l'opacité avec la longueur d'onde. En effet, il est possible de déduire

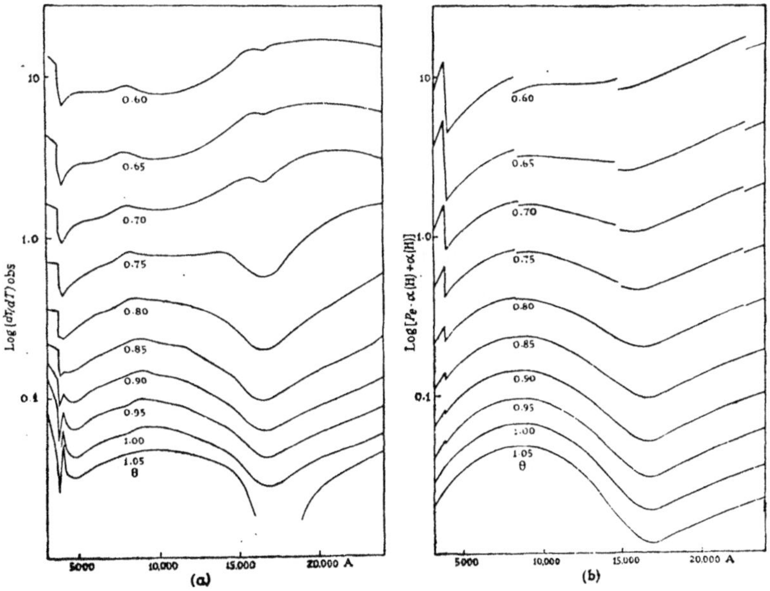

Fig. 11. A gauche (a), le logarithme observé de l'inverse du graient de température à droite (b), le logarithme du coefficient d'absorption théorique. Noter les différences du côte ultraviolet, ainsi qu'aux grandes profondeurs. (D'après GOLDBERG et PIERCE, 1959.)

des mesures la quantité $dT/d\tau_\lambda$ proportionnelle à l'opacité. La figure 11 reproduite d'après GOLDBERG, PIERCE (1959) montre la comparaison faite enter cette quantité ainsi calculée et l'opacité calculée directement à partir des coefficients d'absorption de Chandrasekhar relatifs à H^-. L'accord est évidemment très satisfaisant sur ce schéma.

Il faut pourtant tempérer cet accord de plusieurs remarques: (a) Les courbes des figures 11a et 11b ne coïncident évidemment qu'à une constante près; (b) Même ainsi, aux longueurs d'onde plus courtes que la discontinuité de BALMER, on observe une absorption supplémentaire non identifiée.

En ce qui concerne le premier point, il est facile d'écrire le rapport de proportionnalité entre le rapport $dT/d\tau_\lambda$ et l'opacité κ_λ par gramme de matière, c'est (en utilisant l'équation d'équilibre hydrostatique):

$$\frac{1}{\kappa_\nu}\frac{d\tau_\lambda}{dT} = \varrho(x)\frac{dx}{dT} = \frac{1}{g}\frac{dp}{dT}\,. \tag{6}$$

Des mesures – et de la connaissance de κ_ν – on peut donc déduire dp/dT, donc, par une intégration consécutive, $p(T)$ à une constante près (c'est ce qui a été fait, notamment, par PEYTURAUX, 1952). Cette détermination repose d'ailleurs sur l'emploi de l'équation de l'équilibre hydrostatique, d'une part, de la relation $T(\tau)$ d'autre part; elle ne doit pas donc différer des déterminations classiques de la pression. Fixer la constante dans la comparaison des courbes de la figure 11 ne sert donc rigoureusement à rien, en principe.

Examinons brièvement maintenant les différences observées dans l'UV; nous pouvons suivre MICHARD (1953) et rendre responsable de ces différences l'absorption métallique et la diffusion RAYLEIGH. Nous pouvons aussi suivre la suggestion de ZWAAN (1962) selon laquelle l'édifice $H + H(^3\Sigma_u^+)$ serait responsable de l'effet en question. Mais il est hors de doute que toute conclusion serait bien mieux étayée par de bonnes mesures dans l'UV lointain. C'est ce que le service d'aéronomie du C.N.R.S. est en train de faire, en fusée, en obtenant l'image photographique du soleil en plusieurs longueurs d'onde de l'UV lointain (BLAMONT et PECKER 1961). Les instruments sont aujourd'hui prêts pour le tir; ils utilisent les filtres à bande passante étroite décrits par BONNET et COURTES (1962).

Remarques

1. Discussion préliminaire des effets de l'hypothèse a (sphéricité). A la suite de REDMAN (1943) et d'EVANS (1947), l'auteur a proposé jadis de faire intervenir une certaine *rugosité* des couches photosphériques (voir aussi MINNAERT, 1953). À cette rugosité de caractère géométrique (rugosité des isobares) et qui prêtait à des critiques de nature hydrodynamique, l'auteur en est venu (voir LEFÈVRE et PECKER, 1961) à substituer la notion de rugosité optique, la profondeur optique 1 correspondant à une profondeur géométrique variable d'un point à l'autre. Les effets sont identiques: au bord du soleil, les régions "creuses" sont cachées. Toutefois, Y. CUNY (1963) a montré que cet effet ne pouvait influencer quantitativement les observations du continu comme nous l'avions cru antérieurement. En revanche, on le verra, les raies sont affectées par cet effet.

2. *Effets de l'hypothèse* (*b*). Là encore, Y. CUNY (1961) a attaqué le problème des populations des niveaux de l'atome d'hydrogène, hors ÉTL, dans la photosphère solaire. La figure 21 (voir ci-dessous) reproduit son résultat. Nous noterons que H^- est en ÉTL, et que le doute pourrait seulement venir de la discontinuité de Balmer et du continu de Balmer. Le calcul montre (Y. CUNY) que tout se passe pour cette région du spectre continu photosphérique comme si l'ÉTL était réalisé. Une amélioration notable de la précision de ces mesures et des calculs devrait cependant peut-être entraîner une révision de l'hypothèse.

B. PRESSION ET PRESSION ÉLECTRONIQUE

1. Nous avons vu (Équation 6) que les hypothèses de l'équilibre hydrostatique et de l'ÉTL permettent de déterminer la pression. Nous rappelerons ici tout d'abord (voir GOLDBERG-PIERCE, 1959, p. 26 et suivantes) que la gravité effective doit en principe tenir compte de l'existence de la pression de radiation et de la pression turbulente; mais ces effets sont négligeables dans le cas solaire, si bien que la valeur de g qui intervient est la valeur classique $g = 10^{4.4}$. Il est aussi simple, et équivalent en principe, d'intégrer tout simplement l'équation:

$$\mathrm{d}p = \frac{g}{\kappa_\nu}\,\mathrm{d}\tau_\nu\,. \tag{7}$$

La relation $T(\tau_\nu)$ est connue – ce à n'importe quelle longueur d'onde. La relation $\kappa_\nu(T, p_e)$ est connue également, si l'on considère l'étude de l'assombrissement comme ayant convenablement prouvé la valeur des déterminations usuelles. Une relation entre T, p_e et p est fixée par l'hypothèse ÉTL donc par la relation de SAHA-BOLTZMANN. Le système donne alors la valeur de p et de p_e en fonction de la profondeur optique. On peut vérifier que ces valeurs ne dépendent pas de la longueur d'onde (PIERCE et WADDELL, 1961).

2. Il est possible également de déterminer p_e directement, sans utiliser l'hypothèse ÉTL, notamment.

On peut en effet, en supposant que l'opacité est due principalement à l'ion H^- et à l'hydrogène atomique, écrire, à toute longueur d'onde:

$$\frac{\mathrm{d}\tau_\nu}{\mathrm{d}T} = \kappa_\nu \varrho\,(x)\,\frac{\mathrm{d}x}{\mathrm{d}T} = \frac{1 - x_{\mathrm{H}}}{m_{\mathrm{H}}}\,[p_e\alpha_\nu(\mathrm{H}^-) + \alpha_\nu(\mathrm{H})]\,\frac{\mathrm{d}x}{\mathrm{d}T} \tag{8}$$

équation qui ne comporte pratiquement aucune hypothèse (celles-ci interviennent dans le calcul de x_{H}) Il en résulte:

$$\frac{\mathrm{d}\tau_{\nu_1}}{\mathrm{d}\tau_{\nu_2}} = \frac{p_e\alpha_{\nu_1}(\mathrm{H}^-) + \alpha_{\nu_1}(\mathrm{H})}{p_e\alpha_{\nu_2}(\mathrm{H}^-) + \alpha_{\nu_2}(\mathrm{H})} \tag{9}$$

d'où

$$p_e = \frac{\alpha_{\nu_2}(\mathrm{H})\, \mathrm{d}\tau_{\nu_1} - \alpha_{\nu_1}(\mathrm{H})\, \mathrm{d}\tau_{\nu_2}}{\alpha_{\nu_2}(\mathrm{H}^-)\, \mathrm{d}\tau_{\nu_1} - \alpha_{\nu_1}(\mathrm{H}^-)\, \mathrm{d}\tau_{\nu_2}}\,. \tag{10}$$

Ce calcul, effectué par Pierce et Waddell (1961), met en évidence une différence considérable avec les résultats d'un calcul de type classique (Figure 12). Quelle en est la raison? À notre avis, il faut d'abord chercher si les *écarts à l'ÉTL* et une certaine sous-ionisation qui en serait la conséquence, peuvent être in-

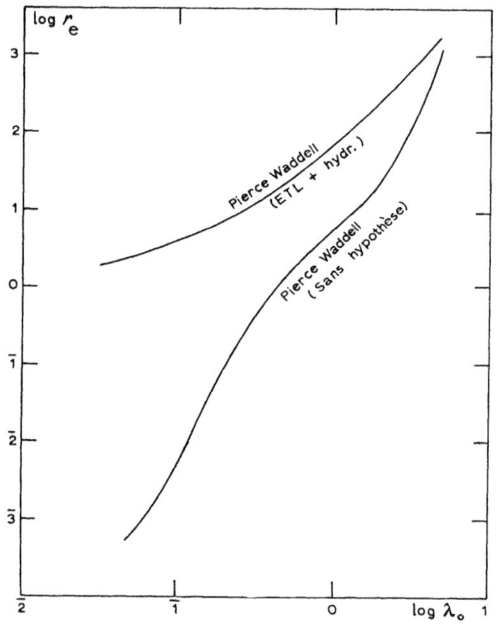

Fig. 12. La courbe inférieure dépend considérablement des erreurs de mesure (voir texte), mais l'effet des écarts à l'ÉTL sur la pression électronique n'est peut-être pas négligeable. (D'après Pierce et Waddell, 1961.)

criminés: Un facteur 1000 est un phénomène qui mérite d'être traité sérieusement! Certes, dans le cas de H^-, et de l'hydrogène atomique (ainsi que l'ont respectivement montré Pagel (1957) et Y. Cuny (1963)), les écarts à l'ÉTL n'entraînent aucun effet appréciable. Mais ne peut-il y avoir un important retard à l'ionisation des métaux, qui fournissent les électrons libres? Cette situation est bien celle de la chromosphère; des calculs comme ceux de Johnson (1963) prouvent que dans un cas comme celui de Na, il en est ainsi (vers $\tau \sim 0.001$); mais l'abondance des éléments ionisés dans la photosphère semble indiquer qu'un facteur 10^3 tel que celui qui résulte des données de Pierce et Waddell est nettement trop élevé pour une interprétation de ce type.

Il est plus vraisemblable d'attribuer l'effet en question à l'extrême sensibilité du rapport (10) aux erreurs de mesures; le dénominateur est voisin de zéro, et serait identiquement nul si H^- était le seul absorbant.

3. Une autre donnée sur les pressions pourrait venir des mesures de *polarisation*. On sait que la polarisation du rayonnement continu a pu être mesurée par LEROY et DOLLFUS (1960) en 4 longueurs d'onde, et en fonction de λ. S. DÉBARBAT (1963) a tenté l'interprétation de ces mesures, en admettant une diffusion due à la fois aux deux processus Thomson (électrons) et Rayleigh (atomes) comme il a été montré par VOIGT (1951). Ce calcul, plus détaillé que celui que l'auteur avait jadis proposé (1950), utilise le modèle de Meudon (pour les températures, cela est sans doute justifié, mais pour les pressions, cela l'est moins). Avec les données usuelles, elle obtient une polarisation beaucoup moins élevée (mais avec le même comportement en fonction de la longueur d'onde et de la position sur le disque) que les mesures (Figures 13). Or la polarisation dépend directement de la quantité

$$x = [p_e\sigma_{Th} + p\sigma_{Ra}]/[\kappa_\lambda + p_e\sigma_{Th} + p\sigma_{Ra}] \,. \qquad (11)$$

Ce fait semble donc aller dans la direction opposée à celle des observations de Pierce et Waddell.*

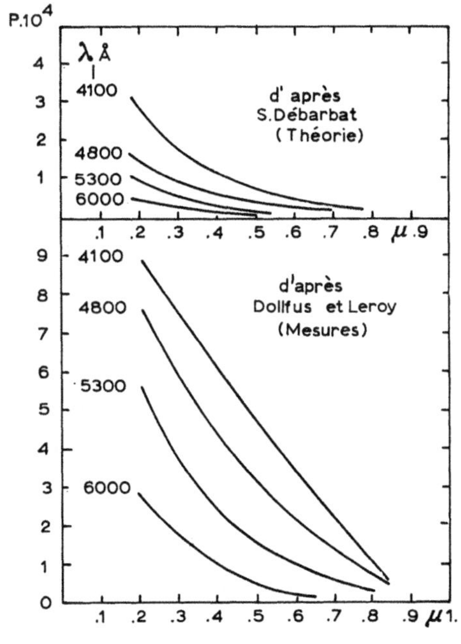

Fig. 13. Pour quatre longueurs d'onde, les mesures et les calculs fournissent, en fonction de μ, le taux de polarisation P. (Figure du haut d'après DEBARBAT, 1963 et figure du bas d'après DOLLFUS et LEROY, 1960.)

* Discutant cette contribution, Dr. Severny a évoqué des mesures de la polarisation, qui s'accordent plus à la théorie que celles de Leroy et Dollfus.

Il semble difficile aujourd'hui de conclure. Il y a certainement des faits à expliquer (polarisation élevée d'une part, d'autrepart variation avec la longueur d'onde de l'assombrissement IR) qui sont étroitement liés à la pression et à la pression électronique. Les opacités en jeu semblent raisonnablement sûres. Écarts à l'ÉTL? Ce serait vraisemblable si on n'avait pas prouvé qu'ils étaient numériquement trop faibles. Inhomogénéités? Aucune étude n'a été faite de leur effet sur les deux phénomènes mentionnés, mais on voit mal comment elles pourraient expliquer les deux effets en question.

Peut-être aussi de nouvelles mesures pourraient-elles réduire considérablement les difficultés? D'autre part, des champs magnétiques peuvent affecter la polarisation (il s'agit de zones spectrales contenant de nombreuses raies)... Les deux problèmes restent posés.

4. Les études des ailes de raies formées par collisions, qu'il s'agisse des ailes STARK, des raies de l'hydrogène, ou des ailes des raies métalliques dues à l'amortissement, ont apporté peu de renseignements valables. Dans le cas des métaux, la théorie des constantes d'amortissement est trop mauvaise (elle en est à la loi de VAN DER WAALS!) pour permettre des déterminations sérieuses, à partir de profils individuels, comme à partir de la seconde partie linéaire de la courbe de croissance.

La théorie des raies de Balmer a progressé considérablement depuis quelques années (nous renvoyons pour une étude générale des aspects astronomiques de cette question, à l'article de BÖHM (1960)) il est évident que le calcul des pressions à partir des ailes des raies de Balmer, tel que celui jadis préconisé par DE JAGER (1952), devrait être repris avec les nouvelles théories; mais on sait depuis ce travail à quel point aussi les ailes des raies de Balmer (comme toutes les raies d'ailleurs) sont influencées par les inhomogénéités de la photosphère. Les deux problèmes devraient être résolus simultanément; ce problème devrait avoir une grande urgence étant donné la qualité des mesures disponibles (voir, par exemple, WHITE (1962)).

5. C'est sans doute dans ce paragraphe qu'il importe aussi de citer la remarquable série de travaux de GIOVANELLI et WILSON (GIOVANELLI, 1959; WILSON, 1960, 1961, 1962ab). Ces travaux ont certes comme but principal de permettre la solution du problème de transfert dans un milieu inhomogène. GIOVANELLI introduit une quantité (nulle dans le cas de l'équilibre radiatif), qui est ε, l'énergie libérée en chaleur par unité de temps, de volume, et d'angle solide. Il écrit l'équation approchée du transfert:

$$\nabla \cdot \left[\frac{1}{\kappa} \nabla J\right] = -12\pi\varepsilon . \qquad (12)$$

Si l'on admet que l'observation (ou tout autre procédé, purement théorique par exemple) fournit J, alors l'équation fournit κ en chaque point, si on admet que $\varepsilon = 0$. C'est par l'intermédiaire de κ qu'on pourrait alors déterminer la pression. Précisément, l'équation-clef des travaux de GIOVANELLI et WILSON est celle qui fixe κ:

$$\kappa\,(z, x) = -\frac{(\frac{2}{3})\,[(\partial J)/(\partial z)]}{J(x_0, 0)\,[(\partial x_0)/(\partial x)]}\,. \tag{13}$$

C'est, essentiellement, la relation (6); et la méthodologie est très voisine de celle décrite ci-dessus.

Dans cette équation, x et z représentent naturellement les mesures des coordonnées dans le plan horizontal et sur la verticale. La relation:

$$\frac{\partial x'}{\partial z'} = \frac{(\partial J)/(\partial x)}{(\partial J)/(\partial z)} \tag{14}$$

définit les courbes du plan (x, z). Le point où la courbe correspondant au point (x, z) intersecte le plan $z = 0$, est le point x_0, 0.

Cette équation (13) n'est qu'approchée, comme on le voit aussitôt en prenant la forme usuelle (pour $x_0 = x$); $J = J_0[1 + {}^3/_2\tau]$. Mais le procédé pourrait peut-être être envisagé avec plus de généralité, l'idée étant de tirer κ des données mesurées relatives à la fonction-source.

Nous reviendrons sur les applications de ces recherches dans le cas des structures inhomogènes.

C. INTERVENTION D'UNE STRUCTURE FINE

L'étude, depuis plus d'un siècle, de la granulation solaire, l'aspect des raies d'absorption, et même les difficultés d'interprétation rencontrées (Figure 14)

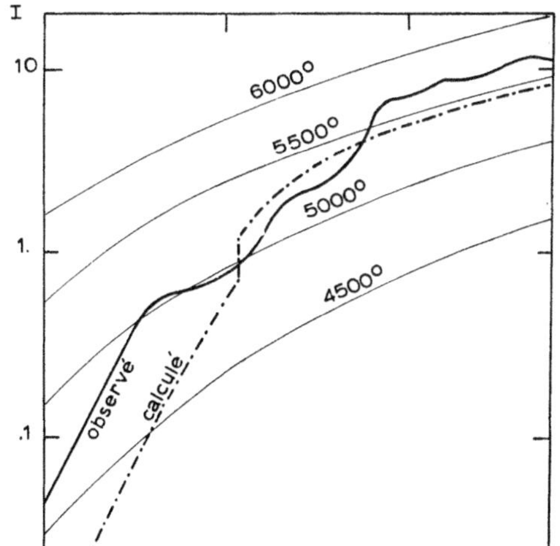

Fig. 14. La courbe en traits pleins correspond aux observations. La courbe calculée (avec un modèle homogène) sous-estime l'influence des régions chaudes. (D'après DE JAGER, 1955.)

dans l'interprétation de certaines caractéristiques du spectre continu, font penser que la photosphère est loin d'être homogène; les quantités déterminées ci-dessus ne sont, c'est bien évident, que des "valeurs moyennes".

L'attitude en face de ce fait a été l'objet dans le passé de deux attitudes différentes, voire contradictoires.

Les uns ont supposé une distribution simple des températures, caractérisée par deux ou trois valeurs discrètes à chaque profondeur (Böhm, 1954; Voigt, 1956; Schröter, 1957; Hubenet, 1961; de Jager, 1959).

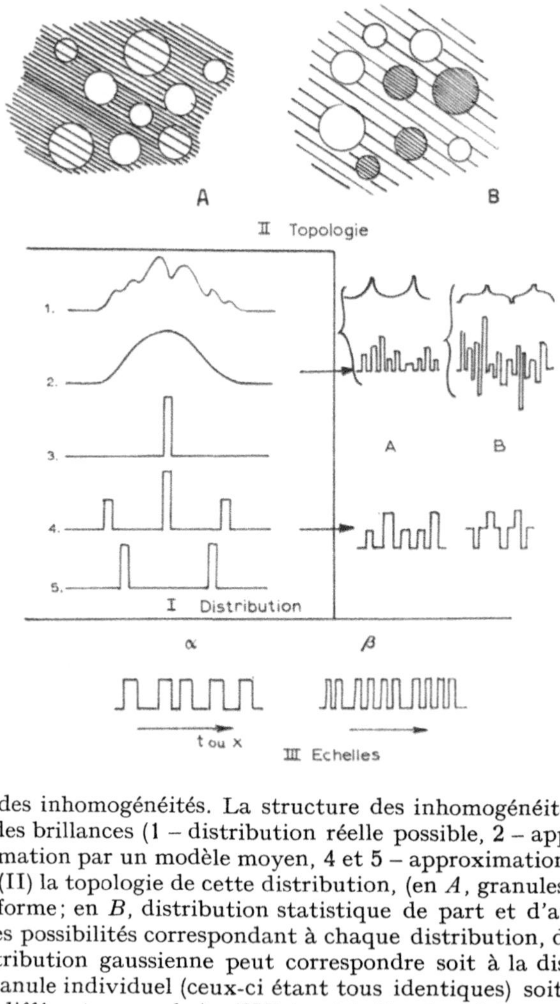

Fig. 15. Structure des inhomogénéités. La structure des inhomogénéités est définie par: (I) la distribution des brillances (1 – distribution réelle possible, 2 – approximation gaussienne, 3 – approximation par un modèle moyen, 4 et 5 – approximations par des modèles à 3 ou 2 colonnes); (II) la topologie de cette distribution, (en A, granules de plusieurs brillances sur fond uniforme; en B, distribution statistique de part et d'autre d'une valeur moyenne). Noter les possibilités correspondant à chaque distribution, données ici à titre d'exemples: la distribution gaussienne peut correspondre soit à la distribution de brillance de chaque granule individuel (ceux-ci étant tous identiques) soit à la distribution des brillances des différents granules); (III) les échelles (spatiales ou temporelles). Au même type de distribution, à la même structure topologique, peuvent correspondre différentes échelles spatiales et temporelles (exemple dans le cas de la distribution 5). *Remarque*: Il est évident que ce schéma reste très simpliste: une véritable théorie statistique des fluctuations des quantités physiques en chaque point de la photosphère est évidemment nécessaire.

Les autres ont plutôt tendance à admettre une distribution à un seul maximum et de genre gaussien (Figure 15). Mais il faut préciser que certains considèrent un modèle à 3 températures comme la représentation schématique d'une distribution en cloche.

Cette opposition s'est aussi manifestée (bien que le problème ne soit pas strictement le même) dans les attitudes diverses en face des clichés de granulation; la controverse dans ce domaine est toujours ouverte. Les uns affirment: "il y a des 'granules', et un 'fond' intergranulaire"; les autres disent: "il y a une distribution quasi-gaussienne des brillances autour d'une brillance moyenne; l'apparence granulaire qui dépend des longueurs caractéristiques de la distribution est associée à certaines propriétés topologiques dont la connaissance s'ajoute à celle de la distribution des brillances". Il est évident que la controverse en ce domaine ne peut s'arrêter que dans un futur encore lointain, lorsqu'on aura obtenu des clichés du soleil avec une résolution et avec une qualité photométrique supérieures, et de beaucoup, aux performances actuelles, déjà marginales.

Ces difficultés sont naturellement compliquées du fait que la déformation instrumentale des objets étale les images, et modifie la distribution réelle des brillances en y superposant une distribution quasi-gaussienne. Dans des mesures de qualité moyenne, c'est même seulement celle-ci qui influence les mesures.

Il est sans doute prématuré de donner ici une conclusion; en fait le problème, à mon avis, se pose autrement.

Toute distribution statistique de températures dans un certain volume est caractérisée:

(a) Par la fonction $P(T)$ (probabilité pour que la valeur de la température soit comprise entre T et $T + \mathrm{d}T$, divisée par $\mathrm{d}T$ – ou fonction de distribution. *Les moments successifs* de la température sont les quantités

$$m_k = \int_0^\infty P(T) T^k \mathrm{d}T. \tag{15}$$

Le moment d'ordre 1 est la valeur moyenne T.

On peut définir aussi les moments "centrés" d'ordre n (le moment d'ordre 2 est la variance; la racine carrée de la variance est l'écart quadratique moyen).

(b) Par la fonction d'autocorrélation spatiale de cette distribution qui fixe l'échelle spatiale du phénomène (*dimensions* caractéristiques de la granulation).

(c) Par la fonction d'autocorrélation temporelle de cette distribution: elle fixe la durée de vie des phénomènes.

(d) Par des caractères qu'on pourrait nommer *topologiques* (ceux qui permettent par exemple de faire la différence entre les deux aspects A et B de la Figure 15) qui dépendent étroitement de la nature physique du phénomène.

Naturellement chacun de ces caractères varie avec la profondeur dans l'atmosphère de l'étoile.

Il est évident qu'à un certain degré d'approximation, deux distributions peuvent être équivalentes: il n'est pas encore évident que nous devions préférer une distribution à deux ou trois pics à une distribution quasi gaussienne à un seul maximum, ou l'inverse. Les premiers moments de ces distributions peuvent coincider, et les observations ne sont pas encore sensibles aux moments suivants. La seule indication que nous ayons à ce sujet provient sans doute de l'étude des clichés obtenus par SCHWARZSCHILD et son équipe grâce au projet Stratoscope I, étude faite par F. EDMONDS (1962). La figure 16, reproduction d'une des figures d'Edmonds, montre la distribution mesurée des intensités. Elle est plus proche d'une représentation gaussienne que d'une représentation à deux ou trois pics; mais c'est le seul exemple que je connaisse d'une preuve analogue, les arguments usuellement utilisés ressortant plus de l'intuition que de la logique. Signalons que DE JAGER et NEVEN sont en faveur d'un modèle à deux colonnes, de préférence à un modèle à trois colonnes (1963).

Nous avons examiné aux paragraphes précédents comment les observations ont conduit à la détermination d'un modèle moyen. La justification de modèles hétérogènes vient de plusieurs arguments:

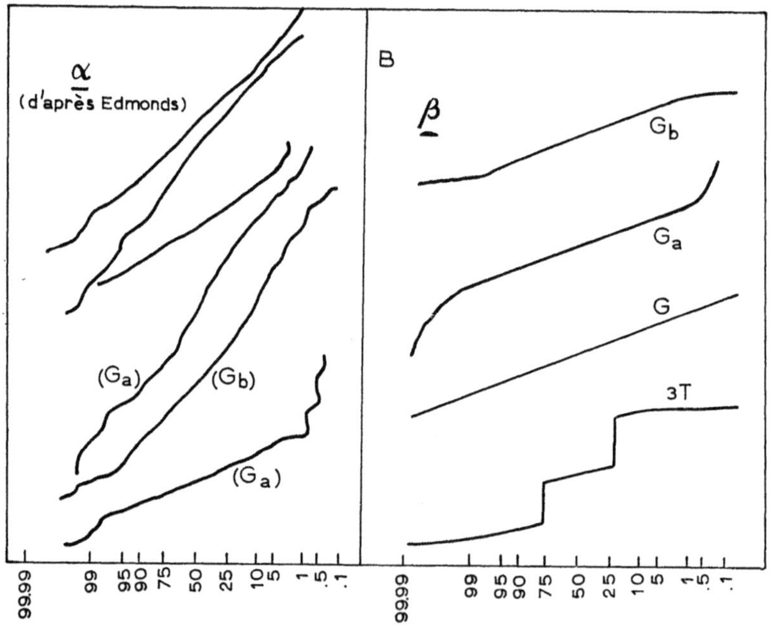

Fig. 16. Distribution des brillances sur les images photosphériques. A gauche (α) (d'après EDMONDS, 1962) analyse des clichés de Stratoscope I. En abscisse, pourcentage du nombre de points où l'intensité a une valeur supérieure à y, porté en ordonnée. Les échelles des ordonnées sont arbitraires. Plusieurs échantillonnages sont analysés. A droite (β), dans le même diagramme, courbes qui seraient obtenus à partir de modèles à 3 colonnes (3T), à distribution gaussienne (G) ; *Ga* et *Gb* correspondent à des distributions quasi-gaussiennes, respectivement à ailes très développées, et à ailes atrophiées.

(a) Existence de la *granulation.*

(b) Difficultés d'interprétation du *spectre continu.* (voir Figure 14). Un exemple bien caractéristique est celui de l'interprétation du spectre UV. On sait que la spectre UV, à cause des propriétés de la fonction de Planck, donnera, relativement, un plus grand poids aux régions chaudes qu'aux régions froides: c'est effectivement une différence entre les mesures, et les prédictions faites à l'aide du modèle moyen déterminé dans le domaine visible qui conduisent à attribuer l'effet aux inhomogénéités (DE JAGER, 1955). Une étude en fonction de la longueur d'onde telle que celle de PIERCE et WADDELL (1961) dépend nécessairement des inhomogénéités. Pourrait-elle être à l'origine des effets anormaux qu'ils ont trouvés? Il faudrait encore le démontrer, et montrer quel type d'inhomogénéités pourrait expliquer de tels effets. Le problème est posé.

(c) Comportement des *raies de Balmer, et de Paschen.* DE JAGER (1959ab). montre comment ces raies sont liées aux inhomogénéités et propose un modèle à deux colonnes, l'une ayant une distribution "rectangulaire" de température, les autres, les chaudes, une distribution "exponentielle".

(d) Comportement des *raies de Fraunhofer.* Les modèles à trois températures ou deux températures de BÖHM (1954), VOIGT (1956) etc., procèdent de cette légitimation: Il s'agit d'expliquer les variations entre le centre et le bord du disque solaire des largeurs équivalentes des raies fortes et faibles du fer. Le modèle de SCHRÖTER (1957) est essentiellement légitimé par l'étude des raies du sodium. Ces modèles sont discutés en détail par HUBENET (1961) et DE JAGER (1959a).

(e) *Intensités centrales des raies de Fraunhofer,* et leurs *variations* entre le bord et le centre du disque solaire. LEFÈVRE et PECKER (1961) ont mis en évidence l'existence d'un effet "d'arête de poisson" (ainsi nommé en raison de l'allure du diagramme obtenu): les fonctions sources représentant, au centre du disque, les intensités centrales des raies de plus d'un même multiplet ne coïncident pas avec les fonctions sources déterminées à partie de la variation centre-bord des intensités centrales des mêmes multiplets. L'analyse de Lefèvre et Pecker lie cette propriété à l'échelle de dimension de la granulation. Nous y reviendrons.

1. Distributions

De différentes analyses issues de ces différents types de données, il faut, semble-t-il, retenir d'abord les faits suivants, dont on peut considérer qu'ils représentent l'état actuel de la discussion relative aux distributions:

(a) *Un modèle moyen ne suffit pas à représenter les observations.*

(b) *Les inhomogénéités sont plus importantes aux grandes profondeurs; elles sont faibles en surface* (SCHRÖTER, 1957; LEFÈVRE et PECKER, 1961). Selon DE JAGER et NEVEN (1963), on peut admettre qu'elles commencent à être notables pour $\tau_0 > 0.1$. Cette limite étant plus profonde que celle qu'admettent SCHRÖTER, LEFÈVRE et PECKER. Toutefois la variation ΔT_b sur le disque solaire, mesurée par EDMONDS sur les clichés "Stratoscope I", ne confirme pas ces modèles. Cet auteur trouve un maximum prononcé à $\mu \sim 0.6$ (Figure 17) qui correspond à un ΔT maximum vers $\tau \sim 0.6$. Il est important de noter que WILSON a

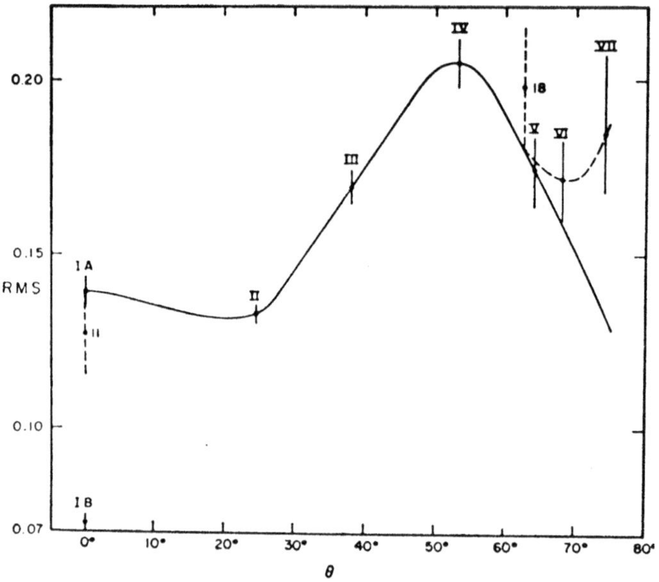

Fig. 17. Distribution des fluctuations d'éclat sur le disque solaire. En ordonnée, valeur de l'écart moyen relatif de la brillance, après correction des effets instrumentaux. Les points 11 et 18 représentent des régions particulières; la courbe en pointillé tient compte de la contribution des structures à grande échelle. (D'après Edmonds, 1962.)

appliqué à l'interprétation de ces mesures d'EDMONDS la méthode décrite plus haut et proposée par GIOVANELLI. Il utilise l'équation (13) sous sa forme plus générale:

$$\kappa(x, z) = -\frac{2}{3}\frac{\partial J}{\partial z}\left[J(x_0, 0)\frac{\partial x_0}{\partial x} - 8\pi\int_z^0 \varepsilon(x', z')\frac{\partial x'}{\partial x}\,\mathrm{d}z\right]^{-1} \tag{16}$$

et pour J et ε des développements (comportant des coefficients indéteiminés) choisis de telle sorte que les moyennes de J, d'ε, et de la valeur de κ résultant de (16) correspondent aux mesures d'assombrissement dans le spectre continu. WILSON prouve par une discussion de la valeur des coefficients indéterminés compatibles avec les mesures d'EDMONDS, que ces mesures ne peuvent s'expliquer que si l'on admet l'existence d'une émission ε issue de régions chaudes localisées.

(c) Ainsi que l'a démontré DE JAGER (1962), les analyses des clichés de granulation de RÖSCH (1962) et celles de BAHNG et SCHWARZSCHILD (1961), coïncident sensiblement; *les ΔT sont assez faibles à une même profondeur optique; au contraire à une même profondeur géométrique, ces différences sont importantes*, ainsi que l'avait remarqué depuis longtemps E. VITENSE (1953). DE JAGER et NEVEN (1963) viennent de préciser ces caractéristiques. On peut admettre que des fluctuations de 300° à 500° (à $\tau_0 = 1$) et de 670° à 1 200° (à la même profondeur géométrique ($\log p_g = 4.98$) sont raisonnables. Il s'agit des différences entre les températures extrêmes (double amplitude des fluctuations) et non de la valeur moyenne des écarts. Soulignons aussi, une fois de plus, que les contrastes de température dans les couches profondes sont masqués par les couches superficielles, plus proches de l'homogénéité thermique; c'est l'effet de "voile" décrit antérieurement par DE JAGER (1948) et DE JAGER et PECKER (1951).

(d) La seule analyse donnant quelque idée des fluctuations d'autres quantités que la température, par exemple des pressions et des pressions électroniques

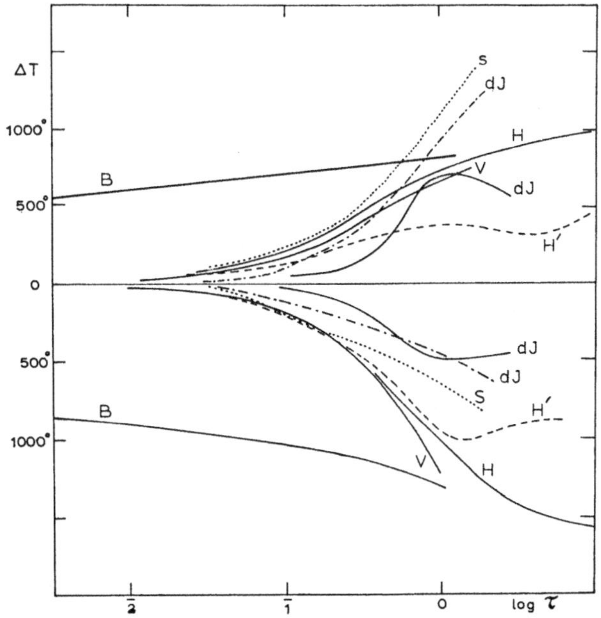

Fig. 18. Modèles inhomogènes de la photosphère. Cette figure porte en ordonnées les différences de températures entre colonnes chaudes (en haut), colonnes froides (d'une part) et les colonnes intermédiaires (modèles à 3 colonnes) ou moyennes (modèles à 2 colonnes). Les courbes en traits pleins correspondent aux modèles à 3 colonnes (B – BÖHM, 1954; V – VOIGT, 1956; H – HUBENET, 1960; dJ – DE JAGER, 1959). Les deux courbes en trait interrompu correspondent aux modèles à 2 colonnes (dJ – DE JAGER, 1954; S – SCHRÖTER, 1957). En abscisses, pour ces six courbes, le logarithme de la profondeur optique à 5000 Å pour le modèle moyen ou intermédiaire. Pour la courbe *H'* en pointillé, les abscisses représentent la profondeur optique à 5000 Å dans chaque colonne. Ces modèles sont inséparables naturellement de la donnée des proportions relatives (en surface traversée) correspondant à chaque colonne, ainsi que des pressions. On se reportera aux références originales.

ou des opacités, est celle de GIOVANELLI et WILSON, déjà citée, mais encore fragmentaire. Aucune des analyses antérieures ne donne rien de lié aux mesures: ce que font les auteurs c'est en général d'admettre que les pressions dans les colonnes chaudes et froides de leur modèle schématique sont égales à la même profondeur géométrique; autrement dit que l'atmosphère est en état d'équilibre quasi-hydrostatique, malgré l'existence de mouvements, considérés comme trop faibles pour affecter notablement cette situation.

(e) La figure 18 représente quelques caractéristiques des modèles inhomogènes proposés.

2. Échelles et structure des inhomogénéités

C'est d'abord l'étude des clichés de granulation qui peuvent évidemment fournir les autres caractères des inhomogénéités (échelles, spatiale et temporelle; topologie).

Ce genre de recherches a été poursuivi depuis des années par de nombreux chercheurs, utilisant des documents plus ou moins bons, des méthodes statistiques plus ou moins élaborées. Si le travail d'Edmonds est certainement non seulement le plus récent, mais sans doute le plus élaboré et celui qui utilise les documents les plus fins, il ne faut pas oublier le travail critique fait antérieurement (STUART et RUSH, 1954; WLÉRICK, 1957; UBEROI, 1955; FELLGETT, 1959). L'auteur a examiné les points essentiels de ces recherches dans une revue bibliographique antérieure (PECKER, 1960). On se reportera aussi utilement aux discussions du colloque de Varenna. Sans entrer les détails, on peut dire que ces travaux ont mis en évidence:

(a) *L'existence d'une structure principale de la granulation, de dimension caractéristique 700 à 1000 km.* Certes, un spectre de dimensions peut être déterminé, limité toujours par l'observation, du côté des petites dimensions (voir ci-dessous Figure 33). Des études comme celle de RICHARDSON et SCHWARZSCHILD (1950) ou de KRAT (1954, 1956) suggèrent une prépondérance de petits granules (150 à 200 km), mais les clichés de Stratoscope I ne conduisent pas à une conclusion si extrême.

Nous verrons ci-dessous quelles indications l'étude des raies fournit sur les échelles des inhomogénités dans des régions moins profondes que celles qui influencent les observations de la granulation dans le spectre continu.

(b) *L'existence d'une structure secondaire 12000 à 15000 km,* et peut-être de structures à plus grande échelle encore. Alors que la structure principale est affectée par les effets de la perspective au bord du disque solaire, les dernières recherches d'Edmonds semblent indiquer qu'il n'en est pas de même pour la structure à grande échelle.

La durée de vie d'un granule est d'environ cinq minutes. L'étude des clichés de Schwarzschild a pleinement confirmé cette ancienne détermination, comme il est indiqué sur la figure 19 (BAHNG et SCHWARZSCHILD, 1961).

La *topologie* de la granulation échappe encore à peu près aux études; les recherches de BLACKWELL, DEWHIRST et DOLLFUS (1959) (en comparant épreuves négatives et positives de granulation) conduiraient presque à conclure à la non-existence de granules en tant que tels; cette conclusion est plutôt différente

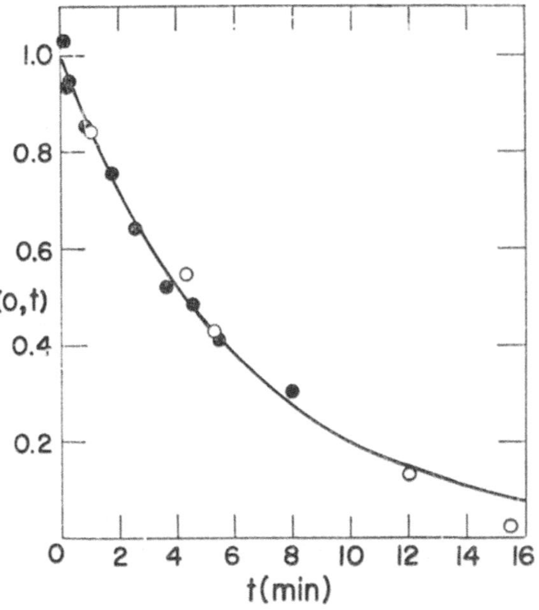

Fig. 19. Autocorrélation, dans le temps, des granules solaires d'après BAHNG et SCHWARZSCHILD. La courbe Q (t) peut se représenter par $Q(t) = Q(00)e^{-(t/\tau)}$, avec $\tau \sim$ 6.3 minutes.

de celle de RÖSCH (1962) ou de SCHWARZSCHILD: on peut d'ailleurs suivre l'évolution complexe de certains granules, objets d'une existence propre pendant leur durée de vie et dont la division ou la modification peut être suivie. On peut lier à la topologie de la granulation les effets dits de rugosité optique (LEFÈVRE et PECKER, 1961). Il est clair en effet que, si à chaque profondeur géométrique, les quantités physiques fluctuent, l'opacité elle-même et donc la profondeur optique (à une longueur d'onde donnée) fluctuent. La couche de profondeur optique $\tau_\lambda = 1$ a donc une forme nettement irrégulière. Il est possible que certaines "aspérités" de cette surface irrégulière portent "ombre" sur d'autres. Au voisinage du bord, on ne verra donc presque pas ceux des points où l'opacité est assez faible (les points chauds par conséquent). Il est évident que cette description est schématique: le long d'un rayon lumineux, tous les points contribuent aux rayonnements observés, et cette notion d'ombre n'a qu'un caractère descriptif. Mais pour schématique qu'elle soit, elle permet de comprendre la nature topologique du phénomène. La quantité de surface cachée en fonction de μ est une mesure de cet effet topologique. Pour le faire comprendre sur un exemple simple, schématisé sur la figure 20, les trois cas A, B et C, pour une inclinaison de 45° des rayons lumineux, qui tous trois correspondent à 50% de régions chaudes et de régions froides, correspondent à une portion des régions chaudes cachées égale à, respectivement, 50%, 0% et 38%.

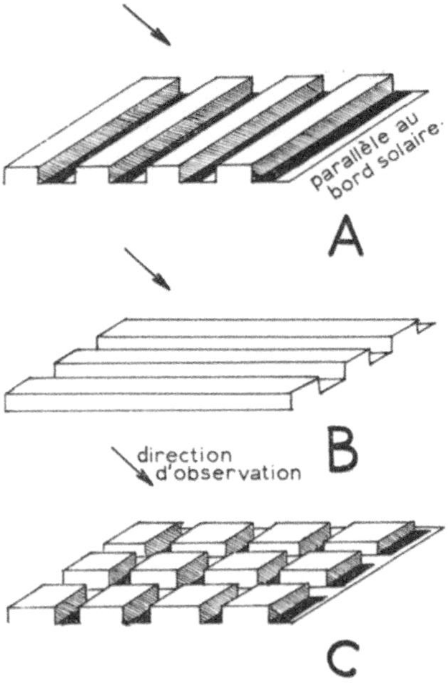

Fig. 20. Calcul de l'effet de rugosité optique.

Nous ne disposons pas de données suffisantes sur cette topologie pour pouvoir associer à une distribution donnée des opacités le degré d'"ombre" en fonction de l'angle θ. En réalité, tout est possible! Notons que les calculs de LEFÈVRE et PECKER (1961), et ceux de JORAND (1962, 1963) utilisent les modèles topologiques de type A, ce qui n'est évidemment pas satisfaisant. D'autre part, ce genre d'effet peut dépendre seulement des composantes de l'irrégularité optique à grande échelle.

D. ECARTS À L'ÉQUILIBRE THERMODYNAMIQUE LOCAL

Parmi les modèles inhomogènes de la photosphère, celui de BÖHM (1954) est particulièrement extrême. Il conduit en surface à des écarts de températures très importants entre éléments chauds et élements froids. On sait que de tels écarts sont considérés comme non réalistes par de Jager et par Schröter. D'où peut provenir cette différence? Böhm lui-même invoque des écarts à l'ÉTL; depuis longtemps de tels écarts étaient déjà soupçonnés dans la photosphère (BRETZ, 1955).

Il semble que ces écarts soient réels, et, aussi faibles soient-ils, ils devient nécessaire d'en tenir compte dans la détermination des inhomogénéités, comme dans celle d'autres caractéristiques de l'atmosphère solaire, les abondances notamment...

Les recherches dans ce domaine, au lieu d'utiliser dans les raies l'équation ci-dessus (4), utilisent l'expression très générale de la fonction-source:

$$S_\nu(\tau_\nu) \equiv B_\nu(T_{\text{exc}\,nm}) \frac{j_\nu}{\varphi_\nu} \tag{17}$$

où la quantité φ_ν représente le profil de l'absorption, et où la quantité j_ν représente le profil de la réémission (c'est dans l'expression de j_ν que s'introduisent les processus non-cohérents). La quantité $T_{\text{exc},\,nm}$ qui dépend naturellement du choix de la transition, s'exprime en fonction de la température électronique et des coefficients b_n et b_m d'écart à l'ÉTL relatifs aux niveaux inférieur et supérieur de la transition:

$$\frac{b_m}{b_n} e^{-(\chi_m-\chi_n)/kT_e} = e^{-(\chi_m-\chi_n)/kT_{\text{exc},\,nm}}\,. \tag{18}$$

Deux façons existent de déterminer les coefficients b_n dans la photosphère.

1. Méthodes théoriques

La première méthode est de nature entièrement théorique. On résout simultanément les équations d'équilibre permanent des niveaux, et les équations du transfert (le modèle est supposé connu) et on en déduit les b_n, donc les T_{exc}. Ce calcul fait en général beaucoup d'hypothèses. Les premiers calculs de Jefferies et Thomas (1959) supposaient des modèles ultra-simples, voire à

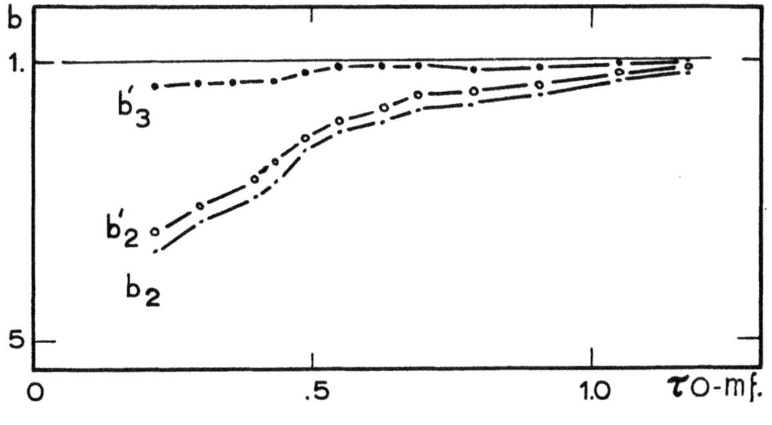

Fig. 21. Ecarts à l'ÉTL dans l'hydrogène photosphérique. En ordonnées, les coefficients b d'écart par rapport à la loi de Boltzmann-Saha; b_2 représente la population du niveau 2, si l'on admet qu'il y a 2 niveaux et un continu; b'_2 et b'_3 représentent la population des niveaux 2 et 3, si l'on admet qu'il y a 3 niveaux et un continu. (D'après Cuny, 1961.)

température constante; de plus ils supposaient valables les approximations d'Eddington; enfin, – c'était le plus grave, – ils supposaient que la quantité ε, rapport de la probabilité de désexcitation par collisions à toutes les autres probabilités de transitions à partir du niveau supérieur, était constante dans l'atomosphère: elle y varie au contraire de plusieurs ordres de grandeur.

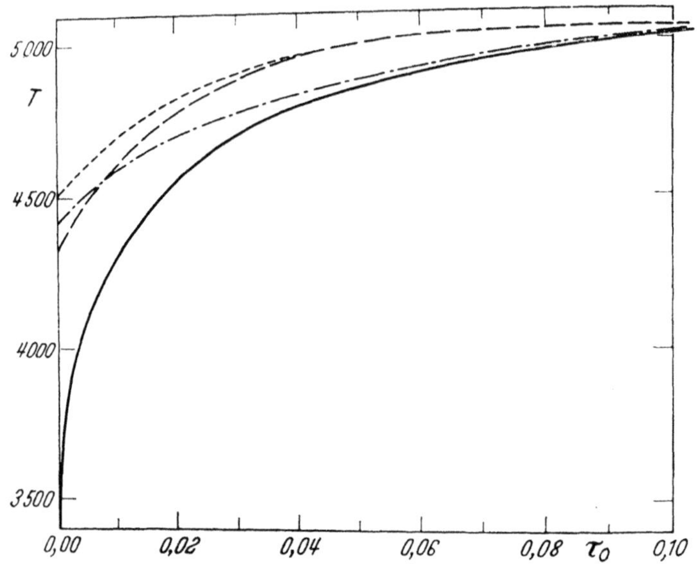

Fig. 22. Calcul de la température d'ionisation pour divers métaux (------: Mg; – – – –: Fe; –·–·–·–·–: Si) comparé au modèle de la photosphère. (D'après MÄDLOW, 1962.)

Ces différentes approximations ont été levées, non sans mal d'ailleurs (voir notamment: DUMONT, 1960, 1961). Mais on continue à mal connaître toutes les quantités physiques en jeu, et on est le plus souvent obligé de représenter un atome réel par un atome "modèle" possédant un petit nombre de niveaux supposés représenter de façon aussi réaliste que possible le grand nombre de niveaux de l'atome étudié (voir notamment DUMONT, 1963b; CUNY, 1963b). De cette façon ont pu être étudiés le cas (sommairement) de H I (CUNY, 1961, 1963b) (voir Figure 21) et le cas de Na I (Hollis Johnson) (Figure 24). Le cas du CaII est en cours d'étude (DUMONT, 1963).

Les résultats, c'est évident, sont très sensibles à la plupart des hypothèses faites, et notamment au modèle, et aux constantes atomiques utilisées (sections de choc, forces d'oscillateur).

Un article récent de M. MÄDLOW (1962) aborde le problème d'une façon simple, en calculant, de façon très approchée, mais suggestive, la température d'ionisation pour plusieurs métaux; il s'en déduit une baisse de l'ionisation, d'où une modification des pressions et de la relation $T_e(h)$ (Figures 22 et 23).

On est encore très loin de posséder des méthodes encore complétement satisfaisantes, aussi ce domaine devrait-il être l'un de ceux où le plus de progrès doivent s'accomplir au cours des prochaines années. Les progrès doivent évidemment intervenir aussi dans le domaine du calcul des constantes physiques, section de choc, forces d'oscillateur nécessaires au développement de la théorie.

Un aspect nécessaire de ces progrès, complémentaire de la théorie déjà citée de WILSON, est sans doute le traitement hors ÉTL des inhomogénéités; autrement dit la prédiction, ou l'interprétation, des fluctuations d'intensité, au centre des raies.

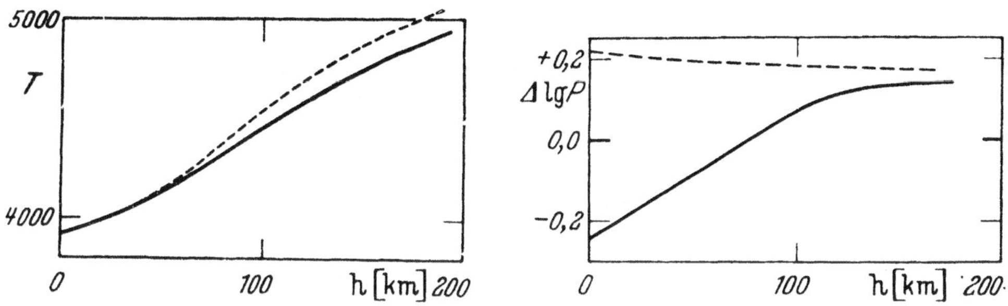

Fig. 23. À gauche, distribution T (h) en ÉTL (----) et hors ÉTL (——) due à la modification de la relation $\tau(h)$; à droite, distribution des pressions électroniques (traits pleins) et gazeux (pointillé): en abscisse, le logarithme de la différence ÉTL–NETL (D'après MÄDLOW, 1962). La comparaison de 22 et 23 permet de se rendre compte des profondeurs optiques ($\tau_0 = 2 \times 10^{-3}$ pour $h = 0$ km, dans le modèle utilisé ici; $\tau_0 = 0.016$ pour $h = 100$ km; $\tau_0 = 0.10$ pour $h = 200$ km.

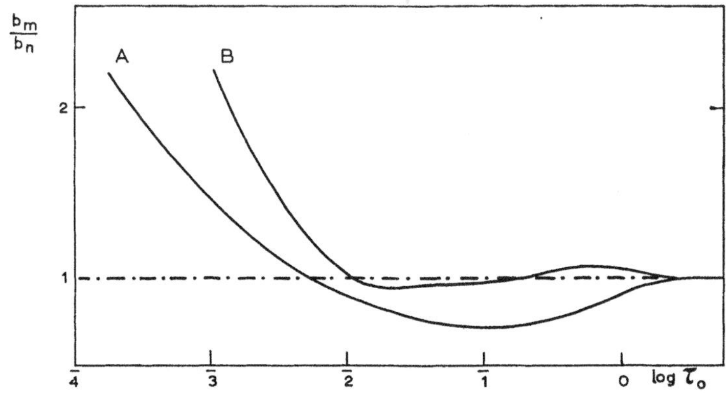

Fig. 24. Écarts à l'ÉTL – Cas de Na I d'après JOHNSON, 1963. La courbe A se refère à la détermination empirique, la courbe B à la détermination théorique. Dans les deux cas, le niveau inférieur de la transition est le niveau 3^2P; mais plusieurs transitions ont été utilisées. On se reportera aux travaux originaux pour une discussion du rôle du niveau supérieur.

2. *Recherches de nature empirique*

Fort controversés ont été aussi les études de caractère empirique ayant abouti à des déterminations des coefficients b_n ou des températures d'excitation. Ces méthodes ont reposé principalement sur le fait que l'intensité mesurée en un point du spectre d'une raie est supposée fournir la fonction-source à la profondeur optique unité (à la longueur d'onde considérée). Cette approximation, bonne sans doute pour les raies intenses, l'est sans doute moins pour les raies faibles. Il me semble inutile de revenir ici sur cette méthode tant de fois décrite (voir PECKER, 1959, 1960b) : mêmes si certains détails peuvent en être caducs, si certains arguments en sont approchés, il nous semble que l'essentiel des raisonnements reste valable, et que le raisonnement physique qui nous a conduit à développer ces méthodes est plus que jamais correct. Le plus étudié des éléments par ces méthodes a été le titane, et nous tirerons de lui nos exemples, dans la suite de cet exposé. Les constituants suivants de l'atmosphère solaire ont été étudiés:

Fe I: GÖKDOGAN *et al.*, 1962	Na I : H. JOHNSON, 1962
Ti I : PECKER 1959; KANDEL, 1960	Ti II: J. ROUNTREE, 1960
V I : PECKER et VOGEL, 1960	CH : F. PRADERIE-PECKER, 1960
Cr I : PECKER et VOGEL, 1960	Be I : M. JOHNSON, 1963b

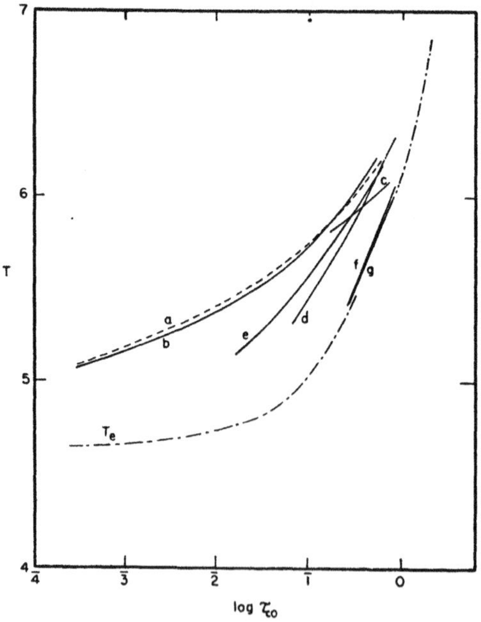

Fig. 25. Écarts à l'ÉTL – Cas de Ti I. Cette figure représente les résultats empiriques (d'apres PECKER, 1959) obtenus pour plusieurs groupes de multiplets, à savoir ceux issus des niveaux a: a^3F (0 v.) ; b: a^5F (0.8 v.) ; c: a^1D (0.9 v.) ; d: a^3P (1.03 v.) ; e: b^3F (1.143 v.) ; f: a^1G; (1.72 v.) ; g: a^5P (1.87 v.).

D'autres métaux sont en cours d'études à Istanbul par les mêmes méthodes.

Dans le seul cas où les deux méthodes (théorique et empirique) aient été appliquées (Na I), un désaccord persiste, que JOHNSON n'arrive pas à lever.

La figure 24 montre le cas du sodium: la figure 25 le cas du titane. Dans le cas de VI également JOHNSON (1963b) montre que le résultat empirique pourrait être prévu théoriquement et est lié à la faible abondance du vanadium.

3. Conséquences des écarts à l'ÉTL

Une première conséquence de tels écarts, c'est que des observations que seules des inhomogénéités permettaient d'expliquer, et des raies qui conduisaient (en ÉTL) à des modèles improbablement froids se trouvent maintenant interprétées de façon naturelle.

Les écarts mis en évidence se réfèrent plutôt à la loi de Boltzmann (excitation). Mais il est évident que l'ionisation est affectée, plus encore peut-être que l'excitation: aussi les modèles usuels pour la pression et la pression électronique sont-ils sans doute caducs. Est-ce à dire que l'on puisse expliquer ainsi les très faibles pressions électroniques mises en évidence par Pierce et Waddell? C'est très peu vraisemblable, on l'a dit.

Une seconde conséquence, et qui découle immédiatement de la première, c'est que pour déterminer les inhomogénéités à partir des raies, ou en tous cas certaines de leurs propriétées, il est nécessaire de tenir compte des écarts à l'ÉTL ainsi obtenus.

Une telle recherche a été faite par CUNY, 1960; LEFÈVRE et PECKER, 1960; et CUNY, LEFÈVRE et PECKER, 1960. Ceux-ci ont utilisé principalement un modèle à trois colonnes, celui de Böhm, par hypothèse, et ils ont montré que ses effets sur la fonction-source apparente au centre des raies était nettement plus faible que celui des écarts. Comme nous l'avons dit ci-dessus ils ont également mis en évidence un effet d'"arête de poisson" (Figure 26), expliquable par les effets de la topologie des surfaces "iso-tau". Une hypothèse sur cette topologie (celle en fait de la figure 20A) conduit alors à fixer l'échelle de dimensions des irrégularités de la surface iso-tau. On trouve $h = 300$ km, ce qui, compte tenu des hypothèses faites, est une surestimation; cette valeur est encore très différente des valeurs obtenues à partir des clichés de granulation. Là encore, une contradiction reste à lever.

Mais on peut noter (Y. CUNY, 1963) que le résultat dépend fort peu du modèle: le seul fait d'observer l'effet "arête de poisson" conduit à une valeur de h de l'ordre de grandeur de l'épaisseur de la photosphère.

Une troisième conséquence des écarts à l'ÉTL, c'est qu'il faudra en tenir compte pour l'interprétation des profils en vue d'obtenir des vitesses. Nous y reviendrons ci-dessous, ainsi que sur l'influence des écarts sur les abondances.

Il est alors évident que la qualité photométrique des données devra être particulièrement soignée. Les excellentes mesures de WADDELL (1960, 1963) montrent bien les difficultés qui apparaissent dans la détermination empirique des fonctions-sources dues aux inhomogénéités et aux écarts à l'ÉTL.

Ce domaine de recherche est évidemment neuf et riche, même dans le domaine photospherique. Est-il besoin de dire que dès lors qu'on admet qu'il y a continuité absolue entre photosphère et chromosphère, avec transition progressive, il est impossible de limiter cette réflexion physique aux couches chromosphèriques?

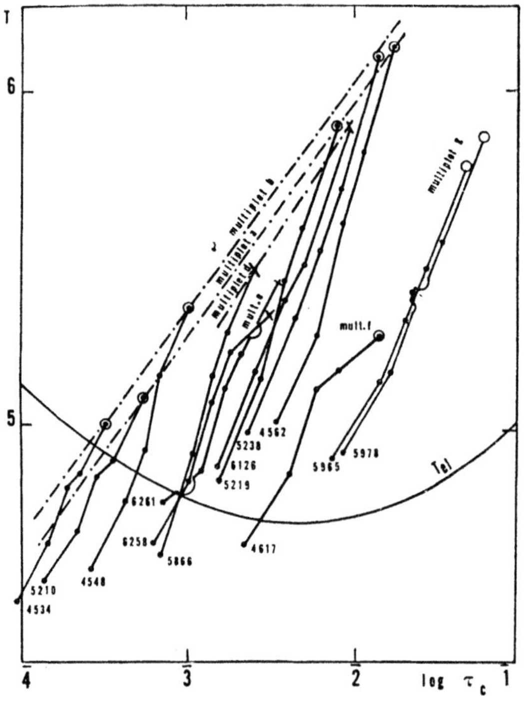

Fig. 26. L'effet des inhomogénéités: "l'arête de poisson" (d'après LEFÈVRE et PECKER, 1961). La figure montre, dans le cas du Titane, comment les fonctions-sources déduites de la variation centre-bord des intensités centrales diffèrent de celles déduites pour les mêmes multiplets de l'étude au centre du disque (les multiplets f et g sont mal placés sur cette figure).

Le fait même que la photosphère soit directement observable prouve que l'équilibre thermodynamique n'y est pas réalisé. Il peut se faire que l'ÉTL le soit, occasionnellement: mais nous estimons qu'il faudra le prouver, et dans chaque cas.

E. La détermination des abondances

Il est évident que la connaissance des abondances des principaux éléments est d'une primordiale importance. Sans entrer ici dans la question au point de vue de la nucléogénèse et de l'évolution stellaire, nous devons remarquer que la connaissance des abondances est nécessaire au calcul de l'opacité (utilisée par exemple pour déterminer la transition photosphère-chromosphère à partir de l'UV), à celui des écarts à l'ÉTL (pour fixer la profondeur de formation) etc.

Les méthodes, usuelles aujourd'hui, de détermination des abondances, reposent principalement sur l'utilisation des courbes de croissance et de leur calcul théorique. Nous n'entrerons pas dans les détails techniques, maintes fois exposés, des méthodes de fonctions de poids et de saturation (mentionnons seulement les intéressants développements qu'en ont récemment proposés Mugglestone et ses collaborateurs (MUGGLESTONE, 1958). Ces auteurs définissent les fonctions de poids en faisant intervenir directement le gradient de température ; mais le principe général de leur analyse reste classique). Les auteurs, suivant en cela une opinion exprimée jadis par UNSÖLD (1948), admettent donc généralement que les raies sont formées par "absorption pure", c'est à dire, dans notre nomenclature actuelle, en ÉTL. Dans ces conditions, il faut encore faire une hypothèse sur le modèle, et notamment le supposer homogène, et disposer de bonnes forces d'oscillateur. L'esprit de ces méthodes est de traiter non pas les raies de tel ou tel métal dans leur ensemble, et de déduire une abondance de la courbe de croissance, mais de traiter chaque raie séparément. Le détail et l'algèbre en ont été maintes fois exposés. Les résultats obtenus par cette méthode appliqués d'une façon extrêmement soigneuse et sûre par GOLDBERG, MÜLLER, ALLER (1960), sont reproduits dans le tableau ainsi que quelques autres valeurs plus anciennes, et plus criticables, et quelques autres valeurs récentes. D'un intérêt particulier est la discussion de l'influence des modèles telle que celle faite par MUGGLESTONE. Cet auteur donne aussi des arguments basés sur la sensibilité du résultat à la précision du calcul de certaines intégrales, pour préférer la méthode dite des "gradients planckiens" à la forme usuelle des fonctions de poids. Certes, de tels travaux sont d'une grande utilité, et leurs résultats sont certainement les meilleurs qu'on puisse obtenir, dans le cadre des hypothèses usuelles. Tout au plus pourrait-on dire qu'une amélioration sensible viendrait de nouvelles mesures des quantités physiques et des raies solaires, de nature à diminuer la dispersion des mesures.

Plus grave à notre avis sont sans doute les critiques qui viennent du fait que ces méthodes ignorent, et les inhomogénéités, et les écarts à l'ÉTL.

En ce qui concerne le second de ces points, des essais ont été tentés par KANDEL (1960), et par l'auteur (1959) dans le cas du titane, le cas du fer étant actuellement en cours d'étude; cependant que Johnson appliquait les mêmes méthodes au cas des éléments comme Be et B. Ces méthodes ont un caractère nécessairement itératif puisque on a besoin des abondances pour fixer les profondeurs de formation des écarts trouvés. Dans les deux derniers cas, les méthodes nouvelles ne changent pas sensiblement le résultat. Mais dans le cas du titane, on trouve que l'abondance doit être multipliée par *un facteur de l'ordre de 5*. Comme les écarts dépendent, en seconde approximation, du degré d'excitation des atomes, nous avions cru pouvoir prédire la variation

TABLEAU

ABONDANCES DES ÉLÉMENTS DANS LA PHOTOSPHÈRE SOLAIRE*

(d'après GOLDBERG, MÜLLER, ALLER, 1960)

ELEMENT		I		II		III	IV
		RUSSELL 1929	GOLDBERG, ALLER, 1943	UNSÖLD 1948	CLAAS 1951	GOLDBERG, MÜLLER, ALLER 1960	*Divers*
1	H	11.5	12.00	12.00	12.00	12.00	
3	Li	2.0	—	—	1.08	0.96	1.26 (GT)***, 0.93 (D) voir (J)
4	Be	1.8	—	—	—	2.36	2.18 (GT)
6	C	7.4	7.56	8.29	—	8.72	9.06 (H), 9.90 (P)
7	N	7.6	8.09	8.61	—	7.98	9.02 (H), 7.86 (M)
8	O	9.0	8.56	8.73	8.65	8.96	9.23 (H), 8.20 (B) 8.73 (CD) 9.01 (M)
11	Na	7.2	6.56	6.28	6.33	6.30	6.13 (W), 5.96 (S), 6.0 (R)
12	Mg	7.8	8.39	7.51	7.57	7.40	7.28 (W), 7.60 (S), 7.25 (R)
13	Al	6.4	6.39	6.33	6.17	6.20	6.13 (W)
14	Si	7.3	7.87	7.29	7.12	7.50	
15	P	—	—	—	—	5.34	
16	S	5.7	7.57	6.92	—	7.30	
19	K	6.8	5.09	5.20	5.01	4.70	5.32 (S)
20	Ca	6.7	6.57	6.23	6.46	6.15	6.17 (W), 6.42 (GT), 6.23 (S)
21	Sc	3.6	—	3.33	—	2.82	
22	Ti	5.2	4.57	4.96	7.56**	4.68	8.15, 5.53 (K)**
23	V	5.0	4.09	4.05	—	3.70	
24	Cr	5.7	4.87	5.58	—	5.36	
25	Mn	5.9	5.09	5.46	—	4.90	
26	Fe	7.2	6.99	7.26	7.16	6.57	6.55 (W)
27	Co	5.6	4.69	5.03	—	4.64	
28	Ni	6.0	6.39	5.95	—	5.91	
29	Cu	5.0	4.39	4.23	4.80	5.04	
30	Zn	4.9	5.57	4.78	4.52	4.40	
31	Ga	2.0	—	—	—	2.36	
32	Ge	3.0	—	—	—	3.29	
37	Rb	1.7	—	—	—	2.48	

38	Sr	3.3	—	3.35	2.88	2.60	
39	Y	2.6	—	3.21	—	2.25	
40	Zr	2.5	—	2.37	—	2.23	
41	Nb	1.0	—	—	—	1.95	
42	Mo	1.4	—	1.78	—	1.90	
44	Ru	1.7	—	—	—	1.43	
45	Rh	0.5	—	—	—	0.78	
46	Pd	1.1	—	—	—	1.21	
47	Ag	1.0	—	—	—	0.14	
48	Cd	2.2	—	—	—	1.46	
49	In	0.0	—	—	—	1.16	
50	Sn	1.2	—	—	—	1.54	
51	Sb	0.8	—	—	—	1.94	
56	Ba	3.3	—	2.95	2.38	2.10	
70	Yb	—	—	—	—	1.53	
82	Pb	1.2	—	2.55	—	1.33	1.25 (Kh), 1.15 (JN)

* Les colonnes se réfèrent aux auteurs cités en tête des colonnes, ou en note 2. Les symboles I, II, III, IV désignent respectivement: I–les méthodes n'utilisant pas la théorie moderne des courbes de croissance; II–les méthodes utilisant la théorie moderne des courbes de croissance; III–le travail fondamental de Goldberg, Müller, Aller; IV–les recherches non systématiques.

Il importe d'attirer le plus l'attention sur l'important ouvrage de L. H. ALLER (1961), consacré entièrement aux abondances des éléments.

** Les raisons des différences entre les différents auteurs sont multiples: modèles différents, méthodes plus ou moins élaborées; forces d'oscillateur de différentes origines: c'est ce dernier facteur qui explique la différence considérable, dans le cas de Ti I, entre CLAAS et GOLDBERG, MÜLLER, ALLER: dans ce cas particulier les deux valeurs de KANDEL se réfèrent à chacun des deux systèmes de forces d'oscillateur.

*** Les abréviations de la colonne IV doivent être interprétées comme suit:

B	BOWEN, 1948	K	KANDEL, 1960
CD	CABANNES, DUFAY, 1948	Kh	KHOKHLOV, 1958
D	DUBOV, 1955	M	MUGGLESTONE, 1958, 1960; FAULKNER et MUGGLESTONE, 1962
GT	GREENSTEIN, TANDBERG-HANSSEN, 1954	P	PECKER, 1951
H	HUNAERTS, 1947	R	RUDKJØBING, 1945
J	JOHNSON, 1963	S	STRÖMGREN, 1940
JN	DE JAGER, NEVEN, 1962	W	WEIDEMANN, 1955

des abondances "apparentes" (cells obtenues par les méthodes classiques") avec l'excitation. De même comme la variation de la fonction-source avec la profondeur n'est pas la même dans la raie et dans le continu, nous avions prédit une variation de l'abondance "apparente" entre le centre et le bord du disque solaire. Ces deux effets n'ont pas été observés, dans la limite de la précision des mesures, malgré quelques tendances que nous avions cru discerner dans les résultats publiés par Goldberg, Müller, Aller: il ne faut pas trop s'en étonner; il ne s'agit là que d'effets de second ordre masqués sans doute par la dispersion des données.

En ce qui concerne les inhomogénéités, il semble difficile actuellement de se prononcer. Il me semble paraitement évident que la juxtaposition de deux étoiles de températures différant de 500° (ce qui est plutôt bas, on l'a vu) ne peut pas correspondre à un continu moyen, et à des raies moyennes, cohérents entre eux... Un calcul rapide, dans ces conditions, peut être fait. En se limitant à faire fluctuer la température (sans se préoccuper de la pression mais ceci ne devrait pas trop changer l'ordre de grandeur ci-après), on peut montrer, pour $\Delta T = 500°$, et dans le cas du fer neutre, que les valeurs usuelles sont sous-estimées *d'un facteur 1.5 à 2*. Il est évident qu'une telle recherche devrait être faite de façon tout à fait systématique, comme l'avait déjà suggéré SCHRÖTER en 1957.

3. Les champs de vitesse dans la photosphère

A. ÉTUDE DES COURBES DE CROISSANCE

On sait (le nombre de bons exposés – UNSÖLD (1955), GOLDBERG-PIERCE (1959), ALLER (1960) par exemple – de cette question nous parait suffisamment élevé, ici encore, pour que nous nous limitions à supposer cela connu) que le palier de la courbe de croissance est sensible aux champs de vitesse que contribuent à l'élargissement des raies par effet Doppler statistique, le long de la ligne de visée. En revanche, la position de ce palier n'est pas sensible aux mouvements qui affectent l'ensemble d'une région des couches photosphériques (mais dont les effets s'additionnent seulement à cause du manque de résolution des images du disque solaire granuleux). On exprime en général ce fait en parlant de "micro-turbulence" et de "macro-turbulence"; les difficultés de cette terminologie ont été mise en évidence bien des fois (colloque de Varenna; DE JAGER, 1959) pour que nous évitions ici ces mots.

Les analyses dépendent évidemment du genre de théorie adoptée pour la courbe de croissance, comme de la qualité des données utilisées.

DE JAGER (1959, p. 97) donne une liste des valeurs obtenues. Elles correspondent à des couches superficielles (comme l'a remarqué VAN REGEMORTER, 1959) d'une profondeur optique de l'ordre de 0.02, et avoisinent des valeurs de l'ordre 1.2 km/s.

À noter que dans le cas du titane ionisé, J. ROUNTREE (1960) a trouvé

4 km/s, valeur correspondant peut être à des couches chromosphériques, mais jamais obtenue par d'autres auteurs: il serait indispensable de procéder à de nouvelles déterminations.

Les méthodes dites de 'courbes de corrélation' introduites par HUANG et STRUVE (1952), qui utilisent simultanément profils et largeurs équivalentes, ont, dans le cas solaire, été appliquées par VAN REGEMORTER (1955), qui a trouveé le résultat suivant dans le cas du fer:

Raies de $\chi_{exc} = 0\text{–}0.15$ volts; $\xi = 4.2$ km/s;

Raies de $\chi_{exc} = 1.4\text{–}1.6$ volts; $\xi = 2.1$ km/s;

Raies de $\chi_{exc} = 2.2\text{–}2.5$ volts; $\xi = 0$ km/s.

Il s'agit encore ici des "micromouvements". Entre le centre et le bord du disque, des recherches de TEN BRUGGENCATE et HOUTGAST (1941) ont montré une diminution de 20% environ du centre au bord (sin $\theta = 0.99$). Plus récemment, M. BRETZ (1955), tout en montrant les difficultés introduites dans ce problème par l'hypothèse ÉTL, retrouve le même résultat (2.5 km/s au centre, 1.5 km/s pour sin $\theta = 0.99$) mais en rend partiellement responsable l'hypothèse en question.

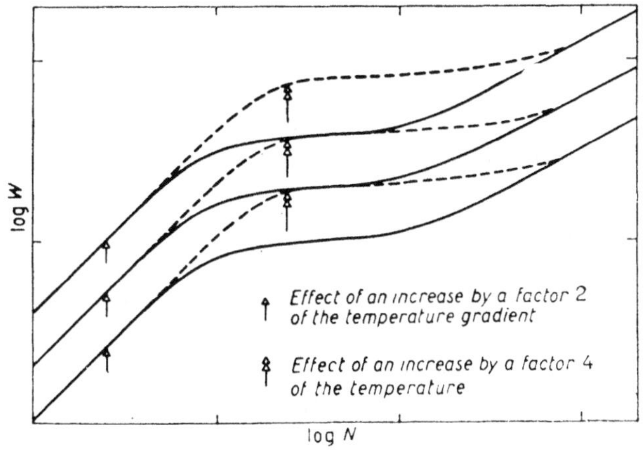

Fig. 27. Effet du modèle sur la détermination des abondances (d'après PECKER et THOMAS, 1961). Effets d'une augmentation d'un facteur 2 du gradient de température (flèche simple), d'un facteur 4 de la température (flèche double).

La validité de ces méthodes a été critiquée souvent. On consultera notamment PECKER et THOMAS (1960): dans la figure 27 reproduite d'après ce mémoire on voit qu'une erreur dans la connaissance de la température ou seulement de son gradient suffit à modifier les résultats; de plus la position du palier de la courbe de croissance dépend des fonctions de poids, et par suite est très sensible aux écarts à l'ÉTL; enfin les données obtenues ont seulement des significations "moyennes". A quelle "profondeur" correspondent-elles? Les définitions usuelles des "profondeurs de formation" n'ont certainement en l'occurrence, et encore, qu'une valeur de première approximation. Si l'on considère toute "profondeur de formation" comme définie (dans la mesure où cela peut avoir un sens) par un théorème de la moyenne, il est évident que la profondeur de formation de la vitesse d'élargissement Doppler τ_v n'est ni celle, τ_D, de l'intensité centrale, ni celle, τ_*, de la largeur équivalente ou de toute autre quantité:

$$I_{r_0} = S(\tau_*) = \int_0^\infty S(\tau) e^{-\tau/\mu} \mathrm{d}\tau/\mu \tag{19}$$

$$I_{r_\nu} = \int_0^\infty f[\xi(\tau)] \mathrm{d}\tau = \int_0^\infty f[\xi(\tau_D)] \mathrm{d}\tau \tag{20}$$

B. ÉTUDE DES PROFILS

1. *Étude classique*

Le profil d'une raie dépend: (1) du profil du coefficient d'absorption; (2) de la variation de la fonction-source avec la profondeur; (3) de la variation de la fonction-source, à une profondeur donnée, avec la longueur d'onde le long du profil. Encore une fois, je renverrai à la communication de PECKER et THOMAS (1961) au symposium de Varenna. Déduire de la variation du coefficient d'absorption les mouvements de matière implique que l'on a pu séparer les deux autres facteurs, et notamment que l'on a su tenir compte des écarts à l'ÉTL. Précisons ici que les micro- et les macro-mouvements ne sont pas séparés par ces études.

La méthode la plus utilisée récemment pour faire cette séparation, proposée par de JAGER (1952), GOLDBERG (1958) et utilisée de façon très systématique par UNNO (1959), est très séduisante: on observe plusieurs raies issues du même niveau fondamental, et avec l'hypothèse que la fonction-source est la même pour ces différentes raies, on compare les points où l'intensité est la même: ils ne se trouvent pas, pour des raies de différentes intensités, à la même distance du centre de la raie. La comparaison entre ces distances au centre de la raie, et les forces d'oscillateur des raies, permet de déduire élégamment le coefficient d'absorption, donc la largeur Doppler, et les vitesses des mouvements. Outre les difficultés de principe signalées déjà (PECKER et THOMAS, 1960), et liées aux variations avec la fréquence de la fonction-source, il faut signaler que lorsqu'il s'agit d'assigner une profondeur de formation aux vitesses ainsi déterminées, UNNO utilise la profondeur calculée en ÉTL. Sur la figure 28, on voit un effet systématique de différence entre Ti, V et Cr, explicable uniquement (et quantitativement) par les écarts différentiels à l'ÉTL relatifs à ces trois éléments, et obtenus indépendamment; il est bien évident que, entre trois éléments de même potentiel d'ionisation, les seules différences de comportement peuvent venir des écarts à l'ÉTL et non pas des inhomogénéités, ni évidemment de quelque différence systématique importante des champs de vitesse. Citons aussi la méthode d'Athay (qui utilise le coude apparent du profil et l'associe aux discontinuités de la dérivée du profil du coefficient d'absorption). Cette méthode (1961) a été sévérement critiquée par THOMAS (1963).

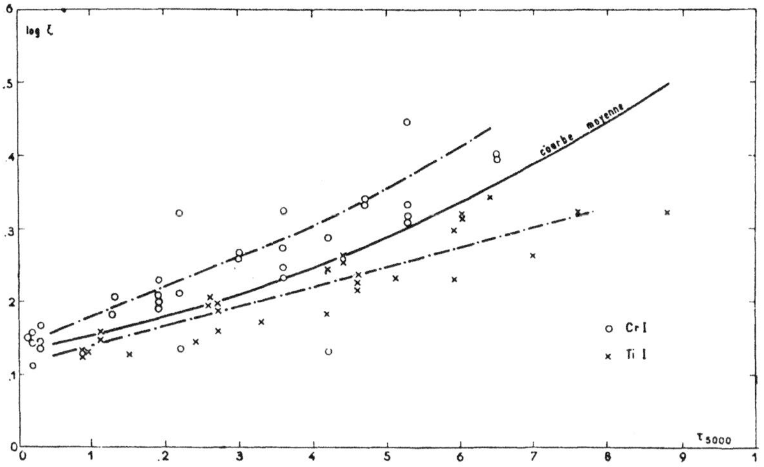

Fig. 28. Effet des écarts à l'ÉTL sur les champs apparents de vitesse d'après UNNO, 1959, et PECKER et VOGEL, 1960. Les points correspondant à l'étude du titane, et ceux correspondant à celle du chrome, sont de part et d'autre de la courbe moyenne. – En ordonnée, logarithme de la vitesse ξ exprimés en km/s.

2. *Recherches hors ÉTL*

Une autre méthode est applicable, utilisant l'ensemble du profil. S'il est possible d'explorer en profondeur une atmosphère en utilisant différentes raies d'un même multiplet (comme on l'a fait pour la recherche des fonctions-sources dans une raie, en vue de la détermination des écarts à l'ÉTL), ou en étudiant un même point d'une même raie entre le centre et le bord du disque solaire (ainsi que nous l'avons fait pour étudier l'effet des inhomogénéités et de leur topologie, comme on l'a montré ci-dessus), on peut aussi faire cette exploration en se déplaçant du centre aux ailes d'une raie. Il sera bien entendu nécessaire de comparer la fonction-source ainsi obtenue avec celle déterminée en utilisant plusieurs raies d'un même multiplet, en leur centre. S'il n'y a pas coïncidence c'est soit que l'opacité (et donc le champ de vitesses) choisie pour le calcul des profondeurs optiques n'est pas correcte, soit que le rapport $j\nu/\varphi\nu$ choisi n'est pas exact. Une étude de ce problème, à l'aide des raies de Ti I et de Fe I, a été entreprise à Meudon par Mmes Dayet-Feautrier, Calmet-Laval et M. Dubois-Salmon. A partir des raies de résonance, et de calcul de $j\nu/\varphi\nu$ effectué au moyen de méthodes antérieurement décrites par Thomas, supposant que la non-cohérence de la raie est due uniquement aux effets Doppler, on trouve pour chaque raie étudiée une courbe de la vitesse en fonction de la profondeur, dans le cas du Ti; dans le cas du fer, un point seulement a pu être obtenu par raie en raison du manque d'information sur l'amortissement dans les ailes. La figure 29 reproduit, d'après la compilation de DE JAGER (1959), les résultats antérieurs, et quelques résultats nouveaux. Nous considérons ceux de Meudon comme préférables dans la mesure ou, seules, ils tiennent compte des écarts à l'ÉTL.

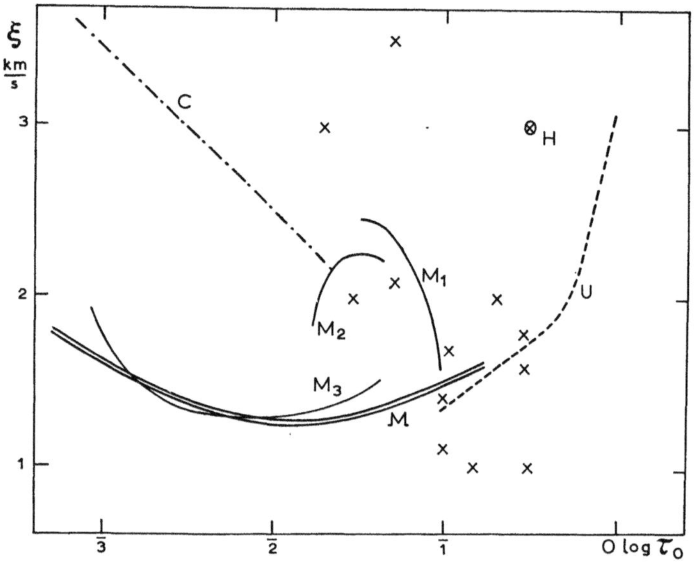

Fig. 29. Champs de vitesses déduits des profils. Les croix représentent des valeurs individuelles rassemblées par DE JAGER (1959). En pointillé, les résultats d'UNNO (1959); en traits mixtes, selon DE JAGER (1959), la moyenne des valeurs chromosphériques. En traits pleins, mesures de Meudon, tenant compte des écarts à l'ÉTL. La valeur notée H correspond aux composantes horizontales des mouvements; toutes les autres à des composantes verticales.

Il est à noter qu'ils correspondent à une valeur de la vitesse plus basse que les valeurs usuelles; une partie de la turbulence apparente est due à un écart systématique par rapport à l'ÉTL, le plus fréquent apparemment pour les métaux neutres, celui qui correspond à $T_{exc} > T_e$, comme dans le cas du titane.

Nous devons citer aussi les arguments selon lesquels (SUEMOTO, 1957) la variation avec la profondeur est telle que ξ décroit vers l'extérieur, parce que les raies faibles ont des largeurs Doppler, toutes choses égales d'ailleurs, plus grandes que les raies moyennes. Ceci n'est pas contradictoire avec le fait découvert (sinon démontré) par ALLEN (1943), et confirmé par WADDELL (1958) que la largeur Doppler des raies étant plus faible au centre du disque qu'au bord; on doit admettre un champ de vitesses anisotrope, la composante horizontale étant plus importante que la composante verticale. Ces conclusions semblent acquises, mais bien difficiles à comprendre.

Les méthodes utilisant les profils ne font pas la différence entre mouvements à petite et grande échelle, dans la mesure où les raies sont mesurées sur le spectre d'une région du soleil suffisamment grande, comme c'est le cas la plupart du temps.

C'est pourquoi il est devenu nécessaire d'étudier les champs de vitesse avec un pouvoir de résolution grandement amélioré sur le soleil.

L'étude de l'asymétrie des raies contient déjà des informations supplémentaires.

3. Profils asymétriques et champs de vitesse

On a pu détecter, à plusieurs reprises (Olson, 1962; Higgs, 1962; Blamont et Roddier, 1961) une certaine dissymétrie des profils des raies de Fraunhofer. Cette dissymétrie, sauf sans doute dans le cas des mesures de Roddier (mais alors une seule raie a été mesurée), est à la limite de la précision des mesures.

Il faut tout d'abord noter que Schröter (1957) a prédit une telle dissymétrie, liée à l'existence dans les modèles à plusieurs colonnes d'une corrélation entre les vitesses et les brillances, les colonnes "chaudes" étant également des colonnes "ascendantes". Le profil des raies correspondant aux différentes colonnes n'étant pas le même, une dissymétrie caractérise nécessairement la raie résultante.

La variation de cette dissymétrie avec la position sur le disque solaire doit donc permettre de déterminer la variation du champ de vitesses avec la profondeur, si l'on suppose connu le modèle, supposé hétérogène – ou vice-versa. Les auteurs cités n'ont publié encore aucune conclusion issue de leurs données. Mais il est hors de doute qu'il s'agit là d'une information très riche.

On doit rattacher à cette dissymétrie le fait que les déplacements apparents des raies, par rapport à la longueur d'onde théorique, sont mesurables et mesurés. Miss Adam (1952) notamment et d'autres auteurs (Findlay-Freundlich et Forbes, 1956) ont publié de nombreuses déterminations de longueur d'onde. Roddier (1963), dans le cas de la raie de résonance du Sr I, a montré comment ces déplacement sonts compliqués par l'existence des dissymétries.

Les mesures de déplacement sont certes d'abord influencés par l'effet Einstein. Mais l'effet Einstein est strictement indépendant du point du disque où l'on se place. En dehors de l'effet Lindholm, négligeable en général, seuls peuvent donc intervenir des effets Doppler. Si l'on considère que les éléments chauds montent et que les éléments froids descendent, la vitesse radiale moyenne apparente n'est pas nulle car les raies "chaudes" et les raies "froides" n'ont pas même intensité. Sans doute faut-il faire intervenir aussi l'effet de la rugosité optique "et de sa topologie": Jorand a montré qu'au bord du disque, les mouvements ascendants, correspondant aux régions chaudes, cachées partiellement, influencent moins les observations que ceux qui correspondent aux régions froides, par rappotr à ce qui se passe au centre du disque. Il est évident également (JORAND, 1963) que les mouvements horizontaux, qui se produisent peut-être dans les couches intermédiaires, entre les régions chaudes et froides, sont cachés par les régions froides, donc de façon préférentielle quand ils correspondent à des déplacements vers le bleu. La figure 30 montre comment (qualitativement) interviennent ces différents effets.

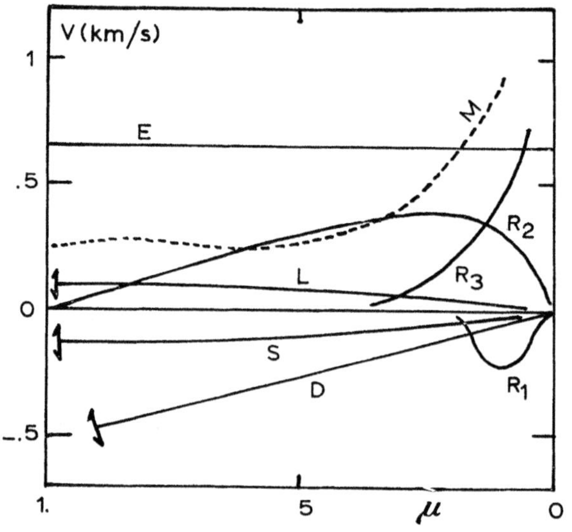

Fig. 30. Effets divers agissant sur la longueur d'onde des raies au bord du disque solaire selon JORAND (1962). La courbe M représente les mesures. Les courbes relatives aux effets calculés sont les suivantes: E – effet Einstein, L – effet Lindholm (amplitude mal fixée mais faible), S – effet Spitzer (critique de l'effet Lindholm), D – effets Doppler des mouvements verticaux (amplitude non déterminée par la théorie); R_1 – effet de la rugosité géométrique; R_2 – effet de la rugosité optique (sur les manifestations des mouvements radiaux); R_3 – effet de la rugosité optique (sur les manifestations des mouvements non radiaux). La vérité correspond peut-être à la superposition des courbes E, S, D, R_2, R_3.

L'étude de la dissymétrie les complète par une étude de ces effets Doppler différentiels en fonction de la profondeur.

Il faut ici faire une remarque; nous faisons ces différentes interprétations en nous servant de modèles à plusieurs températures dont nous savons cependant qu'ils sont très schématiques, et même peut-être fondamentalement erronés. Pour ces modèles nous admettons une corrélation étroite entre vitesse et température, les colonnes chaudes étant en réalité des colonnes ascendantes. Or les observations fournissent une information suffisante pour que l'on puisse dépasser cette conception simpliste. La corrélation (sur les clichés de granulation) entre brillance dans le spectre continu et vitesse a été souvent mesurée (voir BÖHM-VITENSE, 1960) notamment par RICHARDSON et SCHWARZSCHILD (1950) et discutée, notamment par FELLGETT (1959) et SERVAJEAN (1961). Après avoir cru démonter que cette corrélation était élevée, il semble maintenant que les méthodes de mesure étaient fortement sélectives, et que la corrélation en question est plus apparente que réelle ou en tous cas, plus qualitative que quantitative. Nous y reviendrons au paragraphe suivant. Mais d'ores et déjà, il est évident qu'une révision de toutes les déductions que nous avons pu faire, s'impose, dans un cadre d'une théorie plus cohérente des inhomogénéités.

C. ÉTUDE DES MOUVEMENTS À TRÈS GRANDE RÉSOLUTION ANGULAIRE

Depuis quelques années, il est apparu dans la littérature une grande profusion d'excellents spectres, faisant apparaître la structure fine de la granulation dans les raies spectrales. Aux spectres de Michigan (voir GOLDBERG et PIERCE, 1959), du Mont Wilson (RICHARDSON et SCHWARZSCHILD, 1950), du Pic du Midi (SERVAJEAN, 1961) s'est ajouté récemment l'admirable collection de Sacramento Peak (EVANS, MICHARD, 1962a).

Il est évident qu'une étude des vitesses, et des intensités, à cette échelle, doit fournir une information beaucoup plus riche que toute étude du rayonnement sur une surface intégrée.

Avant d'en donner les résultats, précisons cependant que cette étude est encore loin d'avoir fourni toutes les informations possibles: il est évident qu' obtenir point par point des courbes de croissances, des profils, qu'y déterminer les écarts à l'ÉTL, les abondances, est une nécessité qui permettra de se débarasser partiellement des inconvénients signalés tout à l'heure dans l'analyse des données, et dus aux inhomogénéités. Mais de toutes façons, même lorsqu'on disposera de telles données, on n'aura fait qu'une partie du chemin. En effet, on aura étudié la composante à grande échelle des inhomogénéités. Mais il reste des échelles d'inhomogénéités plus petites, celles notamment qui influencent les courbes de croissances (micro-inhomogénéités,) et qui sont hors d'atteinte de la mesure, en raison de la transparence des éléments de ces inhomogénéités.

C'est d'abord dans les recherches de LEIGHTON et de ses collaborateurs (1960), puis dans la série d'articles de Evans et Michard, et de leurs collaborateurs (EVANS et MICHARD, 1962abc; EVANS, MEIN, MICHARD et SERVAJEAN, 1962; BERNIÈRE, MICHARD, et RIGAL, 1962) qu'il faut sans doute voir le progrès le plus important accompli dans la connaissance de la physique de la photosphère depuis de nombreuses années.

Le travail de Leighton est une véritable acrobatie technique; je regrette de ne pas être en mesure de projeter l'un de ses clichés. En superposant des clichés du soleil obtenus par un balayage d'est en ouest puis d'ouest en est, (ces clichés résultant eux-mêmes de la superposition de clichés obtenus dans deux points symétriques par rapport au centre non déplacé de la raie), on fait apparaître un gris général dans les régions qui correspondant à des corrélations nulles entre les champs de vitesse, au même endroit du disque, à deux passages successifs de la fente. Certaines régions au contraire font apparaître des structures nettes, montrant un rétablissement de la corrélation, et ce à plusieurs reprises: pouvait-on mieux prouver l'existence d'une période dans les champs de vitesses macroscopiques, et suggérer l'idée de *mouvements oscillatoires*?

Cette idée fut confirmée nettement par l'ensemble des recherches d'Evans, Michard *et al.*, montrant, pour chaque raie étudiée, l'existence de mouvements oscillatoires amortis, de période 200 à 300 secondes. Pour différentes raies, ces mouvements ne sont pas en phase. Les études de corrélation (Evans, Mein, Michard et Servajean) font apparaître le même phénomène que les études point par point (Figures 31 et 32).

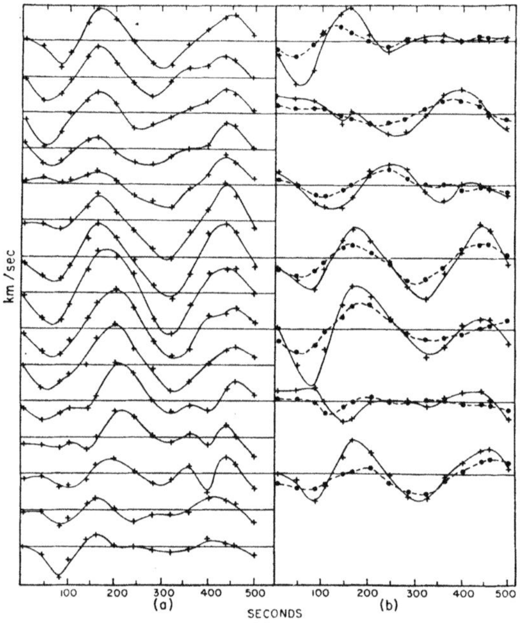

Fig. 31. (D'après Evans, Michard, 1962c.) Vitesse en fonction du temps en différents points du disque: a) pour 15 points distants de 1000 km (raie b_2 de Mg I); b) pour 7 points indépendants pour la raie b_2 de Mg I (traits pleins) et la raie Ti I 5173 (pointillé).

Il n'est pas facile de connaître exactement la vitesse de propagation des ondes, car les méthodes d'estimation des profondeurs de formation sont sujettes à caution, comme l'a fait remarquer Athay (1963). Mais les effets différentiels, d'une raie à l'autre, sont très marqués. Des raies assez superficielles comme la raie b_2 du Mg I montrent des oscillations de période 235 sec, les raies plus pro-

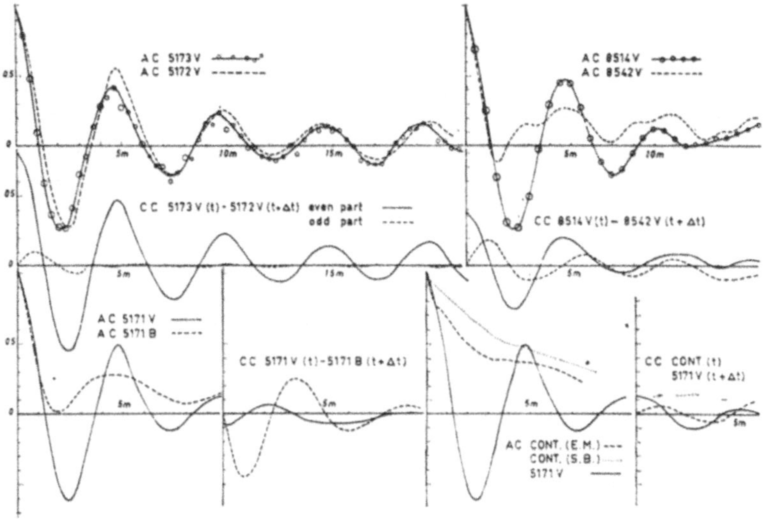

Fig. 32. Courbes d'autocorrélation (*AC*) et de corrélations croisées (*CC*) relatives aux variations des vitesses (*V*) avec le temps, des brillances (*B*) du spectre continu, et de la brillance au centre de la raie 5171 Fe I. Les raies utilisées sont les suivantes: Fe I 8514; Ti I 5173; 5172 Mg I; 5177 Fe I; 8542 Ca II. On notera tout particulièrement les deux courbes du mileu, qui montrent le délai entre les oscillations des deux raies, et le rattrapage progressif de ce retard (voir publication originale pour l'explication des aspects multiples de cette figure très riche, dont le présent article ne met en évidence que certains des aspects). (D'après EVANS, MEIN, MICHARD, SERVAJEAN.)

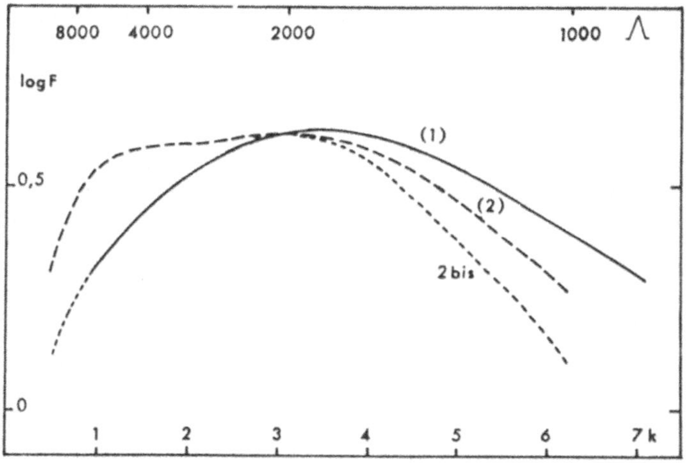

Fig. 33. Répartition spectrale bimensionnelle de l'énergie des fluctuations. (1) Granulation selon BAHNG et SCHWARZSCHILD (1961); (2) Vitesses radiales selon BERNIÈRE, MICHARD et RIGAL (1962); (2') Correction de la courbe (2) pour tenir compte de l'influence du grain photographique. Ces courbes sont corrigées du filtrage des fréquences spatiales dues aux effets instrumentaux et atmosphériques.

fondes (Ti I, 5173) une période de 249 sec, la différence étant très significative; les raies de Mg I sont en avance de huit secondes environ sur celles de Ti dans leur oscillation.

L'échelle de ces mouvements est très remarquable: Si les clichés en question sont assez bons pour détecter les granules individuels de 1000 km environ, en revanche on s'aperçoit que les mouvements sont en phase sur des structures de l'ordre de 3000 à 50000 km. Observation confirmée par la comparaison du "spectre" d'énergie des fluctuations, pour la brillance et pour les vitesses radiales (Figure 33).

D'autre part, quelques aspects de la corrélation entre le spectre continu et les champs des vitesses dans les régions où se forment les raies sont mis en évidence; notamment la remarquable association entre l'apparition de régions brillantes dans le continu, et des oscillations dans les raies, débutant par un déplacement vers le violet.

La vitesse des ondes peut se déduire de l'intervalle (40 secondes) entre les phénomènes du spectre continu, et l'oscillation correspondante, dans la région où se forme la raie: elle est de l'ordre de 5 km/s et proche de la vitesse sonique. Le phénomène oscillatoire diminue grandement avec la hauteur dans la photosphère: ceci est montré par la variation entre le centre et le bord, et par les comparaisons entre les différentes raies; des raies chromosphériques étudiées par Leighton (Ca II, Hα) ne présentent plus le phénomène. Les fortes oscillations sont localisées à la région de transition entre chromosphère et photosphère.

D. REMARQUES

Naturellement, ces vitesses oscillatoires constituent un système compliqué. Leur superposition fait que, parfois, aucune oscillation n'est détectable, le champ de vitesses semble purement statistique. Ce champ a été décrit par plusieurs auteurs (Evans et Michard, 1962a; Goldberg *et al.*, 1960; Bernière *et al.*, 1962). Ses caractéristiques obtenues à partir de spectres à grande résolution angulaire devraient être beaucoup plus précises que ce qu'on obtenait avec les méthodes antérieures. En résumé, ξ augmente avec l'intensité de la raie (c'est à dire vers l'extérieur de la photosphère, en tous cas dans des régions relativement profondes); l'étude des variations du centre au bord montre que le champ de vitesses pourrait consister en une composante verticale qui augmente vers l'extérieur et une composante horizontale indépendante de la profondeur. N'oublions cependant pas que les mesures au bord peuvent être affectées par la rugosité optique.

De plus un champ de vitesses horizontales à grande échelle (naturellement mieux visible près du bord) se développe sur des éléments de surface solaire importants: c'est ce que Leighton appelle la "supergranulation", pour laquelle Leighton, Noyes et Simon trouvent un champ de vitesses centrifuges horizontales, sur des éléments de l'ordre de grandeur de 15000 km.

E. CONCLUSION

L'interprétation de ces données est sans doute la suivante: Il semble que l'on puisse séparer des mouvements oscillatoires progressifs (où la phase dépend de la profondeur) et vite amortis, des mouvements oscillatoires persistants, pour lesquels l'ensemble des couches photosphériques oscille en phase.

Les deux types de mouvements oscillatoires correspondraient peut-être à des ondes de période 300 sec (mouvements persistants) et plus courtes (ondes progressives de pression). La transition d'un type à l'autre s'effectuerait progressivement pendant la courte durée de vie des ondes progressives, comme si leur passage jouait le rôle d'un archet excitateur pour l'ensemble de l'atmosphère, archet mu par la granulation de la zone convective.

4. Le retour à la théorie

Au début de cet exposé, nous avons rejeté comme désuets les modèles de la photosphère en équilibre radiatif; il s'agit d'une représentation physique de la photosphère d'un caractère que l'on peut aujourd'hui tenir pour primitif, naïf même. Les contradictions, entre ce modèle et la réalité, sont devenues flagrantes: l'existence d'une chromosphère, les inhomogénéités, les écarts à l'ÉTL, et ces mouvements plus ou moins systématiques de la photosphère qui la ramènent à son état de zone de transition, tout cela élimine les modèles en équilibre radiatif, hydrostatique, thermodynamique. Les théories de ces modèles telles qu'elles sont décrites dans les ouvrages fameux de CHANDRASEKHAR, (1950), et KOURGANOFF (1951); de SOBOLEV, le dernier en date (1963), ont désormais d'autres champs d'application que la photosphère, sauf peut-être l'étude de problèmes de détail.

Il ne faudrait pas en déduire pour autant que la photosphère est, comme l'atmosphère terrestre ou jovienne, d'une complexité telle que toute théorie y soit impossible.

Les principes essentiels des théories qui devraient rendre compte des observations spectrographiques et autres sont en effet assez simples. Ils doivent certainement d'abord ne négliger l'étude d'aucun phénomène de couplage. Au lieu d'essayer de séparer les variables jusqu'à l'absurde, comme ce fut le cas longtemps, il s'agit au contraire de traiter unitairement la physique de l'atmosphère solaire dans son ensemble.

Deux problèmes fundamentaux se posent. Le premier est un aspect de la théorie de *l'information*. Il s'agit, compte tenu, dans chaque domaine, de la précision des mesures, de déterminer quelles sont les mesures "utiles", celles qui apportent de l'information, et pour lesquelles il importe donc de rassembler des données complètes aussi exactes que possible; il s'agit aussi de déterminer quelles sont les mesures faisant double emploi. N'est-il pas vrai que mesurer l'assombrissement à 5000 et à 5010 Å est inutile, à la précision à laquelle on peut le faire? N'est-il pas vrai que de nombreuses mesures de largeurs équivalentes de raies sont inutiles, et n'apportent rien de plus que d'autres? Il s'agit aussi, une fois établi l'inventaire nécessaire des données nécessaires et suffisantes de déterminer, en nombre ni trop grand ni trop petit, les paramètres physiques, distribution des températures, des pressions et populations atomiques, des vitesses, etc... jusqu'aux moments de ces distributions pour lesquels ils ne seraient plus déterminés physiquement par les mesures, elles d'une précision bien définie.

L'autre problème, c'est de *comprendre* les distributions ainsi obtenues, de les prévoir même.

Il est difficile de présager aujourd'hui de ce que pourrait être la théorie de demain. Certaines de ses idées majeures peuvent cependant être prévues. Tout d'abord, on sait que dans le cours des années récentes, les résultats remarquables sur la propagation des ondes dans la photosphère, avaient été prédits par une série de travaux, dans le détail desquels nous n'entrerons pas ici (voir PECKER, 1960). BIERMANN (1948), SCHATZMAN (1949, 1954), THOMAS (1954), SCHWARZSCHILD (1958) voyaient déjà dans la granulation l'origine profonde d'ondes dont la dissipation dans le chromosphère en déterminerait le chauffage. Des théories comme celle de WHITNEY (1955ab, voir aussi LEDOUX et WHITNEY, 1960) utilisant les granules comme pistons élémentaires, et suivant la propagation des ondes issues de ces pistons, fournirait sans doute l'armature mécanique de ce genre de théorie (voir aussi SCHATZMAN, 1960). Cependant, à cette énergie mécanique, dont le transfert doit être étudié, se superpose, le plus traditionnellement du monde, un flux radiatif. Les échanges d'énergie à toute échelle, l'intervention des phénomènes dépendant du temps, et de leur amortissement, les couplages entre mouvements locaux, champ de rayonnement, état local de la matière, tout cela traité de la façon la plus générale compatible avec la précision des données... Naturellement, il s'agit d'un programme littéralement "formidable" (au sens étymologique de ce mot!). Aussi quelques problèmes individuels doivent d'abord être traités. L'un de ceux qui nous apparaissent urgents est celui de l'équilibre radiatif sans ÉTL; la population des niveaux, qui fixe entièrement l'opacité, commanderait les équations de conservation du flux, donc le "modèle", qui servirait ensuite de base à une détermination des populations des niveaux, par solution simultanée des équations du transfert et d'équilibre statistique; on voit l'itération qui s'amorce... Un autre problème est celui de l'étude des échanges d'énergie associés aux inhomogénéités: les écarts à l'ÉTL, leur relaxation, sont naturellement en cause, la solution des problèmes de transfert dans un plan horizontal également.

On remarque que les études comme celle des écarts à l'ÉTL (c'est à dire des couplages entre matière et rayonnement) sont déjà dépassées; il s'agit maintenant déjà de coupler ces écarts aux inhomogénéités, à l'équilibre d'ensemble des couches photosphériques, etc...

La richesse du sujet (sa difficulté aussi) augmente sans cesse. La photosphère, domaine de recherche qui semblait le refuge de la perfection, et qui apparaissait voici dix ans comme l'exemple type d'un domaine mourant de son propre achèvement, semble aujourd'hui l'un des domaines les plus vivants de la physique solaire. S'y combinent les effets *mécaniques* dus au passage des ondes, des couches de la zone convective à celles de la chromosphère, et les effets *radiatifs* du rayonnement issu de couches internes. C'est une transition entre l'intérieur et l'extérieur: les *écarts à l'ÉTL* y apparaissent. Dans ce milieu en agitation mécanique, les *inhomogénéités* s'entretiennent. De nouveau la discussion est ouverte...

Appendice

L'INVERSION NUMERIQUE DE LA RELATION DE LAPLACE

L'inversion de la relation

$$I(\mu) = \int_0^\infty S(\tau) e^{-\tau/\mu} \frac{d\tau}{\mu} \tag{A1}$$

a été maintes fois effectuée, dans le contexte astronomique, par différentes méthodes numériques approchées.

1. Représentation approchée de la fonction-source S ou de l'intégrale I

Si la fonction $S(\tau)$ est représentable par une somme d'un nombre *limité* de fonctions *simples* (linéaires, exponentielles, integroexponentielles, polynômiales comme suit:

$$S(\tau) = \sum_{i=1}^{N} a_i \varphi_i(\tau) \tag{A2}$$

alors la fonction $I(\mu)$ sera représentable par une somme des transformées, supposées connues, de ces fonctions:

$$I(\mu) = \sum_{i=1}^{N} a_i \Phi_i(\mu) \ . \tag{A3}$$

Si, grâce à la méthode des moindres carrés par exemple, on calcule les coefficients a_i de façon à représenter les observations le mieux possible au moyen d'une telle somme des fonctions $\Phi_i(\mu)$ en nombre limité, alors le retour aux fonctions $\varphi_i(\tau)$ correspondantes permettra, par la relation (A2), de calculer $S(\tau)$ et le problème posé sera résolu.

De très nombreuses représentations ont été utilisées dans le passé. Les plus fréquentes sont les suivantes:

$S(\tau)$	$I(\mu)$	*Références*
$a + b\tau + c\tau^2$	$\rightarrow a + b\mu + \frac{1}{2}c\mu^2$	MILNE (1922), LUNDBLAD (1923), CHALONGE-KOURGANOFF (1946)
$a + b\tau + c\tau^2 + d\tau^3$	$\rightarrow a + b\mu + \frac{1}{2}c\mu^2 + \frac{d}{6}\mu^3$	BARBIER (1946)
$a + b\tau + ce^{-k\tau}$	$\rightarrow a + b\mu + \frac{c}{1 + k\mu}$	KOURGANOFF (1949)
$a + b\tau + cE_2(\tau) + dE_3(\tau)$	$\rightarrow a + b\mu + c\left[1 - \mu \log_e (1 + \frac{1}{\mu})\right] + d\left[\frac{1}{2} - \mu + \mu^2\left\{\log_e (1 + \frac{1}{\mu})\right\}\right]$	KOURGANOFF (1949, 1951), PIERCE, ALLER (1951), PIERCE, WADDELL (1961) (avec $d = 0$ dans ces deux dernières références)

On trouvera dans DUMONT et PECKER (1960) un essai systématique de ces méthodes sur un exemple de fonction-source à dérivée discontinue.

De plus, l'article de MINEUR et PEYTURAUX (1952) fournit une comparaison de la précision de différentes formules à$(n + 1)$ paramètres, différentes valeurs étant utilisées pour le $n$ième paramètre $d<$ (faisceaux de solutions). Ces représentations sont des types suivants pour $S(\tau)$; on se reportera à l'article original pour les formes de $I(\mu)$:

$$a + (b + c\tau + \mathrm{d}\tau^2)e^{-\alpha\tau} \; ; \qquad \frac{a\tau^2 + b\tau + c}{\tau + \alpha} \; ; \qquad a + \frac{b}{\tau + \alpha} + \frac{c}{(\tau + \alpha)^2} .$$

Chacun des faisceaux enveloppe deux points (points de passage obligé) et on utilise en somme une optimisation combinée des différentes représentations en interpolant entre ces points la solution finalement choisie.

PAGEL (1960) a également essayé de rendre compte de l'existence d'une chromosphère (minimum de température) par l'utilisation de la représentation :

$$S(\tau) = a + b\tau + cE_2(\tau) + de^{-\alpha\tau}[1 - \beta\tau]$$

qui correspond à :

$$I(\mu) = a + b\mu + c\left[1 - \mu \log_e \left(1 + \frac{1}{\mu}\right)\right] + \frac{d}{1 + \alpha\mu^2}[1 + (\beta + \alpha)\mu] .$$

Cette solution a des avantages, mais la diminution des résidus n'est peut être due qu'à l'augmentation du nombre de paramètres, non pas à la qualité intrinsèque de la forme de la représentation.

Citons aussi la forme utilisée par Miss BURBRIDGE (1941)

$$S = \frac{b^{-a}}{t_a} \int_0^T t^{a-1} e^{(b-1/b)t} dt \; ; \qquad I(\mu) = \left(1 - b + \frac{b}{\mu}\right)^{-a} .$$

Une méthode comme celle de SYKES (1953) relève de la même optique. Au lieu de représenter la fonction $S(\tau)$ par une somme de fonctions simples, SYKES opère un changement de variable, et utilise les variables $\xi = \log_e \mu$ et $\eta = \log_e \tau$ au lieu de μ et τ; les développements utilisés sont des polynômes d'ordre 3 en ξ et η. Sans fournir une meilleure solution, cette formulation permet sans doute (BÖHM 1961) une discussion plus réaliste de l'effet des erreurs et des approximations.

Il faut noter toutefois qu'il n'y a aucune raison pour que $S(\tau)$ soit représentable par une somme de fonctions simples. C'est une fonction continue, mais elle peut présenter des points où ses dérivées soient discontinues. Une façon de résoudre le problème est alors celle utilisée par PLASKETT (1936, 1964) puis par MINEUR et PEYTURAUX (1952) et préconisée par REIZ (1943), à l'aide d'une formule de quadrature aussi précise que l'on voudra. Nous pouvons en effet écrire:

$$I(\mu) = \sum_{i=1}^{m} S(\tau_i) e^{-\tau/\mu} \frac{P_i}{\mu} \tag{A4}$$

où les τ_i sont les points de division choisis, et les P_i des poids connus d'intégration. L'équation (A4) peut s'écrire pour autant de valeurs de μ =soit n valeurs) que les données en fournissent. On résout alors un système d'équations. Si $m = n$, on en déduit les m valeurs de $S(\tau_i)$. Si $m < n$ on obtient les $S(\tau_i)$ par les moindres carrés; si au contraire $m > n$, on devra (grâce à des mesures nouvelles, ou à des interpolations) augmenter le nombre n de points utilisables. En réalité, cette méthode ne diffère pas essentiellement de la précédente. Elle revient en effet à développer $I(\mu)$ en utilisant des fonctions:

$$\Phi_i(\mu) = \frac{1}{\mu} e^{-\tau_i/\mu} , \tag{A5}$$

donc B suivant les fonctions φ_i dont ces Φ_i sont les transformées. Au voisinage de $\mu = 0$, tel développpement de $I(\mu)$ n'est d'ailleurs pas valable, car $I(\mu)$ tendrait vers zéro pour toute valeur non nulle de τ_i, et vers l'infini pour une valeur nulle de τ_i. Outre cet inconvénient, la méthode conduit le plus souvent à un système indéterminé, en raison du trop grand nombre de points choisis.
Sous la forme en question, cette méthode est donc à éliminer – essentiellement comme étant équivalente à l'emploi d'une mauvaise représentation de S. On peut naturellement utiliser de nombreuses formules de quadrature, certaines étant meilleures que d'autres par un choix, plus approprié au problème en

question, des points de division et des poids. On en trouvera notamment une discussion dans la seconde partie du mémoire de SYKES (1953).

2. Méthode itérative des profondeurs de formation.

Une forme limitée de cette dernière méthode a néanmoins conduit à un succès relatif. On peut en effet, en la limitant à son premier terme, écrire:

$$I(\mu) = S(\tau^*) \qquad \text{(A6)}$$

Si au lieu de considérer cette relation comme une formule d'intégration où les valeurs $\tau_i = \tau^*$ et $P_i e^{-\tau^*} = 1$ sont fournies par quelque considération a priori sur la forme de $S(\tau)$, nous pouvons nous en servir comme d'une définition de τ^*, profondeur effective de formation:

$$S(\tau^*) = \int_0^\infty S(\tau) e^{-\tau/\mu} \frac{\mathrm{d}\tau}{\mu} \qquad \text{(A7)}$$

Il est bien connu que, si S est supposée être une fonction linéaire de τ, alors τ^* est égal à μ. C'est la relation bien connue d'EDDINGTON et BARBIER (voir KOURGANOFF, 1951), dont l'utilisation dans ce contexte a été suggérée par UNSÖLD (1948). De façon générale, si l'on fait une hypothèse sur la forme de S considéré comme la somme d'une fonction linéaire et d'une somme de N fonctions simples, affectées de coefficients a_i:

$$S(\tau) = \alpha + \beta\tau + \sum_{i=1}^{N} a_i \varphi_i(\tau) \qquad \text{(A8)}$$

alors pour que la formule (A6) soit une formule d'intégration exacte, il faut que τ^* prenne une valeur particulière, dépendant de μ et des N coefficients en question. Des tables (et graphiques) représentants τ^* dans le cas de différentes représentations ont été dressées par DUMONT et PECKER (1960–1961). Ces tables permettent de dépasser les résultats de la méthode (a), pour chaque cas (par exemple méthode (a) appliquée avec $S = a + b\tau + cE_2(\tau)$ et méthode des τ^* appliquée avec la même forme pour S). En effet, elles supposent essentiellement que la représentation en question et la fonction réelle S ont des propriétés intégrales communes, mais non qu'elles sont identiques. Dans le cas linéaire, le modèle obtenu avec l'hypothèse $\tau^* = \mu$ correspondante, est nettement meilleur que le "meilleur modèle linéaire possible", étant entendu que la qualité d'un modèle est, en gros, mesurée par l'intégrale:

$$\int_0^1 [I(\mu)_{\text{calc}} - I(\mu)_{\text{vrai}}]^2 \mathrm{d}\mu$$

ou par une intégrale de signification semblable (il serait bon toutefois que l'on s'entendit sur la notion de "qualité" d'un modèle!). Cette méthode n'est donc pas mauvaise mais elle est limitée dans certains cas par les définitions multiples de τ^* entre lesquelles le choix sera ambigu. Sa qualité est démontrée par la figure no. 2 du mémoire cité. Précisons qu'il s'agit évidemment d'une méthode itérative.

3. Expressions explicites de la fonction-source

On peut avoir l'idée d'utiliser aussi une modification de la méthode (a) en se servant d'un développement infini pour I et S, développement qui naturellement devra être convergent. C'est le calcul des coefficients a_i (ici, en très grand nombre) qui rendrait cette méthode difficile, même avec une machine électronique puissante, en raison de haut degré d'indétermination qu'aurait naturellement un tel système. On peut alors songer à l'artifice suivant. Si les fonctions $\Phi_i(\mu)$ sont orthogonales sur l'intervalle (0.1), on peut écrire:

$$I(\mu) = \sum_{i=1}^{n} a_i \Phi_i(\mu) \tag{A9}$$

$$S(\tau) = \sum_{i=1}^{n} a_i \varphi_i(\tau) \tag{A10}$$

avec

$$\int_0^1 \Phi_i \Phi_i d\mu = 0 \ (\text{si } i \neq j) \qquad = \psi_i \ (\text{si } i = j) \tag{A11}$$

Il en résulte que

$$\int_0^1 I(\mu)\, \Phi_i(\mu)\, d\mu = a_i \psi_i \tag{A12}$$

et donc que les coefficients a_i peuvent se calculer par la relation:

$$a_i = \frac{1}{\psi_i} \int_0^1 I(\mu)\, \Phi_i(\mu)\, \mathrm{d}\mu \tag{A13}$$

Des essais ont été faits en utilisant des polynômes de Legendre pour les fonctions Φ_i. Les fonctions $\varphi_i(\tau)$ sont alors aussi des polynômes qu'il est aisé de calculer. Ces essais, faits à la machine de bureau, sont négatifs. Effectuées jusqu'à l'ordre 16, il ne fournissent de bonnes valeurs de S que pour des petites valeurs de τ. La raison en est certainement la représentation polynômiale. Si l'on voulait reprendre cette question, il faudrait trouver un système infini de fonctions orthogonales dont les transformées $\varphi_i(\tau)$ ne croîtraient pas plus vite qu'une fonction linéaire. À notre avis, l'apparition de moyens puissants de calcul électronique permettrait aujourd'hui l'utilisation fructueuse d'une telle méthode, peut-être même avec des polynômes.

Il est à noter que ce genre de méthodes, qui fournissent I par une expression exacte et explicite, sont du même type, essentiellement, que la solution suivante, solution générale de l'équation intégrale en question:

$$S(\tau) = \frac{1}{2\pi i} \int_{-i\alpha}^{+i\alpha} e^{-\tau/\mu} \mu I(\mu) d\mu \qquad \text{(A14)}$$

mais le calcul de l'expression (A14) est difficile directement en raison de l'intervention de quantités complexes; $I(\mu)$ doit être défini sur tout le plan, alors qu'il n'est donné que sur l'axe réel entre 0 et 1. Cette considération explique d'ailleurs les échecs des différentes méthodes.

Signalons aussi, que des méthodes analogues, utilisant des développements en série de Fourier, peuvent également être envisagées, comme Jefferies l'a fait remarquer lors de la discussion du présent exposé.

4. Méthodes de minimisation des corrections.

On devrait également pouvoir appliquer une méthode analogue à celle proposée déjà par DUMONT et PECKER (1960), afin d'annuler progressivement tous les résidus $I(\mu)_{\text{calc}} - I(\mu)_{\text{mes}}$. Si en effet on disposait des transformées S de fonctions I nulles partout sauf autour d'une valeur μ_0 de μ, la caractère linéaire de la transformation permettrait une correction progressive. Il faudrait par exemple dresser une table de fonctions S, chacune correspondant au choix de trois paramètres définissant la position et la forme de la perturbation.

Il n'a pas été possible de dresser cette table à l'aide de fonctions dont les transformées soient calculables simplement. Aussi une telle méthode ne peut-elle être utilisée qu'après la construction d'une telle table, au moyen cette fois de méthodes de calcul puissantes; encore ne serait-elle alors utile qu'aux chercheurs ne disposant pas de ces moyens de calcul puissants, et c'est pourquoi cette direction de recherche n'est sans doute plus très actuelle. Citons seulement les intéressantes recherches de PRZYBYLSKI (1957) qui procède par correction d'un modèle primitif, le calcul de la "correction" étant fait alors par une méthode voisine de celle de Plaskett, en remplaçant l'intégrale par une somme, et en calculant la correction ΔS en sept points représentatifs.

5. Autres méthodes

Enfin (PECKER, 1950) des dispositifs utilisant des analogies électriques peuvent être conçus. Celui auquel s'est intéressé l'auteur, et qui utilisait une machine à "courants carrés" ne peut malheureusement pas fonctionner, en raison du fait que les fréquences les plus élevées sont absentes, que les courants ne sont donc pas vraiment carrés, et que la transformation a les mêmes défauts qu'une transformation utilisant des polynômes de degré élevé, mais non infini.

Sans vouloir exagérer l'importance, pour les astrophysiciens, de ce problème, nous devons, en conclusion, constater qu'il n'est pas encore résolu de façon suffisante. Aucune méthode appliquée jusqu'à present n'a permis encore de diminuer les résidus $I_{\text{calc}}-I_{\text{mes}}$ en deça des erreurs d'observation; aucune non plus ne permet d'évaluer les erreurs commises sur la fonction solution $S(\tau)$. Il serait donc souhaitable d'arriver à une meilleure compréhension numérique du problème.

Bibliographie

Cette bibliographie, pour longue qu'elle soit, est naturellement incomplète et peut-être injuste pour certains mémoires que l'auteur s'excuse d'avoir méconnus; y figurent notamment très peu de mémoires relatifs à des résultats de mesures. Les astérisques se réfèrent à des mises au point récentes sur des problèmes intéressant la structure photosphérique et comportant souvent de nombreuses références bibliographiques.

Adam, M. G.: 1952, *Mon. Not. Roy. Astron. Soc.* **112**, 548.

Allen, C. W.: 1934, *Mem. Commonwealth Solar Obs. Canberra* **1**, 5.

* Aller, L. H.: 1960, in *Stellar Atmospheres* (ed. by Greenstein) Chicago Univ. Press, p. 156.

* Aller, L. H.: 1961, *The Abundance of the Elements*, Interscience, New York.

Athay, R. G.: 1961, *Astrophys. J.* **134**, 765.

Athay, R. G.: 1963, *Astron. J.* **68**, 272 (*Astrophys. J.* sous presse).

Bahng, J., Schwarzschild, M.: 1961, *Astrophys. J.* **134**, 337.

Barbier, D.: 1946, *Ann. Astrophys.* **9**, 173.

Barbier, D.: 1951, *Ann. Astrophys.* **14**, 316.

Bernière, G., Michard, R., Rigal, G.: 1962, *Ann. Astrophys.* **25**, 279.

Biermann, L.: 1948, *Z. Astrophys.* **25**, 161.

Blackwell, D. E., Dewhirst, D. W., Dollfus, A.: 1959, *Mon. Not. Roy. Astr. Soc.* **119**, 98 (voir planche 4).

Blamont, J. E., Pecker, J.-C.: 1961, Liège Symp. (The Ultra-violet), p. 45.

Blamont, J. E., Roddier, F.: 1961, *Phys. Rev. Letters* **7**, 437.

Böhm, K. H.: 1954, *Z. Astrophys.* **34**, 182.

* Böhm, K. H.: 1960, 'Basic Theory of Line Formation', in *Stellar Atmospheres* (ed. by Greenstein), Chicago Univ. Press, p. 88.

Böhm, K. H.: 1961, *Astrophys. J.* **134**, 264.

Böhm-Vitense, E.: 1954, *Z. Astrophys.* **34**, 209.

Böhm-Vitense, E.: 1960, 'Aerodynamical Phenomena in Stellar Atmospheres' IAU Symposium **12**, p. 330.

Bonnet, R., Courtes, G.: 1962, *Ann. Astrophys.* **25**, 337.

Bowen, I. S.: 1948, *PASP* **60**, 16.

Bretz, M.: 1955, *Z. Astrophys.* **38**, 259.

Bruggencate, P. ten, Houtgast, J.: 1941, *Z. Astrophys.* **20**, 149.

Burbridge, I. W.: 1941, *Mon. Not. Roy. Astron. Soc.* **101**, 26.

Cabannes, J., Dufay, J.: 1948. *Compt. Rend.* **226**, 1569.

Chalonge, D., Kourganoff, V.: 1946, *Ann. Astrophys.* **9**, 69.

* Chandrasekhar, S.: 1950, *Radiative Transfer*, Oxford Univ. Press.

Cuny, Y.: 1960, *Compt. Rend.* **250**, 3117.

Cuny, Y.: 1961, *Compt. Rend.* **252**, 4111.

Cuny, Y.: 1963a, Comm. personelles.

Cuny, Y.: 1963b, *J. Quant. Spectr. Rad. Transfer* **3**, 129.

Cuny, Y., Lefèvre, J., Pecker, J.-C.: 1960, *Ann. Astrophys.* **23**, 923.

Claas, W. J.: 1951, *Rech. Astron. Obs. Utrecht* **12**, 1.

Debarbat, S.: 1963, En préparation pour publication.

Detwiler, C. R., Purcell, J. D., Tousey. R.: 1961, Liège Symposium (The Ultra-violet). p. 253.

DOLLFUS, A., LEROY. J.-L.: 1960, *Compt. Rend.* **250**, 665.
DUBOV, E. E.; 1955, *Astron. Cirkular* **159**, 11.
DUMONT, S.: 1960, *Ann. Astrophys.* **23**, 879.
DUMONT, S.: 1961, *Compt. Rend.* **252**, 2670.
DUMONT, S.: 1963a, *J. Quant. Spectr. Rad. Transfer* **3**, 125.
DUMONT, S.: 1963b. Comm. personelle.
DUMONT, S., LALOUX, R., MOUELLIC, L., PECKER, J.-C.: 1961, *Ann. Astrophys.* **24**, 328.
DUMONT, S., PECKER, J.-C.: 1960, *Ann. Astrophys.* **23**, 655.
EDMONDS, F. N. (Jr.): 1962a, *Astrophys. J.* **136**, 507.
EDMONDS, F. N. (Jr.): 1962b, *Astrophys. J.* (Suppl.) **6**, 357.
EUGÈNE-PRADERIE, F., PECKER, J.-C.: 1960, *Ann. Astrophys.* **23**, 622.
EVANS, D. S.: 1947. *Mon. Not. Roy. Astron. Soc.* **107**, 433.
EVANS, J. W., MICHARD, R.: 1962a. *Astrophys. J.* **135**, 812.
EVANS, J. W., MICHARD, R.: 1962b, *Astrophys. J.* **136**, 487.
EVANS, J. W., MICHARD, R.: 1962c, *Astrophys. J.* **136**, 493.
FAULKNER, D. J., MUGGLESTONE, D.: 1962, *Mon Not. Roy. Astron. Soc.* **124**, 11.
FELLGETT, P.: 1959, *Mon. Not. Roy. Astron. Soc.* **119**, 475.
FINDLAY-FREUNDLICH, E. F., FORBES, E. G.: 1956, *Ann. Astrophys.* **19**, 185 et 215.
GIOVANELLI, R. G.: 1959, *Australian J. Phys.* **12**, 164.
GÖKDOGAN, N., HOTINLI, M., PECKER, J.-C.: 1962, *Ann. Astrophys.* **25**, 324.
GOLDBERG, L.: 1958, *Astrophys. J.* **127**, 308.
GOLDBERG, L., ALLER, L. H.: 1943, *Atoms, Stars, Nebulae*, Blakiston.
GOLDBERG, L., MOHLER, O., UNNO, W., BROWN, J.: 1960, *Astrophys. J.* **132**, 184.
* GOLDBERG, L., MULLER, E. A., ALLER, L. H.: 1960, *Astrophys. J.* (Suppl.) **5**, 1.
* GOLDBERG, L., PIERCE, A. K.: 1959, 'The Photosphere of the Sun', *Handbuch der Physik* **52**.
GREENSTEIN, J. L., TANDBERG-HANSSEN, E.: 1954, *Astrophys. J.* **119**, 113.
HIGGS, L. A.: 1960, *Mon. Not. Roy. Astron. Soc.* **121**, 31.
HUANG, S. S., STRUVE, O.: 1952, *Astrophys. J.* **116**, 410.
* HUBENET, H.: 1960, *Rech. Astron. Obs. Utrecht* **16**, 1.
HUNAERTS, J.: 1947, *Ann. Astrophys.* **10**, 237.
DE JAGER, C.: 1948, *Proc. Acad. Sci. Amsterdam* **51**, 731.
* DE JAGER, C.: 1952, *Rech. Astron. Obs. Utrecht* **13**, 1.
DE JAGER, C.: 1954, *Nature* **173**, 680.
DE JAGER, C.: 1955, *Ann. Géophys.* **11**, 330.
* DE JAGER, C.: 1959a, 'Structure and Dynamics of the Solar Atmosphere', *Handbuch der Physik* **52**.
* DE JAGER, C.: 1959b, *Bull. Astron. Neerl.* **13**, 133.
DE JAGER, C.: 1961, Liège Symp. (The Ultra-violet), p. 578.
DE JAGER, C.: 1962, *Med. Kon. Vlaamse Acad. Wet. België* **26** (9).
* DE JAGER, C.: 1963, Liège Symposium (The Infrared), p. 151.
DE JAGER, D., NEVEN, L.: 1962, *Bull. Astron. Neerl.* **16**, 307.
* DE JAGER, C., NEVEN, L.: 1963 Liège Symposium (The Infrared), p. 213.
JOHNSON, H. R.: 1961, *Compt. Rend.* **253**, 1403.
JOHNSON, H. R.: 1962, *Ann. Astrophys.* **25**, 30.
JOHNSON, H. R.: 1963a, en préparation pour publication.
JOHNSON, H. R.: 1963b, *Astrophys. J.* **137**, 858.
JORAND, M.: 1962, *Ann. Astrophys.* **25**, 57.
JORAND, M.: 1963, Comm. personelle.
KANDEL, R.: 1960, *Ann. Astrophys.* **23**, 995.
KHOKHLOV, M. Z.: 1960ab, *Izv. Krimsk. Astrophys. Obs.* **22**, 118, 128.
KOURGANOFF, V.: 1948, *Comptes Rend.* **227**, 895.
* KOURGANOFF, V.: 1951, *Basic Methods in Transfer Problems*. Oxford University Press.
KRAT, V. A.: 1954, *Izv. Gl. Astron. Obs. Pulkovo* **152**, 1.
KRAT, V. A.: 1956, *Izv. Gl. Astron. Obs. Pulkovo* **155**, 17.
* LEDOUX, P., WHITNEY, C.: 1960, 'Aerodynamical Phenomena in Stellar Atmospheres', IAU Symposium 12.
LEFÈVRE, J., PECKER, J.-C.: 1961, *Ann. Astrophys.* **24**, 328.

LEIGHTON, R.: 1960, 'Aerodynamical Phenomena in Stellar Atmospheres', IAU Symposium 12, p. 321.
LEIGHTON, R., NOYES, R., SIMON, G.: 1962, *Astrophys. J.* **135**, 474.
LUNDBLAD, R.: 1923, *Astrophys. J.* **58**, 113.
MÄDLOW, M.: 1962, *Z. Astrophys.* **55**, 29.
MICHARD, R.: 1953, *Ann. Astrophys.* 16, 217.
MICHARD, R.: 1962, *Compt. Rend.* **225**, 1494.
MILNE, E. A.: 1922, *Phil. Trans. A* **223**, 214.
MINEUR, H., PEYTUREAUX, R.: 1952, *Ann. Astrophys.* **15**, 383.
* MINNAERT, M. G.: 1953, 'The Photosphere' in *The Sun* (ed. by G. KUIPER), Chicago Univ. Press.
MUGGLESTONE, D.: 1958, *Mon. Not. Roy. Astron. Soc.* **118**, 432.
MUGGLESTONE, D.: 1960, *Mon. Not. Roy. Astron. Soc.* **120**, 193.
OLSON, E. C.: 1962, *Astrophys. J.* **136**, 946.
PAGEL, B.: 1955, *Mon. Not. Roy. Astron. Soc.* **115**, 493.
PAGEL, B.: 1956, *Mon. Not. Roy. Astron. Soc.* **116**, 608.
PAGEL, B.: 1957, *Astrophys. J.* **115**, 298.
PAGEL, B.: 1960, *Astrophys. J.* **132**, 790.
PAGEL, B.: 1963, *Plan. Space. Sci.* **11**, 333.
PECKER, J.-C.: 1950, *Compt. Rend.* **231**, 947.
PECKER, J.-C.: 1950, *Ann. Astrophys.* **13**, 319
PECKER, J.-C.: 1951, *Ann. Astrophys.* **14**, 115.
PECKER, J.-C.: 1959, *Ann. Astrophys.* **22**, 499.
* PECKER, J.-C.: 1960a, Liège Symposium (Modèles d'étoiles et Ev. Stellaire), p. 343.
PECKER, J.-C.: 1960b, *Ann. Astrophys.* **22**, 366.
PECKER, J.-C.: 1961, Liège Symposium (The Ultra-violet), p. 487.
PECKER, J.-C.; DE JAGER, C.: 1951, *Compt. Rend.* **232**, 1645.
PECKER, J.-C., THOMAS, R. N.: 1960, IAU Symposium 12, Aerodynamical Phenomena in Stellar Atmospheres, p. 401.
PECKER, J.-C., VOGEL, L.: 1960, *Ann. Astrophys.* **23**, 594.
PEYTURAUX, R.: 1952, *Ann. Astrophys.* **15**, 302.
PIERCE, A. K., ALLER, L. H.: 1951, *Astrophys. J.* **114**, 145.
PIERCE, A. K., WADDELL, J. H.: 1961, *Mem. RAS* **68**, 89.
PLASKETT, H. H.: 1936, *Mon Not. Roy. Astron. Soc.* **96**, 402.
PLASKETT, H. H.: 1941, *Mon. Not. Roy. Astron. Soc.* **101**, 1.
PLASKETT, H. H.: 1962, *Mon. Not. Roy. Astron. Soc.* **123**, 541.
PRZYBYLSKI, A.: 1957, *Mon. Not. Roy. Astron. Soc.* **117**, 600.
REDMAN, R. O.: 1943, *Mon. Not. Roy. Astron. Soc.* **103**. 173.
REGEMORTER, H. van: 1955, *Compt. Rend.* **240**, 1968.
REGEMORTER, H. van: 1960, *Ann. Astrophys.* **22**, 249, 341, 363, 681.
REIZ, A.: 1943, *Ark. Mat. Astr. Fys. Suède* **29**, 1.
RICHARDSON, R. S., SCHWARZSCHILD, M.: 1950, *Astrophys. J.* **111**, 351.
RODDIER, F.: 1963, Communications personelles.
RÖSCH, J.: 1959, *Ann. Astrophys.* **22**, 571, 584.
ROUNTREE, J.: 1960, *Ann. Astrophys.* **23**, 633.
RUDKJØBING, M.: 1945, *Publ. København Obs.* **140**.
RUSSELL, H. N.: 1929, *Astrophys. J.* **70**, 11.
SCHATZMAN, E.: 1949, *Ann. Astrophys.* **12**, 203.
SCHATZMAN, E.: 1954, *Bull. Acad. R. Belg. Class. Soc.* **40**, 139.
SCHRÖTER, E. H.: 1957, *Z. Astrophys.* **41**, 141.
SCHWARZSCHILD, M.: 1948, *Astrophys. J.* **107**, 1.
SERVAJEAN, R.: 1961, *Ann. Astrophys.* **24**, 1.
* SOBOLEV, V. V.: 1963, *A Treatise on Radiative Transfer*, Van Nostrand, New York.
STRÖMGREN, B.: 1940, *Publ. Medd. København Obs.* **127**.
STUART, F. E., RUSH, J. H.: *Astrophys. J.* **120**, 245.
SUEMOTO, Z.: 1957, *Mon. Not. Roy. Astron. Soc.* **117**, 2.
SYKES, J. B.: 1953, *Mon. Not. Roy. Astron. Soc.* **114**, 198.

THOMAS, R. N.: 1954, *Bull. Acad. R. Belg. Cl. Sc.* **40**, 621.
THOMAS, R. N.: 1963, *Astrophys. J.* **137**, 38.
* THOMAS, R. N., ATHAY, G. R.: 1961, *Physics of the Solar Chromosphere*, pp. 124 et sq.
TOUSEY, R.: 1961, Liège Symposium (The Ultra-violet), p. 211.
TOUSEY, R.: 1963, *Space Research* **4**.
UBEROI, M. S.: 1955a, *Astrophys. J.* **121**, 400.
UBEROI, M. S.: 1955b, *Astrophys. J.* **122**, 466.
UNNO, W.: 1959, *Astrophys. J.* **129**, 338.
* UNSÖLD, A.: 1937, *Physik der Sternatmosphären*, Springer, Berlin.
UNSÖLD, A.: 1948, *Ann. Phys. (Allemagne)* **3**, 124.
UNSÖLD, S.: 1948, *Z. Astrophys.* **24**, 306.
* UNSÖLD, A.: 1955, *Physik der Sternatmosphären*, Springer, 2ème éd.
VAN 'T VEER, F.: 1963, *Ann. Astrophys.* **26**, 185.
* VARENNA 1960, désigne l'ensemble du Symposium IAU, No. 12, 'Aerodynamical Phenomena in Stellar Atmospheres'.
VITENSE, E.: 1951, *Z. Astrophys.* **28**, 81.
VITENSE, E.: 1953, *Z. Astrophys.* **32**, 135.
VOIGT, H. H.: 1951, *Z. Astrophys.* **28**, 176.
VOIGT, H. H.: 1956, *Z. Astrophys.* **40**, 157.
WADDELL, J. H.: 1958, *Astrophys. J.* **127**, 284.
WADDELL, J. H.: 1960, *Ann. Astrophys.* **23**, 917, 928.
WADDELL, J. H.: 1963, *Astrophys. J.* **137**.
WHITE, O. R.: 1962, *Astrophys. J.* (Suppl.) **7**, 333.
WHITNEY, Ch.: 1955a, *Ann. Astrophys.* **18**, 375.
WHITNEY, Ch.: 1955b, *Astrophys. J.* **122**, 385.
WILSON, P. R.: 1960, *Australian J. Phys.* **13**, 461.
WILSON, P. R.: 1961, *Australian J. Phys.* **14**, 57.
WILSON, P. R.: 1962a, *Mon. Not. Roy. Astron. Soc.* **123**, 387.
WILSON, P. R.: 1962b, *Mon. Not. Roy, Astron. Soc.* **124**, 383.
WILSON, P. R.: 1963, *Astrophys. J.* **137**, 606.
WLÉRICK, G.: 1957, *Smithsonian Astrophys. Contr.* **2**, 26.
ZWAAN. C.: 1962, *Bull. Astron. Inst. Netherl.* **16**, 225.

THE CURRENT REVISION OF THE SOLAR SPECTRUM TABLE λ 2935 Å TO λ 8770 Å

CHARLOTTE E. MOORE-SITTERLY

National Bureau of Standards, Washington, D.C., USA.

Abstract. The current edition of *The Solar Spectrum Table* is nearing completion. All wavelengths have been corrected to conform to the 1928 scale of International Standards. For some 6500 lines, new measurements have been used in deriving the final weighted means recorded in the Solar Table.

Equivalent widths have been determined by Minnaert and his staff for all lines listed, except 152 atomic lines that are absent from the disk spectrum, but present in the sunspot spectrum.

The intensity behavior of individual lines in the spot spectrum as compared with the disk spectrum is briefly described in a separate column under letters that indicate whether the line is strengthened, weakened, unchanged, or obliterated in the spot spectrum.

The revised identifications for atomic lines have been made from the data assembled for the program on *Atomic Energy Levels*. They are based on a study of series and multiplet structure as far as possible. The lower excitation potential and multiplet number are given for the individual lines.

A revision of identifications for molecular lines has also been made for the leading molecules present in the disk spectrum. Tables of approximate counts of these lines in the solar spectrum are given for OH, NH, CN, CH, C_2, and MgH. The rotational line and vibration band are indicated in respective columns for each molecular line present in the solar spectrum. The electronic transition can be found in an auxiliary table under the note number in the last column of the Solar Table.

The present total number of lines listed, including the atomic spot lines, is about 22000 (over a longer range) as compared with about 20000 from Rowland's Table, listed in the 1928 edition. To date 71 percent are wholly or partially identified, as against 57 percent in 1928.

A general summary of elements present in the sun is discussed. There are 62 present without question, and 4 whose presence is subject to question. Two additional elements, B and F, whose presence has been reported from evidence afforded only by molecular spectra, require further study.

Observations of the solar spectrum today extend the spectral range from the X-ray to the microwave region. In one sense it may appear that the subject of this paper is an old-style one, but it is not one that has outlived its usefulness. With all the years of study since Rowland's classical work in 1895, there are still unsolved puzzles regarding the most familiar span of this interesting stellar spectrum.

At the Zurich meeting of the International Astronomical Union in 1948 it was proposed that a revison of the 1928 edition of Rowland's Table[1] be undertaken by Prof. Minnaert, Dr. H. D. Babcock and the author. At this time the project on *Atomic Energy Levels* had been started at the National Bureau of Standards in Washington, D.C., and a wealth of material on laboratory atomic spectra was being assembled for these volumes.[2]

The plan devised for this solar program placed the responsibility for the

All rights reserved.

measurement of equivalent widths of the solar lines exclusively with Prof. Minnaert. Dr. H. D. Babcock has worked out the corrections to the wavelength scale of the 1928 edition,[3] and has furnished his unpublished notes on individual lines throughout the range mentioned above. The revision of identifications and assembly of the final press copy for the current edition has been entrusted to the present author.

Table 1 is a sample taken from the new Solar Table. The procedure in handling the data in the various columns will be described below in some detail.

The Wavelengths (Column 1): In the 1928 edition Rowland's scale of wavelengths was converted to the International System by means of a correction curve based on the International Wavelength Scale of 1922. In making this conversion, three classes of lines resulted: (1) those that were remeasured for use in correcting the scale, designated by "m"; (2) lines marked "w", which were weighted means of remeasured lines and of Rowland's corrected wavelengths; (3) the great majority, taken from Rowland's original table, but corrected to the International System by the correction curve mentioned above.

Further work on the International Standards has indicated that a small running correction is required to fit the wavelengths published in 1928 to the

TABLE 1

SOLAR SPECTRUM WAVELENGTHS

Wave-length (Å)	*Equivalent Width $\Delta\lambda$ mÅ*	*Reduced Width $\Delta\lambda/\lambda$*	*Spot*	*Solar Identification*	*Low E P or Rot. Line*	*RMT No. or Vib. Band*	*Notes*
4037.548	4.5	1.1					
4037.688	22	5.4	*u*	Fe I	2.28	118	
4037.913a	4.5	1.1					
4037.976	4.5	1.1					
4038.124	13	3.2	*s*				
4038.276	3	0.7	*uN*	Ni I	3.65	150	
4038.479r	3.5	0.9					
4038.627	26	6.4	*u*	Fe I	3.30 3.41	600 728	
4038.795	64	15.8	*u*	Fe I			
4038.946r	2.5	0.6					
4039.096	42	10.4	*s*	Cr I	3.85	251	
4039.296	7	1.7	*u*	Cr I–	3.85	251	
4039.432r	0.5	0.1					
4039.574	9	2.2	*u*	V II	1.82	32	
4039.743	8	2.0					
4039.864	44	2.2	*s*	Y I	0.00	5	
4039.946		8.9	*u*	Fe I	2.73	276	
4040.097m	61	15.1	*u*	Fe I			
4040.270	31	4.0		Zr II–	0.93	54	
4040.310		4.0	*s*	Ti I	2.12	185	
				CH	P1	1.1	2

TABLE I (continued)

Wave-length (Å)	*Equi-valent Width Δλ mÅ*	*Re-duced Width Δλ/λ*	*Spot*	*Solar Identi-fication*	*Low E P or Rot. Line*	*RMT No. or Vib. Band*	*Notes*
4040.514	27	6.9	*u*	–CH	Q3	1.1	2
4040.647m	78	19.3	*u*	Fe I	3.30	655	
4040.790	39	9.6	*s*	Ce II	0.44	138	
				Nd II	0.18	30	
4040.951r	4	1.0					
4041.066	8	2.0	*u*				
4041.283r }	168	16.3	*u?*	Fe I	3.30	603	
					3.30	654	
4041.374 }		28.4	*s*	Mn I	2.12	5	
4041.656m	58	14.3	*u*				
4041.914	13	3.2	*u*	Fe I	3.30	602	
4041.989r	3.5	0.9					
4042.05 a	3	0.7					
4042.149r	3.5	0.9					
4042.245	10	2.5	*s*	Cr I	2.54	36	
4042.363	6	1.5	*u?*				
4042.443r	3.5	0.9					
4042.592	13	3.2	*u*	Ce II	0.49	140	
4042.758	10	2.5	*u*	Fe Ip	3.28	556	
4042.907	18	4.4	*u*	La II–	0.92	66	
				(Sm II)	0.10	9	
4042.997r	1.5	0.4					
4043.346	24	5.9	*uN*	CH	Q4	1.1	2
4043.47 a	2.5	0.6					
4043.608	6.5	1.6	*s*	Zr I	0.52	32	
4043.695	17	4.7	*u*	Cr I	4.45	306	
				Fe Ip	2.48	122	
4043.768	27	7.4	*u*	–Ti I	2.16	208	
4043.906 }	146	22.0	*u*	Fe I	2.73	276	
					3.24	557	
4043.988 }		14.1	*u*	Fe Ip	3.24	559	
4044.145	16	4.2	*s*	K I	0.00	3	
4044.274r	0.5	0.1					
4044.380	6.5	1.7		CH	P2	1.1	2
4044.497	40	14.3	*u*	Fe Ip	4.10	1073	
4044.615	105	28.0	*u*	Fe I	2.83	359	
4044.843r	2.5	0.7					
4044.959r	1	0.4					
4045.115	50	17.3	*u*	‖ Fe I	3.02	425	
				Mn I	4.33	48	
4045.217	19	7.4	*uN?*	Mn I			
4045.390	48	25.0	*u*	Co I	1.04	31	
4045.508	9	8.2					
4045.600	29	26.4	*u*	Fe Ip	3.21	559	
				∣ Zr II	0.71	30	
4045.715r	7.5	13.3					
4045.825m	1174	*309.9*	*S*	Fe I	1.48	43	
4045.968	14	23.0					

TABLE I (continued)

Wave-length (Å)	*Equi-valent Width Δλ mÅ*	*Re-duced Width Δλ/λ*	*Spot*	*Solar Identi-fication*	*Low E P or Rot. Line*	*RMT No. or Vib. Band*	*Notes*
4046.074	44	36.1	*u*	Fe Ip—	3.26	557	
4046.341	9	4.4		Ce II	0.55	81	
				V II	3.12	177	
4046.459	19	7.4	*u*	Fe Ip	4.15	1075	
4046.615r	1	0.4		Fe I	3.07	487	
4046.760	7.5	2.3	*u*	| Cr I	2.54	36	
				Ni I	4.09		
4047.016	10	2.7	*u*				
4047.189r	5	1.4					
4047.23 a	5	1.4	*S*	K I	0.00	3	
4047.310	41	10.9	*u*	Fe I	{2.28	117	
					{3.63	853	
4047.407r	3.5	1.0					
4047.673	21	5.4	*s*	Y I—	0.00	6	
				CH	Q5	1.1	2

scale adopted in 1928 by the International Astronomical Union. The corrections range from −0.0012 Å at 2995 Å to −0.016 Å at 7100 Å.[3] In the present edition, the wavelengths from 6600 Å to 8770 Å are taken directly from the 1947 Monograph.[4] Similarly, from 2935 Å to 3062 Å Babcock has re-observed the spectrum[3] and his values are used with a correction of −0.002 Å. For the intermediate range, the running correction described above, applies.

In the course of the work many questions have arisen regarding the reality of faint lines, the resolution of close pairs, and the inclusion of new lines. A set of spectrograms taken from the Utrecht-Mt. Wilson collection made for the preparation of the Minnaert Atlas[5], has been measured by the author in order to clarify some of these questions. The International Standards and remeasured (corrected) wavelengths from the 1928 edition have been used to derive the correction curves for obtaining the wavelengths of the newly measured lines. In all, about 4785 lines have been measured on one or more exposures in the range 3948 Å to 6600 Å. More than 1700 of the fainter Rowland lines have also been confirmed by Zalubas[6], who has measured high-dispersion plates taken for this purpose, in the interval 5000 Å to 6000 Å.

It has not been possible to remeasure all of the faint lines recorded in Rowland's Table. Many of them are clearly visible but need to be measured with a low-power eyepiece. For approximately 6500 lines, however, new measurements have been included in the final weighted means entered in the Solar Table. More than 190 faint lines have been rejected as unreal, and a number of new lines measured to two decimal places from the Minnaert Atlas have been added. It is felt that all entries represent real lines, and that as time goes on, many more faint solar lines will be recorded.

The Intensities (Columns 2 and 3.) From 2935 Å to 3062 Å it has been impossible to measure meaningful equivalent widths. Here, estimated intensities[3] are entered in brackets. From 3062 Å to longer waves, the equivalent widths of the lines expressed in mÅ have been measured by Prof. Minnaert and his staff at Utrecht[7] (Column 2). Similarly, Column 3, the Reduced Equivalent Width $\Delta\lambda/\lambda$ has been furnished by Minnaert. These data will be discussed separately, by the authors who provided them.

Atomic Lines in the Sun-Spot Spectrum, Column 4: The next column is headed "Spot". It deals with general intensity behavior of atomic lines in the spot spectrum as compared with the disk spectrum. The symbols used are as follows: *S* – greatly strengthened in the spot spectrum; *s* – strengthened in the spot spectrum; *u* – unchanged in the spot spectrum; *w* – weakened in the spot spectrum; *ob* – obliterated in the spot spectrum.

For those atomic lines appearing only in the spot spectrum, no disk intensity is given, and the wavelengths have been measured by the author[8] or else they are means of these measures and those by R. S. Richardson.[9] These lines, numbering about 152 in all, have been included to supplement evidence as to the correctness of the identification assignments. The spot behavior of lines in multiplets as a whole is striking. Frequently it furnishes reliable information regarding both blended and unblended lines. If the multiplet as a whole gives strong spot lines and weak disk lines, the additional information furnished by the spot lines alone should not be rejected or omitted. The same comment applies to lines that are weakened or obliterated in the spot spectrum.

The Solar Identifications (Column 5). For atomic lines, the identifications have been assigned on the basis of multiplet behavior, as far as possible. The data assembled for the project on *Atomic Energy Levels*[2] have been utilized to the fullest extent. Series have been studied for the less complex spectra. Lacking these general guides, the solar intensities of the strongest laboratory lines of a given spectrum, have furnished reliable information as to the probable presence of weaker lines.

For unblended lines, the entries are self-explanatory. In the case of blends, i.e. cases where there is more than one contributor to the solar intensity, the symbols "||" and "|" denote, respectively, predominant and appreciable contributors. In Table I, for example, the solar line at 4045.115 Å is accounted for primarily by Fe I, but Mn I is also a contributor to the line. Dashes follow or precede the chemical symbol in cases of blends in which the laboratory wavelength is respectively shorter or longer than that of the solar line.

Along with the identifications of atomic lines in the solar spectrum, there are also selected diatomic molecules known to be present in the disk spectrum. A general report on molecules in the solar spectrum was published in 1957.[10] The spot spectrum is a more favorable source for molecular spectra, but at least

TABLE 2

OH; $A\,^2\Sigma^+ - X\,^2\Pi$

Laboratory			Sun					
Vib. Band	*Wavelength Range*	*Total No. Lines*	*Strongest Solar Intensity* $\Delta\lambda/\lambda$	*Summary of Counts*				
				Present	*Blend*	*Masked*	*Absent*	*Total*
0,0	3021 Å to 3362 Å	283	79.5	108	63	69	43	283
1,1	3109 Å to 3378 Å	231	58.0	52	56	81	42	231
2,2	3184 Å to 3372 Å	159	14.7	15	5	23	15	58
		673		175	124	173	100	572

TABLE 3

NH; $A\,^3\Pi_i - X\,^3\Sigma^-$

Laboratory			Sun					
Vib. Band	*Wavelength Range*	*Total No. Lines*	*Strongest Solar Intensity* $\Delta\lambda/\lambda$	*Summary of Counts*				
				Present	*Blend*	*Masked*	*Absent*	*Total*
0,0	3280 Å to 3487 Å	260	58.1	94	71	81	14	260
1,1	3307 Å to 3485 Å	197	20.5	64	39	76	18	197
		457		158	110	157	32	457

TABLE 4

CN VIOLET

Laboratory				Sun					
					Summary of Counts				
Electronic Transition	*Vibrational Transition*	*Wavelength Range*	*Total No. Lines*	*Strongest Solar Intensity*	*Present*	*Blend*	*Masked*	*Absent*	*Total*
$B\,^2\Sigma^+ - X\,^2\Sigma^+$	0,0; 1,1; 2,2; 3,3;	3583 Å to 3883 Å	1070 41*	62.5	327	157	113	102	699
	Undesignated				90	38	129	114	371
			1741		417	195	242	216	1070
	0,1; 1,2; 2,3; 3,4	4073 Å to 4216 Å	918	12.4	86	56	194	119	455

CN RED

$A\,^2\Pi i - X\,^2\Sigma^+$	0,0	10871 Å to 11559 Å			X				
	1,0	9106 Å to 9930 Å			?		(Atm)		
	2,0	7850 Å to 8298 Å	336	3.3	141	51	118	26	336

* Two laboratory lines of CN contribute to one solar line.

TABLE 5

CH

Laboratory				Sun					
					Summary of Counts				
Electronic Transition	*Vibrational Transition*	*Wavelength Range*	*Total No. Lines*	*Strongest Solar Intensity* $\Delta\lambda/\lambda$	*Present*	*Blend*	*Masked*	*Absent*	*Total*
$A\,^2\Delta - X\,^2\Pi$	0,0	4133 Å to 4413 Å	309	39.6	155	92	46	16	309
	1,1	4185 Å to 4446 Å	235	24.3	70	68	72	15	225
	2,2	4238 Å to 4468 Å	188	26.6	54	36	61	8	159
	0,1	4726 Å to 4941 Å	198	0.6	5	3	14	9	31
			930		284	199	193	48	724
$B\,^2\Sigma^- - X\,^2\Pi$	0,0	3871 Å to 4084 Å	106	49.4	43	41	20	2	106
	1,0	3627 Å to 3710 Å	54	13	18	10	22	4	54
	1,1	4025 Å to 4119 Å	50	7.6	16	11	22	1	50
			210		77	62	64	7	210
$C\,^2\Sigma^+ - X\,^2\Pi$	0,0	3086 Å to 3219 Å	119±	53.0	31	41	18	6	96
	1,1	3119 Å to 3222 Å	50	23.1	7	14	24	5	50
			169±		38	55	42	11	146

TABLE 6

C_2

Laboratory				*Sun*					
				Strongest Solar Intensity $\Delta\lambda/\lambda$	*Summary of Counts*				
Electronic Transition	*Vibrational Transition*	*Wavelength Range*	*Total No. Lines*		*Present*	*Blend*	*Masked*	*Absent*	*Total*
$A\,^3\Pi_g - X\,^3\Pi_u$	(0,0) ; (1,1)	4932 Å to 5165 Å	480	5.6	139	62	98	181	480
	(1,0)	4652 Å to 4737 Å	650	3.8	69	22	175	227	493
	(0,1) ; (1,2) ; (2,3)	5490 Å to 5635 Å	415	2.1	106	14	85	145	350
	Undesignated				21				21
			1545		335	98	358	553	1344

TABLE 7

MG H

Laboratory				*Sun*					
				Strongest Solar Intensity $\Delta\lambda/\lambda$	*Summary of Counts*				
Electronic Transition	*Vibrational Transition*	*Wavelength Range*	*Total No. Lines*		*Present*	*Blend*	*Masked*	*Absent*	*Total*
$A\,^2\Pi_{\frac{3}{2},\frac{1}{2}} - X\,^2\Sigma$	0,0	4956 Å to 5210 Å	201	3.5	72	35	42	52	201

six molecules can be identified with certainty in the disk spectrum: CO, OH[11], NH, CN[12], CH[13] and C_2. The first occurs to longer waves than the span of the present Table. One other molecule, MgH, which is clearly present in the spot spectrum, is probably represented, also, by faint lines in the disk spectrum. Six Tables, numbered 2 through 7 summarize the approximate counts of the respective molecular spectra entered in the present Solar Table. These Tables are self-explanatory. For each molecule they give the electronic and vibrational transition, the range of wavelengths, the strongest solar intensity and the counts of lines considered to be in each of four categories: "Present (unblended), Blend, Masked, and Absent" as regards the Solar Spectrum.

To longer waves the numerous atmospheric lines, Atm O_2 and Atm H_2O mask a large part of the real solar spectrum. More work is needed on the Doppler shift of the solar lines to disentangle them from terrestrial lines. This was stressed by MEGGERS as early as 1918.[14] Just as for the molecules present in the solar spectrum, the band designations are being given, also, for the atmospheric lines. Improved atlases, based on observations made at the Jungfraujoch[15], are being prepared in this region. The present work serves to provide a generous overlap with the new observations to longer waves.

In assigning the solar identifications of these six diatomic molecules, the author has attempted to use the rotational structure as a guide, similar to the multiplet structure used for atomic lines. The intensity behavior along the rotation bands has been taken into account, as well as the agreement in wavelength, $\Delta\lambda$ (⊙–lab.).

Short of λ6600 Å a tolerance exceeding $\pm$ 0.030 Å in $\Delta\lambda$ has seldom been permitted for either atomic or molecular lines except for conspicuous blends. Judgment enters constantly in assigning identifications, and no two people produce identical results. In general, the assignments are conservative.

The Low Excitation Potential or Rotation Line (Column 6): For atomic lines, the lower excitation potential has been calculated with the conversion factor from cm^{-1} to eV equal to 0.00012395. This factor has been used to derive all ionization potentials in the Volumes on *Atomic Energy Levels*.[2]

This column indicates, also, the rotational line, giving the branch and quantum number for the molecular entries in the identification column.

The RMT No. or Vibrational Band (Column 7). This column is arranged similarly to the preceding one. For atomic lines the multiplet number is entered when known.[16,17] It is planned to replace the present Multiplet Tables by a new one covering the range from the X-ray region of the spectrum to the microwave region, arranged by elements. Extensive revisions of the published tables are needed, but there is an insistent demand to retain the old Multiplet Numbers in the new table. This will be done by a method that permits insertion of new multiplet data along with the retention of the old multiplet numbers.

In this column, the Vibration Band is indicated for molecular lines.

Notes (Column 8). The numbers in this column refer to special notes regarding individual lines. The electronic transitions of the molecular lines entered in the Solar Table are given in auxiliary tables for the respective bands. Note numbers refer to these Tables.

Lines that were formerly considered double and are listed as single here, have a note number. Conversely, those listed as single that may be unresolved pairs, are so indicated.

The atomic spot lines mentioned above have a special note. Atomic solar lines whose spot behavior may be influenced by the presence of a molecular line are also noted.

All special cases of importance are handled by these Auxiliary Notes.

Summary. Approximate counts of the lines in the present revision as compared with similar counts in 1928, are tabulated in Table 8. Short of 3062 Å and long of 6600 Å new observations replace those by Rowland. Consequently, the counts have been made by regions in order to emphasize the comparison over the long span from 3062 Å to 6600 Å, where the earlier table has not yet been superseded. In 1928, fifty-seven percent of the 20000 lines recorded in Rowland's Table were wholly or partly identified as to chemical origin. The corresponding percentage in the present compendium of nearly 22000 lines is seventy one.

If many of the familiar spectra such as Fe I and Ni I could be re-observed in the laboratory with electrodeless lamps as sources, the analyses could be extended, and more solar identifications could easily be made. As a result of

TABLE 8

THE SOLAR SPECTRUM; APPROXIMATE COUNTS

1928 Edition

Range	*Total*	*Ident.*	*Unident.*	*Spot lines*	*% Ident*
2975 Å — 3062 Å	479	241	238		
3062 Å — 6600 Å	18301	10390	7911		
6600 Å — 7330 Å	1247	742	505		
2975 Å — 7330 Å	20027	11373	8654	75	57

1964 Edition

Range	*Total*	*Ident.*	*Unident.*	*Spot lines*	*% Ident*
2935 Å — 3062 Å	666	512	154		
3062 Å — 6600 Å	17134	11917	5217	132	
6600 Å — 7330 Å	1919	1484	435	10	
2935 Å — 7330 Å	19719	13913	5806	142	71
7330 Å — 8770 Å	2260	1694	566	10	
2935 Å — 8770 Å	21979	15607	6372	152	71

recent observations of C I, Mg I, Ca I, and Si I a number of new identifications have been made in the solar spectrum.[18]

The entire subject of molecules in the sun is still in an early stage of development. Reliable and extended observations and analyses of diatomic molecular spectra of abundant elements are urgently needed. An excellent start has been made on red CN by DAVIS and PHILLIPS in Berkeley.[19] Doubtless many of the faint solar lines at present unidentified are of molecular origin.

No discussion of the solar spectrum would be complete without mention of the number of elements detected in the solar atmosphere. Sixty-four are present without doubt. Four others need additional confirmation: Tc II, Tb II, Er II, and Au I. Two more, B and F, have been classed as present from a study of the molecular spectra BH and MgF. Further study of these molecular spectra is desirable to confirm these identifications.

When high-dispersion ultraviolet solar spectrograms have been reliably measured, a new field of study will provide answers as to the presence of Tc II, Au I, B I and perhaps other elements. The *raies ultimes* of these spectra are in the short wavelength region. The element neon was not found until the discovery the *raies ultimes* of Ne VIII in the vacuum ultraviolet solar spectrum. Similarly, argon was first identified from the presence of Ar X in the solar corona.

It is possible that one third spectrum may be represented in the present Solar Table. The strongest observed line of Ce III is at 3055.589 Å. A faint unidentified solar line has been observed at 3055.594 Å. This is the first evidence of the presence of a third spectrum in the region discussed here, where all lines are absorption lines. TOUSEY[20] has reported the presence of S III in absorption, to shorter waves. Further study of rare earths in the sun is urgently needed as soon as sufficient laboratory observations become available.

Finally, a statement of progress toward publication is in order. The present volume will contain approximately 610 ledger sheets of tabular data. Of these, the final copy has been completed for 390, and 350 have been released for publication. Galley proof has been received for 199 sheets.*

Acknowledgments. The large amount of material discussed here could not have reached its present stage of completion without the splendid assistance of many colleagues. Thanks are extended to all who generously furnished unpublished data for inclusion in the Solar Table.

H. P. Broida and W. S. Benedict have provided valuable help with the work on molecular spectra. The author is indebted to her husband and to Mrs. Doris Sykes for their assistance in obtaining the statistical data. Mrs. Isabel D. Murray has prepared the press copy of the solar ledger with meticulous care and skill. It is a pleasure to express cordial thanks to all who have so willingly contributed to this extensive solar program.

* Note added in proof Sept. 1964: 504 released for publication; galley proof received for 449 sheets.

References

1. St. John, C. E., Moore, C. E., Ware, L. M., Adams, E. F., and Babcock, H. D.: 1928, *Carnegie Inst. Wash. Publ.* 396, *Papers Mt. Wilson Obs.* **3**, 238 pp.
2. Moore, C. E.: 1949, Atomic Energy Levels, *Circ. Nat. Bur. Stand.* 467, Vol. I 309 pp.
Moore, C. E.: 1952, Atomic Energy Levels, *Circ. Nat. Bur. Stand.* 467, Vol. II, 227 pp.
Moore, C. E.: 1958, Atomic Energy Levels, *Circ. Nat. Bur. Stand.* 467, Vol. III, 245 pp.
3. Babcock, H. D., Moore, C. E., and Coffeen, M. F.: 1948, *Contr. Mt. Wilson Obs.* 745; *Astrophys. J.* **107**, 287.
4. Babcock, H. D. and Moore, C. E.: 1947, *Carnegie Inst. Wash. Publ.* 579; *Papers Mt. Wilson Obs.* **7**, 95 pp.
5. Minnaert, M., Mulders, G. F. W., and Houtgast, J.: 1940, *Photometric Atlas of the Solar Spectrum 3332 Å–8771 Å*, I. Schnabel, Amsterdam.
6. Zalubas, R.: 1955, Dissertation Georgetown University, unpublished.
Zalubas, R.: 1956, *Astrophys. J.* **123**, 544.
7. Minnaert, M.: 1960, *Rech. Astron. Obs. Utrecht* **15**, 168 pp. and unpublished material.
8. Moore, C. E.: 1933, *Atomic Lines in the Sunspot Spectrum*, Princeton, 47 pp.
Moore, C. E.: 1932, *Contr. Mt. Wilson Obs.* **446**; *Astrophys. J.* **75**, 222, 298.
9. Richardson, R. S.: 1932, unpublished material, see Ref. 8.
10. Broida, H. P. and Moore, C. E.: 1957, *Mém. Soc. Roy. Sci. Liège* **18**, 217.
11. Moore, C. E. and Broida, H. P.: 1959, *J. Res. Nat. Bur. Stand.* **63A**, 279.
12. Moore, C. E. and Broida, H. P.: 1957, *Mém. Soc. Roy. Sci. Liège* **18**, 252.
13. Moore, C. E. and Broida, H. P.: 1959, *J. Res. Nat. Bur. Stand.* **63A**, 19.
14. Meggers, W. F.: 1919, *Publ. Alleghery Obs.* **6**, No. 3, 13–44.
15. Delbouille, L. and Roland, G.: 1963, *Photometric Atlas of the Solar Spectrum from λ 7608– λ 12016 Å, Mém. Soc. Roy. Sci. Liège*, Special Volume **4**.
16. Moore, C. E.: 1959, The 1945 Edition of the Princeton 'Multiplet Table of Astrophysical Interest'. *Technical Note* **36**, PB 151 395, Office of Technical Services, U.S. Dept. of Commerce.
17. Moore, C. E.: 1950, 'An Ultraviolet Multiplet Table', *Circ. Nat. Bur. Stand.* 488 Sec. **1**, 78 pp.
Moore, C. E.: 1952, 'An Ultraviolet Multiplet Table', *Circ. Nat. Bur. Stand.* 488 Sec. **2**, 115 pp.
Moore, C. E.: 1962, 'An Ultraviolet Multiplet Table', *Circ. Nat. Bur. Stand.* 488 Sec. **3** (94 pp.), **4** (65 pp.), **5** (30 pp.).
18. Johansson, L. and Risberg, G.: 1963, unpublished material.
Johansson, L.: 1963, *Ark. Fys.* **25**, No. 31, 425.
19. Davis, S. P. and Phillips, J. G.: 1963, *Berkeley Analyses of Molecular Spectra* **1**, 214 pp.
20. Tousey, R.: 1963, private communication.

DISCUSSION ON LINE FORMATION AND PHOTOSPHERIC STRUCTURE

A. UNSÖLD
Chairman
University of Kiel, Kiel, Germany

WASABURO UNNO AND YASUMASA YAMASHITA*

On Curves of Growth for Line Profiles Truncated at Artificial Continuous Levels

Abstract. Areas on the photometric tracing enclosed by line profiles and an artificially truncated continuous level is studied. With these areas (fictitious equivalent widths), fictitious curves of growth can be constructed either theoretically or empirically. From the vertical and the horizontal shift between the theoretical and empirical curves, the Doppler width and the population of atoms are estimated for each truncated continuous level. Thus, the depth dependence of the Doppler width can be obtained empirically. For the solar photosphere, the result shows good agreement with the earlier determination by Unno.

1. The determination of Doppler widths at the photospheric level has been made by many authors. The investigations may be classified according to the method employed. The first method is due to the curve of growth, the second due to the line profile, and the third is the multiplet method. These three methods are different to each other in the degree of inclusion of the large scale component however, the macro-turbulence is not predominant, and the three methods give almost identical results.

The latter two methods were used to derive the depth dependence of the Doppler width (e.g., SUEMOTO[1] and UNNO[2]), but the curve of growth method gives only the average Doppler width and is difficult to reveal the depth dependence. The generalization is possible, however, so as to take off this drawback and to retain some of the advantages of the curve of growth method. A study along this line was made and details were published elsewhere.[3] In the present paper, discussions will be confined to the depth dependence of the Doppler width in the solar photosphere.

2. The equivalent width is an area on the photometric tracing enclosed by a line profile and the continuum in units of continuous intensity level. Fictitious equivalent width can be defined similarly, if an artificial intensity level which is placed between the continuum and the zero-level is used in place of the real continuum.

* Department of Astronomy, University of Tokyo, Tokyo.

C. de Jager (ed.), The Solar Spectrum. All rights reserved.

Based on ALLEN's compilation[4] of data by various authors, the Planck function at $\lambda = 5000$ Å is approximately expressed in terms of the optical depth $\tau (= \tau_{5000})$ as

$$B_{5000}(\tau) = B_0(1 + 2.06\tau - 0.75e^{-7.4\tau}) \, . \qquad (1)$$

Line profiles at the center of the solar disc in the pure absorption model of Milne-Eddington are computed for various values of η_0, the line absorption coefficient at the line center divided by the continuous absorption coefficient. The limiting residual intensity given by Equation (1) is $1 - \gamma_c = 0.082$, in accordance with the observed value. Fictitious continua are drawn at levels of 80% ($\gamma^* = 0.2$) and 50% ($\gamma^* = 0.5$) of the real continuum. These levels correspond to $\eta^* = 0.30$ and 1.50, respectively. (The asterisks indicate the quantities concerned with the truncation of the continuum.) Theoretical fictitious curves of growth as well as the real one are shown in Figure 1, where W_λ^* and γ_c^* denote the fictitious equivalent width and the limiting residual intensity in units of the fictitious continuum.

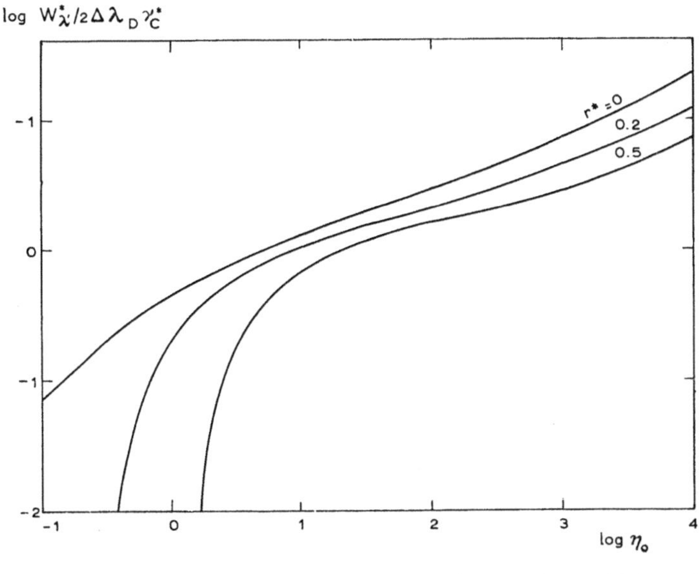

Fig. 1. The theoretical curve of growth for the truncated equivalent width for the intensity in the solar spectrum. (Note: the lines are labeled with the parameter γ^*.)

Corresponding empirical curves are obtained as shown in Figure 2 with the use of the Utrecht Atlas for W_λ^* and of values of log X given by WRIGHT.[5] Inferiority of fictitious curves of growth to the real one is apparent as seen from Figure 2 that the correction for the instrumental width (vertical distance from an observed point to a solid curve) is quite appreciable for lower fictitious continuum and for weaker lines. The same is true for the macro-turbulence and the rotation. High resolution spectrograms should, therefore, be used for a detailed study.

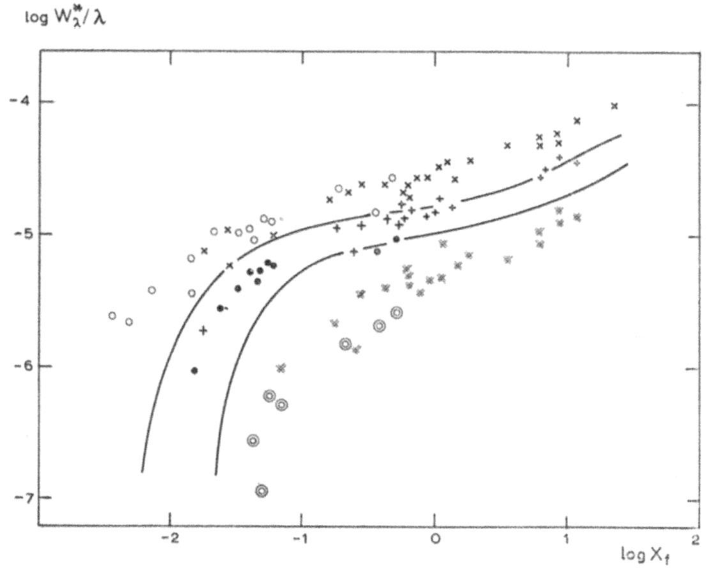

Fig. 2. Empirical curve of growth for the truncated equivalent width in the solar spectrum. Ordinary equivalent widths of lines of Fe (×) and Ti (○), equivalent widths truncated at the level of 80% of the continuum (Fe, +; Ti, ●) and equivalent widths truncated at 50% of the continuum (Fe, #; Ti, ◎) are plotted against log X_f secured by Wright. Two solid lines represent the mean curves corrected for the instrumental blurring.

Similarity is remarkable between the theoretical curves in Figure 1 and the empirical curves in Figure 2. This may indicate that the macro-turbulence is unimportant in the solar photosphere. Closer inspection reveals that the successive distances between the three curves ($\gamma^* = 0$, 0.2, 0.5) is smaller for the theoretical curves than for the empirical curves. In other words, different vertical and horizontal shifts should be applied to every empirical curves to bring them in reasonable superposition on corresponding theoretical curves.

The Doppler width $\Delta\lambda_D$ is obtained for each fictitious curves of growth from the amount of the vertical shift, while the quantity ($\Delta \log \eta_0 + \Delta \log \lambda_D$) is obtained from the amount of the horizontal shift. The latter quantity should be approximately equal to $[\chi_1 - \chi_{0\cdot5} - \chi(\mathrm{H}^-)]\Delta\theta$, according to the Saha-Boltzmann formulae for the atom and the negative hydrogen. This condition is found to be approximately satisfied, after the effective depths are estimated. The Doppler widths obtained are 1.8, 1.7 and 1.3 in units of km/sec for $\gamma^* = 0$, 0.2 and 0.5, respectively. Unfortunately, the depths to which these velocities should be attained are somewhat uncertain.

The fictitious equivalent width may be very roughly interpreted as the equivalent width for the model atmosphere so modified that B_λ is set to be constant ($B_\lambda(\tau_\lambda^*)$) for $\tau_\lambda > \tau_\lambda^*$. Here, τ_λ^* is the effective optical depth ($\sim 1/(1+\eta^*)$) of a line having the residual intensity $\gamma(\eta^*)$ equal to γ^*. Therefore, lines with $\eta < \eta^*$ practically disappear and only stronger lines survive. In this model, the intensity saturation should be appreciable for $(1 + \eta) > \alpha(1 + \eta^*)$, of which α may be taken to be 2. We assume further that widths of strong lines are determined practically by the part of the line wing where $(1 + \eta)$ is equal to $2(1 + \eta^*)$. Then, the effective depth for the fictitious curve of growth is given by

$$\tau_{\text{eff}}^* = 0.5/(1 + \eta^*) \tag{2}$$

with

$$\gamma(\eta^*) = \gamma^* \tag{3}$$

For $\gamma^* = 0, 0.2$ and 0.5, we have $\tau_{\text{eff}}^* = 0.50, 0.37$ and 0.17, and the corresponding Doppler velocities are 1.8 km/sec, 1.7 km/sec and 1.3 km/sec, respectively. The result is in good agreement with an earlier determination by UNNO[2] which was based on the multiplet method with the use of spectrograms taken at the McMath-Hulbert Observatory.

Detailed discussion is not given for the horizontal shift between the theoretical and empirical fictitious curves of growth in this paper. It is noted here that the study of the model atmosphere may be made from the study of the horizontal shift. This kind of study seems to be useful especially for late type stars of which theoretical models are not reliable and the continuum is heavily disturbed by overlapping absorption lines.

References

1. SUEMOTO, Z.: 1957, *Mon. Not. Roy. Astron. Soc.* **117**, 2.
2. UNNO, W.: 1959, *Astrophys. J.* **129**, 375.
3. YAMASHITA, Y. and UNNO, W.: 1963, *Publ. A. S. Japan* **15**, 230.
4. ALLEN, C. W.: 1955, *Astrophysical Quantities*, pp. 133–134.
5. WRIGHT, K. O.: 1944, *Astrophys. J.* **99**, 249.

DONALD H. MENZEL AND BRUCE W. SHORE*

Generalized Tables for the Calculation of Line and Multiplet Strengths in LS Coupling

The radiative transition-probability for a transition between two atomic levels, each comprising several degenerate states, is expressible simply as the product of several factors, when the atomic states are pure *LS*- or *jj*-coupling states.

* Harvard College Observatory, Cambridge, Mass. U.S.A.

Apart from factors of universal physical constants and the density oı radiating atoms, the energy emitted is proportional to the square of a reduced matrix element of a tensor operator, first introduced by Racah. For "allowed transitions" in LS coupling, this tensor operator is the dipole-moment operator of all the electrons, and we use the methods of Racah to evaluate the reduced matrix element.

The reduced matrix element of the dipole operator breaks into a line factor, a factor depending upon the multiplet structure, and a transition integral. The seniority scheme of Racah satisfactorily solves the problems that arise when the transition involves equivalent d electrons. The multiplet factor of the reduced matrix element then includes a coefficient of fractional parentage. The seniority scheme permits the separation of terms of the same type, such as the two ^2D terms of d^3 and d^7, which occur with equivalent d electrons. The seniority states, however, are not eigenstates of the coulomb interaction. The *actual* ^2D term, for example, is a linear combination of the two *seniority* ^2D terms; the specific linear combination depends, for a given atom, on the values of Slater-type integrals.

In the transformation from LS-coupling to intermediate-coupling, the sum of reduced matrix elements between basic LS-seniority terms replaces the single reduced matrix element. For this transformation we require relative phases (signs) of each reduced matrix element. The existing dipole line-strength tables omit these signs. The multiplet tables separate terms of the same type. We have tabulated the several factors required for calculating the reduced dipole matrix element in an LS-seniority scheme in transitions of the type $(Isp - Is^2)$, $(Il - Il')$, $(l_0^{n-1}l - l_0^n)$, and $(l_0^{n-1}sl - l_0^n s)$ for s, p, and d electrons. The form of the tabulation provides both the spectral line strength in LS-coupling and the reduced matrix element used with intermediate coupling.

This investigation was supported by a grant from Air Force Cambridge Research Laboratories.

R. Cayrel*

Sur la température d'équilibre radiatif d'une couche de gaz optiquement mince dans le champ de rayonnement de la photosphère solaire

On admet généralement que dans un modèle d'atmosphère stellaire en équilibre radiatif la température varie de façon monotone, dans le sens d'une décroissance continue avec l'altitude, la température tendant asymptotiquement vers une valeur limite T_0 dite température de surface de l'étoile.

* Observatoire de Paris, Section de Meudon

Nous montrons que cette conception n'est pas valable:

(1) Même dans l'hypothèse de l'équilibre thermodynamique local (ÉTL en abrégé), lorsque par suite de la décroissance de la pression avec l'altitude l'absorbant change de nature dans les couches supérieures de l'atmosphère;

(2) Dans l'hypothèse plus réaliste de l'équilibre statistique (pas d'ÉTL à priori). On peut avoir alors un changement important de température même si la source de l'opacité continue reste la même (ion H^- par exemple) mais si les équilibres contrôlant la concentration de l'ion absorbant sont dominés par des processus différents dans les couches proprement photosphérique ($N_h = 10^{16}$ à 10^{17} atomes d'hydrogène par cm^3) et dans les couches supérieures de densité "chromosphérique" 10 à 1 000 fois plus faible.

CHANGEMENT DE TEMPÉRATURE DE SURFACE DU AU CHANGEMENT DE L'ABSORBANT EN ÉTL

Considérons par exemple la photosphère solaire. Dans le champ de rayonnement à la surface ($\tau_\lambda \simeq 0$) la température d'équilibre radiatif est déterminée par la condition:

$$\int_0^\infty k_\lambda^* J_\lambda(0) \mathrm{d}\lambda = \int_0^\infty k_\lambda^* B_\lambda(T_e) \mathrm{d}\lambda \tag{1}$$

où k_λ^* est le coefficient d'absorption continue de la matière stellaire calculé en supposant l'ÉTL. Les autres notations sont usuelles.

Lorsque la pression électronique est plus grande que quelques dixièmes de dyne l'absorption continue est principalement due à l'ion négatif hydrogène et:

$$k_\lambda^* = k_\lambda^* (H^-) \,. \tag{2}$$

On peut résoudre (1) et (2) par rapport à la température; on trouve:

$$T_e = 4800°K \,.$$

Lorsque la pression électronique est beaucoup plus faible que quelques dixièmes de dyne l'absorption due à l'ion hydrogène négatif est au contraire négligeable devant celle de l'hydrogène atomique. On a alors:

$$k_\lambda^* = k_\lambda^* (H) \,. \tag{3}$$

En résolvant (1) et (3) par rapport à T_e on trouve $T_e \simeq 4950$ °K.

Le champ de rayonnement $J_\lambda(0)$ est le même dans les deux cas puisque l'épaisseur optique continue entre les deux altitudes considérées est négligeable.

CHANGEMENT DE TEMPERATURE DE SURFACE DUE À UN CHANGEMENT DU MECANISME CONTROLANT LA CONCENTRATION DE L'ION ABSORBANT.

Tant que $N_H \gg 10^{16}$ atomes d'hydrogène par cm^3 la réaction d'*associative detachment* $H^- + H \rightleftarrows H_2 + e$ (Pagel, 1959) est beaucoup plus fréquente que la photoionisation de l'ion négatif hydrogène. D'après Unsöld (1963) la concentration en hydrogène moléculaire est elle-même contrôlée par les collisions triples entre atomes d'hydrogène ce qui fait qu'au total la concentration en ions hydrogène négatif est déterminée uniquement par des processus collisionnels.

Il en est ainsi dans pratiquement toute la photosphère où l'on peut admettre que la concentration en ions H^- est correctement donnée par l'hypothèse de l'ÉTL.

Mais aux densités chromosphériques, quand $N_H \leqslant 10^{16}$ atomes/cm³ la conclusion précédente n'est plus valable. Les ionisations par collisions électroniques jouant un rôle négligeable (PAGEL, 1959) et l'émission "free-free" étant faible par rapport à l'émission "bound-free", les photoionisations et les recombinaisons radiatives deviennent alors les processus dominants. Les équations déterminant la concentration N_{H^-} de l'ion hydrogène négatif et la température électronique T_e sont alors:

$$4\pi N_{H^-} \int_0^\infty a_\lambda J_\lambda(0) d\lambda = 4\pi N^*_{H^-} \int_0^\infty a_\lambda B_\lambda(T_e) d\lambda \tag{4}$$

$$4\pi N_{H^-} \int_0^\infty \frac{a_\lambda J_\lambda(0)}{(hc/\lambda)} d\lambda = 4\pi N^*_{H^-} \int_0^\infty \frac{a_\lambda B_\lambda(T_e)}{(hc/\lambda)} d\lambda \tag{5}$$

où a_λ est le coefficient d'absorption atomique de H^-, $N^*_{H^-}$ la concentration fictive en ions H^- calculée comme s'il y avait ÉTL et où (hc/λ) est l'énergie du photon de longueur d'onde λ.

L'équation (4) exprime le bilan d'énergie et l'équation (5) le bilan statistique entre le nombre de photoionisation et le nombre de recombinaisons.

Les équations (4) et (5) déterminent à la fois N_{H^-} et T_e on peut éliminer N_{H^-} entre (4) et (5) pour obtenir l'équation déterminant T_e:

$$\frac{\int_0^\infty a_\lambda J_\lambda(0) d\lambda}{\int_0^\infty \frac{a_\lambda J_\lambda(0)}{(hc/\lambda)} d\lambda} = \frac{\int_0^\infty a_\lambda B_\lambda(T_e) d\lambda}{\int_0^\infty \frac{a_\lambda B_\lambda(T_e)}{(hc/\lambda)} d\lambda} \tag{6}$$

L'équation (6) montre que lorsque $J_\lambda(0)$ peut être assimilé à un champ de rayonnement dilué de température de couleur T_C:

$$J_\lambda(0) = W.B_\lambda(T_C)$$

la solution est:

$$T_e = T_C$$

quelque soit le facteur de dilution W. Cette propriété, bien connue en physique des nébuleuses, est en contraste avec le cas où les collisions maintiennent H^- en ÉTL et où T_e dépend fortement de W.

Nous avons effectué le calcul numérique du premier membre de (6) pour le champ de rayonnement de la photosphère solaire d'après ALLEN (1963). On trouve:

$$T_e = 5\,600°\text{K}$$

et en portant dans (4) ou (5):

$$N_{\text{H}^-}/N^*_{\text{H}^-} = 1.8$$

Nous voyons donc que dans un modèle en pur équilibre radiatif où l'absorption continue est due principalement à la photoionisation de H^- la température a une remontée "chromosphérique" de 4800° K à 5600° K localisée dans la région où la densité passe de la valeur $N_{\text{H}} = 10^{16}$ à une valeur d'un ordre de grandeur inférieur. Cela correspond dans le cas solaire à des profondeurs optiques $\tau_0 = 0.01$ à 0.0001 environ.

Bien entendu la situation réelle est plus complexe que ce que nous venons de décrire à cause: (1) de la dissipation d'énergie mécanique qui seule peut expliquer les températures bien plus élevées de la haute chromosphère et de la couronne; (2) de l'influence des fortes raies de résonance des ions abondants Ca^+ et Mg^+ qui jouent un rôle important dans le bilan d'énergie comme l'a montré un travail récent (Hélène PIRON, 1963). Cependant ce phénomène de remontée "radiative" de la température ne parait pas négligeable dans la basse chromosphère solaire et est peut-être prépondérant dans certaines étoiles de population II où la turbulence est faible et l'abondance de Ca et Mg beaucoup plus petite que dans le cas solaire.

D'autre part on a souvent considéré la remontée de température chromosphérique comme une preuve d'un apport d'énergie par dissipation. L'analyse qui précède montre que ce point de vue n'est pas rigoureux puisque la théorie prévoit une remontée de température purement radiative.

References

ALLEN, C. W.: 1963, *Astrophysical Quantities*, second edition, The Athlone Press, London.
PAGEL, B. E. J.: 1959, *Communications de l'Observatoire Royal de Belgique* **157**, p. 28.
PIRON, Hélène: 1963, *Diplôme d'Etudes Superieures*, Université de Paris.
UNSÖLD, A.: 1963, *Z. Phys.* **171**, 44.

DISCUSSION ON NEW INSTRUMENTAL TECHNIQUES

A. UNSÖLD
Chairman
University of Kiel, Kiel, Germany

F. RODDIER*

High Resolution Profiles and Wavelength Measurements with an Atomic Beam Spectrometer

Our experiment has yet been described in a *Phys. Rev. Letter.*[1] – The light to be analyzed excites the optical resonance of the vapor of a metal. The intensity of the light re-emitted by the vapor, proportional to the incident intensity, is measured at right angle to the exciting beam by a photomultiplier. The scattering frequency is shifted by applying a magnetic field and by observing one σ component (through a circular analyzer). The variation of intensity of re-emitted light as a function of the field gives the profile of the exciting light.

We have operated on an atomic beam in order to minimize the width of the absorption line of the vapor which gives the resolution of our spectrograph (between 1 and 2 mÅ). Therefore no correction of instrumental profile is necessary. For each profile the error on the wavelength deduced after correction of relative motion of the Earth and the Sun, does not exceed ± 0.5 mÅ. The angle used on the Sun is about 30″. The time necessary to obtain one profile is 20 min.

The first line we have studied is the Strontium I resonance line at $\lambda = 4607.3$ Å. Figure 1 shows an example of profiles obtained at the centre of the disk and near the limb. Dotted line represents the axis of the profile (middle of horizontal chords). The wavelength is shifted toward the red but the shift is less than the Einstein red shift. At the north limb the shift is very near the predicted shift.

At the center of the disk, the line is asymmetrical. Figure 2 shows wavelengths measured on 18 different profiles at 6 different residual intensities. The mean asymmetry appears clearly and is very similar to that observed by OLSON and HIGGS[2,3] but here the wavelength is known for each point. The fluctuations of about 2 mÅ observed exceed the experimental error and seem to be real as it has been yet pointed out by different observers.[5,6]

To compare our results with M. G. ADAM's wavelength measurements[4,5], we have first measured our wavelengths at the bottom of the profiles for different values of cos θ (Figure 3a). For cos $\theta > 0.7$ our values are nearer the Einstein shift. We have then measured our wavelength for a 50 mÅ width (Figure 3b).

* Observatoire de Marseille, France.

All rights reserved.

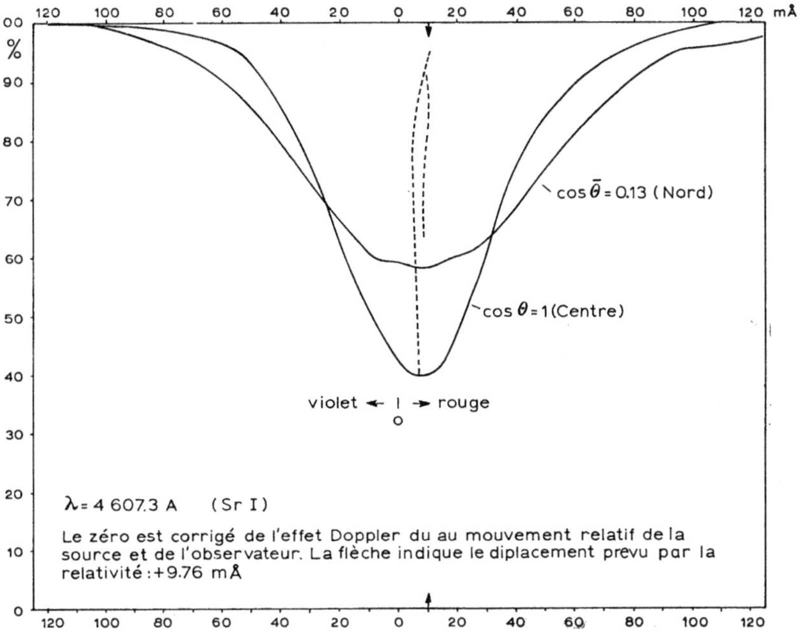

Fig. 1.

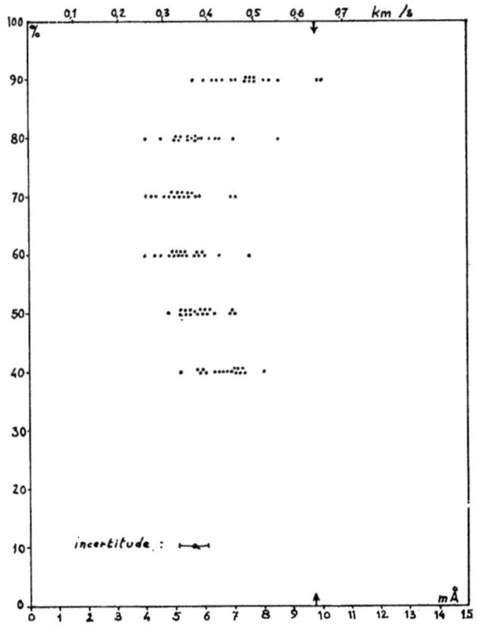

Fig. 2. Asymmetry of profiles at the center of the disk.

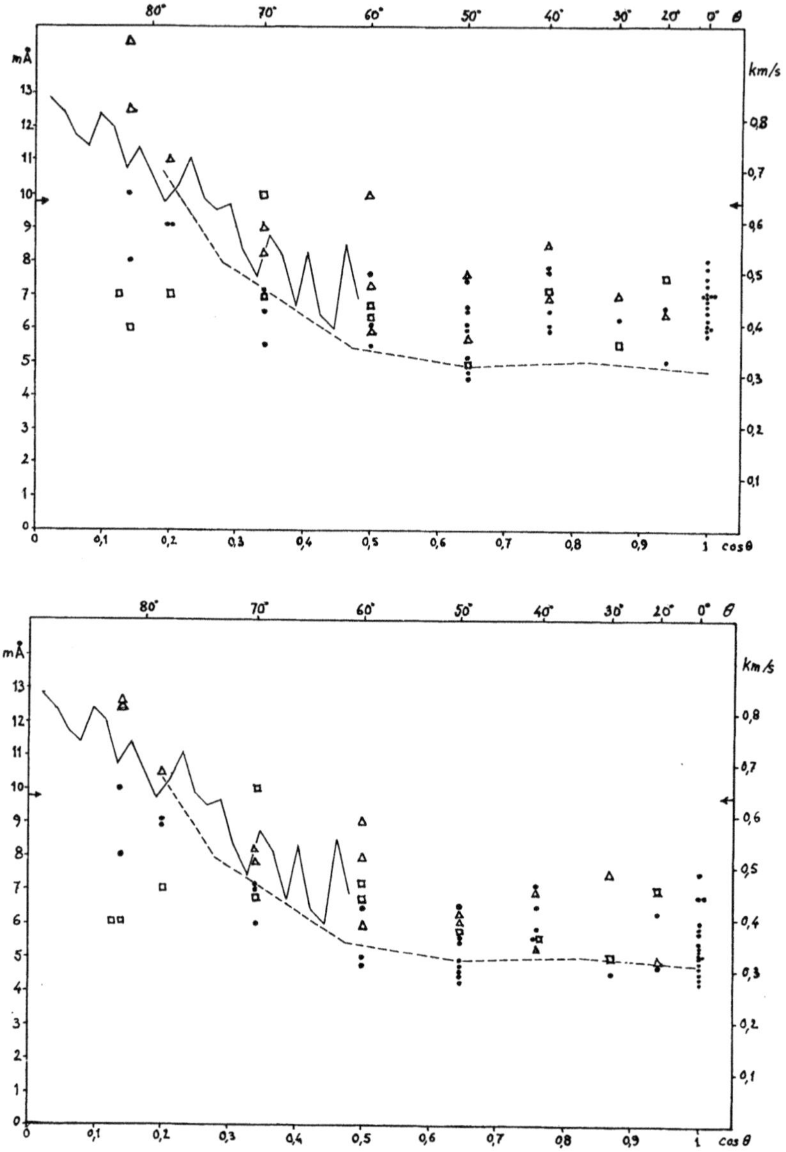

Fig. 3. Center-limb variation of wavelength (a) at the bottom of the profiles; (b) for a 50 mÅ width. Dots – measured toward the Northern limb; Triangles – measured toward the Eastern limb; Squares – measured toward the Western limb;

The agreement is much better. This result shows the necessity of complete knowledge of profiles to give correct interpretation of center-limb variations of wavelength.

Eastern and Western measurements have been corrected from the Doppler effect of solar rotation directly measured on our profiles. For the equatorial synodic rotation we find 1.985 km/s or 14°15 per day which gives a time of revolution of 25.5 days with an error of about 2%. This is in good agreement with previous measurements[7] but for cos $\theta \leqslant 0.2$ we find a much slower rotation. We did not take this into account, so that Eastern and Western wavelengths are very different near the limb. The same effect has been observed

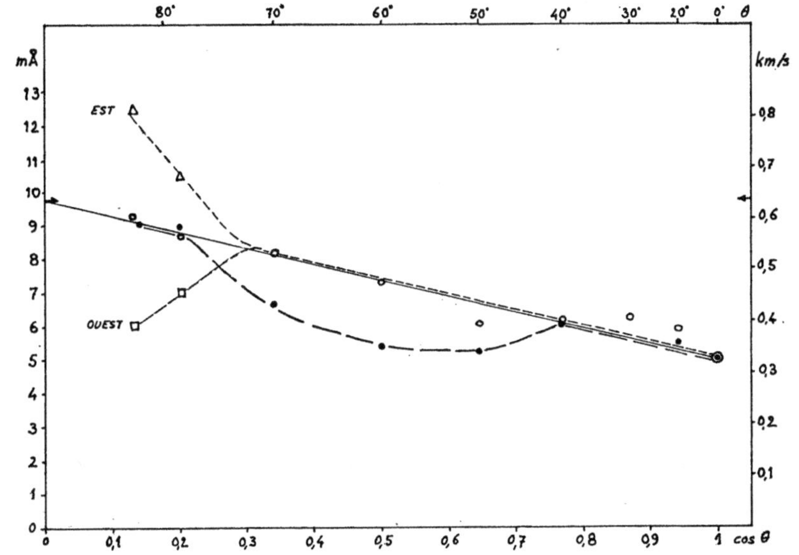

Fig. 4. Mean values of Figure 3b and suggested laws.

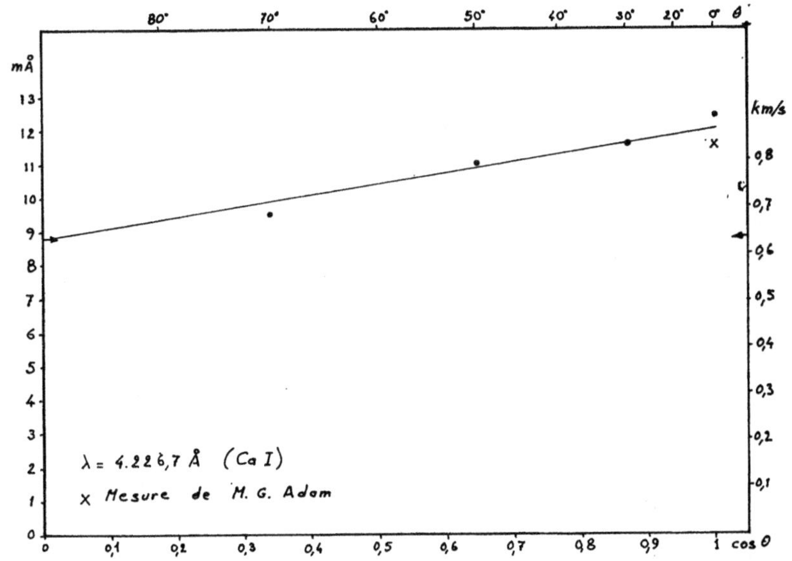

Fig. 5. Center-limb variation of wavelengths for Calcium resonance line toward the Northern limb.

by M. G. Adam. We think it to be real but we did not find any satisfactory explanation. The average of Eastern and Western wavelengths gives the Northern wavelength.

Figure 4 shows the mean wavelengths. Eastern-Western measurements seem to follow a linear law, but Northern measurements show a systematic difference which can be interpreted by meridional currents toward the equator.

Recently we have obtained Calcium I profiles at $\lambda = 4226.7$ Å. Figure 5 shows the wavelengths measured. They all exceed the Einstein value but approach this value at the Northern limb.

References

1. Blamont, J. E., Roddier, F.: 1961, *Phys. Rev. Letters* **7**, 437.
2. Olson, E. C.: 1962, *Astrophys. J.* **136**, 946.
3. Higgs, L. A.: 1962, *Mon. Not. Roy. Astron. Soc.* **124**, 51.
4. Adam, M. G.: 1948, *Mon. Not. Roy. Astron. Soc.* **108**, 446.
5. Adam, M. G.: 1959, *Mon. Not. Roy. Astron. Soc.* **119**, 460.
6. Hart, A. B.: 1954, *Mon. Not. Roy. Astron. Soc.* **114**, 17.
7. de Jager, C.: 1959, *Handbuch der Physik* **52**, 338.

J.-C. Pecker. I cannot see to the Eastern-Western asymmetry noted by Roddier any other interpretation but a distorsion of the corrugated iso-tau surfaces, due to solar rotation, and which would produce an asymmetry of the optical rugosity effect. But this should be proved on the basis of some physical theory!

M. Migeotte*

Progress on the Observation of the Solar Spectrum at the Jungfraujoch (Switzerland)

Since 1958, the spectrometer of the Pfund type used to map the solar spectrum between 2.8 μ and 23. 7 μ[1] has been replaced by a large grating instrument of the Ebert-Fastie type coupled to an appropriate telescope. A brief description of the new installation as well as part of the results obtained up to the present have been published recently by L. Delbouille and G. Roland[2] in a *Photometric Atlas of the Solar Spectrum from λ 7488 to λ 12016***. In this region, 10450

* Institut d'Astrophysique, Cointe-Sclessin (Belgium).

** As this work was partly supported by the "Air Force Cambridge Research Laboratories", Office of Aerospace Research, U.S.A.F., Laurence G. Hanscom Field, Bedford, Mass. (U.S.A.), copies of the Atlas will be distributed by this office. Other ones will be sent directly from our institute. Copies of the Atlas may also be ordered to Dr. H. Sauvenier, General Secretary of the "Société Royale des Sciences de Liège" (7, Place du XX août, Liège, Belgium).

absorption lines – solar and telluric – have been found as compared with 4765 for the same part of the spectrum in the table of BABCOCK and MOORE.[3] The lengthy work of identifications is underway with the help of W. S. BENEDICT (John Hopkins University, Baltimore, Maryland, U.S.A.), O. C. MOHLER (University of Michigan, U.S.A.) and J. W. SWENSSON (Institut d'Astrophysique, Cointe-Sclessin).

The records reproduced in the Atlas have been obtained by using the spectrometer in single pass. However, during these last months, the grating has been used in double pass and a slit has been placed in the intermediate spectrum. These improvements gave the possibility: (1) to increase the resolving power; (2) to use the grating, blazed at 2.7 μ, in high orders because of the practical elimination of the Rowland ghosts; (3) to reduce the scattered light.

The first results obtained for $\lambda > 7000$ Å have been published by L. DELBOUILLE, L. NEVEN and G. ROLAND.[4] The new records have been compared with the data published in the Utrecht and Göttingen Solar Atlas[5,6] as well as with the best solar spectra obtained in single pass, at the McMath Observatory, Pontiac (Mich.), U.S.A.[7] As a result of these comparisons, it has been decided to remap systematically the Solar Spectrum at the Jungfraujoch, from 7500 Å to the shortest ultraviolet wavelengths reaching our observing station. The necessary equipment is now set up to record simultaneously analogous tracings and digital data on punched paper tape.

References

1. MIGEOTTE, M., NEVEN, L., SWENSSON, J.: 1956, 1957; *The Solar Spectrum from 2.8 to 23.7 Microns*; Part I, Photometric Atlas, Part II, Measures and Identifications; *Mém. Soc. Roy. Sci. Liège*, Special Vol. **1** and **2**.
2. DELBOUILLE, L. and ROLAND, G.: 1963, *Photometric Atlas of the Solar Spectrum from λ 7488 to λ 12016, Mém. Roy. Soc. Sci. Liège*, Special Vol. **4**.
3. BABCOCK, H. D. and MOORE, C. E.: 1947, 'The Solar Spectrum 6600 to 13495', *Carnegie Inst. Publ.* **579,** Washington.
4. DELBOUILLE, L., NEVEN, L., and ROLAND, G.: 1963, *J. Quant. Spectr. Rad. Transfer* **3,** 189.
5. MINNAERT, M., MULDERS, G., and HOUTGAST, J.: 1940, *Photometric Atlas of the Solar Spectrum from λ 3612 to λ 8771*, Utrecht.
6. BRUCKNER, G.: 1960, *Photometric Atlas of the Near Ultraviolet Solar Spectrum 2988 Å to 3629 Å*, Göttingen.
7. MOHLER, O.: Private Communication.

DISCUSSION ON PHOTOSPHERIC MOTIONS

A. UNSÖLD
Chairman
University of Kiel, Kiel, Germany

J. EVANS, R. MICHARD ET R. SERVAJEAN*

Analyse de Fourier des oscillations de l'atmosphère solaire

Cette communication résume certains des résultats déduits de séquences de spectres solaires réalisées par J. EVANS à Sacramento Peak, grace à une étude statistique objective poursuivie à Meudon.

De nombreux paramètres peuvent être mesurés simultanément sur les spectrogrammes: fluctuations de vitesse radiale pour différentes raies, fluctuations de brillance dans le continuum et au centre des raies, fluctuations de largeurs équivalentes. Ceci permet, au moins en principe, d'aborder sérieusement les problèmes de corrélation (*cohérence* et écarts de *phase*) entre les fluctuations des divers paramètres physiques à différents niveaux de l'atmosphère.

Les fluctuations de la quantité $a(t)$ peuvent être représentées par leur spectre de puissance (power spectrum)

$$G_a(\nu) = 2\int_0^\infty C_a(t)\cos 2\pi\nu t\,\mathrm{d}t$$

où C_a est la fonction d'autocorrélation. Les relations entre les deux grandeurs $a(t)$ et $b(t)$ peuvent être décrites en fonction de la fréquence ν par leur cohérence

$$P_{ab}(\nu) = \left[\frac{R_{ab}^2 + S_{ab}^2}{G_a G_b}\right]^{\frac{1}{2}}$$

et leur différence de phase

$$Z_{ab}(\nu) = \operatorname{arc\,tg} S_{ab}/R_{ab}$$

expressions dans lesquelles les fonctions R_{ab} et S_{ab} sont les parties réelle et imaginaire de la transformée de Fourier de la fonction de corrélation croisée

$$R_{ab} + iS_{ab} = G_{ab}(\nu) = \int_{-\infty}^{+\infty} C_{ab}(t)\exp 2\pi i\nu t\,\mathrm{d}t$$

Dans l'utilisation des fonctions G_a, P_{ab}, Z_{ab} on doit garder présentes à l'esprit des restrictions importantes:

* Observatories of Sacramento Peak, and Paris-Meudon.

C. de Jager (ed.), The Solar Spectrum. *All rights reserved.*

(1) Elles représentent des propriétés moyennes des fluctuations de l'atmosphère solaire, ces moyennes étant prises sur un intervalle suffisant de temps et d'espace, mais elles ne décrivent pas les propriétés des perturbations individuelles qui composent le champ fluctuant;

(2) La précision avec laquelle on peut les déterminer est limitée par l'étendue des données disponibles;

(3) La fonction de cohérence n'a plus de sens quand l'une ou l'autre des puissances G_a, G_b est petite, et la fonction de phase n'a pas de sens quand la cohérence est faible.

CHAMP DES VITESSES

a. Spectres de puissance des vitesses verticales

Les spectres de puissance des fluctuations de vitesse radiale au centre du disque sont données dans la figure pour les raies

8514.08 Fe I	W =	110 mÅ
5173.75 Ti I		70 mÅ
5172.70 Mg I		(b_2)
8498.06 Ca II et		8542.13 Ca II

et finalement Hα.

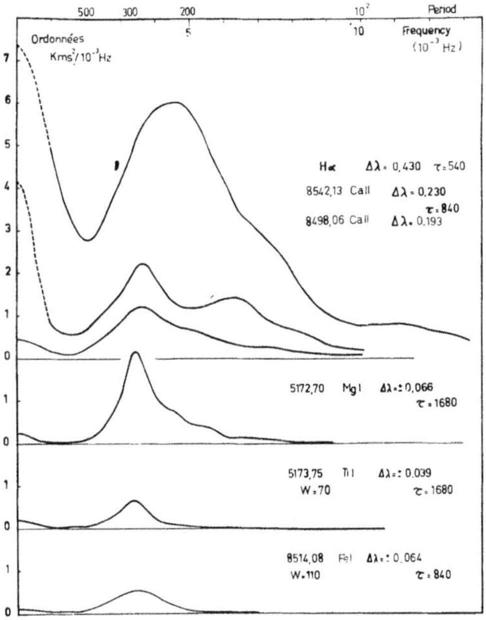

On constate que tous ces spectres de puissance comportent un maximum marqué le *pic de résonance*; il est accompagné d'une *aile de hautes fréquences* pour les raies les plus fortes formées aux niveaux chromosphériques. La fréquence du pic de résonance augmente lentement avec l'altitude de formation des raies ($T_r = 280$ *s* pour les raies fortes de Ca II).

L'aile de haute fréquence est totalement absente pour les raies "photosphériques" de Fe et Ti. Elle apparaît faiblement pour la raie b_2 de Mg et son importance augmente ensuite avec la hauteur de formation pour les raies chromosphériques. Pour Hα, les deux composantes du spectre de puissance semblent se fondre pour donner une large bande unique avec un maximum au voisinage de 220 sec.

b. Phases relatives

Dans la région spectrale du *pic de résonance*, les oscillations observées dans une raie forte sont en retard sur celles d'une raie faible. Par exemple l'écart de phase moyen entre les champs de vitesses de 5172.70 Mg I et 5173.75 Ti I est de 5°; il atteint 35° à 40° entre les raies 8542.13 Ca II et 8514.08 Fe I.

Compte tenu d'ordres de grandeur raisonnables pour les différences de "niveaux effectifs" entre les observations des différentes raies; on constate que ces écarts de phase sont compatibles avec les calculs de BAHNG et SCHWARZSCHILD (*Astrophys. J.* **137**, 901) pour les oscillations de période cinq minutes de modèles schématiques de l'atmosphère solaire.

Dans *l'aile de haute fréquence* l'écart de phase entre les raies 8498.06 et 8542.13 Ca II est nul (à l'approximation actuelle de nos statistiques). Ce résultat est surprenant, car on s'attendrait à ce que les oscillations de fréquence supérieure à celle du pic de résonance s'identifient à des ondes sonores.

L'aile de haute fréquence observée pourrait être liée à la *distorsion* des oscillations photosphériques: on a vu en effet (J. EVANS, R. MICHARD, 1962, *Astrophys. J.* **136**, 493) que ces oscillations ne sont pas exactement sinusoïdales, leur période et leur vitesse de phase apparentes variant au cours de la "vie" de la perturbation.

Cependant diverses autres possibilités doivent être examinées: ondes de gravité à propagation horizontale, et surtout l'effet des interférences entre perturbations voisines.

c. Oscillations "fortes" et champ général des vitesses

Dans le mémoire déjà cité, nous avons étudié les propriétés individuelles d'oscillations *choisies* pour leur grande amplitude et leur "pureté". Ces oscillations fortes ont des propriétés assez distinctes de celles du champ général des vitesses: elles occupent la région des plus courtes périodes dans le spectre de puissance, tout en restant à l'intérieur de la bande de résonance. Les écarts de phase entre les niveaux de b_2 et de 5173.75 Ti sont plus grands d'un facteur 2 pour les oscillations fortes que pour le champ général, et vont d'ailleurs en décroissant au cours de la vie de l'oscillation.

Nous supposons que les oscillations fortes sélectionnées dans notre première étude, sont en moyenne les plus *jeunes*. Une oscillation solaire type, commence par une poussée verticale se propageant à la vitesse du son (?); aprés réflexion, cette perturbation initiale évolue vers une oscillation quasi-stationnaire à la fréquence de résonance.

OSCILLATIONS ET FLUCTUATIONS DE BRILLANCE

a. Granulation

Les spectres de puissance de la granulation et des vitesses radiales diffèrent de manière radicale car le spectre de la granulation décroît de manière monotone des basses vers les hautes fréquences. Les fluctuations de brillance du continuum et celles de vitesses, sont partiellement corrélées (cohérence 0.35), les vitesses ascendantes étant en retard de 29° sur les brillances positives, ceci au voisinage de la fréquence de résonance des oscillations.

Les résultats statistiques sont compatibles avec la relation "génétique" entre granules et oscillations, déjà signalée dans le mémoire de J. EVANS et R. MICHARD.

b. Brillances au centre des raies

Le spectre de puissance des fluctuations de brillance au centre des raies fortes (5172.70 Mg I et 5171.62 Fe I) présente deux caractéristiques intéressantes:

(1) un domaine de très basses fréquences correspondant à des structures persistantes non corrélées avec la granulation photosphérique (*coarse mottling* des spectrohéliogrammes?);

(2) un maximum au voisinage du pic de résonance des vitesses.

Dans ce dernier domaine spectral, brillances et vitesses sont bien cohérentes et sont approximativement *en quadrature.*

Les fluctuations de brillance au centre des raies sont dues aux variations d'épaisseur optique de la "couche renversante", provoquées par les changements de pression associés à des oscillations aérodynamiques essentiellement isothermes.

Les résultats détaillés de ce travail sont en instance de publication aux *Annales d'Astrophysique.*

R. W. NOYES. It is possible to reproduce many of the observed properties of the oscillatory velocity field via a numerical integration of the equations of hydrodynamics and radiative transfer for a one-dimensional solar atmosphere. A study presently under way at the Smithsonian Astrophysical Observatory involves introducing perturbations into such an atmosphere in the region $\tau \geqslant 1$, observing the resulting behavior in higher layers. It is found that the velocity response in these layers to any reasonable perturbation of either pressure of radiative flux below $\tau = 1$ is characterized by (a) oscillations with the observed period (about 300 sec), and (b) the observed decrease of period with increasing altitude.

However, the associated temperature field depends very much on whether the perturbation is in the pressure below $\tau = 1$ or in the radiative flux from below. If the perturbation involves only the pressure (or velocity) field below $\tau = 1$, a compressional oscillation is set up at all altitudes, but the concomitant temperature fluctuations are significant only at high altitudes, where the atmosphere is adiabatic. (At optical depths between $\tau = 0.05$ and $\tau = 1$, the radiative relaxation time is very short compared, so the time scale of any reasonable perturbation, preventing any temperature build-up due to wave compressions.) Thus temperature fluctuations increase steadily in size with increasing altitude, and as a result the relative contrast (i.e. the root mean square percentage fluctuation of $I(O, \mu)$) increases steadily with decreasing μ.

If, however, the perturbation is a change in the radiative flux from below, a presume pulse is induced with the above-mentioned variation of compressional heating with altitude, but there is in addition a direct heating of the lower layers by the increased flux. This extra source of heating yields a contrast which first increases, then decreases toward the limb, in better agreement with observations. Furthermore, in this second case, the correlation between temperature and velocity at different altitudes agrees well with the observations.

Thus it seems probable that the sudden appearance of a rising hot granule at the level $\tau = 1$, followed by the rapid radiation of its excess energy into the overlying layers, may be the event which initiates the oscillatory motions, rather than a "mass aging of the atmosphere from below by more or less isothermal pistons".

R. HOWARD. From recent line-shift observations made with the solar magnetograph at Mount Wilson, the 5-minute oscillation is easily apparent in the line λ 5250 of Fe I. These observations have extended to periods of more than 5 hours, the 10-second square aperture being held at one spot on the solar disk. On each case the autocorrelation function which results shows a sinusoidal character. After an initial decrease in the amplitudes of this sinusoidal curve, they increase again after approximately 45 minutes. This increase occurs again after twice and three times this period. This indicates that the 5-minute oscillation may show a tendency to return in phase.

WASABURO UNNO*

Influence of Macro-turbulence on Chromospheric Doppler Velocities Determined from the Multiplet Method

Abstract. The multiplet method (L. GOLDBERG, *Astrophys. J.* **127** (1958) 308) to determine the Doppler width is generalized to include the effect of the macro-turbulence. The method

* Department of Astronomy, University of Tokyo, Tokyo.

gives the total Doppler width for faint lines. Contribution of macro-turbulence decreases as the line width or the saturation effect increases. The observational evidence that the Doppler width of chromospheric lines measured by the multiplet method for a single pair of lines decreases towards the line center in contradiction to the general increase of the width with height is explained by the presence of macro-turbulence. Velocities of micro and macro-turbulence are about equal and they increase steadily with height.

A. INTRODUCTION

Very recently, SUEMOTO[1] has obtained from his eclipse data that the turbulent velocity increases steadily with height up to 2000 km from the base of the chromosphere and above this height the velocity is about 6 km/sec within a spicule and is about 20 km/sec for the relative motion between spicules. These different classes of motion combined with the inhomogeneity constitute the characteristics of the chromosphere which should be taken into account in the study of the line formation of chromospheric lines.

Some years ago, the author[2] studied the chromospheric turbulence with the use of multiplet method developed by GOLDBERG.[3] By this method, the height dependence of the Doppler width could be obtained in principle from two sources. Since widths of a pair of lines within a multiplet are compared at the same residual intensity to derive the Doppler width, we can obtain the change of the Doppler width by changing the residual intensity in the measurement of a single pair on one hand, and by choosing multiplets of various intensity on the other. The results of these two methods are, however, contradictory to each other; the former gives decreasing velocities with height, while the latter gives increasing velocities. It was concluded in the previous paper[2] that the former should be misleading, since the inhomogeneity in turbulence and the effect of non-coherent scattering might upset the conclusion of the first method but not that of the second.

In this paper, the spicule-to-spicule motion is considered to act partially as the macro-turbulence and partially as the micro-turbulence in the line formation, and the study is made to examine its effect on the measured Doppler width by the multiplet method. Since the velocity of that motion is about three times that of the turbulence within a spicule, the above approximation is legitimate, if the superposition of spicules along the line of sight is incomplete down to the depth at which the optical depth reaches unity at the core of the line. This assumption seems to be true for the strong lines of which the core is practically formed within the chromosphere.

B. THE MULTIPLET METHOD

The multiplet method developed by GOLDBERG[3] is the manifestation of the idea that if the absorption coefficient at the wave length $\Delta\lambda_A$ from the line center of a line A of a multiplet is the same as the absorption coefficient at $\Delta\lambda_B$ of a line B of the same multiplet, residual intensities should be the same between the wave lengths, $\Delta\lambda_A$ and $\Delta\lambda_B$, of respective lines. For the Doppler core, this idea is expressed as

$$1 - r_A(\Delta\lambda_A) = f(\Delta\lambda_A^2) \tag{1}$$

and

$$1 - r_B(\Delta\lambda_B) = f(\Delta\lambda_B^2 + \ln K \cdot \Delta\lambda_D^2), \tag{2}$$

where $(1 - r_A)$ and $(1 - r_B)$ denote residual intensities of the lines A and B, $f(\Delta\lambda_A^2)$ represents the profile of the line A, ln K the natural logarithm of the ratio of the gf-values,

$$K = (gf)_A/(gf)_B, \tag{3}$$

and $\Delta\lambda_D$ the Doppler width. Since the equality of absorption coefficients is given by

$$\Delta\lambda_A^2 = \Delta\lambda_B^2 + \ln K \cdot \Delta\lambda_D^2, \tag{4}$$

equations (1) and (2) show that $1 - r_A(\Delta\lambda_A) = 1 - r_B(\Delta\lambda_B)$. The underlying assumptions are (i) the source function is common for both lines, (ii) the profiles are not disturbed by the macro-turbulence and the rotation, and (iii) the atmosphere is not heterogeneous and the turbulence is homogeneous. The non-coherent scattering, the Stark and the Zeeman effects violate the condition (i), but their effects are not large and in some respect similar to the effect of the macro-turbulence on the multiplet method.

The Doppler width $\Delta\lambda_D$ is easily obtained observationally with the help of Equation (4). If the condition (iii) is not satisfied, the value of $\Delta\lambda_D$ in this method will be some weighted mean of non-uniform Doppler widths.

An interesting consequence of Equations (1) and (2) is that the profile of the line B is completely determinate from the profile of the line A and the Doppler width as far as the Doppler core is concerned. If this is not the case, the Doppler width varies with height very rapidly or at least one of the conditions (i), (ii) and (iii) is not satisfied.

It was found that the value of $\Delta\lambda_D$ with the use of Equation (4) increases with $\Delta\lambda_A$ or $\Delta\lambda_B$. This may imply that the turbulent velocity might decrease with height in contradiction to the evidence that stronger pairs of multiplet lines give larger values of $\Delta\lambda_D$. In this paper, the macro-turbulence is assumed to be dominant in the chromosphere, and an attempt is made to interprete the apparent contradiction by means of reasonable values of the macro-turbulent velocity.

C. THE EFFECT OF MACRO-TURBULENCE

The profiles of a pair of lines undisturbed by the macro-turbulence are given by Equations (1) and (2) which may be explicitly expressed as

$$1 - r_A = e^{-(\Delta\lambda_A/D)^2}[a_0 + a_1(\Delta\lambda_A^2/D^2) + \dots] \tag{5}$$

and

$$1 - r_B = K^{-(\Delta\lambda_D/D)^2} e^{-(\Delta\lambda_B/D)^2}[a_0 + a_1\{(\Delta\lambda_B^2 + \ln K \Delta\lambda_D^2)/D^2\} + \dots]. \tag{6}$$

Here, D, a_0, a_1, etc. are parameters depending on the atmospheric model and the mechanism of the line formation. Then, the profiles blurred by the macroturbulence are approximately given by

$$1 - R_{A,B} = \frac{1}{\sqrt{\pi}\,\delta} \int_{-\infty}^{+\infty} [1 - r_{A,B}(x')]\, e^{-(x' - \Delta\lambda_{A,B})^2/\delta^2}\, dx', \tag{7}$$

or more explicity by

$$1 - R_A = \left[e^{-\Delta\lambda_A^2/(XD)^2}/X\right]\left[a_0 + a_1\left\{\frac{1}{2X^2}\left(\frac{\delta}{D}\right)^2 + \frac{1}{X^4}\left(\frac{\Delta\lambda_A}{D}\right)^2\right\}\right] \tag{8}$$

and

$$1 - R_B = \left[K^{-(\Delta\lambda_D/D)^2}\, e^{-\Delta\lambda_B^2/(XD)^2}/X\right]$$
$$\left[a_0 + a_1\left\{\frac{1}{2X^2}\left(\frac{\delta}{D}\right)^2 + \ln K\left(\frac{\Delta\lambda_D}{D}\right)^2 + \frac{1}{X^4}\left(\frac{\Delta\lambda_B}{D}\right)^2\right\}\right], \tag{9}$$

where

$$X^2 = 1 + (\delta/D)^2\,. \tag{10}$$

The equality of $(1 - R)$ of both lines no longer holds with Equation (4). We define, therefore, the Goldberg width by

$$\Delta\lambda_G^2 = (\Delta\lambda_A^2 - \Delta\lambda_B^2)/\ln K\,, \tag{11}$$

where $\Delta\lambda_A$ and $\Delta\lambda_B$ are measured at same intensities. From Equations (8) and (9), we have

$$\Delta\lambda_G^2 = X^2\left\{\Delta\lambda_D^2 - \frac{D^2}{\ln K}\ln\left[1 + \frac{a_1 \ln K(X^4\Delta\lambda_D^2 - \Delta\lambda_G^2)}{a_0 X^4 D^2 + a_1(X^2\delta^2/2 + \Delta\lambda)_A^2}\right]\right\}, \tag{12}$$

Qualitative discussions can be made with Equation (12). If $a_1 = 0$, we have

$$\Delta\lambda_G^2 = X^2\Delta\lambda_D^2 = [1 + (\delta/D)^2]\,\Delta\lambda_D^2\,. \tag{13}$$

For weak lines, in particular, $D = \Delta\lambda_D$ and, therefore,

$$\Delta\lambda_G^2 = \Delta\lambda_D^2 + \delta^2\,. \tag{14}$$

This means that the Goldberg width gives the total Doppler width for weak lines, while it gives practically the Doppler width of small scale for strong lines (δ/D $<$ 1). On the other hand, strong lines show saturated profiles with flat bottoms. This is qualitatively represented by positive values of a_1. Then, it can be shown that the solution $\Delta\lambda_G$ of Equation (12) increases with increasing value of $\Delta\lambda_A$ for reasonable values of parameters. In this case, $\Delta\lambda_G$ may possibly be smaller than $\Delta\lambda_D$ for small $\Delta\lambda_B$ depending on the values of model parameters, a_0, a_1 and D.

D. OBSERVATIONAL ESTIMATION OF $\Delta\lambda_D$ AND δ

Eqs. (8) and (9) are written as

$$1 - R_A = e^{-\Delta\lambda_A^2/(XD)^2} [A_0 + A_1 \Delta\lambda_A^2] \tag{15}$$

and

$$1 - R_B = K^{-(\Delta\lambda_D/D)^2} e^{-\Delta\lambda_B^2/(XD)^2} [B_0 + A_1 \Delta\lambda_B^2] \,. \tag{16}$$

Unknown parameters are A_0, A_1, B_0, (XD) and $(\Delta\lambda_D/D)$, and therefore five measurements along the cores of A and B lines, e.g., $(\Delta\lambda_A)_i$ at $(1 - R_A)_i$ $(i = 0, 1, 2)$ and $(\Delta\lambda_B)_i$ at $(1 - R_B)_i$ $(i = 1, 2)$, can determine these unknowns. From the comparison of Equations (8) and (9) with Equations (15) and (16), we have

$$D^2 = [A_1/(B_0 - A_0)]\, (XD)^4 \ln K(\Delta\lambda_D/D)^2 \,. \tag{17}$$

Values of $\Delta\lambda_D$ and δ are determined with the help of Equations (10) and (17) with the values of (XD) and $(\Delta\lambda_D/D)$.

It is, however, obvious that Equation (5) is not the only type of explicit representation of Equation (1). We can use a simpler form as follows,

$$1 - r_A = a_0' + a_1'(\Delta\lambda_A)^2 + a_2'(\Delta\lambda_A)^4 + \ldots \tag{18}$$

with which we have

$$\begin{aligned} 1 - R_A &= a_0' + a_1'(\Delta\lambda_A^2 + \delta^2/2) + a_2'(\Delta\lambda_A^4 + 3\delta^2\Delta\lambda_A^2 + 3\delta^4/4) \\ &= A_0' + A_1'(\Delta\lambda_A)^2 + A_2'(\Delta\lambda_A)^4 \end{aligned} \tag{19}$$

and

$$\begin{aligned} 1 - R_B = {} & a_0' + a_1'(\Delta\lambda_B^2 + \ln K\Delta\lambda_D^2 + \delta^2/2) \\ & + a_2'\{(\Delta\lambda_B^4 + 3\delta^2\Delta\lambda_B + 3\delta^4/4) + 2\ln K\Delta\lambda_D^2(\Delta\lambda_B^2 + \delta^2/2) + \\ & + (\ln K\Delta\lambda_D^2)^2\} \\ = {} & B_0' + B_1'(\Delta\lambda_B)^2 + A_2'(\Delta\lambda_B^4) \,. \end{aligned} \tag{20}$$

Unknowns A_0', A_1', A_2', B_0' and B_1' are to be determined observationally, and then $\Delta\lambda_D$ and δ are given by

$$\Delta\lambda_D^2 = (1/2\ln K)\,[(B_1 - A_1)/A_2] \tag{21}$$

and

$$\delta^2 = [(A_1 + B_1)/4A_2] - (B_0 - A_0)/(B_1 - A_1) \,. \tag{22}$$

The essential advantage of the Goldberg method is still conserved in the fact that the detailed knowledges of the model and of the mechanism of line formation are immaterial in the analysis.

Analysis has been made with the same data as in the previous paper[2]

TABLE

Element	λ	*ln K*	*Equations (15), (16)*		*Equations (15), (21)*		*adopted*		ξ_t*(km/sec)*		*height (km)*
			$\Delta\lambda_D$ *(mA)*	δ *(mA)*	$\Delta\lambda_D$ *(mA)*	δ *(mA)*	$\Delta\lambda_D$ *(mA)*	δ *(mA)*	*micro*	*macro*	
Fe I	4271.8 4202.0	1.02	38	31	42	40	40	35	2.6	2.5	350 h
Fe I	4383.6 4407.8	0.66	49	38	54	45	50	40	3.2	2.7	500 h
Mg I	3838.3 3832.3	0.65	58	52	50	53	55	50	3.9	3.9	850 h
Ca II	3933.7 3968.5	0.69	140	120	136	111	140	115	10.5	8.7	2600 h

ln *K* denotes the natural logarithm of the ratio of (*gf*)-values; for the first pair from King and King (*Astrophys. J.* **87** (1938) 359) and for the others from the relative strength for *LS* coupling.

(Michigan data due to O. C. MOHLER). The results are listed in the table. Discrepancies between the fourth and sixth column and between the fifth and the seventh column indicate that the result depends rather sensitively on the form of the analytic approximation for the profiles. This does not necessarily mean that the number of fitting parameters should be increased, since the result is also sensitive to observational errors. It seems, therefore, to be important to choose an analytic form which fits well with the accurately measured profiles with a small number of model parameters. In this respect, the present results are merely preliminary. From the degree of deviation in the fitting, the weight of the determination is estimated for the two cases and the weighted average is listed in the eight column for $\Delta\lambda_D$ and in the ninth column for δ. The thermal component is substracted from $\Delta\lambda_D^2$, and the micro and macro-turbulent velocities are given in the tenth and the eleventh column.

The height above the photosphere for each measured Doppler width was tentatively estimated in the previous paper.[2] Its value was different for different chromospheric models. The average is adopted here and a small correction due to the difference in $\Delta\lambda_D$ between the present and the previous determination is applied. The result is listed in the last column of the table. A scale factor h is attached to the values of the height in order to account for the spicule structure. In this respect, discussions will be made in the following section.

E. DISCUSSIONS

In the preceding analysis, variations of $\Delta\lambda_D$ and δ with height and other disturbing influences such as non-coherent scattering are neglected. Their effects may probably be small or at most smaller than the effect of the heterogeneous structure represented by spicules.

MORIYAMA and SUEMOTO[4,5] presented a model of the chromosphere composed merely of spicules having transition layers to the corona. According to them, the interspicular region is nothing but the corona which almost directly contacts with the photosphere with a transition layer of negligible thickness. The model can account for eclipse and radio observations as well as the solar ultraviolet radiations.

If their model is applicable also to the formation of strong Fraunhofer lines at the disc center of the sun, even the cores of the lines are formed practically in the photosphere and the contribution from spicules diminishes slightly the depression in the core because of large turbulent velocities between spicules. Then, the Goldberg widths derived from strong absorption lines may have little physical significance.

Their model, however, seems to be an idealization of the real situation. In fact, the essential feature is reserved, if spicules have observable transition layers only in their active phases and inactive spicules or the quiet chromosphere covers the most part of the solar surface. If this is the case, strong absorption lines are really formed within the chromosphere and, therefore, information on the chromosphere can be derived from the observation of these lines. Then, the values of $\Delta\lambda_D$ and δ obtained in the preceding section may give their representative average values within the chromosphere.

A major consequence of the spicule structure is that the observed Doppler widths should be attributed to a variety of heights above the photosphere. Since the values in the last column of Table 1 are based on the uniform model of the chromosphere, they should be multiplied by a scale factor h, depending on the extension of each spicule. The value of h may be as large as two for spicules and less than one for the quiet region.

F. ACKNOWLEDGEMENT

The author expresses his thanks to Professor Z. Suemoto for stimulating discussions.

References

1. Suemoto, Z.: 1964, *Publ. A. S. Japan*, in preparation.
2. Unno, W.: 1959, *Astrophys. J.* **129**, 388.
3. Goldberg, L.: 1958, *Astrophys, J.* **127**, 308.
4. Moriyama, F.: 1961, *Ann. Tokyo Astr. Obs.*, 2nd ser., **7**, no. 3.
5. Suemoto, Z. and Moriyama, F.: 1962, *Third COSPAR Symposium*, p. 800.

SESSION II

THE QUIET CHROMOSPHERE

THE THEORY OF THE CHROMOSPHERIC SPECTRUM

JOHN T. JEFFERIES*
Joint Institute for Laboratory Astrophysics, Boulder, U.S.A.

1. Introduction

The richness and variety of the emissions of the solar chromosphere, together with its complex geometric forms present a difficult problem for analysis. It is, therefore, hardly surprising that a thoroughly satisfying chromospheric model has not so far emerged from the various analytical attempts, though many of these have brought out very pertinent points on individual requirements on the model. Our purpose here is to present the theoretical bases for the analysis of the line spectrum of the chromosphere. In fact the presentation is more general than this may suggest; it should be as applicable to any glowing gas as to those in the solar chromosphere.

This is in no sense intended as an historical survey but rather as the statement of a personal view of the present situation and of some possible directions towards the solution of outstanding problems in the formation of, and the analysis of, line spectra.

It is important to distinguish these two aspects of the problem. Anyone wishing to derive the physical properties of a gas from knowledge of its emitted spectrum faces a dual problem. He must, firstly, formulate a theory for the formation and transfer of radiation in a gas of given properties and secondly he must devise an analytic procedure which can turn this theory around – so to speak – to infer the physical properties of the gas from a knowledge of its spectrum. While both aspects will be treated here, the primary emphasis is to be given to the former since its elucidation is essential before an adequate discussion of the interferential problem can be attempted.

The specific intensity of radiation emerging in a particular direction from an atmosphere is given by the well-known equation

$$I_\lambda(0,\sigma) = \int_0^S \varepsilon_\lambda(s) \exp\left[-\int_0^s \varkappa_\lambda(s)\mathrm{d}s\right]\mathrm{d}s \qquad (1)$$

where S is the total thickness in the line of sight – whose direction is $\boldsymbol{\sigma}$ – and ε_λ, $\varkappa_\lambda$ are respectively the monochromatic emission and absorption coefficients. We may conveniently distinguish two cases of this equation, according to whether

* Present address: Hawaii Institute of Geophysics, Honolulu 14, Hawaii.

All rights reserved.

$$\int_0^S \varkappa_\lambda(s)\mathrm{d}s$$

is negligible or is not negligible. In the first, or optically *thin*, case we have

$$I_\lambda(0,\boldsymbol{\sigma}) \simeq \int_0^S \varepsilon_\lambda(s)\mathrm{d}s\,. \tag{2}$$

In principle we cannot, of course, solve this integral equation unless we have some independent method of determining just how the emission coefficient varies with position. If we *assume* some functional form for the variation of ε with position then knowing S we may invert (2) to determine ε_λ(s). The problem of interpreting the observed line intensity then reduces simply to that of interpreting the emission coefficient which is, of course, simply proportional to the product of the population of the upper state of the transition and the line shape. We shall not discuss the shape; it is frequently Gaussian, or closely so, with a width which is not always clearly interpretable; our interest here lies in understanding the state populations.

In the second case above, where the atmosphere is optically thick, and so self absorption is important, it is convenient to rewrite Equation (1) in the form

$$I_\lambda(0,\boldsymbol{\sigma}) = \int_0^{\tau_1(\lambda)} S_\lambda(t_\lambda)e^{-t_\lambda}\mathrm{d}t_\lambda \tag{3}$$

where $t_\lambda = \int \varkappa_\lambda \mathrm{d}s$ is the optical path length in the direction $\boldsymbol{\sigma}$, S_λ is the source function and τ_1 is the total path length through the atmosphere, also measured in the direction $\boldsymbol{\sigma}$.

The dual problem mentioned above is clearly illustrated in Equation (3) – which includes Equation (2) as a special case. To interpret the observed I_λ we must be able, firstly, to infer $S_\lambda(t_\lambda)$ – the analytic problem – and secondly to determine, from this, the physical properties of the atmosphere. Thus the joint problem of the inference of, and the interpretation of the source function S constitutes the fundamental problem of spectral line analysis.

How is all this to be related to the chromospheric spectrum? I think it would be generally agreed, now, that there are two ways of observing the chromospheric spectrum, namely by direct observations at the limb, as at an eclipse, or by observations of strong lines like H and K of Ca II on the disk. These give different sorts of information; the limb spectrum contains many lines which are weakly absorbed, if at all, and which correspondingly may be analyzed in terms of Equation (2). Of these, the visible lines of helium are an excellent example. On the other hand, an analysis of the center to limb intensity distribution in the early Lyman lines, or in the cores of H and K, can only proceed through the self absorbed case, embodied in Equation (3).

These two different observational methods frequently require different analysis; often in the flash spectrum only the total line intensity can be obtained, while for center-limb studies a less urgent situation allows careful observations of line profiles. Consequently rather different analytical tools are used and they will be discussed in turn in this article. In Section 2 the 'thin atmosphere' case, as embodied in Equation (2), is studied and an application of the general ideas is made, in Section 3, in an attempt to interpret some aspects of the helium flash spectrum. In the following Section 4, we turn to the more general problem contained in Equation (3) and outline some of the current problems associated with its solution.

2. Interpretation of the Emission in Optically Thin Lines

Any discussion of the interpretation of the *shape* of a line formed *in an optically thin gas* would lead us away from the immediate objective of this paper. This is not to say, of course, that the shape may not contain vital information. Consider for example the overlapping of high lines in the Balmer series which gives, through the theory of the Stark effect, information on the electron density at the region where the emission originates. Many lines, however, seem to be broadened by mass motions and since our primary interest lies in studying the excitation of the gas we shall consider only the *total* line intensity when dealing with optically *thin* lines.

For the total emission integrated over wavelength in the line, Equation (2) becomes

$$I = \int_0^S \varepsilon(s)\mathrm{d}s \tag{4}$$

where I and ε are wavelength integrated quantities. Our interpretive problem may then be stated as follows: Given the observed quantity I, we wish to determine $\varepsilon(s)$ and hence to determine a compatible atmospheric model.

Consider the determination of $\varepsilon(s)$ from a slitless spectrogram of an eclipse. In that case we determine the total line emission above the moon's limb as a function of time – or equivalently of the moon's distance, h, above the solar limb. Thus at a point on the crescent we determine, in fact,

$$E(h) = \int_h^\infty I(y)\mathrm{d}y = \int_h^\infty \mathrm{d}y \int_{-\infty}^{+\infty} \varepsilon(x,y)\mathrm{d}x \tag{5}$$

where x is a distance in the line of sight and y is perpendicular to x and to the solar limb.

A simple differentiation of $E(h)$ then gives the integrated specific intensity

$$I(h) = \int_{-\infty}^{+\infty} \varepsilon(x, h) \mathrm{d}x \,. \tag{6}$$

As already mentioned, we cannot invert this equation unless we make some assumption on the variation of ε with position. The usual practice is to suppose that ε is a function only of the radial distance of the element above the limb, in which case the inversion is fairly straightforward. Essentially the same result is obtained if we suppose the material to be concentrated in thin radial columns occupying a fraction a of the total volume. In this case, however, the derived emission coefficient, which refers to a unit volume, must be correspondingly increased by the factor a^{-1}. However, in principle a can have any value less than unity and we are faced with a fundamental limitation only resolvable by appeal to a requirement of consistency with some other observation.

Suppose however that, on the basis of some reasonable assumption, we have an estimate for ε. Knowing the Einstein transition probabilities we can then determine the number of atoms in the upper state of the line. The interpretation of the emitted spectrum in this optically thin case then comes down to the interpretation of the excitation of the gas, i.e., to the distribution of atoms among the various possible states.

In a stationary situation this distribution is determined by a set of equations of statistical equilibrium which state the requirement of equality, at each position in the gas, of the rates of exit from and entrance to any given state. This is simply a stationary condition; it includes local thermodynamic equilibrium as a special case. All that is needed to solve these equations – and so to interpret the spectrum – is a knowledge of the collisional and radiative transition rates between each pair of states in the atom.

Collisional rates present no problem in principle; one simply averages the product of cross section and velocity over a Maxwellian electron velocity distribution. In practice one needs values of the cross sections and thanks largely to the work of Burgess, Seaton, and van Regemorter – see for example Seaton (1962) – good semi-empirical formulae for these exist, at least for optically allowed excitation and for ionization. There remain, however, some outstanding problems – for example that of the cross sections for intersystem transitions.

The rates of spontaneous downward transitions are atomic constants whose values have generally been well estimated, theoretically or experimentally, for many important transitions. The rate of radiative absorption including stimulated emission as a negative term is, however, governed by the local value of the radiation intensity. For those lines for which the chromosphere is optically thin this is simply the corresponding intensity of the photospheric radiation. For lines whose strengths are mainly controlled by chromospheric excitation – like the resonance lines of H and He I – we cannot have recourse to this simple solution. Since such lines are formed in the chromosphere, their strengths clearly depend on those atomic populations which we are trying to determine.

Before considering a way out of this problem we shall examine a practical difficulty involved in solving the equilibrium equations. Recall that there is one equilibrium equation for each state and that the number of states identifiable in the spectral emission is very large. We shall certainly need to find a way to limit the number of states if we are not to become involved in prohibitive labor. At the same time we must keep sufficient states that the excitation can be meaningfully described.

A reasonable place to terminate the discrete level structure is at a level N such that, in the given atmosphere, the probabilities of an upward and of a downward transition from N are equal. Since, in general, the closer the level lies to the continuum the greater the chance of an upward rather than a downward transition, an atom in or above state N is more likely to reach the continuum before going downward again. Since, then, levels above N are well mixed with the continuum and since the continuum states are assumed to be distributed as in thermodynamic equilibrium, so also will be the levels above N. In this case we may readily show – as would be expected physically – that the levels above N may be effectively regarded as part of the continuum and that in discussing the equilibrium we may, with negligible error, quite literally regard the continuum as starting at the energy corresponding to level N. For an electron density of 10^{11} cm^{-3} we find $N \simeq 8$ for H and He I.

With this out of the way we may return to the problem of specifying the strengths of those lines which are produced in the chromosphere. These can only be determined through the appropriate transfer equations which may, to adequate accuracy, be written in the form due to Eddington,

$$\frac{1}{3}\frac{\mathrm{d}^2 J_\nu}{\mathrm{d}\tau_\nu^2} = J_\nu - S_\nu\,, \tag{7}$$

where J_ν is the mean intensity $1/4\pi \int I_\nu \mathrm{d}\omega$ and τ_ν the optical depth. There is one such equation for each line.

Instead of approaching the problem immediately via the transfer equations, with all the attendant complexity introduced by the interlocking of line with line, we may introduce the unknown radiation intensities parametrically through the 'net radiative bracket' Z defined, for the j–i transition, by (see THOMAS, 1960)

$$A_{ji} Z_{ji} n_j \equiv A_{ji} n_j - 4\pi \int (J_\nu \alpha_\nu \mathrm{d}\nu / h\nu) n_i\,. \tag{8}$$

with α_ν the atomic absorption coefficient. The right hand side of this equation gives the excess number of emission over absorption transitions in the j–i line. At sufficiently great optical depths, $Z = 0$ so that a detailed balance holds in the radiative transition between states i and j. In the opposite extreme, i.e.

$\tau_1 \ll 1$, J_ν is essentially equal to the incident intensity – if this is zero then Z is close to unity. These are just the 'optically thick' and 'optically thin' cases often referenced in the literature.

While representing a set of convenient parameters, the numbers Z_{ji} in the different lines are not independent; their values are governed by the mean intensities in the relevant lines and these in turn are mutually interdependent since, physically, photons may be transferred back and forth between the different lines through radiative and collisional transitions. The relationship between the net radiative rates for the various lines is not properly known. A simple approximate study of the interlocking problem – JEFFERIES (1960) – indicates that, frequently, the relative Z values of resonance lines are roughly inversely proportional to the squares of their absorption coefficients. This is shown in Figure 1 for some sample computations of the simultaneous transfer

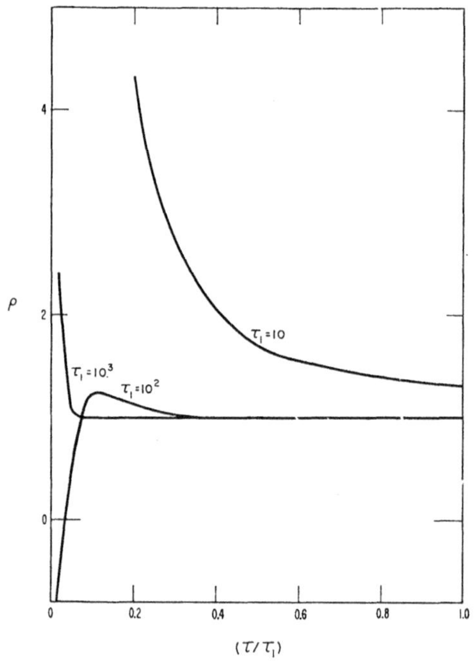

Fig. 1. Variation of ϱ through the atmosphere for the indicated values of total optical depth τ_1.

in the resonance lines of a three level atom of the type discussed by the author (1960) and for various values of the total optical thickness. The quantity ϱ plotted in the Figure is defined by

$$\varrho = \frac{Z_{21}}{Z_{31}} \left[\frac{\alpha_{21}}{\alpha_{31}} \right]^2 \tag{9}$$

and it can be seen that this is close to unity over most of the atmosphere for $\tau_1 > 10$. The detailed shape of the curves in Figure 1 is of no consequence; the essential thing is the convergence of the ratio ϱ to unity.

The physical significance of the result is a little obscure; it seems more likely that an exact solution of the coupled transfer equations would show the relative Z to be inversely proportional to the corresponding absorption coefficient – rather than its square – i.e., that

$$Z_{ji}\alpha_{ji} = \text{Constant} \tag{10}$$

providing the total optical depth in the ji line is greater than about 10^2. Given some such relations among the Z_{ji} we may then solve the equilibrium equations parametrically in one Z and determine which value or values gives the best fit to observation.

Even granting that we can find a set of values for the net radiative rates such that the computed populations are consistent with those inferred from observation, this is simply a parametric solution. We must next solve a transfer equation to establish that the necessary Z values are consistent with the other physical properties of the model.

As an illustration we compute the depth variation of Z for an uncoupled transfer problem. The mean intensity J_ν is given by the equation

$$\frac{1}{3}\frac{d^2 J_\nu}{d\tau_\nu^2} = J_\nu - \tilde{\omega}_0 \int J_\nu \phi_\nu d\nu - \lambda B \tag{11}$$

where $\tilde{\omega}_0 \equiv 1 - \lambda$ is the fraction of the absorbed photons which are subse-

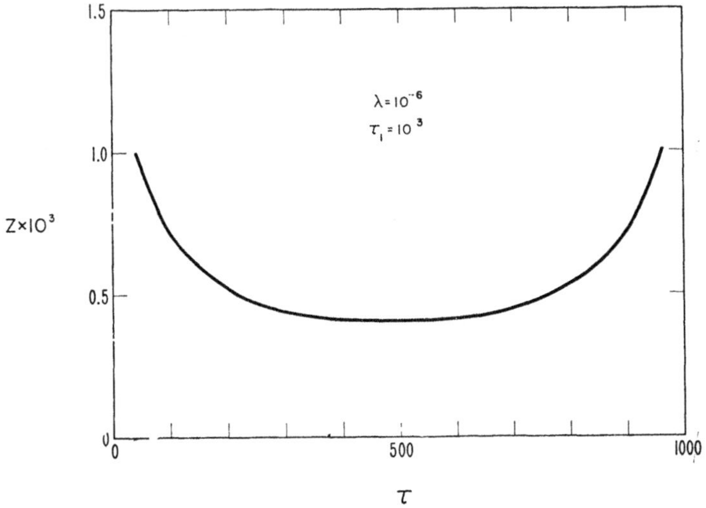

Fig. 2. Depth variation of the net radiative parameter Z for $\lambda = 10^{-6}$ and a total opacity of 10^3 in the line center.

quently re-emitted, ϕ_ν is the normalized profile of the absorption coefficient and B is physically analogous to the Planck function. This equation has been solved by methods given by JEFFERIES and THOMAS (1958), in a ninth order quadrature representation for the integral, for atmospheres of various thicknesses and albedos $\tilde{\omega}_0$. Values of Z obtained from this solution are shown for an atmosphere of total optical thickness 10^3 in Figure 2 – the results are not critically dependent on λ; the Figure is based on $\lambda = 10^{-6}$. The corresponding variation of the source function with depth is shown, for interest, in Figure 3.

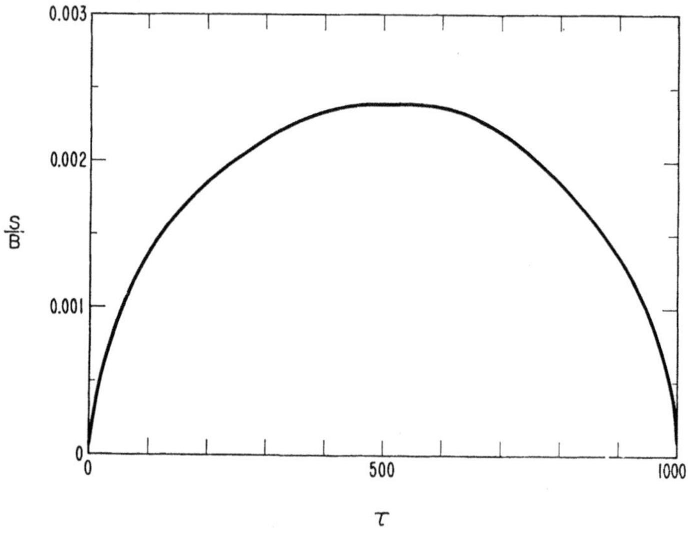

Fig. 3. Depth variation of the line source function for the same conditions as for Figure 2.

In summary, the interpretation of the chromospheric flash spectrum in the unabsorbed lines requires a solution of the statistical equilibrium equations. The coefficients of these equations depend among other things on the radiation strengths. We know these from the disk intensity for the weak lines but not for strong lines formed in the chromosphere itself and we can only find them from the appropriate transfer equations. We find it convenient, for these self absorbed lines, to introduce the *net* radiative transition rates through the parameters Z – one for each self absorbed line. These various Z values need to be related to one another, however, and the form of this relationship is, as yet, unknown. A conjectured form is obtained in Equation (10) and on this, or some other, basis the equations of statistical equilibrium may be solved parametrically in the electron temperature, the electron density, and one Z value. A comparison with observation may then delimit consistent values of these quantities. Finally the derived model must yield, through a solution of the transfer equation, a consistent value of the number Z.

The weak point in this procedure is clearly the uncertainty in the relationship between the various net radiative rates. This can be overcome only when a systematic and reliable method for the solution of coupled transfer equations is available. This problem is central to the whole of the modern theory of line formation and is under active study by many workers, see e.g. the proceedings of the 1962 I.A.U. Symposium of Commission 36.

3. An Illustration: Chromospheric Helium

The detailed analysis of the flash lines of a particular element will be guided by the details or peculiarities of the emission and may well differ from case to case. To some extent, though, all will rely on the general ideas set out above in Section 2. To make these more concrete we shall study some aspects of the chromospheric helium emission taking our data from the 1952 HAO eclipse expedition as given by ATHAY and MENZEL (1956). This is not a full analysis, certain details will be studied, others ignored. The purpose of this Section is illustrative only.

The model atom adopted for this study consisted of twelve levels, namely, the ground state 1^1S, the continuum $\varkappa$, and the excited states $2S$, $2P$, 3, 4, 5 of each of the singlet and triplet series. Fine structure states in $n = 3, 4, 5$ were supposed populated in the ratios of their statistical weights. The limiting level N was set at 6; this is, perhaps, a little low but auxiliary calculations have indicated that the error should be small, particularly for the lower lying states. The continuum was supposed to start at the energy of the state $n = 6$ i.e., at 24.2 eV rather than at the ionization energy 24.6 eV. Transition rates were estimated using cross sections given by SEATON's (1962) formulae and f-values of TREFFTZ *et al.* (1957). The photospheric radiation field, where known, was used to compute radiative absorption transition rates; for transitions caused by absorption in the far infra-red the sun was supposed to emit like a black body at 6000° K.

The statistical equilibrium equations were then solved parametrically in $Z^{1}_{2P},{}^{1}_{1S}$ the Z values for the other resonance transitions being related in accordance with the conjectural Equation (10). The actual quantities solved for were $n_j/n_1 n_e$ and $n_\varkappa/n_1$. We want to determine what limitations can be placed on the chromospheric model from a comparison between the observed and computed line strengths. For this illustration we shall consider only the interpretation of the intensity in $D_3(3^3D - 2^3P)$ and $\lambda\, 6678$ $(3^1D - 2^1P)$. A notable feature of the flash spectrum is the great discrepancy in the intensities of these two lines which imply a population ratio $R \equiv \overset{3}{n_3}/\overset{1}{n_3} \simeq 40$ – see ATHAY and MENZEL (1956) – compared to the thermodynamic equilibrium ratio of close to 3. A possible way to account for this is through a radiative leak in the resonance lines which will reduce the singlet populations but influence much less those of the triplets because of their metastability with respect to the 1^1S state.

The size of the radiative leak is proportional to Z; the observed ratio n_3^3/n_3^1 then allows determination of the value of Z for the chromosphere.

The computed value of R is shown in Figure 4 as a function of $Z^{1}_{2P},{}^{1}_{1S}$ for $T_e = 1.25 \times 10^4$ and 3×10^4 °K and for $n_e = 10^{11}$ cm^{-3}. The results are insensitive to reasonable variations in n_e and can be seen to be little influenced by T_e. The observed ratio is obtained for $Z \simeq 10^{-3.4}$. From Figure 2 it is seen

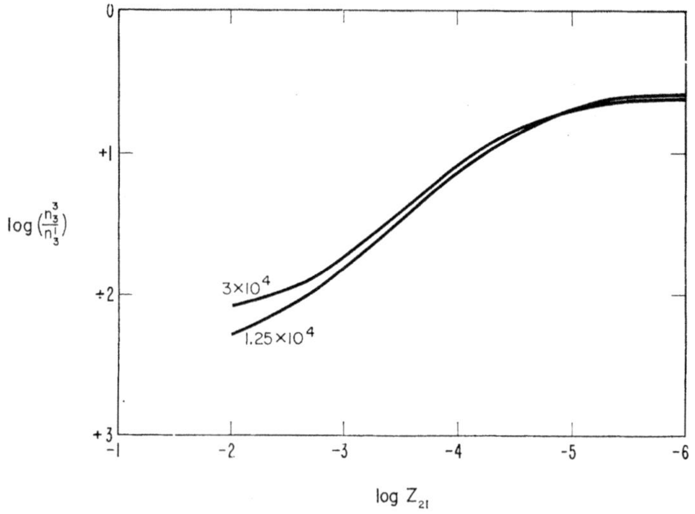

Fig. 4. Variation with Z of the triplet to singlet population ratio for the $n = 3$ states.

that such a value applies over most of an atmosphere whose total optical thickness at the center of $2^1P - 1^1S$ has the substantial value of order 10^3. This requirement, as will be seen, places some rather severe restrictions on the model.

The relative strengths of D_3 and λ 6678 can take us no further than this estimate of τ_1; however we expect that the absolute intensity of D_3 can lead to further information on chromospheric structure. As shown by ATHAY and MENZEL (1956) the observed D_3 emission in the chromosphere requires that, at 1500 km,

$$\mathbf{a} n_{3D}^{3} \simeq 10(\text{cm}^{-3}) \tag{12}$$

where **a** is the fraction of the chromospheric material which emits the helium spectrum. In the appearance of this number we see reflected the basic uncertainty associated with the inversion of Equation (2). The implications of this observational data on the chromospheric model are found as follows. Denote the helium to hydrogen abundance ratio by A and suppose that all the hydrogen is ionized so that n_e is the hydrogen concentration. The equilibrium equations are solved for the ratios $n_j/n_1 n_e$ and $n_\varkappa/n_1$ so that we can immediately determine $n_j/n_e^2 A$. Table 1 shows computed values for $Z_{21} = 10^{-3}$ for $n_3^3/n_1 n_e$ and $n_\varkappa/n_1$ and, in the fourth column, the value of $\mathbf{a} A n_e^2$ which would be required to satisfy the observational deduction $\mathbf{a} n_{3D}^3 = 10$ at 1500 km above the limb. In the center of the $2^1P - 1^1S$ line the optical thickness of an atmosphere of depth z is given by

$$\tau_1 = 7.4 \times 10^{-12} z n_1 T_e^{-1/2} \tag{13}$$

if the broadening is purely thermal. The last column of Table 1 shows the values of $A n_e z$ necessary to make $\tau_1 = 10^{3.4}$, as required to explain the observed

TABLE 1

He I EXCITATION ($Z = 10^{-3}$)

$T(10^4\,°K)$	$n_3{}^3/n_1 n_e$	$n_\varkappa/n_1$	$aAn_e{}^2$	$An_e z$
1.5	1.5×10^{-21}	1.2×10^{-2}	1.2×10^{22}	1.7×10^{16}
2.0	7.5×10^{-20}	8.1×10^{-1}	4.4×10^{20}	3.4×10^{16}
3.0	4.2×10^{-18}	7.0×10^{1}	3.1×10^{20}	1.7×10^{18}

emission scale height of the helium lines is ~ 1000 km; we should make little error in setting $z = 10^8$ cm. This analysis of the ratio of D_3 to $\lambda 6678$ strengths and of the strength of D_3 has thus led to two relations between the

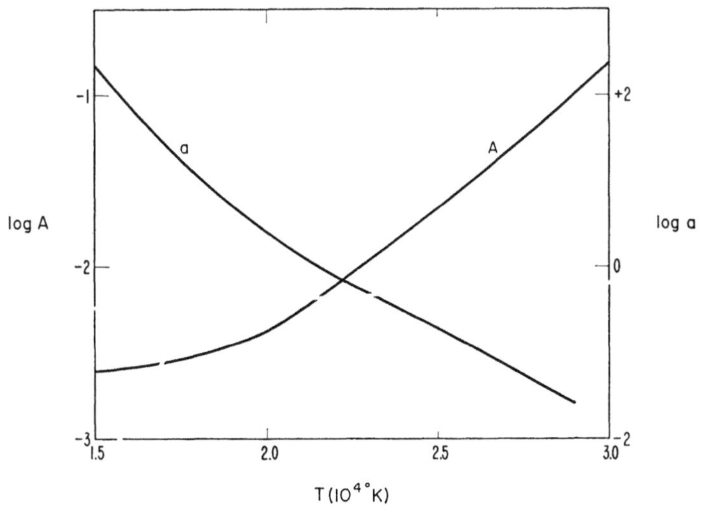

Fig. 5. Consistent values of electron temperature T_e, helium abundance (A) and fractional volume (**a**); case $N_e^2\, T_e^{-3/2} = 1.5 \times 10^{16}$ cgs units.

four unknowns **a**, A, n_e and T_e. In point of fact if $a < 1$ we need to specify electron density and temperature in both components of the model.

To proceed from here we could, for example, use the observed Balmer continuum emission which is found, from a reanalysis of data given by ATHAY

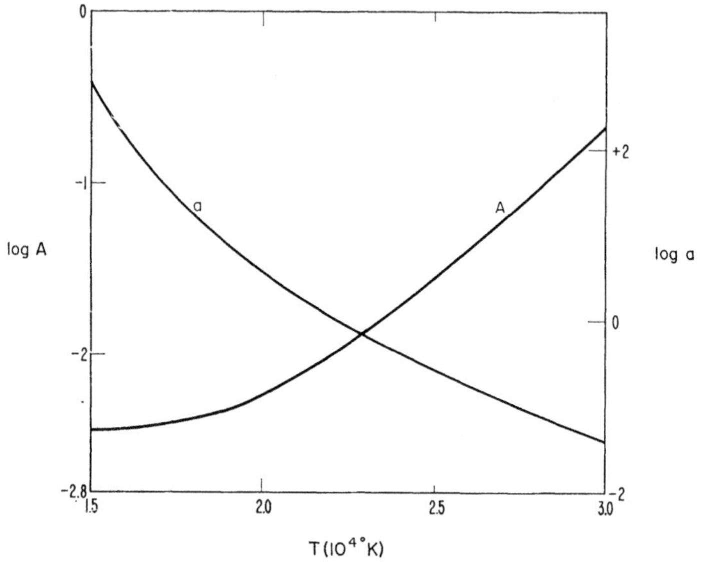

Fig. 6. As for Figure 5 but for $N_e^2 T_e^{-3/2} = 0.75 \times 10^{16}$ cgs units.

et al. (1954), to require $n_e^2 T^{-3/2} \simeq 1.5 \times 10^{16}$ cgs units. Strictly speaking this refers to the average over both hotter (helium emitting) and colder components, if such exist. If we wish to maintain pressure constancy at a given height, the value would be smaller in the hotter elements. We consider two cases, namely, where $n_e^2 T^{-3/2}$ has a value either 1.5×10^{16} or 7.5×10^{15} cgs units. For consistency with the requirements on τ_1 and $\mathbf{a} n_{3D}^3$ as expressed in columns 4 and 5 of Table 1, we then need values of **a** and A given, as functions of T_e, in Figure 5 and 6. Values of **a** greater than unity are meaningless; the inferred values of A, n_e and T_e for **a** $= 1$ and 0.1 are shown in Table 2.

TABLE 2

DERIVED ATMOSPHERIC PARAMETERS

	Case (a)		*Case (b)*	
	$a = 1$	$a = 0.1$	$a = 1$	$a = 0.1$
A	7.0×10^{-3}	3.6×10^{-2}	1.3×10^{-2}	8.5×10^{-2}
T_e	2.2×10^{4}	2.6×10^{4}	2.2×10^{4}	2.8×10^{4}
n_e	2.2×10^{11}	2.5×10^{11}	1.6×10^{11}	1.9×10^{11}

The inferred electron densities and temperatures are reasonably insensitive to variations in the fractional volume **a**, provided that this is held within such limits that the corresponding abundance A is reasonable. However this latter number is so poorly known that it cannot serve as a very useful guide.

In summary, this illustration has attempted to show how one could apply the general analytical procedures outlined in Section 2 to a practical case. The analysis has some interest in its own account suggesting as it does that the helium emitting regions of the chromosphere seem to be well characterized by a temperature of order 2.5×10^4 °K and an electron density at 1500 km of about $1\text{–}2 \times 10^{11}$ cm^{-3}. The value of **a** is not well determined; there is a suggestion that it may be significantly less than unity. A more complete analysis of He I is in progress but unfortunately could not be completed in time for inclusion here.

4. Self Absorbed Lines

A. GENERAL CONSIDERATIONS

A line formed in an optically *thick* region will display, through its *profile*, information on the atmospheric structure. The specific intensity, at wavelength λ, emerging at angle $\cos^{-1} \mu$ to the normal to a plane parallel atmosphere is given by Equation (3), viz

$$I_\lambda(0, \mu) = \int_0^\infty S_\lambda(\tau_\lambda) e^{-t_\lambda/\mu} \mathrm{d}t_\lambda/\mu \tag{14}$$

where

$$\begin{gathered} S_\lambda = \frac{S_l + r_\lambda S_c}{1 + r_\lambda} \\ \mathrm{d}t_\lambda = \mathrm{d}t_c + \mathrm{d}t_l(\lambda) \equiv \mathrm{d}t_l(\lambda)\,(1 + r_\lambda) \\ r_\lambda = \varkappa_c/\varkappa_l(\lambda) = r_0/\phi(\lambda), \quad \phi(\lambda) = \varkappa_l(\lambda)/\varkappa_l(0)\,. \end{gathered} \tag{15}$$

The symbols S_l and S_c represent the line and continuum source functions, $\mathrm{d}t_l(\lambda)$ and $\mathrm{d}t_c$ the elements of optical depth in the line (at wavelength λ) and continuum while $\varkappa_l(\lambda)$ and $\varkappa_c$ are the corresponding absorption coefficients.

The problem of the interpretation of the center-limb variation of the disk spectrum may then stated as follows: Given the observed distribution $I_\lambda(0, \mu)$, can we infer uniquely the distribution of physical parameters (temperature, density) necessary to produce this intensity, and if so, how do we do this? This problem is fourfold. First we must examine the limitations in principle to the inversion of Equations (14) and (15), i.e., whether the inversion is uniquely possible with arbitrary accuracy; second we must study the limitations imposed by observational inaccuracies or incompleteness; third we must devise a method for inferring what information we can on $S_\lambda(t_\lambda)$, and only then can we turn to the question of interpreting this in terms of physical variables. We shall consider each of these in turn.

B. THE ANALYTICAL PROBLEM

It is not difficult to show that Equations (14), (15) possess no unique solution for the unknowns S_l, S_c, r_0, $\phi(\lambda)$ even when the intensity is known with arbitrary accuracy. Thus suppose we were to be given $I_\lambda(0, \mu)$ at l values of λ and m values of μ; we have lm known quantities.

Let us assume, as the most favorable case, that S_l is independent of frequency – as suggested e.g. by THOMAS (1957). We then must determine S_l, S_c, r_0 and $\phi(\lambda)$ at m depths and $\phi(\lambda)$ at $(l - 1)$ frequencies – $\phi(\lambda) = 1$ at the line center – in order to satisfy exactly the given values of $I_\lambda(0, \mu)$. This is generally impossible however since there are then $m(l + 2)$ unknowns and only ml data.

A way out of this difficulty has been suggested theoretically – JEFFERIES (1960), and demonstrated by WADDELL (1962, 1963) from his fine observations, showing that the members of certain multiplets have nearly equal source functions at a given *physical* depth. Observations of the emergent intensities of a *pair* of lines of known relative strength and belonging to the same multiplet then doubles the information with no further increase in the number of unknowns and allows a unique determination in principle of the unknowns S_l, S_c, r_0 and $\phi(\lambda)$.

In practice $I_\lambda(0, \mu)$ is observed only over a limited range in μ – because of seeing and other difficulties – and only with a finite accuracy. It is clear that these observational limitations will severely restrict the information recoverable on the details of $S_\lambda(t_\lambda)$. In his discussion of this problem BÖHM (1961) has been able to derive quantitative estimates for the recoverable information. For example he has shown that with 1% accuracy in I_λ and a range of 1 to 0.2 in μ one can determine no more than three reliable pieces of information on $S_\lambda(t_\lambda)$.

The next step in the determination of $S_\lambda(t_\lambda)$ from Equation (14) has been discussed by many authors. PECKER (1963) has considered most of these in an appendix to his contribution to this Symposium, and it is not appropriate to reconsider them here. It seems that no generally satisfactory procedure is available as yet. In the author's experience, a fairly satisfactory approach has been to use an expansion procedure, for example,

$$S(t_\lambda) = a_\lambda + b_\lambda t_\lambda + c_\lambda t_\lambda^2 \tag{16}$$

in which case we find

$$I_\lambda(0, \mu) = a_\lambda + b_\lambda \mu + 2c_\lambda \mu^2 \tag{17}$$

and the unknown coefficients a_λ, b_λ, c_λ can then be found from a least squares fit of Equation (17) to the observed $I_\lambda(0, \mu)$.

As the final step in this analytical phase we need to determine from the inferred $S_\lambda(t_\lambda)$, the corresponding values of S_l, S_c, r_0 and $\phi(\lambda)$ via Equations (15).

Now consider the derived $S_\lambda(t_\lambda)$ at two equal wavelength separations from the line centers in each of two lines of a multiplet, (e.g. NaD$_1$ and D$_2$). Denote their values at a given *physical* depth by S_{11}, S_{12}, S_{21}, S_{22} the first subscript indexing the line, the second the wavelength separations.

By making use of the fundamental requirements that the line source functions S_l are independent of frequency and are equal at a given *location* – at least for D_1 and D_2 – we may eliminate first S_l and then S_c from the defining equations for the S_{ij} and derive, after some algebra, the following equations for r_{11} and r_{12}

$$\begin{aligned} a_1 r_{11} + a_2 r_{12} &= 1 \\ b_1/r_{11} + b_2/r_{22} &= 1 \end{aligned} \tag{18}$$

where

$$\begin{aligned} a_1 &= (S_{11} - S_{21})/D_1 \\ a_2 &= (S_{22} - S_{12})/D_1 \\ b_1 &= (S_{22} - S_{12})/D_2 \\ b_2 &= (S_{11} - S_{21})/D_2 \end{aligned} \tag{19}$$

and

$$\begin{aligned} D_1 &= (S_{12} - S_{11}) + g(S_{21} - S_{22}) \\ D_2 &= (S_{12} - S_{11}) + \frac{1}{g}(S_{21} - S_{22}) \end{aligned} \tag{20}$$

with $g = \varkappa_2(\lambda)/\varkappa_1(\lambda)$ $(= 2$ for the sodium D_1 and D_2 lines).

These equations only apply if the S_{ij} refer to the same physical location in the atmosphere; however we only derive the run of S_λ with the optical depth t_λ, and so we must determine what values of t_λ (the optical depth in line and continuum) refer to the same physical depth. It seems that this can only be done iteratively assuming e.g. as a first approximation that $r(\lambda)$ is independent of depth, solving the basic system (15), (18, (19), (20), to determine a second approximation and so on. As the Equations (18) for r_{11} and r_{12} are rather sensitive to changes in the S_{ij}, it is probably desirable to solve them for mean values through a least squares set at four or five depth values. In doing this one must, of course, keep within reasonable values of t_λ i.e., between about 0.2 and (say) 1.5 within which $S_\lambda(t_\lambda)$ is fairly well determined.

C. RELATION OF RADIATION PARAMETERS TO PHYSICAL VARIABLES

The most we can possibly hope to determine from the observed intensities are the radiation parameters S_l, S_c, r_0 and $\phi(\lambda)$; the preceding discussion outlines a method for effecting this. As the final stage in the interpretation of the emergent intensity we now attempt to relate these quantities to the atmospheric structure. Though each of the four quantities reflects some aspect of the atmospheric properties, we shall only study here the interpretation of the line source function S_l. We shall always expect to get information on this quantity whereas if r_0 is small both S_c and r will be poorly determined.

Much of the previous work on chromospheric line spectra has supposed the excitation to correspond to a state of local thermodynamic equilibrium in which case S_l is simply the Planck function at the appropriate local temperature. It is recognized now, however, that chromospheric densities are too low to produce, through collisions, a sufficient degree of coupling between the thermal and radiative modes to result in an L.T.E. distribution in the bound state populations.

In the general case the evaluation of S_l requires knowledge of the radiative and collisional transition rates between all levels so that we may determine the relevant population ratios entering S_l. The radiation field in any one line depends in a complicated manner on the gas temperature and density, on the incident flux in all other lines, on the location in and even the thickness of the atmosphere. It is, therefore, going to be a very involved problem indeed to attempt to interpret the derived run of S_l with depth in terms of an atmospheric structure. Indeed one is not entirely sure that a unique solution exists.

We cannot, as yet, make much claim to physical insight in this problem. THOMAS (1957) has made an important qualitative distinction according to whether a line falls under 'collisional' or 'radiative' control. All that is described by these terms is the dominant mechanism for the creation and destruction of photons in the line. For example the most likely way of producing D line photons in the solar chromosphere is through photoelectric ionization from the ground state followed by spontaneous recombination to the upper 3^2P state and then emission of D_1 or D_2. Similarly destruction of a D line photon goes most probably through the inverse transition sequence. In other words the source function for the chromospheric D lines is controlled by the strength of continuum radiation, i.e. by the radiation incident on the chromosphere. The line source function cannot, then, display the thermal structure of the chromosphere since this enters only through collisions and, for these lines, collisions are not significant in creating or destroying photons. For this reason the solar D lines are typical of Thomas' radiatively controlled lines.

In the opposite extreme only collisional processes are important in producing and destroying new photons. These 'collisionally controlled' lines – typified by solar Ca II H and K can and do reflect the thermal structure as is shown, for example, in the occurrence of the well known H and K line 'emission core'.

It is possible to realize states intermediate to the above two extremes but the essential point to recognize is that the chromosphere itself will have only secondary influence on the line excitation if the line is photoelectrically controlled. Correspondingly one will not expect to derive much information on the chromospheric structure from a detailed study of such a line.

While Thomas' dichotomy is very useful and clear for a simple atomic model, it becomes rather less clear when interlocking of comparable strength lines is significant; a resolution of this problem awaits the clarification of the interlocking problem.

In the much misunderstood case of a two level atom interlocking does not arise and the transfer equation is given by equation (11). Corresponding solutions in a ninth order approximation for S as a function of τ in the line center for $\lambda = 10^{-4}$ and 10^{-6} are shown for a semi-infinite atmosphere in Figure 7. The quantity B is assumed constant. The source function increases to its

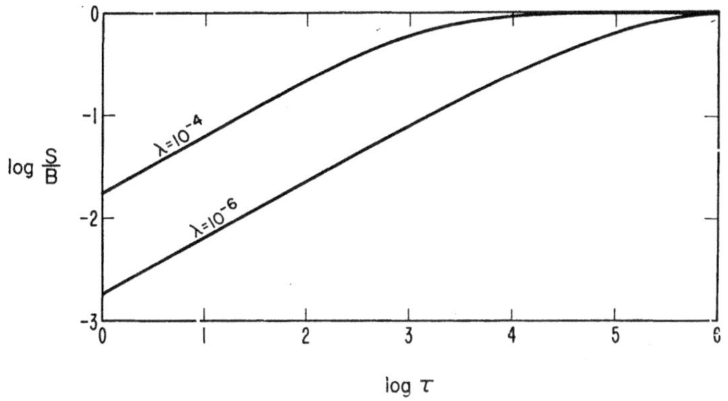

Fig. 7. Depth variation of the line source function for two representative values of the scattering parameter λ. The atmosphere is supposed semi-infinite.

saturation value, B, with a scale length in τ which varies between about $1/\sqrt{\lambda}$ at small τ to $\sim 1/\lambda$ at large τ. It may be shown indeed that asymptotically,

$$S(\tau) \sim (1 + 5 \times 10^{-2}/\lambda\tau)^{-1} . \tag{21}$$

Clearly the emergent line will be seen as an absorption line, while for a finite atmosphere the line is self reversed, as follows immediately from Figure (3). The physical reason for these effects is that the source function, being dependent on the population ratio, depends on the line intensity. Since this decreases towards the surface – from which radiation can escape – so will the population of the upper state of the line decrease and, with it, S_l. Thus we would expect the normal line emitted by a stellar atmosphere to be an absorption line, not because T decreases towards the surface (it is constant in our models) but solely because the atmosphere has a boundary. The long scale length S_l is characteristic of such diffusion problems – the photon has a small probability (λ) of destruction at each absorption so it makes a random walk of $\sim 1/\lambda$ steps (of variable length since its frequency is changed on scattering) covering a distance $> 1/\sqrt{\lambda}$ measured in units of τ in the line center.

This is all well known, but is has been criticized as being of little practical interest because of the limitation to a two level atom. Certainly it is quantitatively limited in application, however the physical basis of the argument can hardly be doubted; in any case it receives more than adequate qualitative support from the observation e.g. of the H and K reversals in stellar chromospheres which are an almost inevitable consequence of the theory.

The extension to interlocked lines has only been discussed in a very simple approximation – JEFFERIES (1960) – for a case resembling L_α and L_β in the sun. If we denote by s_α the relative source functions $S_{L\alpha}(\tau)/S_{L\alpha}(\infty)$ and define s_β similarly, then the computed ratio $R = s_\alpha/s_\beta$ is shown as a function of τ in Figure 8. Because of the strong interlocking due to H_α absorption and emission

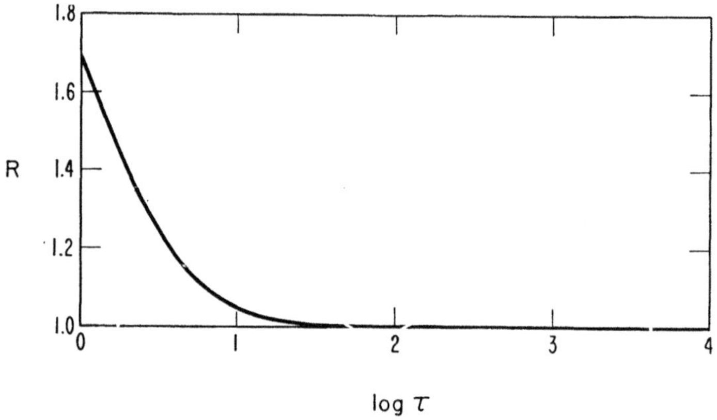

Fig. 8. Relative variation of two resonance line source functions; when $R = 1$ the two functions have the same depth variation.

transitions, the populations of n_1 and n_2 are closely coupled and, with them, are the corresponding source functions. For close lying multiplet lines one might anticipate an equality in the source functions and, as mentioned above, Waddell has indeed demonstrated this.

Before we can properly interpret spectral line profiles, however, we need much more detailed studies of the interlocking problem to relate the S_l in different multiplet lines. This in turn waits essentially on the derivation of fast and reliable methods for solving the general transfer equation either in its integral or differential form. Some of the problems involved in this were discussed at the 1962 IAU Symposium of Commission 36 and are under intensive study but, so far, without complete success.

In summary, then, the pertinent facts seem to be as follows: It has been shown by a series of investigators that the line source function is essentially frequency independent in most situations. This is an enormous advantage as it allows us to characterize a line by a single excitation parameter S at a given depth. Secondly we know that the relative source functions of two lines of a multiplet will, in general, bear a close and definite relationship to one another. Thus we can characterize two (or more) spectral lines by a single parameter S at a given depth. These two facts allow us to make a start, at least, on the analysis of line profiles.

5. Inhomogeneous Atmospheres

One need only look at a spectroheliogram to see that the solar atmosphere is not homogeneous. Any analytical procedure must then ultimately find some way of incorporating the presence of these inhomogeneities into the formalism. How this is to be done depends on whether the analysis is for an optically thick or an optically thin line since in the latter case we only observe the total integrated emissivity along the line of sight and this could be redistributed in any fashion without affecting the observed intensity. Clearly this is not so for the self absorbed or optically thick atmosphere where the emergent intensity depends critically on the arrangement of material along the line of sight. We shall content ourselves here with a few general remarks on the problem.

(a) *Optically thin atmosphere.* A simple way to incorporate inhomogeneities in this case is that used above – in discussing the inversion of Equation (6) – where the atmosphere is regarded as of two components each itself homogeneous with elements of each distributed statistically in the line of sight and occupying fractional volumes $1-\mathbf{a}$ and $\mathbf{a}$. As mentioned above we need not associate a characteristic scale length with the elements in this case since none enters directly into the analysis.

Various other one-parameter ways for describing the inhomogeneities are possible. For example JEFFERIES and ORRALL (1963) have obtained an approximate prominence model using an electron temperature distribution $f(T) \propto T^{-g}$ and using observational data to determine the exponent g.

(b) *Optically thick atmospheres.* The situation is quite different for the analysis of self absorbed lines formed in an inhomogeneous atmosphere. The basic Equation (14) of Section 4 cannot then be used, as the transformation from line of sight to normal optical depth is not a simple geometrical one. One must, presumably, again adopt a statistical distribution for the emission and absorption coefficients and compute the relevant moments of the emission from the atmosphere as a function of the parameters of the statistical distribution. Certain special cases along these lines have been discussed by GIOVANELLI (1963) and WILSON (1962).

Just how the development will proceed is uncertain; it seems more than likely, however, that in many cases the analysis of Section 4 will still be necessary as a first step to establish some sort of mean atmosphere, subsequent refinements to the model being in terms of statistical fluctuations from this mean state.

Clearly we need only as many parameters for the model as will satisfy the observations; it is trite, but necessary, to add that the latitude given by the possibility of including inhomogeneities in the model should not be abused. An observation can be discordant with a theoretical prediction (based on a given model) for three reasons; either the model is wrong, the theory is wrong, or the observations are wrong. A common tendency in astrophysics is to regard the above order as one of probability with very low doubt on the theory and observations. One may be forgiven, perhaps, for wondering whether the consequent multiplicity of atmospheric parameters could not be reduced were as much labor and ingenuity spent in devising an adequate theory as in devising a model.

References

Athay, R. G., Billings, D. E., Evans, J. W., and Roberts, W. O.: 1954, *Astrophys. J.* **120**, 94.

Athay, R. G., and Menzel, D. H.: 1956, *Astrophys. J.* **123**, 285.

Böhm, K. H.: 1961, *Astrophys. J.* **134**, 264.

Giovanelli, R. G.: 1963, *Progress in Optics* (ed. by E. Wolf), Vol. II, Chapter III, North Holland Publishing Co., Amsterdam.

International Astronomical Union Symposium Commission 36: 1962, *J. Quant. Spec. Rad. Transfer* **3**.

Jefferies, J. T. and Thomas, R. N.: 1958, *Astrophys. J.* **127**, 667.

Jefferies, J. T.: 1960, *Astrophys. J.* 132, 775.

Jefferies, J. T. and Orrall, F. Q.: 1963, *Astrophys. J.* **137**, 1232.

Pecker, J.-C.: 1963, this volume, pp. 29–88.

Seaton, M. J.: 1962, *Atomic and Molecular Processes* (ed. by D. R. Bates), Chapter 11, Academic Press Inc., London.

Thomas, R. N.: 1957, *Astrophys. J.* **125**, 260.

Thomas, R. N.: 1960, *Astrophys. J.* **131**, 429.

Trefftz, E., Schlüter, A., Dettmar, K. H., and Jorgens, K.: 1957, *Z. Astrophys.* **44**, 1.

Waddell, J. H.: 1962, *Astrophys. J.* **136**, 223.

Waddell, J. H.: 1963, *Astrophys. J.* **137**, 1210.

Wilson, P. R.: 1962, *Mon. Not. Roy. Astron. Soc.* **124**, 383.

THE CHROMOSPHERIC SPECTRUM AT ECLIPSE

R. GRANT ATHAY
High Altitude Observatory Boulder, Colorado

1. Evolution of Concepts and Techniques

An evaluation of our present understanding and knowledge of the chromospheric spectrum depends heavily upon our concept of the thermodynamical and geometrical properties of the chromosphere. Our knowledge of the spectral properties of the chromosphere still derives largely from observations made at total solar eclipse. Our concepts of geometrical structure, however, derive mainly from observations made outside of eclipse. The spectral and geometrical properties must be combined in order to formulate a meaningful thermodynamical picture of the chromosphere. The thermodynamical picture, in turn, modifies our concepts of the relevant geometrical and spectral properties.

Early observers of the chromospheric spectrum seen at eclipse regarded the chromosphere as an uncharted "no-man's land". Their observations were limited accordingly and usually consisted of a single spectogram obtained with an objective grating spectrograph. In this way they learned that the chromospheric spectrum in some ways is very similar to the photospheric spectrum and in other ways is very different; that it exhibited strong emission lines of helium; and that scale-heights of intensity differ from line to lines and are generally large relative to photospheric density scale-heights.

The realization that the chromospheric spectrum changes markedly with height, inspired a second era of observers to obtain several spectrograms in a time sequence or a single spectrogram with a moving image as the moon progressively covered (or uncovered) more and more of the chromosphere. The first successful attempt to obtain a sequence of spectra led PANNEKOEK and MINNAERT (1928) to the important discovery that the chromospheric helium line at λ 4 472 exhibits a maximum intensity in the low chromosphere, as subsequently has been discovered for other helium lines and the Balmer continuum as well.

Improvements in instrumental design and film speeds have made possible in chromospheric eclipse studies the introduction of both high dispersion spectrographs for the study of line profiles and fast spectrographs capable of much shorter exposure time intervals and correspondingly greater height resolution in the chromosphere. Recent trends in eclipse observations of the chromosphere have tended to take advantage of these improvements in instruments and films. Photoelectric techniques are just beginning to be used effectively in eclipse studies and hold great promise for the future.

C. de Jager (ed.), The Solar Spectrum. All rights reserved.

It is perhaps fair to say that existing eclipse observations of the chromosphere have not been fully exploited. In view of the sparsity of data and the difficulties surrounding eclipse observing, this is indeed an unfortunate situation. There exist many sets of eclipse spectra all of which contain far more information than has been extracted from them. This is particularly true of faint features of the spectrum. For example, we note that although many earlier eclipse expeditions successfully recorded the continuum spectrum of the chromosphere, this feature of the spectrum was first studied seriously at the 1952 eclipse. It is imperative that we make fuller use of eclipse data and make it more readily available to all who are interested.

Although the central theme of this paper is concerned with eclipse observation of the chromospheric spectrum, a few general remarks seem appropriate.

The chromosphere is visible in projection against the solar disk in only a relatively few lines, and, then, only near the centers of these lines. Nevertheless, this type of observation of the chromosphere is far simpler to execute and can be accomplished far more frequently than successful eclipse observations. In spite of this advantage in relative ease of observing, disk studies have not been vigorously pursued for purposes of defining the chromospheric spectrum. Much of this inactivity can be attributed to the inherent limitations on the number of Fraunhofer lines that can be identified as chromospheric and to the inherent difficulties of separating out the photospheric contribution that is superposed on the chromospheric line. Part of the inactivity in this type of observing however must be attributed to the lure of eclipse chasing.

Disk observations of the chromosphere are characterized by a complexity of structures distinguished by their different brightnesses and their motions. Structures revealed by brightness fluctuations have been recognized for many decades, but have been studied only rarely. Structures in the velocity field have been known to exist for many years. Only in the last few, however, has their study been pursued.

A large number of chromospheric lines can be observed at the limb by the simple expedient of placing a spectrograph slit just beyond the limb on a solar image formed under conditions of good seeing with an ordinary solar telescope. Again, this technique of studying the chromospheric spectrum has not been pursued vigorously, in this case, because of the complexities introduced by stray photospheric light scattered into the chromospheric image by the earth's atmosphere and by the telescope. Coronagraphs offer some improvement but not so much as might be hoped for.

Limb observations made outside of eclipse make their greatest contribution to chromospheric studies in isolating and defining the properties of the elusive spicules, which gave rise to the "burning prairie" described by Secchi in the last century. The recognition of spicules and their implications in terms of chromospheric geometry are forcing a completely new interpretation on the spectrum and thermodynamics of the upper and middle chromosphere, the full impact of which is only now beginning to be understood.

The terrestrial atmosphere is opaque to wavelengths less than about $\lambda 2800$, and many of the more interesting features of the chromospheric spectrum have long been shielded from our observation. Rocket and satellite spectroscopy from above the atmosphere are revealing a wealth of new features of the far ultraviolet chromospheric spectrum, as observed against the disk, that offer unique opportunities for understanding the thermodynamics of the chromosphere. Indeed, it seems safe to conclude that these new observations will stimulate a new interest in chromospheric physics already partially aroused during the past two decades by the discovery of spicules and by the advancing concepts of departures from thermodynamic equilibrium in the outer solar atmosphere.

Still another advancement in observational techniques has and will continue to stimulate interest in chromospheric problems. Solar radio noise at wavelengths less than a few centimeters originates mainly in the chromosphere. This extension of the spectrum beyond the photographic regions into radio wavelengths offers still more opportunities for studying unique features of the chromosphere and their associated effects.

Finally, we note that chromospheric radiation includes energetic particles emitted in association with flares. The advent of "man in space" programs and the discovery of polar proton events has greatly increased interest in corpuscular radiation from the sun. Emphasis must be placed on the proper understanding of such transient chromospheric phenomena as flares and plages. This understanding must await a rather complete knowledge of flare and plage spectra, which currently are known only partially and for only small isolated spectral bands. We recognize that flare and plage spectra will be difficult to interpret even when information is plentiful, and that the spectra may vary considerably from point to point within the flare or plage. This difficulty should not deter our attempts to obtain such spectra, however. It may well be that the study of these transient features will aid materially in our understanding of the quiet solar atmosphere.

Spectra obtained by rockets in the far ultraviolet, at radio wavelengths, of active phenomena and of energetic particles form the subject of other papers and will not be further elaborated upon here.

2. Chromospheric Geometry and Boundaries

A. THE BASE OF THE CHROMOSPHERE

For convenience and for explicitness, we define the lower boundary of the chromosphere as the spherical surface for which the tangential opacity, τ_t, through the solar atmosphere is unity at $\lambda 5000$. The radial opacity, τ_r, and geometrical height of this surface with respect to a reference surface at, say,

$\tau_r = 1$ depend upon the opacity gradient in the atmosphere. If we adopt a mean scale height of 60 km for τ_r at $\lambda 5000$, we find $\tau_r = 0.0037$ at $\tau_t = 1$ and a height separation of 335 km between the surfaces $\tau_r = 1$ and $\tau_t = 1$.

At heights above the base of the chromosphere it is convenient to relate eclipse data directly to the geometrical height above some point defined with respect to the continuum intensity and/or its derivatives. If the scale height of τ_r and the continuum source function are both constant near the surface $\tau_t = 1$, we find that $\tau_t = 1$ at $\lambda 5000$ corresponds to

$$\frac{d^3 E_{5000}}{dh^3} = 0 ,$$

where

$$E_\lambda = \int_h^\infty \int_0^\tau S_\lambda e^{-\tau_t} d\tau_t dh \tag{1}$$

and

$$\tau_t = \tau_t(0) e^{-\beta h} . \tag{2}$$

Hence, in a plot of E versus height, as shown in Figure 1, the point of inflection marks the surface $\tau_t = 1$ if the foregoing assumptions are satisfied. The heights in Figure 1 correspond to those adopted from preliminary data and have not been altered to fit the data plotted. We note from Figure 1 that the point of inflection clearly falls below $h = 100$ km and is probably in the interval $0 \leqslant h \leqslant 100$ km.

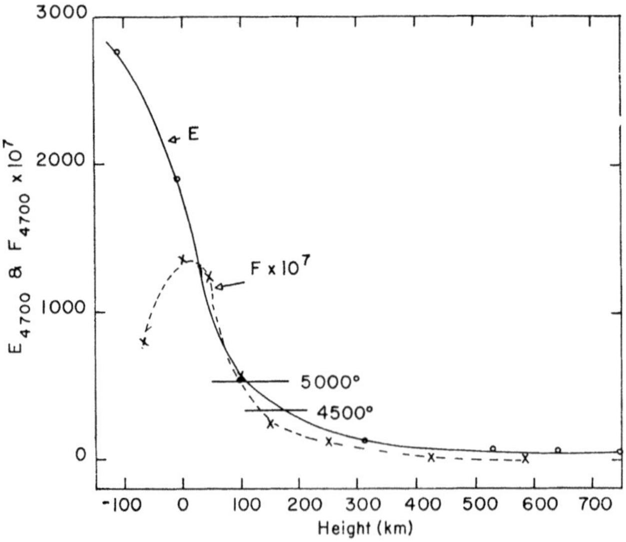

Fig. 1. Determination of the surface $\tau_t = 1$ from the point of inflection in plots of E versus height and from the magnitude of F (1952 HAO data).

A further check on the location of $\tau_t = 1$ can be made by computing expected values of

$$F_\lambda = -\frac{dE_\lambda}{dh}. \tag{3}$$

If S_λ is constant with τ,

$$\begin{aligned} F_\lambda &= S_\lambda(1 - e^{\tau_t}) \\ &= 0.632\ S_\lambda \quad \text{at} \quad \tau_t = 1\,. \end{aligned} \tag{4}$$

A plot of the empirical F_{4700}, which is derived from E_{4700}, is given in Figure 1 along with computed values of $F_\lambda(\tau_t = 1)$ assuming $S_\lambda = B_\lambda$. At heights below $h \leqslant 50$ km, the empirical F_λ lie systematically above the computed values for 0.632 B_λ (5000°) and 0.632 B_λ (4500°) by factors of about two and three, respectively. Within about 120 km below the height where $\tau_t = 1$, we expect F_λ to increase by about a factor $1/0.632 \approx 1.6$. Since the temperature at $\tau_t = 1$ is almost certainly below 5000°, we may conclude that the absolute intensities given by the 1952 eclipse data are too high by a factor of about two, as has been previously suggested (THOMAS and ATHAY, 1961, Chap. 2) from other indications.

If we allow for a factor of two reduction in the empirical F_{4700}, both the magnitude of F_{4700} and the sharp drop in F_{4700} in Figure 1 suggest that the best location of $h = 0$ is within the height range designated as 50–100 km. In summary, then, absolute intensities in the 1952 data seem to be too high by a factor of about two and the assigned heights seem to be in error by about +75 km. Similar checks on the precision of other chromospheric eclipse data are lacking unless one makes the perhaps dubious assumption that chromospheric intensities are independent of the sunspot cycle.

Data for the 1962 eclipse obtained by the joint expedition of the Sacramento Peak Observatory, High Altitude Observatory, and National Bureau of Standards will soon be available. These data will be more complete in wavelength coverage than any existing data of comparable height resolution and more accurate photometrically than the 1952 data. A more precise determination of the reference height $h = 0$, should be possible with the 1962 data.

B. THE UPPER CHROMOSPHERIC BOUNDARY

The upper boundary of the chromosphere is not conveniently defined in terms of either τ or h. Instead, we define it as the upper limit of observable emission in the spectral lines of either neutral or singly ionized atoms, exclusive of prominences. Most of the spectral lines fade into the continuum below $h =$ 1000 km. At 3000 km, there are only some few dozen lines observed in the visual spectrum and these consist mostly of hydrogen and helium lines. Only the very strongest of the metal lines persist to this height. At 5000 km, only about a dozen lines of hydrogen, helium and ionized calcium can be observed in the visual spectrum.

At about 3000 km, the chromosphere is relatively easily resolved into discrete structures, and at 5000 km and higher spicule structure is readily recognized. A few spicules persist to some 15000 to 20000 km, whereas, at a given time, others may terminate near 3000 km.

We consider the spicules as part of the chromosphere and draw the upper boundary of the chromosphere accordingly. So far as is known, there is no observable chromospheric-type emission originating between spicules. On the other hand, there is good evidence that coronal line emission occurs down to at least 7000–8000 km and perhaps lower (Athay and Roberts, 1955). We visualize the spicules, then, as penetrating the corona. In fact, there is reason to believe that all interspicule spaces above about 3000 km are coronal in character, and should not be included in our definition of the chromosphere. If we adopt this point of view, then the chromosphere above about 3000 km consists solely of spicules.

C. GENERAL COMMENTS ON GEOMETRICAL IRREGULARITIES

Chromospheric irregularities other than spicules are known to exist. All strong Fraunhofer lines exhibit some form of brightness mottling on the solar surface that most probably are related, in some way, to inhomogeneities in temperature and density. These brightness mottles, in most cases, are too large and too numerous to be spicules. Furthermore, radio data at mm wavelengths require an irregular surface in the lower chromospheric layers where they reach optical depths near unity, and, again, these irregularities must be more numerous than spicules. (Coates, 1958; Thomas and Athay, 1961, Chap. 8). In view of these indications of inhomogenieties, it is helpful to consider the circumstances under which discrete chromospheric features can be observed at the limb.

Let us assume that there is a spherical surface concentric with the sun and that discrete features protrude from it into the higher atmosphere. We assume also that these discrete features have the following properties: (a) their vertical extent above the reference sphere is at least 1000 km; (b) they have similar spectral properties; and (c) they are numerous and either uniformly or randomly distributed. If we then specify the total number, N, and lateral dimensions, D, of the features, we can determine the average number, n, of the features intersected by a given ray tangent to the reference sphere.

A cylinder tangent to the reference sphere will intersect all features of at least 1000 km vertical extent that are located on the solar surface within 38000 km of the circle of tangency. It follows that over 5.4% of such features will be intersected by the tangent cylinder. The number of such features in a uniform distribution intersected by a ray tangent to the reference sphere will be the same as the number contained in an arc whose length is the same as the diameter of the features. Hence, we will have

$$n = \frac{0.054\,ND}{\pi R_\odot}\,. \tag{5}$$

If the features are randomly distributed over the solar surface, there will be some statistical clumping and spreading of features so that averages are not particularly meaningful. For this case, we compute the number of arcs of length D along the solar limb for which no features appear. This will correspond to the number of groups of features, S, that will appear at the limb as distinct entities, and is given by (THOMAS and ATHAY, 1961, Chap. 3)

$$S = 0.054\,N\,e^{-0.108\,ND/\pi R}\,. \tag{6}$$

Values of n and S are given in Table 1 for assumed values of N and D. If the distribution is uniform we would not expect to observe features when $n \geqslant 2$, and if the distribution is random we will not observe features resembling those actually present unless $S/0.054\,N \gtrsim 0.1$. We note from Table 1, then, that

TABLE 1

VISIBILITY OF CHROMOSPHERIC FEATURES AT LIMB

Fraction of Sun Covered by Features	*N*	*D* (*km*)	*n*	*S*
0.001	10^4	900	0.22	430
	10^5	300	0.7	1200
	10^6	90	2.2	470
0.01	10^4	2700	0.7	120
	10^5	900	2.2	47
	10^6	270	7	1
0.1	10^4	9000	2.2	4.7
	10^5	3000	7	0
	10^6	900	22	0

features covering 0.1 % of the solar surface and having a true diameter of more than 0.5″ of arc are likely to be observed as such. On the other hand, features covering 1 % of the surface are not likely to be observed unless their true diameter exceeds about 3″ of arc and features covering 10 % of the surface are not likely to be observed unless their diameter exceeds about 15″ of arc.

We may infer from this that at heights where we first clearly recognize such phenomena as spicules the total area of the sun covered by these features is less than 1 % of the solar surface area. If we adopt a diameter of 1″ of arc for spicules, they must number fewer than 10^5 at the lowest heights where they are distinctly observed. At 5000 km in the chromosphere, for example, spicule statistics give $S \approx 1200$, $N \approx 5 \times 10^4$, and a fraction 0.002 of the solar surface covered by spicules.

One consequence of the foregoing considerations is that major irregularities in chromospheric structure in many cases will not be observable at the limb. The types of mottling observed in the Fraunhofer lines, for example, are necessarily obscurred in limb observations unless they form a very minor component of the structure.

Center-to-limb observations at 8.6 mm wavelengths show only small changes in brightness temperature for $R < 0.98R_{\odot}$. In a spherically symmetric solar atmosphere the opacity is proportional to $\cos^{-1}\theta$ and to λ^2. Hence, at $R = 0.98R_{\odot}$, $\cos^{-1}\theta = 5$, the brightness temperature at 8.6 mm should be the same as the brightness temperature at the center of the disk at 1.9 cm. Observations at 1.9 cm are not available, but at 3 cm the central disk brightness temperature is about 16000–19000° compared to 8500° at 8.6 mm. It seems necessary then, that we drop the assumption of spherical symmetry. When this is done, we find that in the layers where the 8.6 mm waves originate the chromosphere must be about equally divided between hot, low density structures and cold, high density structures (THOMAS and ATHAY, 1961, Chap. 8). These structures will therefore not be observable at the solar limb as discrete features.

D. SPECIFIC COMMENTS ON GEOMETRICAL IRREGULARITIES

From the base of the chromosphere to about 750 km, current models (THOMAS and ATHAY, 1961, Chap. 6) indicate that hydrostatic equilibrium, and spherical symmetry are good first approximations. Geometrical irregularities undoubtedly exist, as evidenced by strong Fraunhofer lines, but they evidently do not seriously influence our interpretations of the chromospheric spectrum.

Near 1000 km helium line data (ATHAY, 1963) and Mg line data (ATHAY and HOUSE, 1962) clearly indicate that marked departures from spherical symmetry are present. The helium lines appear quite suddenly in the spectrum with strong intensity. Yet no other lines in the visual spectrum show either a pronounced decrease or increase in intensity associated with this sudden appearance of helium lines. This must mean that near 1000 km helium emission arises in volume elements that form a small fraction of the total volume and that emission from other elements arises, at these heights, in the remaining volume. In this same height range near 1000 km, hydrostatic equilibrium apparently breaks down sharply, as applied to a spherically symmetric atmosphere (THOMAS and ATHAY, 1961, Chap. 6). This apparent break down of hydrostatic equilibrium, evidenced by an artificial sharp rise in temperature in models derived from continuum data, is not eliminated by introducing departures from spherical symmetry, which are known to exist. In fact, an atmosphere departing from spherical symmetry in temperature and density cannot be in strict hydrostatic equilibrium.

In the height range 1500 to 3000 km, strong departure from spherical symmetry occur, but their character remains largely uncertain. SUEMOTO (1962) reports that spicule structure is distinguishable in Sr II lines on 1958 eclipse plates as low as 1500 km. If this is true, then the combined fractional areas of the polar surface covered by spicules and any other small scale features with spectral properties similar to spicules cannot exceed more than about 1% at 1500 km. Hence, Suemoto's observations become of crucial importance in the interpretation of chromospheric spectra in the height range 1500–3000 km.

At 3000 km and above, DUNN's (1958) observations in H-α show unmistakable spicule structure and, again, we must conclude that spicules cover less than 1% of the solar surface at these heights. Since the interspicule regions above 3000 km show no evidence of a chromospheric type spectrum, we infer that eclipse spectra of the chromosphere above 3000 km refer solely to spicules.

Interpretations of absolute and relative intensities of lines, of their profiles and of their intensity gradients must be made in terms of the geometrical properties of the chromosphere. Hence, below about 750 km present evidence indicates that we may interpret the spectra in terms of a spherically symmetric atmosphere with reasonable expectations of success. From about 1000 km to 3000 km an essentially unknown geometry must be introduced. Above 3000 km we may rely on spicule statistics to define the chromospheric component of the atmosphere.

Intensity gradients of chromospheric lines above about 1000 km reflect geometrical properties of the atmosphere as well as thermodynamic properties. Above 3000 km, the fractional volume of the atmosphere occupied by spicules decreases, and this factor alone may account for part or all of the observed decrease of line intensities with height (cf. ATHAY, 1963).

In equatorial regions of the sun, to which most eclipse data apply, spicule numbers decrease by a factor of four in the height range 3000 to 6000 km. Line intensities observed at eclipse are proportional to the total spicule volume exposed beyond the lunar limb if the spicules are optically thin. Since spicules have considerable vertical extension and, in general, some inclination to the radial direction, their volume decreases considerably faster with height than does their number. Although accurate statistics are not available from which we can evaluate these effects on spicule volume, the available statistics suggest that in equatorial regions the volume of spicule material decreases about 1.5 to 2 times as fast as spicule numbers. Thus, in the interval 3000 to 6000 km, the volume of spicule material decreases by a factor of 6 to 8. By comparison, the intensity changes by a factor of about 25 in Hα, 80 in Hβ, 40 in D_3 and 8 in the K-line. The spicules are probably thin in Hβ and D_3, but they have significant opacity in Hα and relatively high opacity in the K-line (ATHAY, 1961). Thus, we conclude that the changes in Hβ and D_3 intensity over this 3000 km interval are about a factor 8 larger than can be accounted for in terms of volume of spicule material exposed beyond the moon. This would correspond to a true intensity scale height of about 1400 km in these lines, as compared to an observed scale height of about 700 km.

3. The Chromospheric Spectrum at Eclipse

A. GENERAL REQUIREMENTS ON DATA

Eclipse observations of the chromosphere, upper photosphere and corona are made in a variety of ways and for a variety of purposes. The ultimate aim of all such observations is to provide accurate values of the quantity

$$F(\lambda, h, \phi) = \int_0^{\tau_{max}(\lambda, h, \phi)} S(\lambda, h, \phi) e^{-\tau(\lambda, h, \phi)} d\tau(\lambda, h, \phi) \tag{7}$$

where ϕ is the solar latitude. With slitless spectrographs, one measures

$$E(\lambda, h', \phi) = \int_{h'}^{\infty} F(\lambda, h, \phi) dh \tag{8}$$

where h' is the projected height of the lunar limb. With a slit spectrograph, one measures

$$E(\lambda, h_1 - h_2, \phi) = E(\lambda, h_1, \phi) - E(\lambda, h_2, \phi) = \int_{h_1}^{h_2} F(\lambda, h, \phi) dh \tag{9}$$

where h_1 and h_2 are the projected heights of the slit jaws on the solar image.

Scale heights of total intensities in chromospheric lines range upwards from about 200 km, and scale heights for significant features in line profiles may be even smaller. Eclipse observers usually experience seeing of a few seconds of arc at best. Hence even with observations obtained with a slit spectrograph reliable values of $F(h)$ can be obtained only by differentiating the observed quantities $E(h_1 - h_2)$.

When spicule-like structure is present and the scale size in the h, ϕ plane is small, additional corrections must be made for the diffusion disk caused by seeing. Such corrections have not been made in eclipse data thus far obtained. Instead, we have averaged data over several chromospheric features and introduced the ϕ dependence through an average parameter that is later evaluated from other considerations or, more frequently, guessed at.

Several observations of $E(\lambda, h, \phi)$ have been made for different values of ϕ corresponding to large scale features either on the lunar limb or on the solar limb. These data seem to indicate that at a given time in solar equatorial regions the chromospheric spectrum has essentially a unique character when averaged even about 1° of solar latitude or, equivalently, over about 20″ of arc subtended at the earth in the ϕ coordinate. Several outstanding exceptions to this generalization have been noted in the literature, however. (cf. THOMAS and ATHAY, 1961, Chap. 3 and 7.) These exceptions are clearly evident to visual inspection of eclipse spectrograms.

Almost all eclipse data on the chromospheric spectrum refer to latitudes near the solar equator. Very little is known either quantitatively or qualitatively about the chromospheric spectrum at the poles.

Attempts to measure $F(\lambda, h)$ with sufficient precision to give reliable line profiles still leave a great deal to be desired. In fact, we have only a few observations of $E(\lambda, h)$, which lack the necessary accuracy to derive $F(\lambda, h)$.

It would seem that the determination of $F(\lambda, \phi)$ in the chromosphere observed at the limb can be very greatly improved through observations made outside of eclipse with large solar telescopes and spectrographs now available. Eclipse observations must be continually stressed, however, because much of the chromospheric spectrum is beyond the observing capabilities of even the best telescopes at permanent installations.

Finally, there is a great need for tying chromospheric eclipse data solidly to disk data. We have already alluded to this with respect to the continuum data in Figure 1, where we noted that absolute intensities in 1952 eclipse data seemed to be too high relative to disk by about a factor of two. The problem can be further illustrated as shown in Table 2. Columns 3 and 4 contain values

TABLE 2

COMPARISON OF CENTRAL INTENSITIES* ON DISK AND ABOVE THE LIMB

Line	*Mean ht. km*	$E(\lambda)$	$F(\lambda)$	$\Delta\lambda$ *(approx)*	$F_0(h)/F_0(\mu=1)$	μ	$F_0(h)/F_0(\mu)$
Hα	1500	4.2×10^{14}	1.3×10^{6}	2	1.5	0.2	2.2
Hβ	1000	3.2×10^{14}	2.0×10^{6}	1.5	2.4	0.2	4.5
H	1000	8×10^{14}	4.8×10^{6}	2	18	0.2	22
λ8542 Ca II	1000	8.8×10^{13}	1.0×10^{6}	2	1.4	—	—
λ4078 Sr II	1000	1.2×10^{13}	1.4×10^{5}	0.5	0.6	—	—
λ5184 MgI	1000	1.4×10^{13}	1.6×10^{5}	0.4	1.1	0.04	2.2

* Intensity units are ergs sec^{-1} $d\omega = 1$ $d\lambda = 1Å$. F_0 $(\mu = 1)$ and F_0 (μ) obtained from WHITE (1962); GOLDBERG, MOHLER, and MÜLLER (1959); WADDELL (1963).

of $E(\lambda)$ and $F(\lambda)$ at heights given in Column 2. Values of $F(\lambda)$ are obtained by differentiating curves of $E(\lambda, h)$ given by 1952 eclipse data. Column 5 gives an estimate of the approximate width of the line profile, assuming that they are rectangular in shape. Columns 6 and 8 give ratios of intensities at line center at the heights indicated in Column 2 to the intensities at $\mu = 1$ and the values of μ given in Column 7.

Chromospheric opacity is expected to be high in each of the lines included in Table 2. Thus, we expect a central intensity in the chromosphere not too different from that on the disk. We note that with the exception of the λ4078 line the central intensities in the chromosphere are consistently above those near the edge of the disk. If we reduce the chromospheric intensities by a factor of two, there is rather good agreement with disk intensities, except for the H line results, which appear to be anomalous. Indeed, if we compare 1936 eclipse data with 1952 data (ATHAY, MENZEL and ORRALL, 1957) there is a clear indication that the H line intensity is too high in 1952 data by about a factor of 10.

We again conclude, then, that absolute intensities in the chromospheric spectrum given by High Altitude Observatory's 1952 eclipse data are systematically high by about a factor of two. By comparison, HOUTGAST's (1957) data for the 1952 eclipse are about a factor 1.25 below the HAO intensities. Absolute intensities for 1932 and 1936 eclipse data derived by ATHAY, MENZEL and ORRALL (1957) are about a factor of 3 to 4 below the HAO 1952 intensities. Data given by VYAZANITSYN (1951, 1952) for the 1941 and 1945 eclipses, while showing considerable scatter, give absolute intensities that are about a factor of 2 below the HAO 1952 intensities and, presumably then, are in good agreement with disk data.

As noted in the foregoing discussion, the accuracy in eclipse data – that is of first importance – is the accuracy with which $F(\lambda, h, \phi)$ can be derived. There are no eclipse data yet available for which one could justifiably claim that F is known at a given λ, h, and ϕ to an accuracy better than a factor of 2. Such accuracy requires considerable precision in the measurement of absolute intensities and even greater precision in the measurement of relative intensities. In addition, the observational data must have a height resolution that is smaller than the effective scale length of the feature being observed. Data from the 1958, 1961 and 1962 eclipses offer hope for substantially improving both the precision of F at a given λ, h and ϕ and the variation of F with these three variables.

B. CONTINUUM

The continuum spectrum of the low chromosphere arises from a variety of sources including the free-bound and free-free emission of H^- and neutral hydrogen, and scattering of photospheric radiation by hydrogen atoms and free-electrons. The relative contributions of each component vary markedly with height and wavelength. As a result, the continuum spectrum of the low chromosphere is rich in useful information. In spite of the fact that the chromospheric continuum is relatively easily observed at eclipse, eclipse observers have paid little attention to it. Reliable quantitative data are still scarce. However, recent expeditions have concentrated on this problem and hold promise of considerable improvement in the data available.

Throughout the visible spectrum the chromospheric continuum resembles the photospheric continuum in color. This information, however, is based on only one observation of $E(\lambda)$ at a single height (530 km) in the chromosphere. Reliable values of $F(\lambda)$ can be obtained only if good data are available for

$E(\lambda, h)$. At present the only detailed data available for $E(h)$ are at wavelengths near $\lambda 4700$ and near the Balmer series limit at $\lambda 3647$. Even at these wavelengths, the data are much in need of improved accuracy in both absolute and relative intensities.

Below about 500 km, the continuum near $\lambda 4700$ is dominated by H^- free-bound emission. At about 800 km, H^- free-bound, electron scattering and Paschen free-bound emission each contribute about one-third of the total emission at $\lambda 4700$, and above 1 500 km electron scattering dominates. Free-free emission and Rayleigh scattering by neutral hydrogen atoms contribute less than 10 percent to the observed continuum at any height in the chromosphere (cf. THOMAS and ATHAY, 1961, Chap. 6).

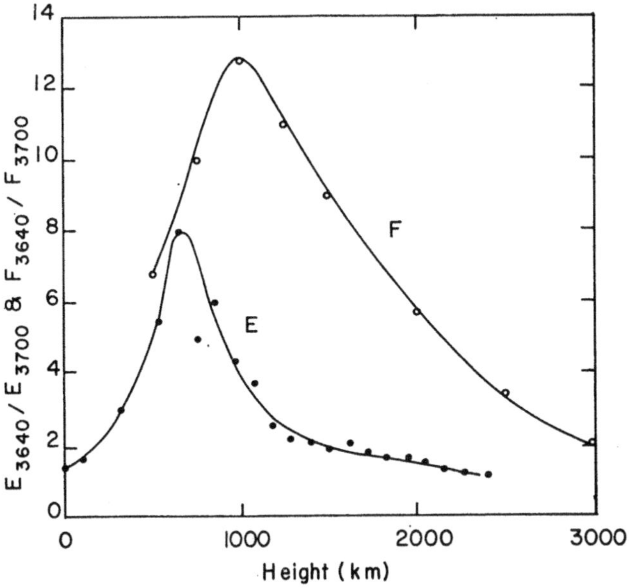

Fig. 2. The Balmer jump as observed in E and F at different heights in the chromosphere (1952 HAO data).

Balmer free-bound emission is added to the foregoing sources of continuum for $\lambda < 3647$, and is a conspicuous feature of the spectrum in the low chromosphere. Figure 2 exhibits plots of E_{3640}/E_{3700} and F_{3640}/F_{3700} as a function of height. We note that the Balmer jump, when observed in E, has a pronounced maximum of about 6–8 in the height interval 600–800 km, and when observed in F a maximum of about 13 occurs at a height near 1 000 km. The curve of F_{3640}/F_{3700} is obtained from smooth curves fitted to the observed values of $E(h)$.

Below 100 km and above 2000 km the Balmer jump as evidenced by

E_{3640}/E_{3700} is weak. The Paschen jump is relatively inconspicuous at all heights in the chromosphere and has not been reported by eclipse observers. The reason for this follows directly from the observed Balmer jump. The ratio E_{8200}/E_{8400}, which measures the Paschen jump, is given by

$$\frac{E_{8200}}{E_{8400}} = \frac{E_{3640}}{E_{3700}} \frac{E_{3700}}{E_{8400}} \frac{E_{8200}}{E_{3640}} + 1 . \tag{9}$$

For E_{3700}/E_{8400}, we may use the observed value for the photospheric spectrum of about 0.2, given in ergs per unit frequency (cf. MINNAERT, 1953), and for E_{8200}/E_{3640} we may use the theoretical ratio at the series limits given by $2^3/3^3$. With these values, a Balmer jump of 6 corresponds to a Paschen jump of 1.3. Thus, we expect the continuum intensity to jump a maximum of only 30 percent at the Paschen series limit. Throughout most of the chromosphere the Paschen jump will be still smaller.

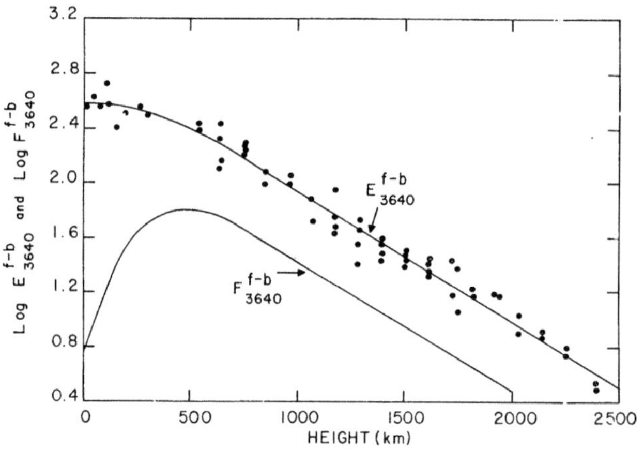

Fig. 3. The maximum in F^{f-b}_{3640} (arbitrary units) in the low chromosphere as derived from observed values of E^{f-b}_{3640} (1952 HAO data).

By subtracting F_{3700} from F_{3640} we obtain just the Balmer free-bound contribution to the continuum at $\lambda 3640$, which we label F^{f-b}_{3640} and plot in Figure 3. Near 400 km F^{f-b}_{3640} exhibits a sharp maximum, so sharp, in fact, that values of F^{f-b}_{3640} obtained below 400 km are unreliable. Since

$$F^{f-b}_{3640} \propto \int_0^\infty n_e n_p T_e^{-3/2} \mathrm{d}x \tag{10}$$

where n_e and T_e are electron density and temperature, n_p is proton density and x is the line-of-sight coordinate, the decrease in F^{f-b}_{3640} above 400 km is readily explained by a decrease of n_e and n_p with height. The rapid increase of $F^{\text{f-b}}_{3640}$ from 0 to 400 km, however, can only result from a strong increase in n_p with height that off-sets the decrease in n_e and $T_e^{-3/2}$. This can, and will, happen if the ionization of hydrogen increases rapidly with height, which, in turn, requires an increase of temperature with heights.

The electron temperature at 500 km can be obtained from the continuum intensities at $\lambda 3640$ and $\lambda 4700$. The H^- emission, which dominates at this height at $\lambda 4700$, is a strong function of temperature, so the determination of T_e is relatively precise. We obtain from this method $n_e = n_p = 5 \times 10^{11}$ cm^{-3} and $T_e = 6250°$ (THOMAS and ATHAY, 1961, Chap. 6). An independent determination of T_e at 500 km has been made from the spectral distribution of the continuum intensity. This method gives $T_e \approx 6000°$ (ATHAY, 1959). By comparison, at $h = 0$, the temperature is near 4500°, $n_e = 2 \times 10^{12}$ and $n_H = 2 \times 10^{16}$ cm^{-3}. This yields $n_p = 4.5 \times 10^9$. The product $n_e n_p T_e^{-3/2}$ is therefore 3×10^{16} at $h = 0$, and 5×10^{17} at 500 km. Hence, the absolute intensities of the Balmer continuum and the continuum at $\lambda 4700$ at 500 km confirms the indicated increase of $F^{\text{f-b}}_{3640}$ with height in the lowest layers of the chromosphere. It is noteworthy, however, that the maximum in $F^{\text{f-b}}_{3640}$ has been reported in only the 1952 eclipse data.

The acquiring of carefully calibrated values of $E(\lambda, h, \phi)$ and $F(\lambda, h, \phi)$ for the continuum in the chromosphere and at the extreme edges of the disk certainly forms one of the most pressing problems of eclipse observing.

In the higher layers where spicule structure is pronounced electron scattering is the main source of continuum and the observed continuum may be weighted heavily in favor of the inter-spicule regions. Spicules occupy less than one percent of the emitting volume, and unless their electron density exceeds that in the inter-spicule regions by more than a factor 10^2, which is unlikely, they will contribute less than half of the observed emission.

C. HYDROGEN

Above the first few hundred km of the chromosphere the hydrogen Balmer lines and continuum are the outstanding features of the spectrum. A few lines of helium and Ca II are their only competitors. It is also true that no other lines have been studied in as much detail as have the hydrogen lines. Even so, there is still much to be done to improve our knowledge of the chromospheric hydrogen spectrum.

Prior to the 1962 eclipse data, which will be available shortly, we have had no reliable data for intercomparing Balmer and Paschen lines. Or, for that matter, we have not had a completely reliable comparison of early and late members of the Balmer series. The wavelength intervals are large and the intensity range through the Balmer series approaches a factor of 10^4. Both of these factors tend to make accurate photometric comparisons difficult.

Chromospheric opacities at line centers range from very high at Hα to very low in the later Balmer lines. Hence, the Balmer lines cover the transition region from high to low opacity, and efforts to evaluate the chromospheric opacity have centered around the Balmer lines. To a first approximation, the Balmer line intensities can be represented in the form

$$E(h) = E_0 e^{-\beta h} . \tag{11}$$

With the exception of the intensities of the first few Balmer lines in the low chromosphere, no demonstrable departures from this simple form for $E(h)$ have been noted. Yet, the complex chromospheric geometry and the strong effects of self-absorption in the Balmer lines almost demand that this simple form cannot hold for all lines. Furthermore, the maximum in $F_{3640}^{\mathrm{f-b}}$ near 400 km requires a similar maximum in the later Balmer lines. Data that are adequate to exhibit this maximum are still not available.

Empirical values of β decrease steadily from the Balmer series limit to Hα, an effect that has been attributed mainly to self-absorption. Attempts to infer the quantitative effects of self-absorption and to arrive at chromospheric opacities in the hydrogen lines show widely divergent results and additional data and analyses are necessary. The problem would be much more easily solved if a simultaneous treatment of the Paschen lines was possible. At the 1952 eclipse we attempted such a comparison, but unfortunately without success. The Paschen line data were poor and the intensities were weaker relative to the Balmer lines than is expected in even a thin atmosphere. Self-absorption should weaken the Balmer lines relative to the Paschen lines, which is opposite to the effect observed. The 1962 data should provide a greatly improved set of Paschen and Balmer line data, and a useful intercomparison of these lines will be possible.

Departures from thermodynamic equilibrium in the populations of energy levels in chromospheric hydrogen are clearly evident in the Balmer lines. At heights above 2000 km the decrement of intensities through the later members of the Balmer series ($n \geqslant 10$) is steeper than predicted by thermodynamic equilibrium in a thin atmosphere. This can be explained only by a super population of the lower energy levels relative to the higher levels, i.e., the b_n parameters measuring departures from thermodynamic equilibrium are greater than unity and increase as n decreases.

Separation of the quantitative effects of self-absorption and departures from thermodynamic equilibrium is a difficult problem that has not been adequately solved. Again, the Paschen line data will aid in this quantitative evaluation.

Profiles of chromospheric hydrogen lines have been studied both at eclipse and outside of eclipse. Eclipse observations have been seriously handicapped by the difficulty of determining the effective height of the observations and the appropriate limits of integration effected by the finite size of the spectrograph slit and the seeing excursions of the image. Extra eclipse studies of the early hydrogen lines (cf. SMITH, 1957) show pronounced changes with height in the profiles of the early Balmer lines that place rather difficult requirements on eclipse observations.

For the later Balmer lines where intensities are low, only eclipse data on profiles are available. These data show the composite effects of broadening by self-absorption in the earlier lines and broadening by the Stark effect in the later lines. The result is a minimum width near H15, which probably still exhibits residual effects of both self-absorption and Stark broadening in the low chromosphere (REDMAN and SUEMOTO, 1954).

Studies of line profiles for individual spicules indicate that a single spicule has enough opacity to broaden the Hα and perhaps Hβ profiles by self-absorption. Profile observations for the later Balmer lines in spicules could be accomplished at eclipse and would provide valuable data.

In general, the profiles of chromospheric lines observed at the limb are somewhat broader than the cores of the corresponding Fraunhofer line.

Interpretation of the chromospheric hydrogen line intensities and profiles in terms of temperatures and densities in the chromosphere have so far ended in frustration. Self-absorption and stark effect in the profiles are difficult to evaluate with sufficient precision to have confidence in the residual profile for purposes of deriving thermal motions.

The Balmer lines observed at the limb have a large contribution from scattering of radiation filtering up through the photosphere and chromosphere. In addition, their intensities are closely coupled to the radiation field in the Lyman series. Small departures from radiative detailed balance in the early Lyman lines may produce greater effects on the Balmer line intensities than relatively large changes of temperature. Hence, an interpretation of the Balmer line intensities will require a careful treatment of the radiative transfer problems in both the Lyman and Balmer series.

D. HELIUM

The chromospheric spectrum is rich in lines of neutral helium, and their study is perhaps the most fruitful of all chromospheric lines. Some 20 lines from the singlet and triplet series $2S$–nP, $2P$–nS and $2P$–nD are observed in the visual regions of the spectrum. Those that are unblended with other lines show a maximum in $F(h)$ at about 1 100 km, as shown in Figure 4. Below the maximum $F(h)$ drops rapidly and has not been reliably determined. Above the maximum, both $F(h)$ and $E(h)$ decrease rather slowly with increasing height, having an intensity scale height of about 800 km. Only the λ 4 686 line of He II and the heavily self-absorbed lines of Ca II decrease as slowly with height.

All of the unblended lines of neutral helium decrease in intensity with height at the same rate. Hence, the ratios of $E(h)$ and $F(h)$ for different lines remain constant with height. This is a convincing demonstration that self-absorption is unimportant and that we may treat the chromosphere as being optically thin in these lines.

Because there are no effects of self-absorption in neutral helium lines and because f-values are relatively well known, these lines are ideal for searching for departures from thermodynamic equilibrium. Comparison of corresponding

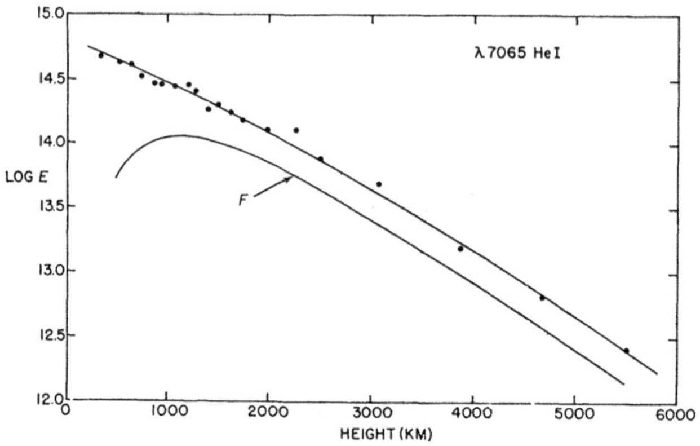

Fig. 4. The maximum in F (arbitrary units) for λ 7065 of He I derived from observed values of E (1952 HAO data).

triplet and helium lines do, indeed, reveal large departures from thermodynamic equilibrium, which provide a strong boundary condition on chromospheric models along with a challenging problem of interpretation. The theoretical ratio of $E_{5876} : E_{6678}$ is 3.5 : 1 and the observed ratio in the chromosphere is 40 : 1. For lines originating on higher levels the theoretical and observed ratios are more nearly equal, but even lines originating on levels as high as $n = 6$ and $n = 7$ still exhibit recognizable departures. In evaluating the significance of these departures from the theoretical intensity ratios, one must remember that we are comparing ratios of populations of energy levels and that the actual departures of populations from their theoretical value may be considerably greater than indicated by the above ratios.

Wherever we see helium line emission in the chromosphere, whether it be at 1 000 km where the chromospheric structure is complex or in a single spicule, the ratios of line intensities are essentially the same and the singlet levels are underpopulated with respect to the triplets.

Populations of the triplet levels of He I in the chromosphere are governed mainly by the intensity of the radiation field incident from the photosphere at the wavelengths of the triplet lines and continua and by the local density. Temperature enters in only a secondary way. Populations of excited singlet levels depend upon the same equivalent radiation field for the singlet lines and continua. The temperature and density dependence of the singlet populations, however, enter in a complicated way. The excited singlets become underpopulated with respect to the triplets by electrons draining into the 1^1S level. Such a drainage can occur if part of the photons in the resonance lines escape the local environment so that fewer photons are absorbed than are emitted. We express the photon leak, also referred to as the net radiative bracket, by

$$\varrho_{nn'} = 1 - \frac{B_{n'n} N_{n'} \int J_\nu \phi_\nu \mathrm{d}\nu}{A_{nn'} N_n}, \tag{12}$$

where $B_{n'n}$ and $A_{nn'}$ are the Einstein coefficients, J_ν is the radiation intensity averaged over direction and ϕ_ν is the normalized frequency dependent absorption coefficient. Physically, $\varrho_{nn'}$ measures the divergence of the photon flux. Analysis of the helium line data yields limits on $\varrho_{nn'}$ for the resonance series. Approximate values of $\varrho_{nn'}$ as a function of $\tau_{nn'}$ can be computed from theory.

If the opacity is very high in the resonance lines, no photons will escape and $\varrho_{nn'} = 0$. If the opacity is intermediate, the photon leak becomes a strong function of the opacity and a constant ratio of singlets to triplets would require a relatively constant opacity. As the opacity becomes low the photon leak approaches an asymptotic value and the singlet : triplet ratio approaches a constant value. This, then, is the most logical solution to adopt, viz., that the opacity in the resonance lines of He I has dropped low enough that the photon leak is near its asymptotic limit. It should be noted that since most of the resonance line radiation is produced by scattering, $\varrho_{nn'}$ is always small compared to unity even when τ is small.

The photon leak is a function of τ, which, in turn, is a function of the temperature and density structure of the atmosphere. Hence, one gets temperature information from the helium lines by first obtaining semi-empirical values of $\varrho_{nn'}$ and then relating these by theory to T_e and n_e. Although rigorous solutions have not been accomplished for this latter phase of the problem, it seems clear from the approximate solutions that the temperature in the helium regions of the chromosphere is of the order of 50000° (ATHAY, 1963). A similar temperature is inferred from the widths of helium line profiles.

Observations of the 2^3S–2^3P and 2^1S–2^1P lines at λ 10830 and λ 20582 would add greatly to the empirical determination of $\varrho_{nn'}$ and should be attempted at future eclipses.

A further quantity of importance from helium line data is $n_e n_{\mathrm{II}} T_e^{-3/2}$, where n_{II} is the density of He II. This quantity can be obtained by extrapolating the intensities of helium lines to their series limit. The accuracy of the extrapolation depends upon the manner in which the line intensities approach the series limit. Theory predicts significant differences in the approach to the series limit depending upon the values of n_e and T_e in the helium regions, and it is important that the region of extrapolation be reduced to a minimum. It is by comparison of the quantity $n_e n_{\mathrm{II}} T_e^{-3/2}$ with the quantity $n_e n_p T_e^{-3/2}$

derived from Balmer continuum that we obtain the most reliable estimates of the solar abundance of helium. Hence, we should make a deliberate effort to observe higher members of the helium series.

The quantity $n_e n_{II} T_e^{-3/2}$ is extremely useful in deriving chromospheric models and values obtained at the 1952 eclipse have played an important role in formulating our picture of the chromosphere. Quantitative evaluation of departure from spherical symmetry in the height range 1000–3000 km is based mainly on the quantity $n_e n_{II} T_e^{-3/2}$ in relation to continuum data at λ4700 and λ3640. If we adopt a H : He ratio of 10 : 1, a region of the chromosphere in which helium is predominantly singly ionized would yield $n_e n_{II} T_e^{-3/2} = 0.1\ n_e n_p T_e^{-3/2}$. This is found to be the case for $h > 3000$ km, and we conclude that hydrogen and helium emissions arise in essentially the same volume elements, viz., spicules. Below 3000 km $n_e n_{II} T_e^{-3/2} < 0.1\ n_e n_p T_e^{-3/2}$ and we conclude that either helium is predominantly neutral or that helium and hydrogen emissions originate in separate and distinct volume elements. The latter suggestion seems the more compatible of the two.

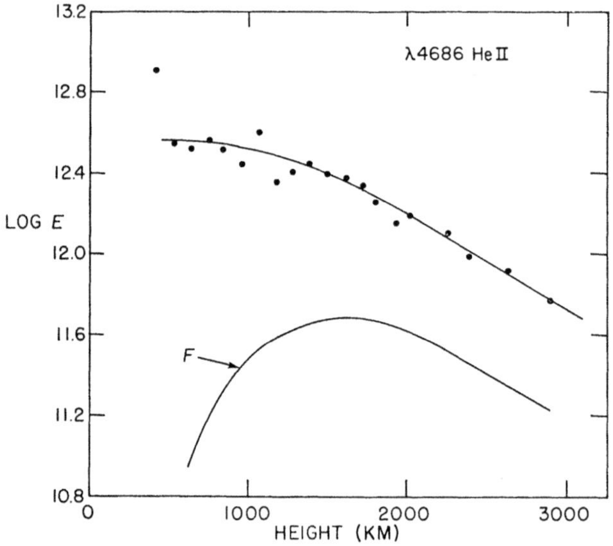

Fig. 5. The maximum in F (arbitrary units) for λ4686 of He II as derived from observed values of E (1952 HAO data).

The λ4686 line of ionized helium corresponds to the Paschen-α line of hydrogen and is the only ionized helium line present in the visual chromospheric spectrum in the normal chromosphere. The behavior of the line is much like that of He I lines. It rises to maximum intensity near 1500 km, as shown in Figure 5, although the maximum seems flatter and more ill-defined than for

He I. The rate of decrease of $\lambda 4686$ intensity above the maximum seems to be somewhat slower than for He I, but the emission is weak and the observational data are not sufficient to firmly establish this point. Profile information is totally lacking.

Temperatures inferred from He I line data are sufficiently high to excite the He II line. The problem is complicated, however, by possible fluorescent excitation of the 2–4 transition in He II with the hydrogen Lyman-α line (ZIRKER, 1956) and by the finite optical thickness of the chromosphere. The latter effect complicates the transfer problem in He II and the former requires a knowledge of the Lyman-α line. The Lyman-α profile emergent from the sun suggests, however, that the fluorescent excitation of $\lambda 4686$ is not of major importance.

E. HEAVY ELEMENTS

So little work has been done with chromospheric lines originating from elements heavier than helium that it is convenient to discuss them as a group. In slitless spectra, by far the strongest of these lines are those of Ca II, but numerous strong lines are also present from such atoms as O I, Na I, Mg I, Al I, Ca I, Sc II, Ti II, Cr I, Cr II, Mn I, Mn II, Fe I, Fe II, Ni I, Sr II, Y II, and Ba II.

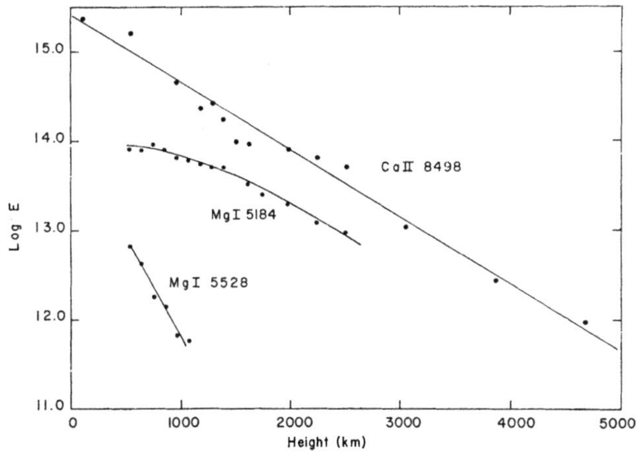

Fig. 6. Extremes of emission scale heights for lines of heavy elements in the chromospheric spectrum (1952 HAO data).

One of the striking features of chromospheric emission in lines of these heavier elements is the great diversity of intensity scale heights. Values of β, defined by Equation (11), range from about 5×10^{-8} cm^{-1} down to about 0.7×10^{-8} cm^{-1}, and in the lowest layers of the chromosphere strong effects of self-absorption give local values of β that are still smaller. Individual atomic species may show almost as much diversity in β from line as the whole spectrum of lines. Thus, the Mg I line at $\lambda 5528$, 3^3P–4^1D, has $\beta \approx 5 \times 10^{-8}$ cm^{-1} and $\lambda 5184$, 3^3P–4^2S, has $\beta = 1 \times 10^{-8}$ cm^{-1} in the same height range in the low chromosphere. In the upper chromosphere β for $\lambda 5184$ increases to about 2.3×10^{-8} cm^{-1} (ATHAY and HOUSE, 1962). These effects are illustrated in Figure 6.

The emission gradients, β, may be influenced by a number of factors. Equations (7) and (8) may be rewritten for a given λ and ϕ in the form

$$E(h) = h\nu\, A_{nn'} n_n \{\text{abs}\} = K\, n_e n_i T_e^{-3/2} b_n e^{X_n/kT_e} \{\text{abs}\}\,, \tag{13}$$

where K is a constant for each line, n_i is the ion density for the stage of ionization just above the one giving rise to the spectral line, b_n measures the departure of the ratio n_n/n_i from its value in thermodynamic equilibrium, X_n is the ionization potential measured from level n to the continuum and {abs} is a factor, less than unity, that corrects for the influence of self-absorption. For convenience, further rewrite Equation (13) in the form

$$E(h) = E(n_e, n_i, T_e, \tau)\,. \tag{14}$$

Then, if we compare emission gradients for different lines we may possibly identify the functional dependence that differs from line to line. The functional dependence on n_e enters through its influence on n_i and b_n as well as through its explicit appearance in Equation (13). To a first approximation, however, we shall neglect changes in the functional dependence on n_e from line to line in favor of the remaining variables. The functional dependence of $E(h)$ on n_i enters linearly as in Equation (13). However, n_i may be strongly influenced by T_e and may have a radically different height dependence from atom to atom or ion to ion. For lines of a given atom or ion, n_i has a common effect, of course. The explicit dependence on T_e in Equation (13) may vary from line to line in a given atom or from atom to atom through the factor exp (X_n/kT_e). This effect of T_e may well be less pronounced, however, than the effect exhibited through the dependence of n_i on T_e. Finally, we note that the influence of τ may vary strongly from line to line and atom to atom. Although τ is coupled to n_e and T_e it is convenient for our purposes to carry it as a separate variable.

ZIRKER (1958) has carried out a systematic study of the influence of τ on the β's for lines of heavy elements in the low and middle chromosphere. He finds a statistical relationship between β and the absolute intensity extrapolated to $h = 0$. Strong lines tend to have smaller β's than the fainter lines, as is expected from self-absorption effects. Furthermore, most of the lines suggesting strong effects of self-absorption have a tendency for β to increase with height relative to fainter lines, and if the line is a member of a multiplet the ratio of its intensity to the intensity of weaker members of the multiplet tends to increase with height also. Both of these latter effects are the expected consequences of self-absorption.

In the case of Mg I, we (ATHAY and HOUSE, 1962) have been able to check

ZIRKER's results directly by independent means. The upper level of the $\lambda 4571$, 3^1S–3^3P, line of Mg I is the absorbing level for the $\lambda 5184$ and $\lambda 3838$ triplets. Since $\lambda 4571$ is a forbidden line, its f-value is extremely small, 1.7×10^{-6}, and there is no possibility for an observable effect of self-absorption. Hence, we obtain the population of the 3^3P level, and therefore τ in the $\lambda 5184$ and $\lambda 3838$ triplets, directly from the observed intensity of $\lambda 4571$. The results give relatively large values for τ in these lines. The height of the surface $\tau = 1$ in the center of $\lambda 5184$ and at the center of the disk, for example, is 1100 km. If we reduce absolute intensities by a factor of 2, as suggested in the foregoing, the height of $\tau = 1$ drops to 800 km in rather good agreement with Zirker's results.

A further evidence of self-absorption is shown in line profiles observed by REDMAN (1942, 1955) and REDMAN and SUEMOTO (1954). Lines found by ZIRKER to be self-absorbed in their total intensities have line profiles that show strong evidence of saturation due to self-absorption. Quantitative comparisons of these two evidences of self-absorption effects are difficult to obtain. It seems well established, however, that in many of the stronger lines of heavy elements the chromospheric opacity is high and the β's are reduced by self-absorption.

Lines listed by Zirker as showing pronounced evidence of self-absorption include 46 of the strong lines arising from O I, Na I, Mg I, Al I, Ca I, Ca II, Ti II, Cr II, Mn II, Fe I and Sr II. When the opacities at the limb are converted to radial opacities it is found that in all of the lines mentioned the height of the surface $\tau = 1$ at line center falls within the chromosphere, generally in the range 300–800 km.

The Mg I lines $\lambda 4571$ and $\lambda 5528$, 3^1P–4^1D, are both free from observable effects of self-absorption. Even a generous over estimate of the populations of the absorbing levels predict low chromospheric opacity in these lines. Their β's, however, differ markedly ($\beta_{5528} = 5 \times 10^{-8}$ cm^{-1} and $\beta_{4571} = 2.3 \times 10^{-8}$ cm^{-1}) and their intensity ratio differs from the thermodynamic equilibrium ratio by about a factor of 3 at 500 km and a factor of about 30 at 1000 km. We conclude, therefore, that strong departures from thermodynamic equilibrium are present. The difference between the β's for the two lines are much too large to be attributed to an effect of T_e through the term exp (X_n/kT_e). Since n_i and n_e are common to the two lines, the only remaining factor is b_n. Very similar effects and in the same direction are noted in comparison of the O I triplet at $\lambda 8446$, 3^3S–3^3P, to the $\lambda 7772$ quintet, 3^5S–3^5P (ATHAY and HOUSE, 1962).

Calculations of b_n for Mg I using a model chromosphere (ATHAY and HOUSE, 1962) yield predicted line intensities and gradients that agree very well with observations. One of the reasons for a large β for $\lambda 5528$ is a strong decrease of b_{4^1D} with height, which, in turn, is caused by decreasing opacity, with a resultant increase of $\varrho_{nn'}$, in the resonance line at $\lambda 2852$. We have, then, a situation in which the emission gradient of one line is strongly influenced by the opacity in a completely different line. The effect of reducing the opacity in $\lambda 2852$ is to allow electrons to drain into the 3^1S level thereby depopulating the excited levels linked to the ground state by permitted transition. Since the transition 3^1S–3^3P is forbidden, the population of the 3^3P level is not reduced by this effect. The same mechanism has been found in the helium singlets and triplets, but the quantity $\varrho_{nn'}$ for helium is apparently rather constant throughout the chromosphere. The observed effects in the O I lines most likely arise from the same mechanism.

ZIRKER's assignment of the surface $\tau = 1$ to the chromosphere for many strong lines carries with it implications of departures from thermodynamic equilibrium in these lines. The radiation temperatures at the center of most of these lines in the Fraunhofer spectrum are in the range 3800–5000°, which are considerably lower than our best values of T_e at the same height. This cannot be the case unless the b_n's differ from unity by substantial amounts. Similar effects where $T_{\rm rad} < T_e$ are noted in many other lines in the spectrum, known to be of chromospheric origin. These include the early Balmer lines, the H and K lines of Ca II, the λ 10830 line of He I and all of the chromospheric lines in the rocket spectrum.

The conclusion that many strong Fraunhofer lines of heavy elements are chromospheric in origin and are formed under conditions departing strongly from thermodynamic equilibrium is important to all of astrophysics. The basic assumptions underlying the determination of abundances of elements, line formation in stellar atmospheres and, in some cases, model atmosphere calculations must be re-examined.

Effects of self-absorption and departures from thermodynamic equilibrium, although demonstrably important in the chromosphere, most likely cannot account for all of the differences in β that are observed. Temperature increases with height in the chromosphere and there must be influences on the β's through changing effects of excitation and ionization. Efforts to correlate β with the excitation potentials in the lines have not been successful. Similarly there does not seem to be any significant relationship between β and the ionization potential of the ion or atom giving rise to the spectral line (cf. ZIRKER, 1958).

If we drop the {abs} term in Equation (13), we note that the line intensity is proportional to n_i. We also note that temperatures in the chromosphere are such that elements whose first ionization potential is less than about 8 volts are predominantly ionized. We also note that many of these same elements have second ionization potentials of about 15 volts and that hydrogen, whose ionization potential is 13.56 volts, is ionizing rapidly in the first few hundred kilometers of the chromosphere. This is demonstrated clearly by both the increase of Balmer continuum intensity with height in the low chromosphere and by the intensity of the Balmer lines and continuum relative to other lines and continua. Analysis of helium line data leads to the conclusion that in the regions producing the helium emission, He II is more abundant than He I.

There is reason to believe, therefore, that the lines observed in the chromospheric spectrum from ionization stage $i-1$ may arise mainly in regions where the i stage of ionization is dominant, or perhaps even giving way to the $i+1$ stage of ionization. In the former case, where n_i is not changing relative to other ionization stages there is no reason to expect a correlation of β with ionization potential. In the latter case, where n_i is giving way to n_{i+1}, n_i will decrease somewhat more rapidly with height than otherwise because of a relative increase in n_{i+1}. If $X(i, i+1)$ is low this added decrease in n_i with height may be a dominant factor. This suggests that a correlation may exist between $X(i, i+1)$ and β.

In attempting to find a correlation between β and X, it is important to recognize that self-absorption effects are present. This may be eliminated to some extent by using the maximum value of β observed for a given ion. A plot of β_{max} (Zirker, 1958; Athay and House, 1962) versus $X(i, i+1)$ is shown in Figure 7. A correlation does seem to exist in the sense indicated, viz., small β associated with large values of $X(i, i+1)$ and vice versa. Similar plots of β_{max} against $X(i-1, i)$ show no recognizable correlation.

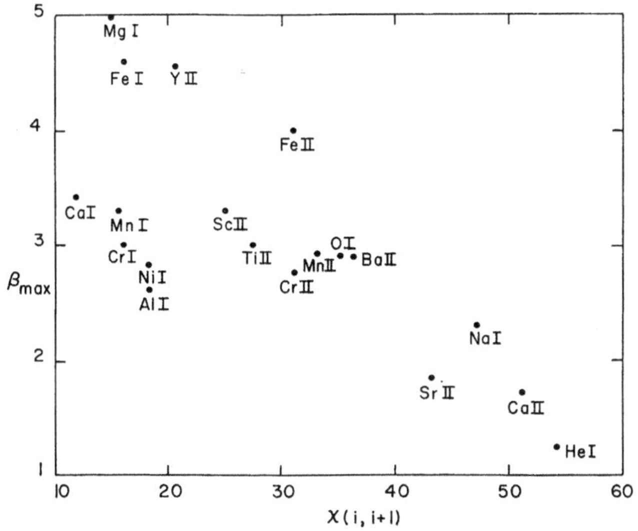

Fig. 7. Correlation between β_{max} and $X(i, i+1)$ for lines arising from ionic species in the $i-1$ stage of ionization as shown in the figure (1952 HAO data).

Some reservation must be used in the interpretation of Figure 7. For example, the only lines of Sr II and Na I for which data are available are the resonance lines and these show evidence of self-absorption. In the case of Ca II, however, the $\lambda 8498$ line is used, and this line is not likely to be strongly affected by self-absorption (cf. Athay, 1961). Also the He I lines are known to be free of self-absorption. In the case of Mg I, only one line is known to have a β as large as that observed, and this is known to be partly due to a decrease of b_n

with height. The next largest β observed for Mg I is 2.3×10^{-8} cm^{-1}. On the other hand, the strongest Y II lines have $\beta = 4.5 \times 10^{-8}$ cm^{-1} and several of the strongest lines of Fe I have $\beta \geqslant 3.9 \times 10^{-8}$ cm^{-1}. REDMAN (1955) has pointed out, along with others, that the lines of singly ionized rare earths in the chromospheric spectrum fade extra rapidly with increasing height. These elements are all characterized by low third-ionization potentials.

Thus, the details of the correlation in Figure 7 may be substantially changed by new data. It does not seem likely that the general correlation will be destroyed, however.

Vastly more data for lines of heavy elements than are available in the literature are contained in existing eclipse spectrograms. The faint lines of heavy elements have been scarcely studied at all. They may well hold the key to many interesting new features of the flash spectrum.

One further point is of interest in connection with the spectrum of heavy elements in the chromosphere. Slit spectra of the low chromosphere have a sharply different appearance from ordinary slitless spectra. In slitless spectra certain lines stand out as the brighter features of the spectrum. In slit spectra, however, a new set of lines dominate. The reason for this is related to the above discussion. Lines such as Y II and the rare earths, for example, have much narrower profiles and their scale heights are much smaller than the lines that are characteristically strong in slitless spectra. The rare earths have profile half-widths of about 0.1 Å compared to values 5 to 10 times this large for lines of H, He and Ca. Also, their local-scale heights may be 10 times smaller. A spectrograph that resolves both the profiles and scale heights may thus see the central regions of the rare earth lines with intensities increased relative to H, He and Ca lines up to 100 fold over the values observed in slitless spectra.

F. MOLECULES

The flash-spectrum is relatively rich in molecular bands for the first few hundred kilometers. In effect, molecules behave as though their second ionization potential were low, i.e., after dissociation occurs the neutral atoms are easily ionized, which prevents the molecule from reforming. Values of β for the CN band at λ3838 are about 4.7×10^{-8} cm^{-1} for both the band head and the total intensity in the band. Thus, if we consider the second ionization potential of CN as being the first ionization potential of C, which is 11.26 volts, the β for CN fits very well into the correlation shown in Figure 7.

PECKER and ATHAY (1955) derived a rotation temperature of 4500° for the λ3838 band of CN at 100 km and computed theoretical band intensities for a model chromosphere in which T_e was near the rotation temperature and in which densities were compatible with continuum data. The results underpredicted the observed band intensities by about a factor of 20. Because of the low dissociation potential of CN, and the strong dependence of CN concentration on density, the local concentration of CN should be very sensitive to inhomogeneities in temperature and density. Thus, the discrepancy found by Pecker and Athay could be readily explained by, and may be taken as evidence for, departures from spherical symmetry in the low chromosphere.

D. V. THOMAS (1958) has rather clearly demonstrated that self-absorption is present near the head of the $\lambda 3838$ CN band in the low chromosphere. Rotational temperatures derived by Thomas after an approximate correction for self-absorption have values of 7500° and 5000° at 0 and 50 km respectively. The heights, however, are admittedly rather uncertain. Rotation temperatures of this magnitude are difficult to understand unless the electron temperature is at least as high, which seems unlikely. Also, if electron temperatures were this high the discrepancy between computed and observed intensities noted by Pecker and Athay would become even more pronounced.

Molecular band spectra in the first few hundred kilometers of the chromosphere can provide much useful information on the geometrical and thermodynamical properties of the chromosphere at these heights. Observational data with improved spectroscopic and height resolution are needed for this purpose.

4. Problems for Future Eclipse Studies

Although the chromospheric spectrum has been observed at eclipse for several decades large gaps still exist in our knowledge of the spectrum. The transition region from the photosphere to the chromosphere has been inadequately observed and even less adequately studied. The behavior of the continuum spectrum, of central and integrated intensities of lines and of line profiles in this region are all problems of pressing importance. The proper understanding of the temperature minimum near the base of the chromosphere and the details of the temperature gradients either side of the minimum is of vital importance in attempts to identify the non-radiative sources of energy that produce the temperature rise in the chromosphere and corona.

The multitudes of faint lines in the spectrum of the low chromosphere, the so-called "flash spectrum", undoubtedly contain a wealth of information. They have been recorded on spectrograms obtained at many eclipses, but never exhaustively studied. We have much to learn concerning the geometrical and thermodynamical properties of this region of the chromosphere, and we cannot continually overlook these important data.

What are the spectral properties of the gases between spicules at heights of 3000–10000 km? At present, we can only conjecture that the spectrum must be like that of the corona. Careful observations made at eclipse seem best suited for answering this important question more precisely.

Spectra of active regions of the chromosphere seen at the limb at eclipse offer great advantages over similar spectra obtained in disk observations. Only a few such eclipse observations exist. One such observation at the 1952 eclipse (cf. THOMAS and ATHAY, 1961, Ch. 3) shows a region in which the chromosphere essentially disappeared. Only the very strongest chromospheric lines bridged faintly across this region, and most of this apparent emission was probably caused by halation in the photographic film. Immediately adjacent to the region of missing chromosphere, was a region of strongly enhanced and extended chromosphere. What is the nature of such regions and how frequently do they occur?

Does the chromospheric spectrum vary from equator to pole and with the phase of the sunspot cycle? There are still insufficient data to answer these questions in any but the grossest terms. We cannot claim to have fully observed the spectrum of the chromosphere until they are fully answered.

Research reported in this paper was supported by Air Force Cambridge Research Laboratories.

References

Athay, R. G.: 1959, *The Observatory* **79**, 12.

Athay, R. G.: 1961, *Astrophys. J.* **134**, 756.

Athay, R. G.: 1963, *Astrophys. J.* **137**, 931.

Athay, R. G. and House, L. L.: 1962, *Astrophys. J.* **135**, 500.

Athay, R. G., Menzel, D. H., and Orrall, F. Q.: 1957, *Smithsonian Contr. to Astrophys.* **2**, 35.

Athay, R. G. and Roberts, W. O.: 1955, *Astrophys. J.* **121**, 231.

Coates, R. J.: 1958, *Astrophys. J.* **128**, 83.

Dunn, R. B.: 1958, Thesis, Harvard College Observatory.

Goldberg, L., Mohler, O. C., and Müller, E. A.: 1959, *Astrophys. J.* **129**, 119.

Houtgast, J.: 1957, *Rech. Astron Obs. Utrecht* **13**, No. 3.

Minnaert, M.: 1953, *The Sun* (ed. by G. Kuiper), University of Chicago Press, Chicago.

Pannekoek, A. and Minnaert, M.: 1928, *Verhandelingen Akad. Wetenschappen, Amsterdam* **14**, No. 2.

Pecker, J.-C. and Athay, R. G.: 1955, *Astrophys. J.* **121**, 391.

Redman, R. O.: 1942, *Mon. Not. Roy. Astron. Soc.* **102**, 134.

Redman, R. O.: 1955, *Vistas in Astronomy* (ed. by A. Beer), Pergamon, London.

Redman, R. O. and Suemoto, Z.: 1954, *Mon. Not. Roy. Astron. Soc.* **114**, 524.

Smith, E. V. P.: 1957, *Astrophys. J.* **126**, 529.

Suemoto, Z.: 1962, *Transactions of the IAU* **11B**, 206.

Thomas, D. V.: 1958, *Mon. Not. Roy. Astron. Soc.* **118**, 458.

Thomas, R. N. and Athay, R. G.: 1961, *Physics of the Chromosphere*, Interscience, New York.

Vyazanitsyn, V. P.: 1951, *Izv. Glavnoi Astron. Obs. Pulkova* **18**, 19.

Vyazanitsyn, V. P.: 1952, *Izv. Glavnoi Astron. Obs. Pulkova* **19**, 40.

Waddell, J.: 1963, *Astrophys. J.* **137**, 1210.

White, O. R.: 1962, Thesis, University of Colorado.

Zirker, J. B.: 1956, *Astrophys. J.* **123**, 536.

Zirker, J. B.: 1958, *Astrophys. J.* **127**, 680.

ROCKET SPECTRA OF THE CHROMOSPHERE

H. E. HINTEREGGER
U. S. Air Force Cambridge Research Laboratories, Bedford, Massachusetts, USA

Introduction

Photographic or photoelectric recordings of solar XUV-spectra* at wavelengths from about 1700 Å to 1 Å can be well-obtained only from altitudes above those reached in mountaintop or even balloon experiments. Therefore, the history of the experimental art of acquiring such spectra begins no sooner than the scientific use of rockets around 1946. Actual spectral resolution of solar emission lines at wavelengths below about 2200 Å, however, was accomplished for the first time about six years later. This accomplishment had to await the successful development of a servomechanism for fairly accurate solar pointing of spectrographs during the high altitude portion of the rocket flight. This development is due to the work of W. A. RENSE and W. B. PIETENPOL and their associates (University of Colorado). In 1952 they obtained the first spectrophotographic evidence (W. A. RENSE, 1953) of the solar hydrogen Lyman alpha line, recovering the record from a grazing incidence grating spectrograph kept properly oriented in a biaxial solar pointing control mounted on an Aerobee rocket.

Historically, the year 1952, merely one solar cycle of 11 years ago, thus marks the first successful beginning of solar chromospheric rocket spectroscopy of the type using actual optical disperison at wavelengths below 2200 Å.

Since then, essentially the same type of biaxial pointing control aboard the same type of vehicles (Aerobee Hi rockets with full instrument payloads have reached altitudes up to about 235 km) have been used extensively by R. TOUSEY, W. A. RENSE, and H. E. HINTEREGGER and their associates of the U. S. Naval Research Laboratory (NRL), University of Colorado (U of C) and U.S. Air Force Cambridge Research Laboratories (AFCRL), respectively. No detailed historical summary of their work will be attempted here since several recent summaries with references to original papers provide an adequate coverage in the published literature (reviews by TOUSEY, RENSE, FRIEDMAN, and HINTEREGGER in *Space Astrophysics* (W. W. LILLER ed., 1961), by Friedman's article on rocket spectroscopy in *Space Science* (D. P. LEGALLEY, ed., 1963) and the latest summary article now appearing in *Space Science Reviews* (R. TOUSEY, 1963) which I had the benefit to obtain in preprint form). Even a representative model of spectrometer for OSO-A (first orbiting solar observatory) prepared by

* The abbreviation XUV has been suggested by Tousey as a brief cumulative designation of the range of wavelengths including both the far vacuum ultraviolet (or extreme ultraviolet, EUV) and soft X-rays.

 All rights reserved.

J. C. LINDSAY and his associates (GSFC) was first subjected to a (successful) test flight using the same type of pointing control and rocket vehicle. It is also being used by L. GOLDBERG and his associates of the Harvard College Observatory (HCO) for testing models of instruments developed for flight on future orbiting solar observatories such as OSO-B. The same is true for the past and current work with various types of solar XUV image-forming devices, truly monochromatic or broad-band (non-dispersive), XUV spectroheliographs, and a white light coronograph (TOUSEY, 1963). This work with biaxial pointing controls on Aerobee rockets, conducted by various groups of U.S. scientists, includes not only preparations for currently projected satellite experiments but also a number of well-known historical achievements accomplished with recoverable photographic rocket instruments. Outstanding examples are the photography of the sun in hydrogen Lyman alpha (R. MERCURE, S. C. MILLER, W. A. RENSE, and F. STUART, 1956; J. D. PURCELL, D. M. PACKER, and R. TOUSEY, 1959), photography of the sun in its X-ray emission (T. A. CHUBB, H. FRIEDMAN, R. W. KREPLIN, R. L. BLAKE, and A. E. UNZICKER, 1961), and most recent NRL spectroheliograph records of the solar disc in other XUV wavelengths such as He II Lyman alpha (304 Å) and even Fe XV (284 Å) (R. TOUSEY, private communication, 1963).

Since an extensive review, 'The Extreme Ultraviolet Spectrum of the Sun', is currently being published in *Space Science Reviews* (R. TOUSEY, 1963), it would be rather worthless to attempt a similarly comprehensive review of the same general scope here. I am indebted to Dr. TOUSEY for his kind permission to use much of his most illustrative material in this symposium lecture and include reproductions in this written version. The utility of the latter as a working copy for convenient reference within this symposium has thus been improved consideraby. Also, the existence of Tousey's excellent review made it possible for me to omit or mention but briefly many points of general interest. The resulting saving in space allowed a number of specific topics and most recent results to be included in this presentation. The afternoon discussion sessions should provide us with further results from most recent experimental work, that could not be included in this summary but may appear in the published proceedings of this symposium.

Technology

INSTRUMENT CARRIERS

The impressive amount of solar research accomplishments using essentially the same type of biaxial solar pointing control with Aerobee or Aerobee Hi rockets clearly labels this combination as a well-proven reliable "work horse". Actually the same facility may remain the only readily available one for U.S. rocket spectroscopy experiments short of those scheduled to be flown on satellites of the OSO-series for at least another year or so.

One can have but little doubt, however, that the addition of one or two types of somewhat different solar experiment carriers of intermediate complexity and cost is well justified at present. The availability of such a facility would indeed have been extremely fruitful during the past few years already. The most obviously desirable type of a work horse No. 2 for solar XUV spectrophotometry experiments and satellite instrument prototype developments would be a carrier providing essentially the same biaxial pointing accuracy of the order of one minute of arc but reaching significantly higher peak altitudes. The higher peak altitudes have mainly the two purposes of (a) extending terrestrial atmospheric XUV extinction measurements beyond 220 km as well as eliminating the residual atmospheric absorption effects for measurements of the "incident" solar XUV fluxes, and (b) providing the often crucially needed longer vacuum exposure of instruments, say 10–20 minutes per rocket shot instead of the present 3–6 minutes.

Another perhaps even more acutely desirable variant, say work horse No. 2A, should provide the same as No. 2 except for a minimized roll of the instrument holder around the solar vector. This version would be of obvious benefit for all those solar experiments involving either actual image formation or other forms of spatial resolution or identification of radiation source points over the solar disc and beyond. The conversion of the latter type of carrier (2A) into a more sophisticated one (2B) providing actual three-axial orientation control with significantly better pointing accuracy, say a few seconds of arc, may be recognized as being practical from both technical and economic viewpoints.

A number of projects in geophysics and solar physics could also greatly benefit from the development of a reliable simpler and less costly type (No. 3) of instrument carrier for various experiments requiring but modest solar pointing and less volume and weight than is actually provided by the present rather expensive and not too conveniently launchable carrier type No. 1 (liquid propellant). A solid propellant rocket capable of carrying some 5 to 10 kg of solar instruments to altitudes ranging from about 120 to 400 km with provisions of crude instrumental solar pointing within about one degree of arc, would certainly represent a very useful additional type of carrier (No. 3). It would primarily facilitate currently projected aeronomical applications. However, the availability of such carriers would also allow some exploratory solar measurements, as well as tests of new solar instruments or at least representative critical parts of such experiments in the course of development for flight on advanced orbiting observatories.

The most general task before us appears to be the more or less continuous acquisition of absolute spectrophotometric records of the total solar disc XUV radiation as well as spectroheliograms of good spatial resolution and wide dynamic range of intensities in a considerable number of different XUV wavelengths. One would like to obtain such records routinely for the most interesting emission lines of different stages of ionization of a fair number of elements besides hydrogen and helium in the solar atmosphere. This major task obviously calls for continuously operational orbiting observatories rather than relatively short lived experiments on non-orbiting carriers ("single shots"). The great importance of a simultaneous vigorous program of non-orbiting experiments, however, must not be overlooked.

TYPES OF INSTRUMENTS AND RECORDS

The technological developments leading to our present experimental capability of obtaining solar XUV spectrophotometric records including some XUV line shape measurements and XUV spectroheliograms have been covered in publish-

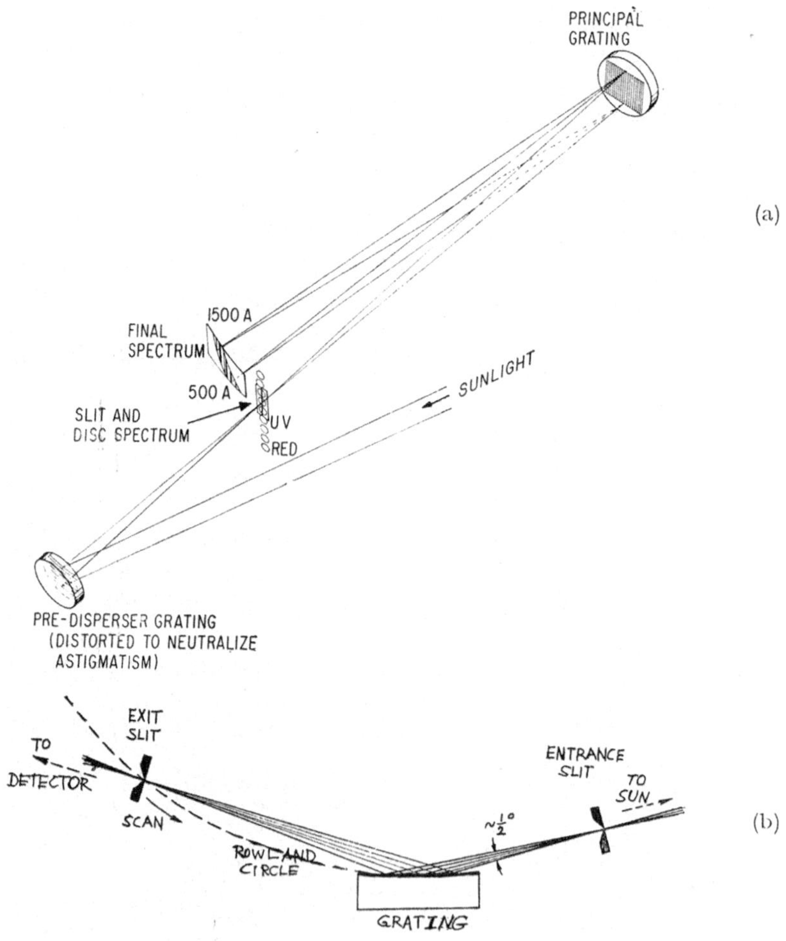

Fig. 1. The current schemes of XUV rocket spectroscopy (a) Double dispersion spectrographs (NRL) (from TOUSEY, 1963, Figure 3); (b) Grazing incidence monochromators (AFCRL, GSFC) (see HINTEREGGER, 1961a). For illustrative purposes the schematic drawing does not show true proportions.

ed reviews practically up to date. Therefore no amplification of such general technological accounts appears necessary here. Familiarity with at least the general basis of currently practicable types of rocket XUV instrumentation will be assumed in the following review of available data. Consequently, the technological part of discussions within this presentation has been reserved for special experimental problems and some new developments not yet covered in the existing literature.

For similar reasons the following selection in reproducing existing solar XUV rocket spectra has been chosen to serve only as a representative summary but by no means as a historically complete one.

All solar XUV rocket spectra have been obtained with diffraction gratings (see Figure 1). The brief designations of spectra as "Normal Incidence" (single dispersion), "Normal Incidence Double Dispersion," and "Grazing Incidence" (single dispersion) therefore provide sufficient identification of the type of optical systems with which the various spectra were acquired. The distinction between photoelectric and photographic detection of the dispersed radiation may be indicated by variously selected terminology. For the sake of brevity, I may use the expressions "*Countograms*" and "*Photograms*" where the existence of a spectrum is clear in the context. Where this is not so, I may speak of "Spectrocountograms" and "Spectrophotograms", respectively. For the written presentation, however, the but slightly longer expressions "Photographic Spectrograms" and "Counter Spectrograms" have been chosen, to avoid the perhaps awkward promotion of new abbreviations.

Although the first photographic solar XUV spectrograms (U of C) and counter spectrograms (AFCRL) were obtained in 1952 and 1959, a reasonably satisfactory spectral definition and more complete coverage of the wavelength range of interesting chromospheric emissions was not achieved until 1958 and 1960, respectively. Existing photographic XUV spectra of 1952 to 1958 (U of C, NRL) as well as the first counter spectrograms of 1959 (AFCRL) are merely of historical interest now. Therefore, they are not reviewed here.

ELECTRONICS VERSUS PHOTOGRAPHY

Nearly all counter spectrograms have been obtained with a relatively rapid wavelength scan. In AFCRL instruments the full wavelength range was covered in about ten seconds for all flights up to May 2, 1963. This high scanning speed was chosen mainly for the sake of aeronomical research objectives (H. E. HINTEREGGER, 1961a, 1961b; K. WATANABE and H. E. HINTEREGGER, 1962; H. E. HINTEREGGER and K. WATANABE, 1962; H. E. HINTEREGGER, 1962). Being of no interest within the scope of this symposium, these areonomical objectives are mentioned here merely as one of the reasons for the generally low spectral resolution of these AFCRL countograms.

This low resolving power of most counter spectrograms may appear rather puzzling in view of the use of gratings of a 2 meter radius of curvature in all

AFCRL monochromators. Besides the aforementioned aeronomical excuse, however, one should note a more basic experimental reason why it is generally difficult to obtain counter spectrograms of equally good resolution as that of photographic measurements for otherwise comparable experimental conditions. These conditions imply use of the same type of grating and the same duration of the photographic exposure as of the duration of complete scan in counting. The basic experimental handicap of scanning monochromator measurements relative to spectrophotography is perhaps best explained in reference to Figure 1. There it is immediately apparent that counter spectrograms are experimentally comparable to photographic records of exposure times much shorter than the scan period of the monochromator. The "effective" exposure times are essentially in the proportion of instrumental monochromator bandwidth ($\Delta\lambda_{\text{instr}}$) to the total scanning range ($\lambda_{\max} - \lambda_{\min}$), that is about $^1/_{100}$ (or $^1/_{1\,000}$) even for a rather poor (or modest) resolution of 10 Å (or 1 Å) of a total range of 1 000 Å.

Although this handicap has been partially offset by improving photoelectric detection techniques, it would be highly imprudent to ignore the obviously persisting value of the photographic emulsion in present and future rocket spectroscopy. In the language of modern electronic space technology, a simple photographic plate would have to be called a fabulously complex array of sensing, computing and data storage systems. Since the currently practicable sensitivities of photoelectric photon counting techniques are variously close to the theoretically expected ultimate limits, a major experimental break-through in photoelectronic rocket spectroscopy may be accomplished only with comparably complex arrays of simultaneously operating detectors. An alternative approach would be that of imaging and intensifying the photoelectrons emitted from a target conforming to a whole section of the Rowland circle. Obviously such a device is conspicuously resembling a photographic instrument in principle but requiring in practice an enormous amount of electronics and a very high information bandwidth of telemetry and data storage.

Representative Solar XUV Spectrograms of 1958–1963

1958/1959 GRAZING INCIDENCE PHOTOGRAPHIC SPECTRA, U OF C

The grazing incidence spectrograph records obtained on June 4, 1958 (U of C), provided the first photographic evidence of the Lyman alpha line of He II (303.8 Å). The records from this flight together with those from a successful repetition on March 30, 1959, already show indications of a remarkably large number of solar XUV emission lines throughout the entire wavelength range from H Ly-α (1 215.6 Å) to 87 Å. The identification of nearly all but the strongest lines from these records, however, was apparently extremely difficult because of disturbing high background of stray light. It was further complicated by the presence of film defects resembling spectral lines after development.

Apparently for these reasons, photographic spectrograms from these 1958/1959 flights were not reproduced in the literature. The published list of observed wavelengths of about 150 lines (T. VIOLETT and W. A. RENSE, 1959) stemming from extensive visual evaluation work, includes many interesting solar identifications as well as visual intensity estimates. This list has stimulated most valuable attention among solar physicists. However, it has also led to a considerable amount of controversy and contraversial use in the subsequent aeronomical and astrophysical literature. A re-evaluation of these controversies is no longer necessary since more recent spectra of better definition are practically free of the aforementioned experimental complications and are more directly applicable to intensity determinations throughout the same wavelength range that was covered in the 1958/1959 U of C experiments.

MARCH 1959 NORMAL INCIDENCE PHOTOGRAPHIC SPECTRA, NRL

The March 13, 1959 flight of the normal incidence spectrograph by NRL appears to have accomplished the most that could be done with a single dispersion normal incidence photographic system (J. D. PURCELL, D. M. PACKER, and R. TOUSEY, 1960) involving the successful use of a selectively reflective coating of the diffraction grating and a collimating mirror made of crystal quartz. The photospheric continuum appeared detectable to wavelengths of about 1700 Å below which it became swamped by instrumental stray-light. In spite of the latter, however, many genuine solar emission lines besides a number of members of the hydrogen Lyman series were clearly recorded in these spectra. They provided the first recordings of coronal emission of Ne VIII at 770 and 780 Å and many other new lines as well as a first indication of the hydrogen Lyman continuum. A reproduction of these already remarkably good spectrophotograms is omitted here only because the adoption of a double-dispersion system within the NRL flight program of normal incidence spectrographs led to truly superior performance soon thereafter.

APRIL 1960 AND AUGUST 1962 DOUBLE DISPERSION NORMAL INCIDENCE PHOTOGRAPHIC SPECTRA, NRL

The successful introduction of double dispersion in rocket XUV spectroscopy (C. R. DETWILER, J. D. PURCELL, and R. TOUSEY, 1961) appears to have been the decisive step in reducing the instrumental stray-light that had been the most serious handicap in all previous photographic observations of the solar XUV spectrum. Solar XUV spectra photographed with this technique on April 19, 1960, cover the wavelength range from 1550 Å to about 500 Å with a linear dispersion of 41 Å/mm. Selected samples, showing a 13 sec. exposure of the 1550–1160 Å region (SWR-C film), a 60 sec. exposure of the 1400–950 Å region (DC-3 film), and a 60 sec. exposure of the 1200–760 Å region are reproduced here as Figures 2b, 3a, and 3b, respectively. Figure 3c is a reproduction of the short wavelength end of the same exposure as shown here in Figure

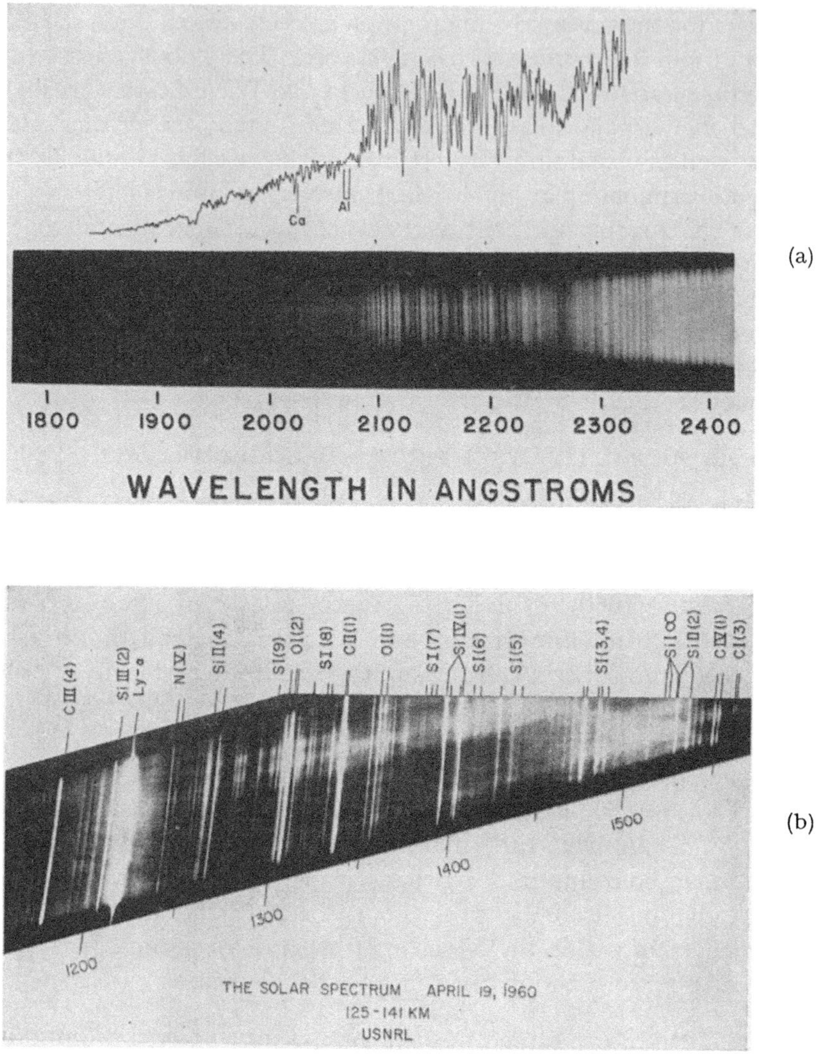

Fig. 2. Illustrating the transition from (a) the typical photospheric UV spectra to (b) the EUV (XUV) spectra of mainly chromospheric emissions observed at shorter wavelengths (from Tousey, 1963, Figures 18 and 19).

3b, produced by the first order spectrum of the predisperser (right) together with that produced by the second order of the predisperser (left). Spectra with a linear dispersion of 20 Å/mm for the 2000–1200 Å region and 10 Å/mm for the 1250–800 Å region were obtained with two similar instruments on August 22, 1962 by Garret, Purcell, and Tousey (R. Tousey, 1963). They are reproduced here as Figure 4.

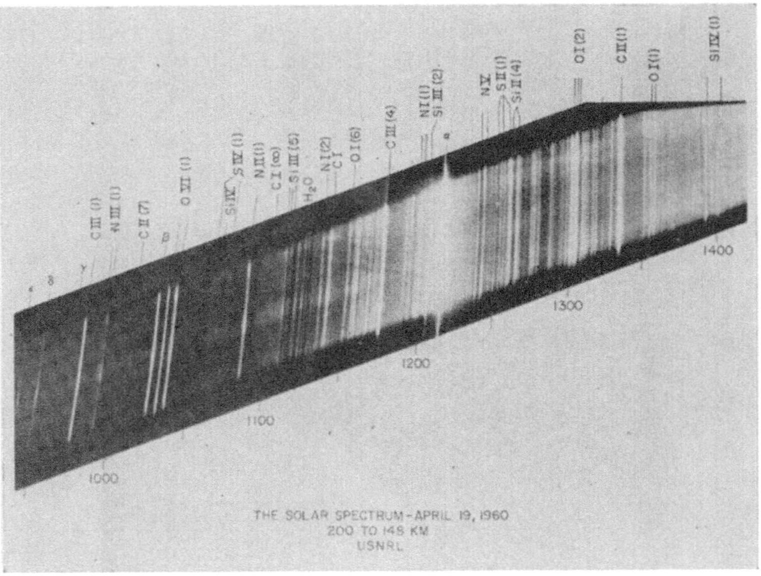
THE SOLAR SPECTRUM-APRIL 19, 1960
200 TO 148 KM
USNRL
1000
1100
1200
1300
1400

Fig. 3a.

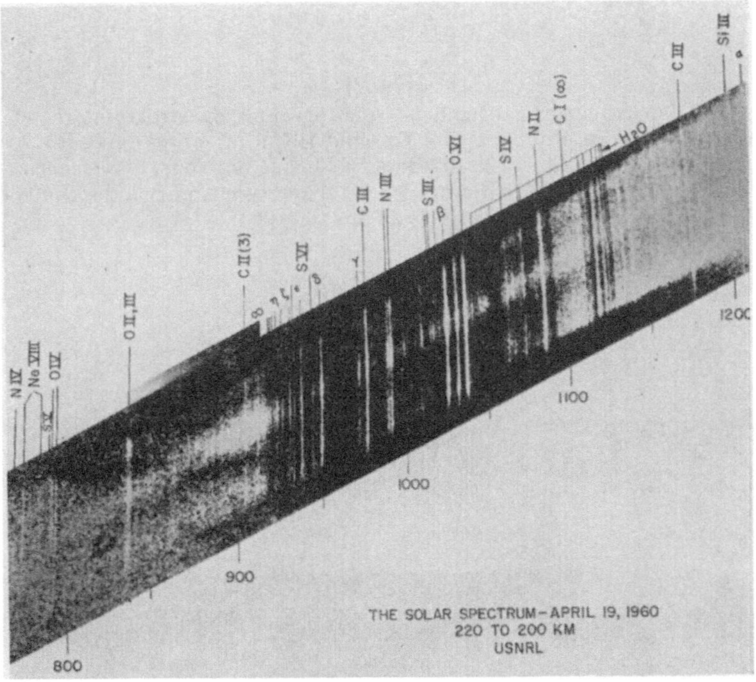
THE SOLAR SPECTRUM-APRIL 19, 1960
220 TO 200 KM
USNRL
800
900
1000
1100
1200

Fig. 3b.

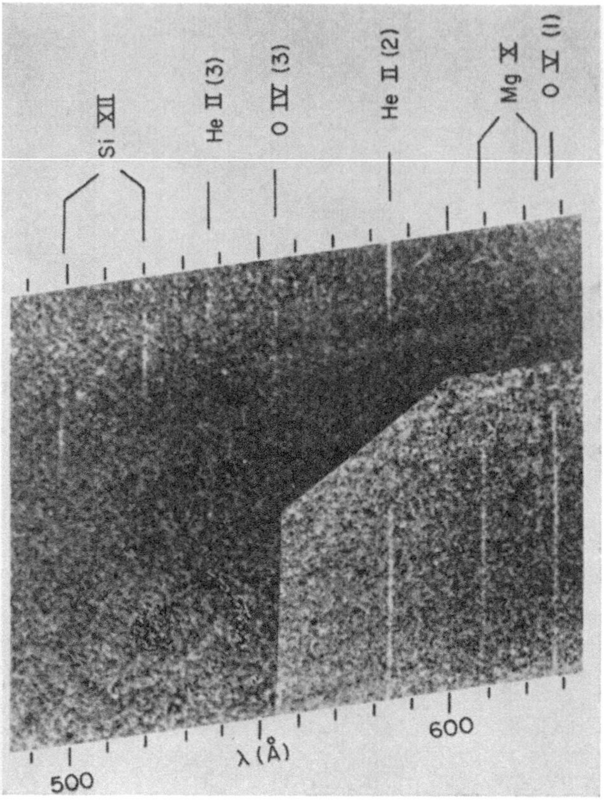

Fig. 3c.

Fig. 3. Double dispersion normal incidence photographic spectrograms, 1400–500 Å, of April 19, 1960 (DETWILER, PURCELL, and TOUSEY, 1961): (a) 60-sec exposure, 200–148 km, DC-3 film; (b) 60 sec exposure, 220–200 km, SC-4 film; (c) short wavelength sections of same exposure as (b), produced by first order pre-dispersion (right) and second order (left), respectively (TOUSEY, 1963).

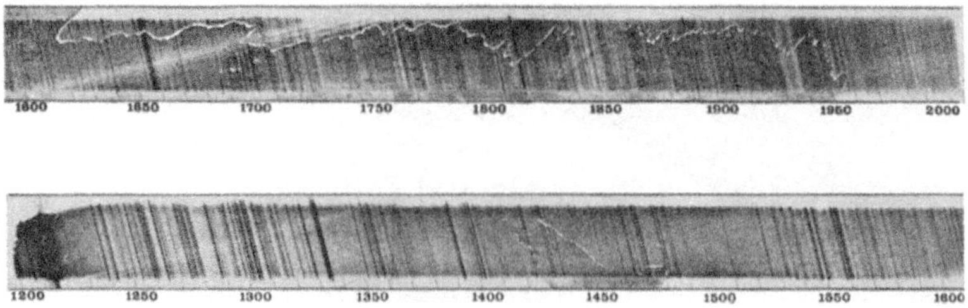

Fig. 4. Double dispersion normal incidence photogram of solar radiation with 0.2 Å resolution (TOUSEY, 1963).

THE INTENSITY DISTRIBUTION OF SOLAR XUV RADIATION FROM THE INTEGRATED DISC

Intensities of the radiation from the whole XUV-visible solar disc are not easily obtained from densitometry of nearly stigmatic solar spectrophotograms. The difficulty is illustrated by the fact that the top and bottom edges of spectra of the type shown in Figures 2 and 3 correspond, roughly, to the limb of the sun and the bright regions traversing the spectrum correspond to plage regions (TOUSEY, 1963). A further complication is introduced by the slight intentional curvature of the slit (TOUSEY, 1963). Reasonably well representative absolute spectral intensity distributions based on densitometry of various NRL spectrograms (C. R. DETWILER, D. L. GARRETT, J. D. PURCELL, and R. TOUSEY, 1961; R. TOUSEY, 1963), however, have been published for the wavelength regions 2100–1750 Å, (1959 spectra) and 1550–860 Å (1960 spectra). These are reproduced here as Figure 5a and the Figures 5b and 6d, respectively. Figures 6a and 6c show 1961 AFCRL monochromator data ("countograms") with overlaid calibration curves (in photons per cm^2 per instrumental wavelength bandwidth $\Delta\lambda$) for comparison. Because of the use of grazing incidence and the omission of any collimating or image-forming devices in front of the entrance slit, the AFCRL countograms are most directly representative of the integrated flux from all XUV-visible parts of the solar atmosphere (HINTEREGGER, 1961a; TOUSEY, 1963).

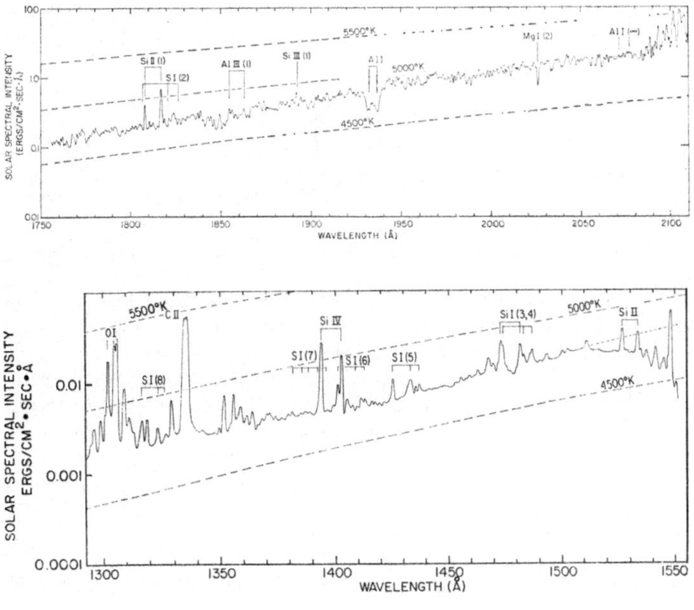

(a)

(b)

Fig. 5. Intensity distributions of sunlight from the integrated disc: (a) from single dispersion spectra (PURCELL, PACKER, and TOUSEY, 1960); (b) from double dispersion spectra (DETWILER, GARRETT, PURCELL, and TOUSEY, 1961).

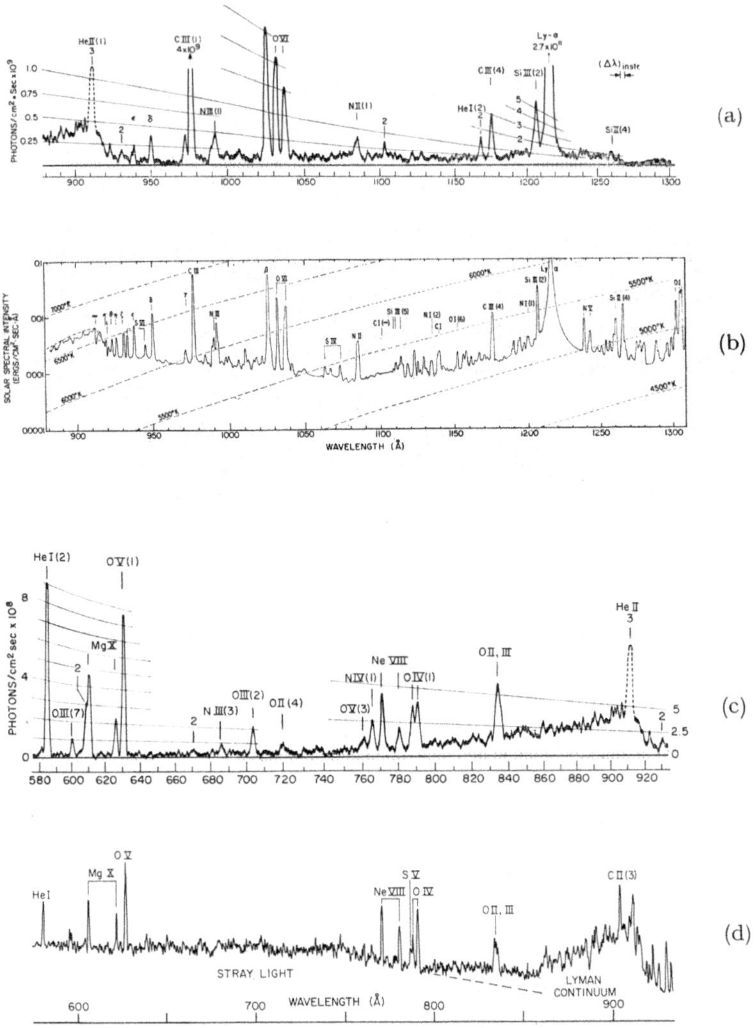

Fig. 6. Solar XUV intensity distribution in the 1300–584 Å region. Note the reasonably good agreement of the grazing incidence counter spectrograms, (a) and (c), with densitometry records, (b) and (d) of double dispersion normal incidence photographic spectrograms (reproduced from Tousey's review (1963)). The grazing incidence records refer most directly to the radiation from the entire XUV-visible solar disc (HINTEREGGER, 1961a; TOUSEY, 1963).

1960/1961 GRAZING INCIDENCE COUNTER SPECTROGRAMS (AFCRL, NRL, GSFC)

Telemetering monochromators transmitting the photoelectron counts produced by the dispersed radiation emerging from a continuously moving exit slit (see Figure 1b) together with signals of the position of the exit slit along Rowland circle have been flown successfully on January 19, 1960 (250–1300 Å scan),

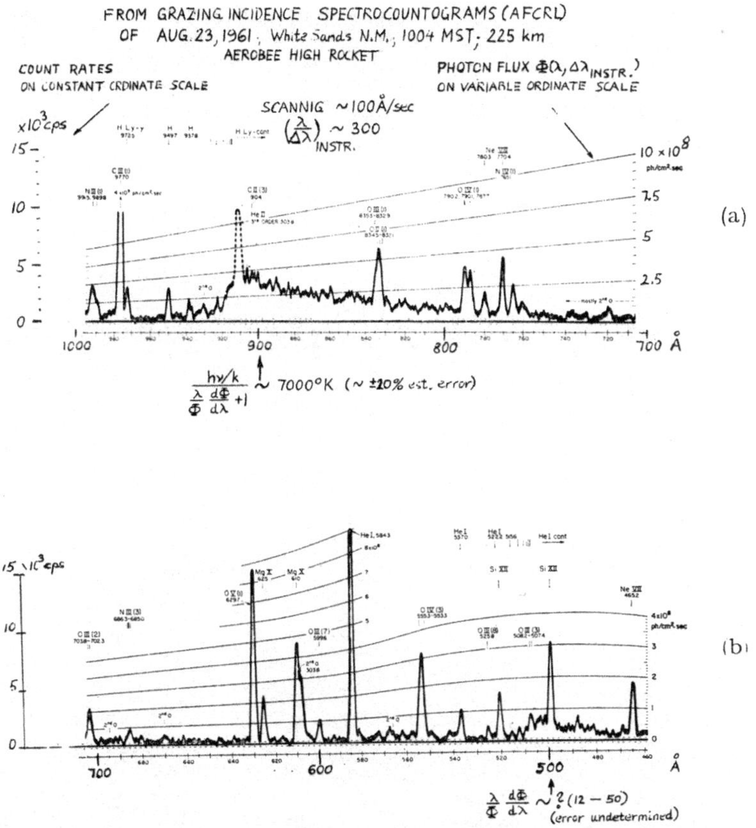

Fig. 7. Fast scan XUV counting spectrograms allowing a fair estimate of hydrogen Lyman continuum intensity distribution. However, a reliable determination of the He I continuum and more accurate evaluations of the H Lyman continuum must await the acquisition of rocket or satellite measurements with much slower scanning speed (OSO-C) and/or increased instrumental optical speed (e.g., collimation in front of entrance slit).

January 29, 1960 (60–310 Å, see Figure 8b), August 23, 1960 (250–1300 Å, see Figure 8b) and August 23, 1961 (250–1300 Å, see Figures 6a, 6c, and Figure 7) by AFCRL (H. E. HINTEREGGER, 1961a; L. A. HALL, K. R. DAMON, and H. E. HINTEREGGER, 1963). All of these flight monochromators were operated with a fast wavelength scan of about ten seconds per full range, since they were "multipurpose" instruments designed for studies of the terrestrial upper atmosphere (H. E. HINTEREGGER, 1961b; K. WATANABE and H. E. HINTEREGGER, 1962; H. E. HINTEREGGER and K. WATANABE, 1962; H. E. HINTEREGGER, 1962) at least as much as for solar physics (H. E. HINTEREGGER, 1960; H. ZIRIN, L. A. HALL and H. E. HINTEREGGER, 1963). As a result the spectral resolution has been considerably lower than might be expected from the use of

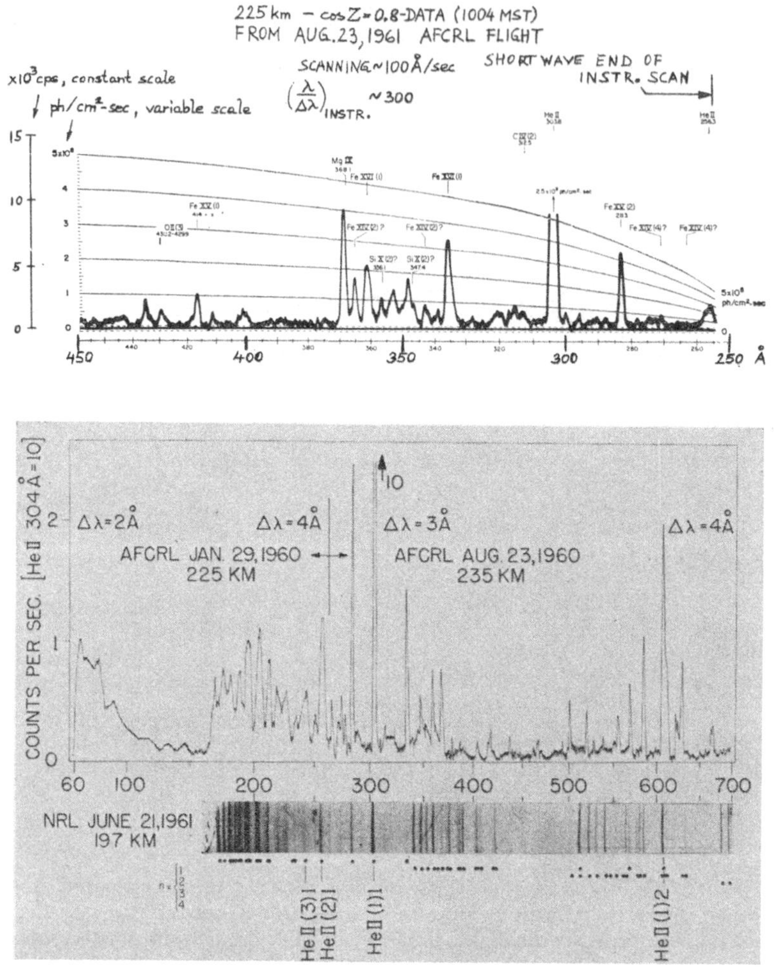

(a)

(b)

Fig. 8. 1960/1961 spectra extending to shortest wavelengths of 170 Å (60 Å) (also see Figure 9c), acquired with *grazing incidence monochromators without filter* (AFCRL), shown in (a) and upper part of (b) and a *grazing incidence spectrograph* with aluminum filter (NRL), shown in lower part of (b) (Tousey, 1963). The conspicuous drop in intensities shortward of 170 Å, though mainly instrumental in Al-filtered exposure has been confirmed as being truly solar in most recent experiments.

diffraction gratings of 2 meter radius of curvature (see previous section on "Electronics versus Photography").

To obtain a notable reduction in the statistical error of the recorded number of photoelectrons per passage of the exit slit over a given emission, a quadratically larger increase of the allotted time per scan is necessary if both instrumental wavelength bandwidth (resolution) and total scanning range are retained. The rate of disturbing background counts, due mainly to XUV stray light from the grating and cosmic rays, on the other hand cannot be lowered below a certain level, at least not within a given state of the experimental art. Therefore the instrumental parameters of scanning time, wavelength range, spectral resolution, and minimum detectable solar emission line intensity cannot be chosen independently, and the determination of the most practical mutual trade-off among partially conflicting instrument capabilities (H. E. HINTEREGGER, 1961b) deserves rather thorough scientific-technological attention. The restriction to a single available type of instrument carrier with a given time of vacuum exposure represents a stringent experimental limitation to present U. S. solar observation experiments of the non-orbiting variety (see section on Instrument Carriers earlier in this paper; also see legend to Figure 7).

The benefit of a slower scanning speed for solar spectroscopy can be seen from a comparison of the notably better resolution of the 60-sec scan of the GSFC monochromator of September 30, 1961 (see Figure 9c) with the 10 sec. scan of the January 29, 1960 data (upper left of Figure 8b) for a comparable total range of scanned wavelengths. The detectors of both the GSFC mono-

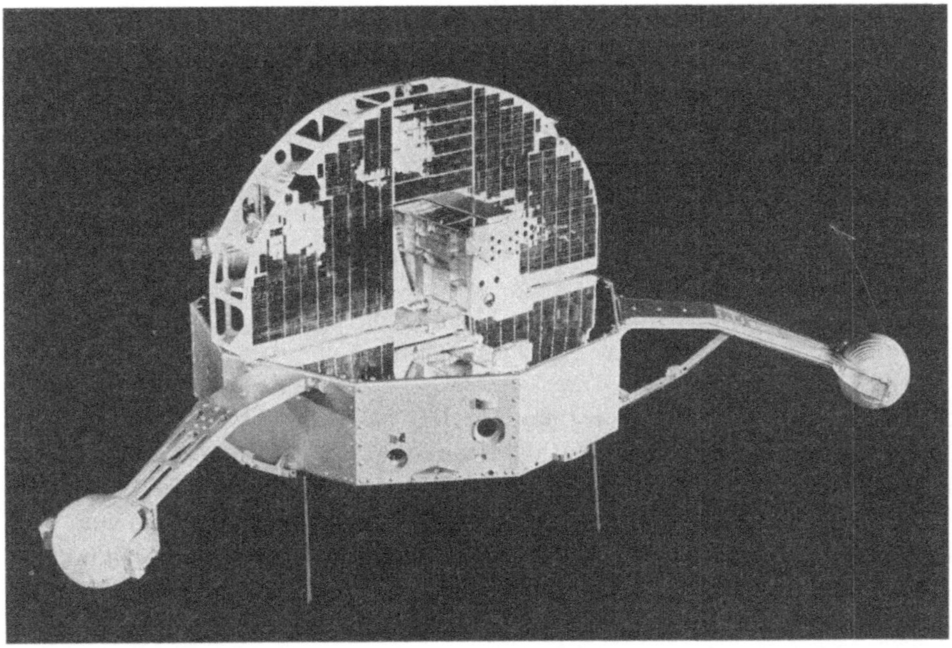

Fig. 9. A grazing incidence monochromator (GSFC) aboard the first orbiting solar observatory (OSO-A, U. S. National Aeronautics and Space Administration) allowed solar XUV spectrocountograms to be acquired over a long time with the same instrument. The spectra (a) and (b) correspond to a relatively quiet and moderately active sun, respectively. The countogram (c) from an earlier non-orbiting flight of a similar GSFC monochromator reveals a higher instrumental efficiency in the first order of diffraction (compare Figure 10a).

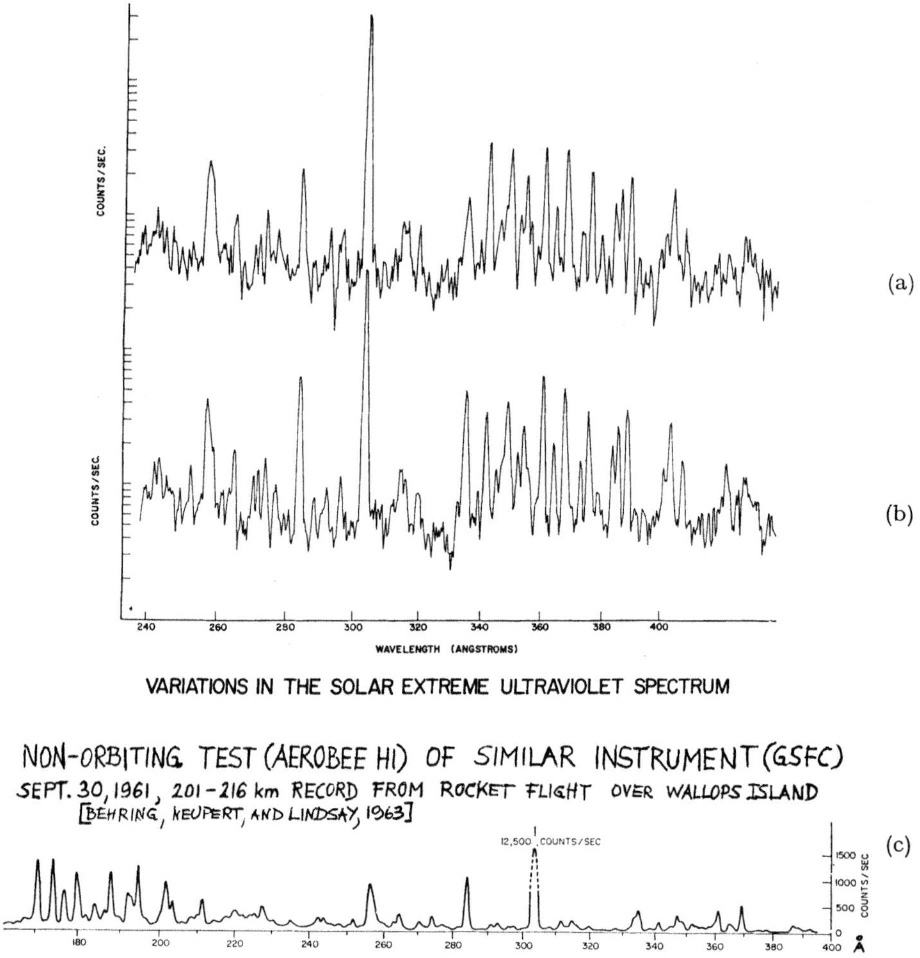

Fig. 9 (continued).

chromator and the January, 1960 AFCRL monochromators used tungsten photocathodes (L. Heroux and H. E. Hinteregger, 1960) whereas the August, 1961 instrument used a non-unipotential cathode of tin oxide baked onto glass (L. A. Hall and H. E. Hinteregger, 1960). Since tungsten was used in both short wavelength monochromator flights (AFCRL, left part of Figure 8b; GSFC, Figure 9c), there remained some doubt as to whether or not the conspicuous drop-off of spectral intensities for wavelengths below about 170 Å, shown clearly in both observations, might be a merely instrumental effect rather than an actual characteristic feature of the solar XUV emission spectrum. Even though the photographic records recovered from a grazing incidence spectrograph flown by NRL on June 21, 1961 (R. Tousey, 1963) again showed a drop-off in observed intensity below 170 Å, this could be attributed readily to the expected absorption effect of the aluminum filter. The latter had to be used to prevent excessive background blackening of the photographic spectrogram in grazing incidence. This NRL spectrogram, though agreeing reasonably well with the photoelectric counting record (see Figure 8b) was therefore neither a confirmation nor a contradiction of the earlier discovered drop-off below 170 Å (upper left of Figure 8b).

1962 GRAZING INCIDENCE SPECTRA (GSFC, NRL, AFCRL)

The most significant event of solar XUV spectroscopy was the successful launch and operation of the first orbiting solar observatory (OSO-A), a photograph of which is reproduced here at the top of Figure 9. Its solar-oriented section contained an XUV scanning monochromator that was provided by J. C. LINDSAY and his associates (GSFC). The instrument was of a design similar to the AFCRL grazing incidence monochromator (H. E. HINTEREGGER, 1961a) except for the use of a smaller radius of the Rowland circle and a different exit slit scanning mechanism. The latter required a moving detector but allowed the use of a smaller version of the Bendix open structure magnetic multiplier with a relatively small tungsten cathode, whereas the AFCRL instruments have used stationary $3\frac{1}{2}$ inch long photocathodes (L. HEROUX and H. E. HINTEREGGER, 1960; L. A. HALL and H. E. HINTEREGGER, 1960). Preliminary results from the successful operation of this GSFC monochromator have been reported at the Third International Space Science Symposium (COSPAR). Expecting a more detailed account of these OSO-A measurements to appear in the literature soon, one can say already that the major achievement was certainly that of providing solar XUV spectrophotometry data for the first time more or less continuously over many weeks covering not only more than the period of a solar rotation but also rather different types of solar activity (see Figure 9a, b). Unfortunately, the ratio of first order diffraction signals to the background in these data is evidently not as high as one should expect from an orbiting instrument of relatively slow spectral scan.

A comparison of the rocket spectrum (Figure 9c) with OSO-A records (Figure 10a) seems to indicate that the grating of the orbiting instrument gave intensities in second order of diffraction appreciably higher than those of either the GSFC Aerobee instrument or the improved NRL grazing incidence spectrograph that was flown on August 22, 1962 (see Figure 10b) (R. TOUSEY, 1963).

A number of remarkably strong solar emission lines in the 205–170 Å region as well as the conspicuous general intensity drop below 170 Å appear again in these 1962 spectra (see both Figures 10a and 10b). Whereas the character of these strong XUV lines as truly solar emission characteristics was thus established beyond any doubt, the reality of the drop around 170 Å as a solar feature was considered to be still questionable (R. TOUSEY, 1963). Even less certain appeared the reality of the spectral features shown by the January, 1960 recordings in the 168 to 60 Å region. Auto-correlation studies of many OSO-A instrument scans for the 170–20 Å are still in progress with preliminary results

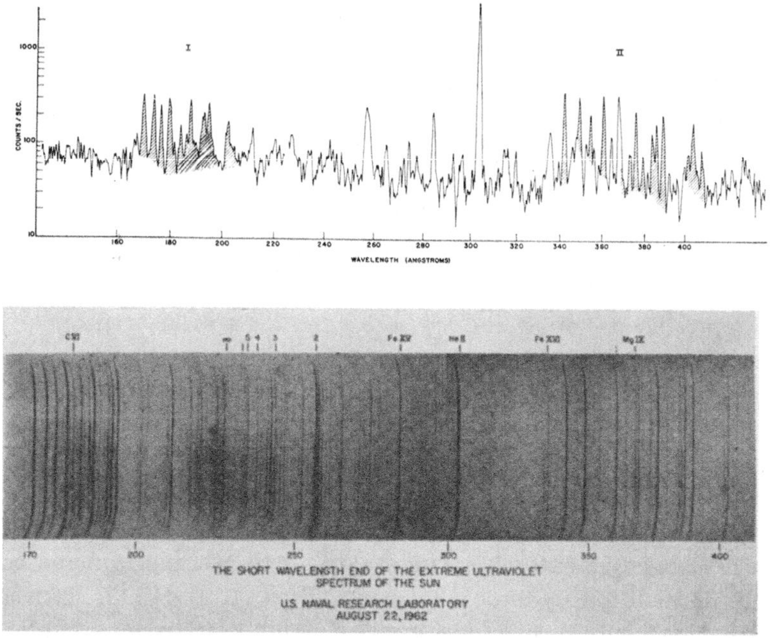

Fig. 10. 1962 spectra of 400–170 Å region of solar XUV emissions. The still largely unidentified observed strong lines between about 170 Å and 204 Å (I) dominate in second order (II) in appearance over the first order lines of considerably actual intensity (see Figure 8a) in both the satellite countogram (a) and Aerobee Hi rocket photogram (b) (Tousey, 1963).

suggesting fair agreement with the 1960 AFCRL data (private communication, J. C. Lindsay, 1963).

The data from two successful AFCRL monochromator flights (June and October, 1962) covering the 1300–250 Å region are not discussed here, since both experiments were conducted for the exclusively aeronomical purpose of comparing the early morning and late afternoon structure of the terrestrial atmosphere (L. A. Hall, W. Schweizer, and H. E. Hinteregger, 1963). This purpose dictated the choice of a large solar zenith angle around 81° resulting in spectra that are significantly modified by atmospheric absorption even near the peak of the rocket trajectory.

An AFCRL monochromator for longer XUV wavelengths was flown on July 10, 1963. However, this instrument was not designed to produce an actual spectrum scan but to monitor the observed intensities at nine fixed wavelengths (ranging from 1206.5 Å to 303.8 Å) as a function of height in the earth's atmosphere with a much better altitude resolution than could be obtained with the previously flown "multi-purpose" monochromators. Since a considerable amount of exploratory information is now available for both solar and terrestrial atmospheric physics, this latest specialization of solar XUV rocket experiments in either exclusively aeronomically or solar physics oriented types appears to be most fruitful. The results from the first specialized AFCRL experiments of May 2 and July 10, 1963 (L. A. HALL, W. SCHWEIZER, and H. E. HINTEREGGER, 1963b) suggest that an even more pronounced differentiation between aeronomical XUV spectrophotometry and solar physics oriented XUV experimentation, including even different experiment carriers, appears to be most desirable from now on.

MAY 2, 1963 GRAZING INCIDENCE COUNTER SPECTRUM FROM 310 Å to 56 Å (AFCRL)

The unfortunate coincidence of various types of rocketry mischief with exactly all those AFCRL monochromators that were flown for the short wavelength

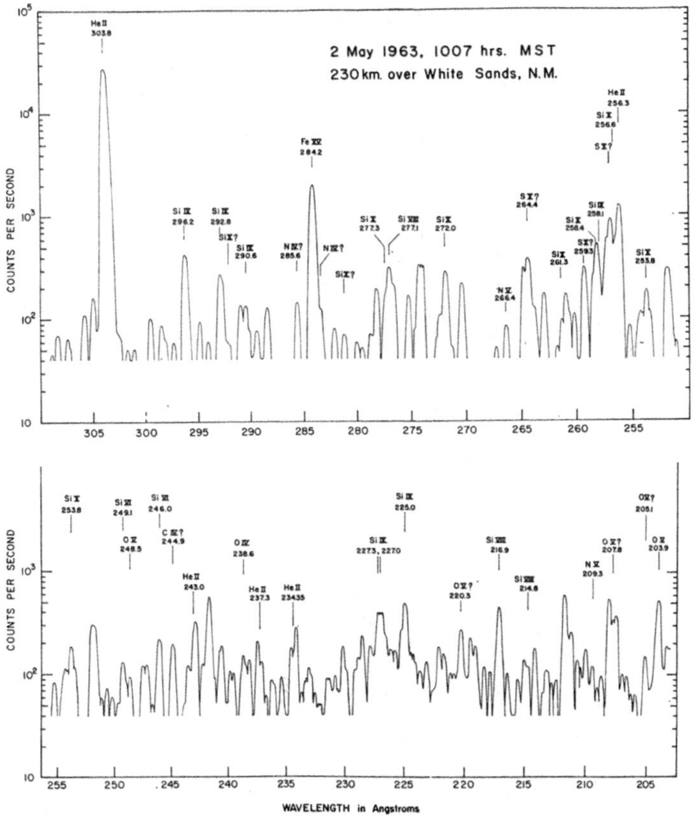

Fig. 11. May 2, 1963 spectrum of 310–56 Å region of solar XUV emissions, from 100-sec scan of grazing incidence monochromator on Aerobee Hi rocket, 10 a.m.; 226–234 km over White Sands, New Mexico (HINTEREGGER, HALL, and SCHWEIZER, 1963). The appearance of the He II Lyman continuum is almost certainly real and the corresponding model temperature determined from the logarithmic slope is approximately 18000 °K (+10000, −3000) without any reliance on the estimated error (compare Figure 7 for H and He I).

region (i.e., with 30000 lines per inch on the grating instead of 7500) since the first successful experiment on January 29, 1960, explains the complete lack of any AFCRL monochromator data of wavelengths below 250 Å for the following period of over three years. Because the primary scientific interest in new data for wavelengths below 250 Å was centered upon the solar emission spectrum as it exists essentially above the absorbing region of the earth's atmosphere rather than atmospheric absorption characteristics, the scanning time of the instrument could be increased from 10 sec to 100 sec. Together with various general instrumental improvements accumulated over three years, this increased scanning time indeed paid off in the form of an incomparably better spectrum (see Figure 11) than the crude first one of 1960 (see left upper part of Figure 8b, where a rather large stray-light background has been subtracted already, whereas Figure 11 represents data without any correction). A more detailed report on this May 2, 1963 flight would not be within the scope of this review and will soon be submitted for publication (H. E. HINTEREGGER, L. A. HALL

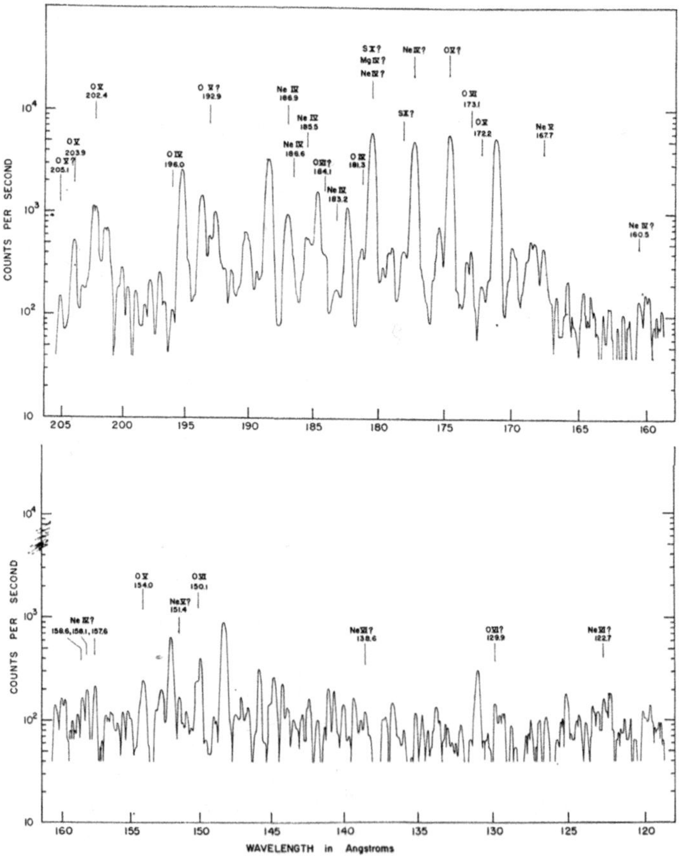

Fig. 11 (continued).

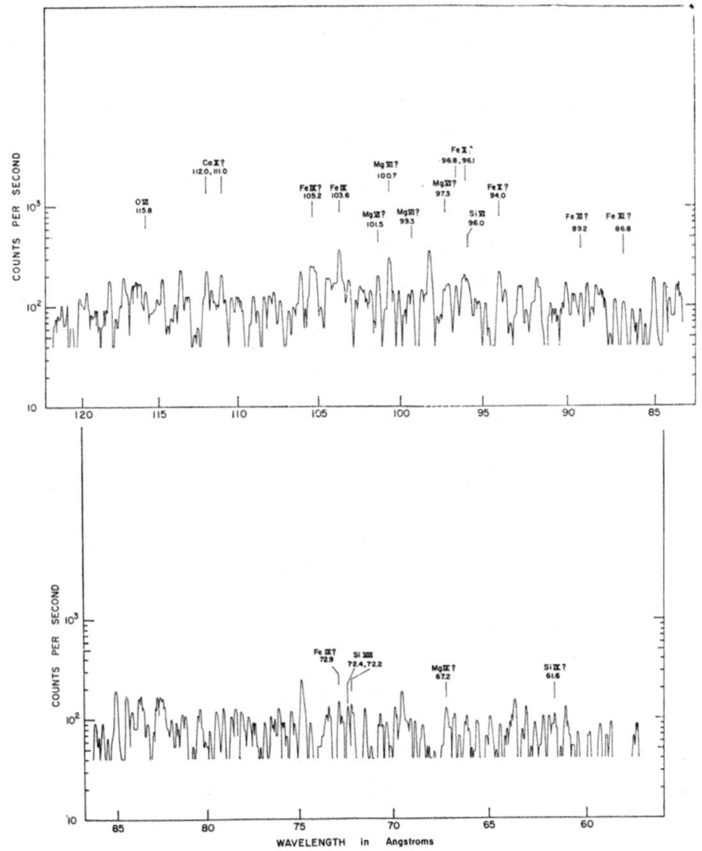

Fig. 11 (continued).

and W. SCHWEIZER, 1963). A number of preliminary working sheets (rolls of large size giving much more detail than could be considered practical for reproduction) have been sent to various interested colleagues in the U. S. and abroad prior to this symposium. Additional copies may be obtained by writing to the author. Considering the recent acquisition of the data, one should consider the solar emission line identifications appearing in the present Figure 11 as merely tentative, included mainly for the purpose of being exposed to discussion at this symposium.

XUV Spectroheliograms

STIGMATIC XUV IMAGING PHOTOGRAPHY

Since the first spectroheliogram in the light of hydrogen Lyman alpha (R. MERCURE, S. C. MILLER, W. A. RENSE, and F. J. STUART, 1956), no truly monochromatic XUV photographs of the sun have been obtained at any wave length in the XUV region of the spectrum other than H Ly-α (1216 A) until 1963 (TOUSEY, 1963), when an NRL spectroheliograph designed by Purcell (R. TOUSEY, 1963) gave images of the sun in He II, 304 Å and apparently clearly distinguishable ones even in Fe XV, 284 Å. Since these exploratory measurements (R. TOUSEY, private communication), have not been submitted for publication as yet, I am showing only briefly two slides of the Lyman alpha spectroheliograph of PURCELL, PACKER, and TOUSEY (1960) and the other of the aforementioned successful NRL rocket spectroheliograph as illustrated in Tousey's review (1963) however without reproduction in the written version of this presentation.

MECHANICAL TWO-DIMENSIONAL DISC SCAN

To avoid the need to recover photographic film from satellite apparatus, it is necessary to resort to photoelectric detection techniques and telemetry. A suitably modified version of the photographic NRL spectroheliograph is shown in a slide from Tousey's review (R. TOUSEY, 1963, Figure 12b, not reproduced here). This NRL instrument is being prepared for flight on the NASA satellite OSO-B. It has been designed to produce three stigmatic images of the sun near the normal to a diffraction grating of lm radius of curvature. The three images are in He II, 304 Å (third order), He I, 584 Å (second order), and H Ly-α, 1216 Å (first order), respectively (R. TOUSEY, 1963). At the place of each image of about 0.6 mm diameter a separate photoelectric detector with a much smaller sensitive area of about 0.02 mm diameter is mounted in a fixed position adjusted to coincide with the center of the solar image when the instrument axis points at the center of the solar disc. This ratio of the diameters of the solar image and the detector aperture area corresponds to an expected spatial resolution of about 1 minute of arc across the solar disc. Telemetry of the photoelectric signals for each of the 1600 steps in the raster scan of the pointed section of OSO-B (40×40 minutes of arc) around the central solar vector then provides the raw data from which the spectroheliograms may be constructed.

NON-MECHANICAL ONE-DIMENSIONAL DISC SCAN

An entirely different method of determining the spatial distribution of monochromatic XUV emission sources across the solar disc has been explored in some detail by BEDO and HINTEREGGER (unpublished AFCRL work 1962/1963; NASA proposal, December, 1962). The principle of this method is simply an application of the diffraction law of planar gratings for appropriately chosen median angles of grazing incidence and suitable instrumental exit beam collimation. The exit collimation is accomplished by a modified type of a Soller collimator (used previously only in X-ray work) for modest resolution, or a grazing incidence collimator mirror for higher resolution.

For a more or less *non-localized* monochromatic source brightness distribution across the solar disc, the solar radiation incident on the grating has an

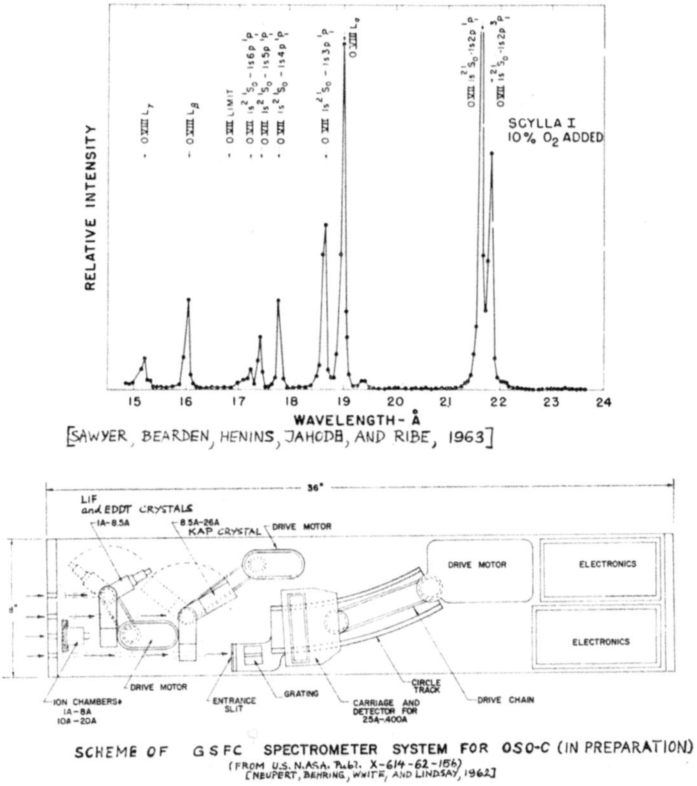

Fig. 12. Steps toward dispersive solar rocket spectroscopy at X-ray wavelengths. An X-ray diffraction countogram (KAP crystal) of the O VII and O VIII spectrum from one of the hottest laboratory plasma sources (Scylla I, Los Alamos Scientific Laboratory) is shown out of context here as (a); Laboratory spectra of this type may soon be compared with solar measurements. In July, 1963, the first successful rocket flight of a KAP crystal spectrometer has been accomplished by scientists of the Naval Research Laboratory (Friedman, private information, 1963). Similar crystal spectrometers are projected to be flown in the pointed section of the NASA satellite OSO-C, within the frame of the GSFC spectrometer system outlined schematically in (b).

angular collimation width of $(\Delta\alpha)_0 \approx 0.01$ radian (α is the angle of incidence relative to the grating normal). The instrumental spectral resolution width is given by the relation

$$(\Delta\lambda)_0 = d\,[(\cos\alpha)(\Delta\alpha)_0 + (\cos\beta)\Delta\beta] \tag{1}$$

where β is the diffraction angle for wavelength λ, $\Delta\beta$ is the angular transmission width of the instrumental exit collimator, and d is the grating spacing.

For a strongly *localized* monochromatic emission from a region on the solar disc subtending a small angle, $(\Delta\alpha)_\bullet \ll (\Delta\alpha)_0$, the corresponding spectral resolution width,

$$(\Delta\lambda)_\bullet \approx d(\cos\beta)\Delta\beta \tag{2}$$

is of course smaller than $(\Delta\lambda)_0$. The ratio of these two instrumental widths approaches the form

$$R \equiv \frac{(\Delta\lambda)_0}{(\Delta\lambda)_*} \approx 1 + \frac{\alpha^*(\Delta\alpha)_0}{\Delta\beta}\sqrt{\frac{d}{2\lambda}} \tag{3}$$

which is valid for sufficiently small glancing angles ($\alpha^* \equiv \pi/2 - \alpha$) and short wavelengths (i.e., $\alpha^* \ll 1$ and $\lambda \ll d$).

This quantity R is actually the most generally descriptive parameter of the entire instrumental operation, since it determines not only the spectral resolution discrimination but also the instrumental spatial resolution. The latter assumes the conveniently simple form

$$(\Delta\alpha)_{\text{instr}} \equiv \left(\frac{\partial\alpha}{\partial\beta}\right)_{\lambda=\text{const}} \Delta\beta \approx \frac{1}{R}\,(\Delta\alpha)_0 \tag{4}$$

for $\alpha^* \ll 1$, $\lambda \ll d$, and $\Delta\beta \ll (\Delta\alpha)_0$. Therefore, this quantity R has a physical meaning similar to but not identical with the ratio of the diameters of the monochromatic solar image and the sensitive detector area in the mechanical scanning scheme of obtaining spectroheliograms mentioned in the preceding section.

The present scheme does not involve any mechanical scan of the instrumental entrance axis around the central solar vector but assumes merely steady pointing; namely, the constancy of the glancing angle α^* of a ray from the center of the solar disc. The scanning consists of merely a rotation, internal to the instrument, of the exit collimator about an axis perpendicular to the plane of dispersion. The measurements are recordings of the dispersed photon flux within the collimator transmission width, $\Delta\beta$, of diffraction angles, β, as a function of angular collimator position that may be labeled in "nominal wavelengths" (instead of β) referring to monochromatic emission localized near the center of the solar disc. A localized monochromatic emission occurring away from the center of the solar disc at an angle ($\Delta\alpha$) measured parallel to the instrumental plane of dispersion therefore appears as a shift in "nominal wavelength". The latter is of course measurable only if one already knows the actual wavelength. This will certainly be the case for the currently projected experiments. Since the size of one such instrument, requiring no entrance optics, will be quite small, it is easy to add at least one more identical instrument with a 90° rotated plane of dispersion. The resulting two-dimensional emission localization detection system will certainly allow a unique evaluation at least for periods when only a few active emission centers are present on the solar disc. Even for relatively complicated actual emission distributions, however, the resulting records are still expected to provide meaningful information.

The present method can obviously not be used for any arbitrarily chosen solar XUV emission wavelengths. The practical restriction of these monochromatic XUV emission localization measurements can be described by saying that one must not pick a wavelength too close to other strong solar emissions. The expression "too close" refers to a spectral separation of $(\Delta\lambda)_0$ for essentially non-localized line emissions and some reasonable multiple of $(\Delta\lambda)$. for localized ones. However, this restriction is apparently no more serious than similar ones inherent in the use of the more conventional projected method of obtaining XUV spectroheliograms.

Solar XUV Spectrophotometric Monitoring

Future work in experimental solar XUV spectroscopy by a still fairly small number of groups of scientists and technologists, will undoubtedly be concentrated on increasingly refined measurements of existing types as well as the exploration of new approaches. Besides this certainly most exciting work, however, we must try to maintain (and to some extent first establish) continuity in relatively simple types of solar XUV spectrophotometry. The importance of this somewhat dull obligation rests on requirements not only of studying the solar chromosphere and corona per se, but also of studying the terrestrial upper atmosphere ("solar-terrestrial relationships"), other planetary atmospheres of our solar system, and the interplanetary medium. These general needs are not reviewed here since they appear sufficiently clearly in various original papers and reviews (e.g., H. FRIEDMAN, 1960; L. GOLDBERG, 1960; D. R. BATES, 1954; H. E. HINTEREGGER, 1961a; H. E. HINTEREGGER, 1961b; H. E. HINTEREGGER and K. WATANABE, 1962; H. E. HINTEREGGER, 1962; I. HARRIS and W. PRIESTER, 1963; A. DALGARNO, 1962; A. DALGARNO, 1963; L. A. HALL *et al.*, 1963) in the English literature as well as a substantial number of publications in Russian journals on astrophysics and aeronomy (e.g., G. S. IVANOV-HOLODNYI, 1959; G. S. IVANOV-HOLODNYI and G. M. NIKOLSKII, 1961; G. S. IVANOV-HOLODNYI and G. M. NIKOLSKII, 1962; G. M. NIKOLSKII, 1962; G. S. IVANOV-HOLODNYI, 1962).

The types of instruments that may be called "non-dispersive" (ion chambers or Geiger counters with various gas fillings and filters, proportional counters, scintillation counters, and photoelectron energy analyzers) have already produced an extremely valuable wealth of information. This has been described in detail most recently in a well organized review by Friedman in the book *Space Science* (H. FRIEDMAN, 1963).

A vigorous extension of these non-dispersive monitoring experiments, which can be conducted rather inexpensively from many different types of rockets and space vehicles without solar orientation control, is of unquestionable value. Over the last few years of rapidly increasing knowledge of details of the solar emission spectrum, however, it has become apparent, that a second type of "solar monitoring" experiments, intermediate to non-dispersive observations and the most advanced types of grating spectroscopy, should be added to the existing monitoring projects. A step in this direction is represented by the solar

XUV spectrophotometer now being prepared at AFCRL for flight on the modestly solar oriented paddle of the NASA satellite POGO (orbiting geophysical observatory to be launched into nearly polar orbit). The instrument promises to achieve a resolving power, $\lambda/\Delta\lambda$, of about 100 throughout the spectral region from about 170 Å to 1700 Å using a stack of planar gratings in a simple non-focusing arrangement with rotating mechanical exit collimators, a matched stack of different photocathodes, and SE multiplication in counting techniques. The principle of operation of this compact "monochromator stack" is similar to that of the aforementioned non-mechanical one-dimensional solar disc scanner except for a considerably larger angular width of the exit collimator transmission, $\Delta\beta \gtrsim (\partial\beta/\partial\alpha)(\Delta\alpha)_0$ in the previous notation. A more detailed discussion of the utility of this or other monitoring types of XUV spectrophotometers of intermediate complexity and small size would be premature at this time.

It seems appropriate to conclude this section with the strongest possible emphasis of the importance of supporting basic experimental laboratory research work. Only a more vigorous pursuit of the latter can be expected to remove a number of embarrassing question marks selected to various important data acquired by space experiments so far. The fortunate recognition of this need has already led to the formation of new organizations dedicated to objectives that may be summarized as "laboratory astrophysics."

Note. High resolution XUV spectroscopy of emission line shapes is not covered in this presentation. Some most recent photographic determinations of the profiles other that H Lyman α e.g. H Lyman β at 1025 Å and O VI at 1032 Å should be reported by J. D. Purcell (NRL) in the afternoon discussion session (see page 212).

References

BATES, D. R.: 1954, *The Earth as a Planet* (ed. by P. KUIPER), University of Chicago Press.
BEHRING, W. E., NEUPERT, W. M., and LINDSAY, J. C.: 1963, *Space Research* 3 (ed. by W. PRIESTER), North-Holland Publishing Co.
CHUBB, T. A., FRIEDMAN, H., KREPLIN, R. W., BLAKE, R. L., and UNZICKER, A. E.: 1961, *Mém. Soc. Roy. Sci., Liège* **4**, 228.
DALGARNO, A.: 1962, private communication.
DALGARNO, A., MCELROY, M. B., and MOFFET, R. J.: 1963, *Plan. Space Sci.* **11**, 463.
DETWILER, C. R., PURCELL, J. D., TOUSEY, R.: 1961, *Mém. Soc. Roy. Sci., Liège* **5**, sér. 4, 254.
DETWILER, C. R., GARRETT, D. L., PURCELL, J. D., and TOUSEY, R.: 1961, *Ann. Géophys.* **17**, 9.
ELWERT, G.: 1963, *Proc. Int. Conf. on The Ionosphere, London,* 1962, Int. of Phys. and Phys. Soc. London Publ.
FRIEDMAN, H.: 1960, *Physics of the Upper Atmosphere* (ed. by P. RATCLIFFE), Academic Press.
FRIEDMAN, H.: 1963, *Space Science* (ed. by LEGALLEY), John Wiley and Sons.
GOLDBERG, L.: 1954, *The Earth as a Planet* (ed. by P. KUIPER), University of Chicago Press.
HALL, L. A. and HINTEREGGER, H. E.: 1960, *AFCRL Publication* TR-60-428.
HALL, L. A., SCHWEIZER, W., and HINTEREGGER, H. E.: 1963a, submitted to *J. Geophys. Res.*

HALL, L. A., SCHWEIZER, W., and HINTEREGGER, H. E.: 1963b, to be submitted to *J. Geophys. Res.*
HALL, L. A., DAMON, K. R., HINTEREGGER, H. E.: 1963, *Space Research* **3** (ed. by W. PRIESTER), North-Holland Publishing Co.
HARRIS, I. and PRIESTER, W.: 1963, *Space Research* **3** (ed. by W. PRIESTER), North Holland Publishing Co.
HEROUX, L. and HINTEREGGER, H. E.: 1960, *Rev. Sci. Instr.* **31**, 280.
HINTEREGGER, H. E.: 1960, *Astrophys. J.* **132**, 801.
HINTEREGGER, H. E.: 1961a, *Space Astrophysics* (ed. by W. LILLER).
HINTEREGGER, H. E.: 1961b, *J. Geophys. Res.* **66**, 2367.
HINTEREGGER, H. E. and WATANABE, K. W.: 1962, *J. Geophys. Res.* **67**, 3373.
HINTEREGGER, H. E.: 1962, *J. Atm. Sci.* **19**, 351.
HINTEREGGER, H. E., HALL, L. A. and SCHWEIZER, W.: 1963, to be submitted to *J. Geophys. Res.*
IVANOV-HOLODNYI, G. S.: 1959, *Izv. AN SSSR, Ser geofiz.* **1**, 108.
IVANOV-HOLODNYI, G. S.: 1962, *Geomagn. i Aeronom.* **2**, 377.
IVANOV-HOLODNYI, G. S. and NIKOLSKII, G. M.: 1961, *Astron. Ž.* **38**, 828.
IVANOV-HOLODNYI, G. S. and NIKOLSKII, G. M.: 1962, *Geomagn. i Aeronom.* **2**, 425.
JOHNSON, F. S., MALITSON, H. H., PURCELL, J. D., and TOUSEY, R.: 1958, *Astrophys. J.* **127**, 80.
LINDSAY, J. C.: 1963, private communication.
MERCURE, R., MILLER, S. C., RENSE, W. A., and STUART, F. J.: 1956.
NICOLET, M.: 1954, *The Earth as a Planet* (ed. by P. KUIPER), University of Chicago Press.
NICOLET, M.: 1960, *Physics of the Upper Atmosphere* (ed. by P. RATCLIFFE), Academic Press.
NIKOLSKII, G. M.: 1962, *Geomagn. i Aeronom.* **2**, 3.
PURCELL, J. D., PACKER, D. M., and TOUSEY, R.: 1959, *Nature* **184**, 8.
PURCELL, J. D., PACKER, D. M., and TOUSEY, R.: 1960, *Space Research* **1** (ed. by H. K. KALLMANN-BIJL), North Holland Publishing Co.
PURCELL, J. D., GARRETT, D. L., and TOUSEY, R.: 1963, *Space Research* **3** (ed. by W. PRIESTER), North Holland Publishing Co.
RENSE, W. A.: 1953, *Phys. Rev.* **91**, 299.
RENSE, W. A. and VIOLETT, T.: 1959, *J. Opt. Soc. Amer.* **49**, 139.
TOUSEY, R.: 1961, *J. Opt. Soc. Amer.* **51**, 384.
TOUSEY, R.: 1963, *Space Sci. Revs.* **2**, 3.
TOUSEY, R.: 1963, private communication.
VIOLETT, T. and RENSE, W. A.: 1959, *Astrophys. J.* **130**, 954.
WATANABE, K. and HINTEREGGER, H. E.: 1962, *J. Geophys. Res.* **67**, 999.
ZIRIN, H., HALL, L. A., HINTEREGGER, H. E.: 1963, *Space Research* **3**, (ed. by W. PRIESTER) North Holland Publishing Co.

DISCUSSION ON THE QUIET CHROMOSPHERE

R. O. REDMAN
Chairman
University of Cambridge, Cambridge, United Kingdom

L. DE FEITER
Secretary
Sterrewacht 'Sonnenborgh', Utrecht

M. J. SEATON. I would like to ask Jefferies two questions:

(1) The chromosphere theory has been expressed in terms of mathematical problems which are difficult but which are nonetheless soluble. But just how much error comes from the assumption of parallel stratified layers?

(2) It has been explained that information on the source function can be obtained from studies of the components of a doublet. Would it not be possible to get good consistency checks using a triplet?

J. T. JEFFERIES. It is difficult to say what error is introduced from the assumption of parallel stratified layers. The analytical procedure which I suggested could perhaps give information on the mean atmosphere. To determine fluctuations about these means we would need to reformulate the transfer equation to include statistical fluctuations in emission and absorption coefficients. In addition of course, one would need to be able to measure the appropriate statistical properties of the limb darkening.

Seaton's second comment is quite correct. We do plan to use a triplet (the Mg b lines) for just the purpose suggested, i.e. to obtain a self consistency check.

E. H. AVRETT. Jefferies has referred to the difficulties that occur in the radiative transfer part of the general line problem. I wish to present accurate solutions that have been obtained for a very simple but characteristic case: that involving a two-level atom with no continuum, and assuming complete redistribution and Doppler broadening. This case has been considered most recently by PECKER and THOMAS (1963), HEARN (1963), and IVANOV (1963). The limited problem is worth some attention because it contains most of the essential mathematical difficulties that are present in the general case. Also, the solutions are useful in testing the applicability of more general methods of solution. Several such methods are now being developed.

We then consider the transfer problem

$$\mu \frac{\mathrm{d}I_x(\tau,\mu)}{\mathrm{d}\tau} = \varphi_x[I_x(\tau,\mu) - S(\tau)]$$

. de Jager (ed.), The Solar Spectrum. All rights reserved.

where

$$S(\tau) = (1 - \varepsilon) \int_{-\infty}^{\infty} \varphi_x J_x(\tau)\,\mathrm{d}x + \varepsilon B$$

φ_x is the normalized Doppler profile $(1/\sqrt{\pi})e^{-x^2}$, τ is the integrated line optical depth, and $x = (\nu - \nu_0)/\Delta\nu_D$. $(l - \varepsilon)$ is the probability that an absorbed photon will be re-emitted in the line; $\varepsilon = C_{21}/(A_{21} + C_{21})$ is the fraction of de-excitations by electron collisions. We neglect stimulated emission. εB is the amount of this collisional energy that reappears at the line frequency. For a slab of total thickness T and no incident radiation, the integral equation for the source function is

$$S(\tau) = (1 - \varepsilon) \int_{0}^{T} K(1t - \tau 1) S(t)\,\mathrm{d}t + \varepsilon B\,,$$

where

$$K(y) = \tfrac{1}{2} \int_{-\infty}^{+\infty} \varphi_x^2 E_1(y\varphi_x)\,\mathrm{d}x\,.$$

ε_{I} is characteristically of order 10^{-6} or smaller. An indication of the severe properties of the integral equation when ε is small is the exact result $S(0) = B\sqrt{\varepsilon}$ for ε and B constant and T infinite.

We have obtained accurate numerical solutions $S(\tau)$ for constant B and constant values of ε as small as 10^{-12}. In the figures 1-3, we show the corresponding emergent normal intensities

$$I_x(\tau = 0, \mu = 1) = \varphi_x \int_{0}^{T} e^{-\varphi_x t} S(t)\,\mathrm{d}t$$

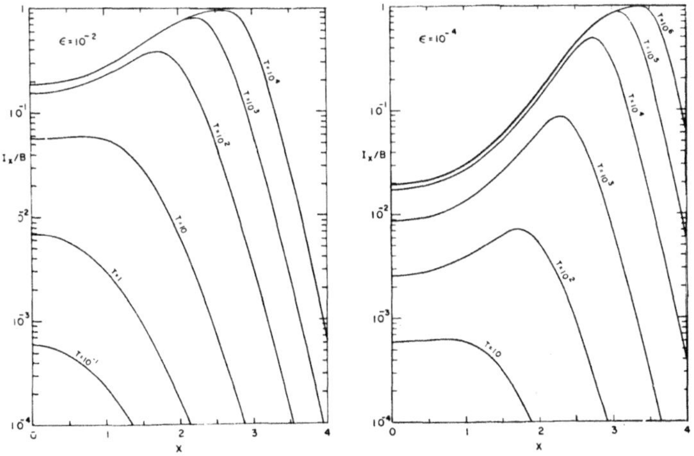

Fig. 1.

for various values of the total thickness T. The self-reversal in the line centre corresponds to the fact that the small numbers of photons that are scattered into the wings then have a much greater mean free path and escape more easily. With coherent scattering, i.e. with the source function given by

$$S_x(\tau) = (1 - \varepsilon)J_x(\tau) + \varepsilon B$$

we would have no reversal.

Thc method used to obtain these results is a specialized one that is not easily adapted to the case of a variable ε. However, we have at present a general computer program that handles any $\varepsilon(\tau)$, but with a great loss of accuracy for $\varepsilon < 10^{-4}$. With certain further developments now in progress we expect this restriction to be removed.

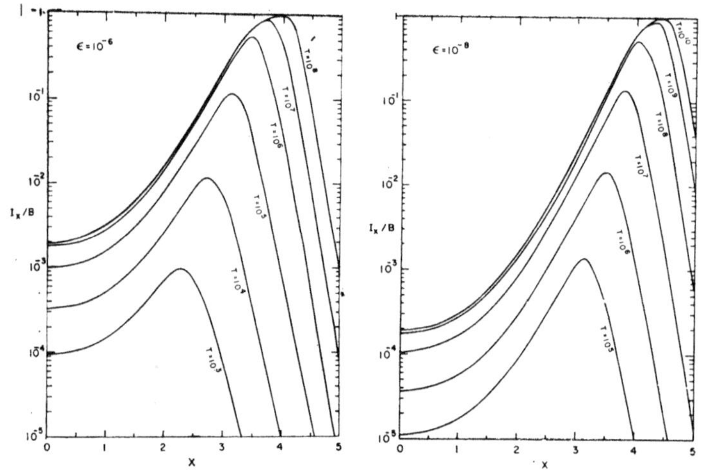

Fig. 2.

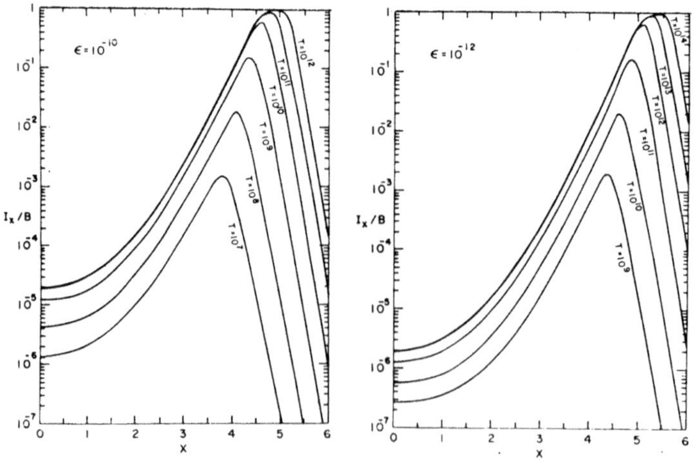

Fig. 3.

This work is being undertaken in collaboration with D. G. Hummer. A complete account will be published in *Monthly Notices of the Royal Astronomical Society*.

References

HEARN, A. G.: 1963, *Proc. Phys. Soc.* **81**, 648.
IVANOV, V. V.: 1963, *Soviet Astronomy-AJ* **6**, 793.
PECKER, C. and THOMAS, R. N.: 1963, *Astrophys. J.* **137**, 967.

J. T. JEFFERIES. It is very encouraging to see the progress made by Avrett and Hummer towards the solution of this problem, so intensively discussed at Herstmonceux last year. However, one should note that before the interlocking problem can be solved it is necessary to obtain solutions to the transfer equation for depth-dependent parameters.

H. ZIRIN. I would support Seaton's point on the necessity of determining the chromospheric structure. There is strong evidence that the helium emission originates in spicules only, and in fact the low optical depth required by Jefferies for the helium resonance lines is only possible in a filamentary structure.

R. N. THOMAS. Zirin asks that we should not waste too much time on detailed non-LTE model calculations until we can be sure the model corresponds to reality. However the only test of reality is to compare the results of the calculations with observations. As Jefferies said there are two parts to the problem: the synthetic and analytic. One needs to understand very clearly that the non-LTE question is not simply whether or not a local value of electron temperature is sufficient to represent distribution functions through the Boltzmann law; rather, it is whether any local set of parameters is sufficient to represent the local distribution function. The big point of the non-LTE calculations in any case where the self-emission of the atmosphere is important is just that the local parameters are indeed insufficient. To obtain the distribution functions the properties of some extended region of the atmosphere must be taken into account as boundary conditions for the solution of the radiation transfer problem.

For example consider a pictorial model of a simple columnar structure:

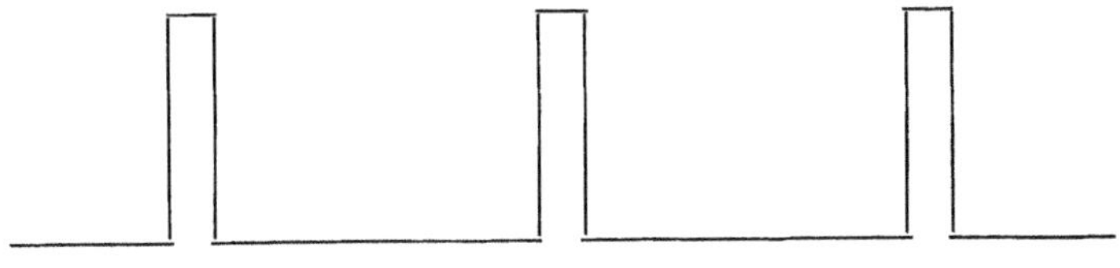

The analytical problem is to invert the inferred total number of atoms along the line of sight, in some energy level, N_n, to obtain n_n the number per unit volume. This n_n is compared with theory, but the theory must also be based on the same geometrical model. In much of our work to date we have indeed considered the atmosphere as a whole in computing distribution functions, and hence considered a non-homogeneous distribution of n_e, T_e, etc. But these have been radial distributions only, not tangential, i.e. we have worked always with the radial optical depth to the "top" of the atmosphere. In a case such as that shown we must include the horizontal distribution and the horizontal optical depth. If the vertical extent of the columns much exceeds the horizontal, then the parameter fixing the character of the atmospheric self-emission may well be the horizontal optical thickness of the column. Only elaborate, detailed, mathematical models of such a configuration can give any answer worth considering. I would emphasize that we will never know what exists until at least we have done these calculations with the correct geometrical boundary conditions.

R. MICHARD. According to a slide shown by Athay the central intensities of some strong lines in the chromosphere would be higher than at the centre of the disk by factors between 2 and 20. This contradicts photometric measurements of these lines made outside eclipses, where I_0 is found to decrease monotonically with μ, and also with increasing height outside the limb.

R. G. ATHAY. The slide showing an increase in central intensities of strong lines on going from the disk to the chromosphere was intended to illustrate the probable errors in chromospheric line intensities and not to establish the reality of the increased brightness in the chromosphere.

J. R. W. HEINTZE. Figure 2 in Athay's paper shows a drop of surface brightness at λ 4700 when going from photosphere to chromosphere, as found by him from flash spectra taken at the 1952 eclipse. Going inward towards the sun's centre from the inflexion point of the curve, which he calls by definition the solar limb, he has three points showing a maximum of surface brightness. His points in this region are rather uncertain, so that he thought it better to draw a line through them, monotonically increasing to the sun's centre.

But perhaps the maximum is real. Reducing flash spectra taken at the 1954 eclipse by Houtgast and myself, I found a remarkable limb-brightening in five wavelengths between 4600 and 6400 Å. My results for λ 4605 Å and 5700 Å are shown in Figure 4.

J. HOUTGAST. In Athay's graphs of line intensities against height in the chromosphere, smooth lines are drawn through the scattered points. For strong lines, however, one has to take into account the fact that after saturation of their cores the development of line wings does increase again the total intensities at low heights in the chromosphere.

Further I would like to stress the need for more and better observations for the very difficult determination of (a) the drop in brightness at the very limb of the sun; and (b) the behaviour of weak lines in the transition from photosphere to chromosphere.

For (a) I would suggest that simultaneous spectrophotographic and photo-

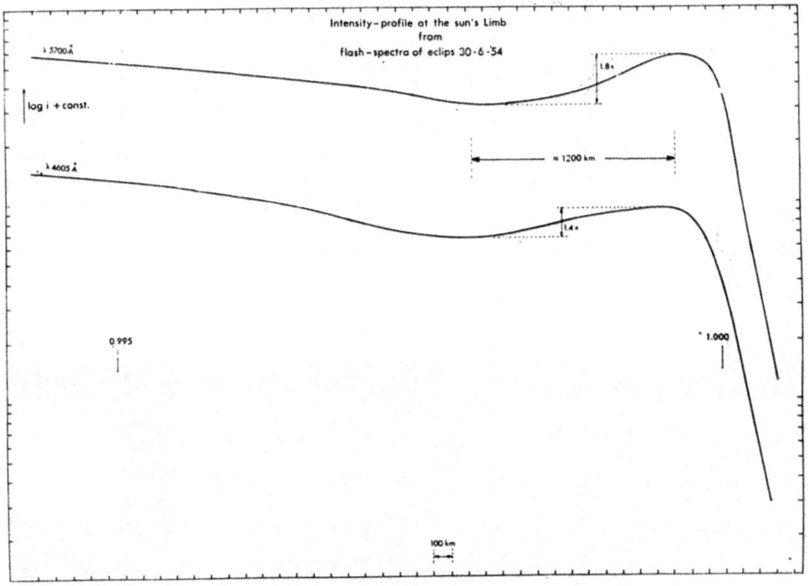

Fig. 4.

electric observations be made. For (b) we have observations from the eclipse of February, 1962, which need to be extended. The theoretical treatment is being made by A. Schadee and myself.

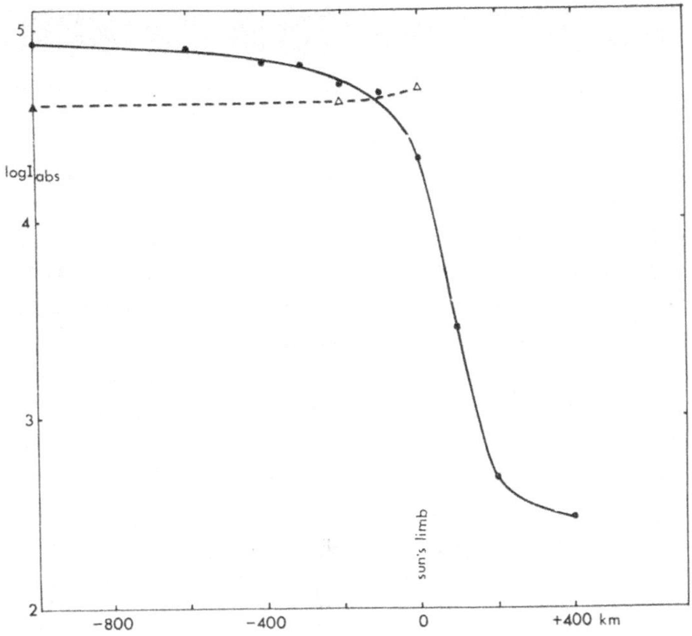

Fig. 5.

A. SCHADEE. The spectrograms from the 1962 eclipse show that ion lines change from absorption to emission about 100 km within the sun's limb. We have computed the behaviour of the Ti II line at λ 4657 Å on a simple model assuming LTE and taking a temperature minimum of 4100 °K at the sun's limb. Figure 5 shows the computed energy in ergs per sec per steradian emitted by a 1 cm² diameter column along the line of sight in 0.1 Å of the continuum near 4657 Å (full curve), and in addition the energy in the same units, within the Ti II line (dotted curve). The computation agrees with the conclusion that the transition from absorption to emission takes place near −100 km.

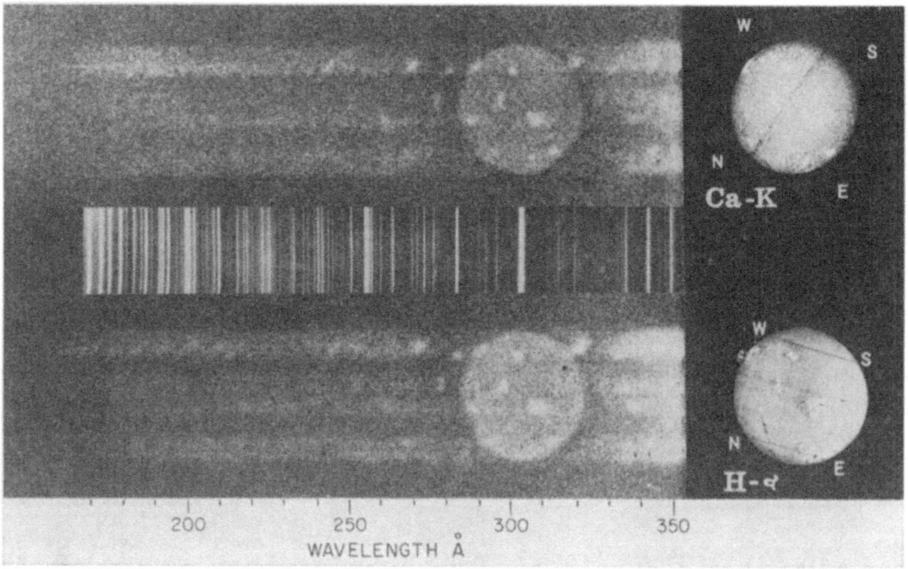

Fig. 6.

J. D. PURCELL. I would like to give some recent results of solar spectroscopy carried out at the E. O. Hulburt Center for Space Research, U. S. Naval Research Laboratory, Washington, D. C. Spectroheliograms in several extreme ultraviolet wavelengths, and grazing incidence spectra down to 40 Å in the soft x-ray region, were obtained by the U. S. Naval Research Laboratory in a rocket flight on 10 May 1963. Photographs in λ 303.8 He II, λ 285 Fe XV, λ 335 Fe XVI and several other nearby lines were obtained with an objective type concave grating spectrograph using an unbacked aluminium film as a filter to reject stray light (Figure 6). In a second double-dispersion stigmatic spectrograph the slit was opened to admit a central 7′ wide band of the sun and images of this band were recorded in λ 977 C III, λ 1025.7 hydrogen Lβ, λ's 1031.9 and 1037.6 O VI and 1215.7 hydrogen Lα (Figure 7). In the several plage regions observed the enhancement of intensity relative to the quiet disk was remarkably similar

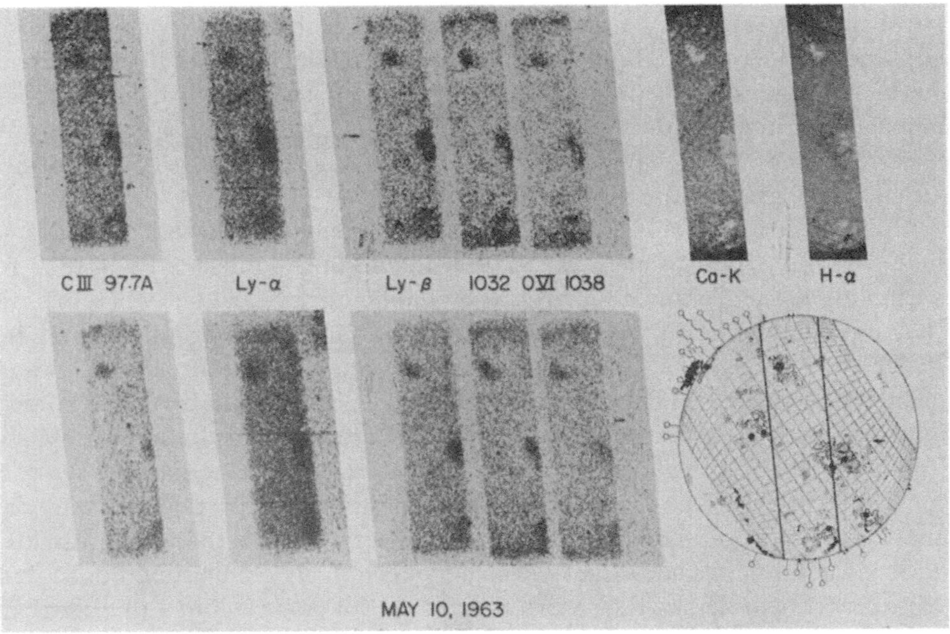

Fig. 7.

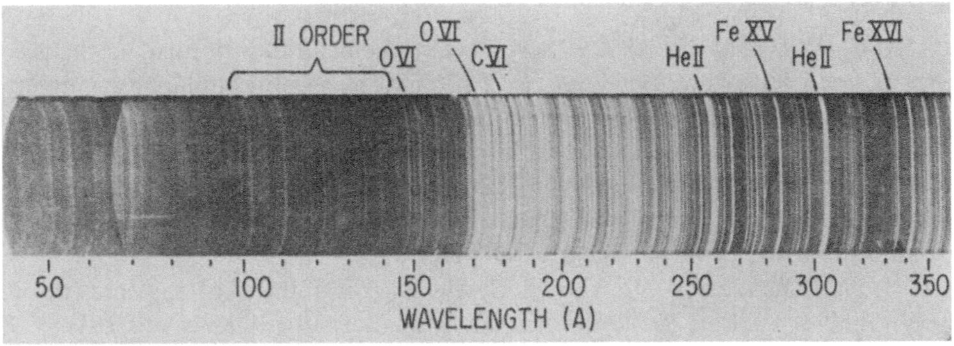

Fig. 8.

in all of the observed wavelengths with the exception of Fe XV and Fe XVI. In the latter wavelengths no discernible disk was present although the plages were recorded with considerable intensity (Figure 6).

The grazing incidence spectrograph employing a thin aluminium filter flown on 1961 June 21 and 1962 August 22 was improved in resolving power and speed. The spectra (Figure 8) show lines beginning at 40 Å, the short wavelength limit of the spectrograph, and extending to about 80 Å in the first order. Many lines are present in this region, together with diffuse emissions, which may be unresolved groups of lines. The presence of a continuum throughout this region is suggested by the spectra although this is not clearly established. Lines in the region 80 Å to 150 Å appear to be higher order lines. At about 150 Å the first order again appears and continues to long wavelengths, overlapping with higher orders above 340 Å. The sharpness of the spectra permits measurement of wavelengths to 0.05 Å or better. Some lines have been identified although the great majority have not.

R. N. THOMAS. Could I ask Purcell to comment on the NRL results on limb darkening measurements in the continuum in the range λ 1200–1800 Å? I can only quote Tousey's remarks at a colloquium in Boulder, where he said that limb-darkening in the λ1800 region turns into limb-brightening at the shorter wavelengths. In such a case, the radiation temperature at those wavelengths where there is neither limb brightening nor darkening gives a very close approximation to the minimum value of electron temperature in the atmosphere. Tousey's values give T_e (min) $= 4700°$. Such a high value has clearly very important implications for the mechanical energy input, particularly implying that this occurs quite low in the solar atmosphere. NRL is to be congratulated on the results and we hope they will soon publish them, eliminating this second-hand report by others of their work. I apologise for doing it, but I think such an important result must be mentioned at this symposium.

E. SCHATZMAN. What is the solar limb, or, more precisely what is the reference level? An optical thickness unity along the line of sight corresponds to what radial optical depth?

R. G. ATHAY. In the 1952 eclipse data the solar limb is defined as the point where $\tau = 1$ for a tangential ray, at λ 4700. If we assume spherical symmetry and adopt a scale height for τ of 60 km this definition places the "limb" at a radial optical depth 0.0037. The validity of the assumptions is open to question; indeed one may question the usefulness of the concept of a solar "limb". However, some reference point must be established for eclipse data and the solar limb seems to be the most convenient and useful.

M. G. J. MINNAERT. We should always keep in mind that the concept of a solar limb is entirely conventional and that actually it does not exist. The transition between the photosphere and the chromosphere is gradual for any parameter that we may consider. In principle a point in that transition layer should be characterized by its distance to the centre of the sun. For practical purposes one may assume any reference point, determined by its optical depth, or by the inflexion point of the brightness curve; but these definitions are arbitrary and will in general not coincide.

H. ZIRIN. Dietz and I have tried to identify lines on Hinteregger's beautiful spectra. Now that we have a resolution of 0.1 Å this has become much easier. We demand wavelength agreement to 0.1 Å ,and also the right multiplet ratios. We have identified lines of Mg IX and X, Si IX and X, Fe X, XI, XIV, XV and XVI, using Edlén's laboratory observations.

The greatest problem lies in the strong lines between 170 and 200 Å, which are as yet unidentified. We suggest that these are due to $3s^2 3p^n \rightarrow 3s\,3p^{n+1}$ and $3s^2 3p^n \rightarrow 3s^2 3p^{n-1} 3d$ transitions in Fe X to XIV. We have made some identifications of such transitions in Fe XIII and Fe XIV with moderately strong lines. In these cases the known splitting of the lower level (coronal line) is used. But the strongest lines are still unidentified and laboratory work is desperately needed.

Mrs. C. MOORE-SITTERLY. This paper emphasizes the urgent need for well-planned *laboratory* programmes of analysis of highly-ionized atomic spectra of abundant elements. Observations in the extreme ultraviolet, made with plasma sources, are required to extend the interpretation of the already existing rocket and satellite solar spectra.

J.-C. PECKER. If we had at our disposal only the UV spectrum of the sun, we would not make any clear distinction between chromosphere and corona. this remark completes Minnaert's comment: it is difficult to speak any more of photosphere, of chromosphere or of corona. There is one continuous solar atmosphere.

C. DE JAGER. It is true that the *brightness temperature* of the continuous solar spectrum reaches a minimum of about 4700° near 1 400 Å. The minimum of the *source function* in the photosphere should correspond to a lower temperature, which turns out to be 4500°. The observed radiation intensity is the result of an integration over the depth and is always greater than the intensity in the region of minimum temperature (published in detail in *Bull. Astron. Inst. Netherlands*, **17** (1963) 209).

On the basis of the intensity measurements in the ultraviolet continuous spectrum, the minimum temperature in the solar photosphere can be determined with fair accuracy. An error of a factor $2^1/_2$ in the measured intensity would correspond to an error of 200° in the resulting brightness temperature. We estimate the actual error of measurement to correspond to $\pm$ 50°. There is in addition an uncertainty in the computation of the kinetic temperature of the gas from the measured brightness temperature. This error must be less than 50°.

I conclude that the minimum temperature of the photosphere is 4500° $\pm$ 100°.

J. T. JEFFERIES. A rather basic difficulty arises from Hinteregger's deduction of a temperature about 20000 °K for the source of the He II continuum, since this is very much lower than the 40000 °K generally conceded to be the minimum necessary to excite this radiation to the observed strength.

H. E. HINTEREGGER. I cannot say any more about estimated accuracies in inferring Planckian model temperatures than I have indicated in discussing the measurements. To reduce our present, not too assessable, uncertainties, we must await experiments of much larger scanning times such as will be available in satellite experiments.

(In answer to a question from the Chairman about the lack of spectrum lines of intermediate ionization): Judging from a probably too simpleminded viewpoint, the two most impressive theoretical arguments have been (a) Suemoto's: that a spherically symmetrical chromosphere should emit considerably less XUV than is observed, and (b) Zirin's: that a spherically symmetrical chromosphere with anything but an extremely thin transition region between chromosphere and corona should emit considerably less XUV than is observed. Both seem to indicate that a spherically symmetrical model of the chromosphere above 1000 km does not make good sense, whereas the assumption that XUV emission comes mainly from a fractional area of the disk, for example from spicules only, does seem to make good sense to me. To some extent, of course, my preference for this picture may be connected with my inability to understand the theory.

H. Zirin. R. Howard and I have been studying the behaviour of spicules on Mt. Wilson high resolution spectroheliograms. We find that the spicules occur in "bushes" above the flocculi making up the chromospheric network. They are most prominent near $r = 0.9$. There are about 2000 "bushes", or 20–30000 spicules. Using this and XUV observations Dietz and I constructed a model of the chromosphere in which the hot spectrum (O II, O III, C III, He) originates in spicules only, while between the spicules (above 500 km) we have corona. To test this model Howard and I have obtained spectroheliograms in the 10830 Å He I line with the 60-ft tower at Mt. Wilson, using the HAO image tube system. We confirm the gross structure found by d'Azambuja. The fine structure shows that the 10830 absorption follows the chromospheric network, i.e. there is absorption only on the flocculi around each network element and the associated "bushes" of spicules. Thus the edges of the network marked by the calcium flocculi indicate the loci of magnetic fields, spicules, the hot regions which absorb 10830 Å; and perhaps the connection of the corona to the photosphere.

C. de Jager. The profile of the infrared He I triplet at 10830 Å has been studied in the quiet sun's spectrum with the Jungfraujoch spectrograph, by Namba, Neven and myself. Accurate photometry enabled us to determine the intensity ratio between the main doublet and the satellite line. The intensity ratio should be 8 for weak lines, but appears actually to be about 3 or 4, so that self-absorption is important for the main doublet, and this notwithstanding the fact that the line is very weak in the solar spectrum, with a central depression of 5% at the centre of the disk, increasing to 10% near the limb.

Hence the line absorption must take place in discrete elements; in each of these the line must be rather deep, with a central depression of about 50%. Hence about 10% of the solar surface must consist of helium absorbing elements, presumably the spicules.

Interestingly, it appears that the physical parameters of the helium absorbing elements of the quiet chromosphere are very similar to those in plages, as studied by Namba. Hence we may conclude that the plages consist essentially of an agglomerate of spicules; the number density of spicules should be greater in the plages than in the quiet chromosphere.

R. G. ATHAY. The picture of spicules that has been presented, in which they are hot relative to their surroundings, creates problems that are difficult to explain. What happens to the spectrum of the interspicule regions? If these regions are relatively cold, their spectra should look more like that of the low chromosphere than that of the corona. The reverse seems to be the case. If one attempts to suppress the interspicule spectrum by keeping the density low, then very large lateral pressure gradients arise. Furthermore, it becomes extremely difficult to construct a stable chromosphere-corona transition region in which pressure does not increase outwards.

H. ZIRIN. I did not make myself clear, so that Athay has misunderstood my point. The spicules contribute the high excitation part of the chromospheric spectrum and are thus the hot part of the chromosphere. Since the spherically symmetric chromosphere does not extend above 1500 km, the spicules themselves are of course immersed in the much hotter corona, which is dense over the "bushes" and less dense elsewhere.

De Jager's result is very encouraging. In our paper Dietz and I placed great weight on the fact that 10830 is a line of depth 0.05 that shows line ratios corresponding to $\tau = 3$. This is only possible if it arises in individual elements of $\tau = 3$ that cover 5 per cent of the surface. It would be very important now to measure the depth of the line where it is weak, i.e. at the centre of the network elements. In these areas its depth is less than 0.01.

H. E. HINTEREGGER. I wonder how much one might expect to learn from XUV observations of larger spicule structures ("bushes") instead of trying the much more difficult task of resolving individual spicules?

H. ZIRIN. The "bushes" are about 10000 km long and 4000 km high; but they might be separated by about twice that amount, so that they might be distinguished with 20 secs of arc resolution.

T. A. CHUBB. During midsummer of 1963 an X-ray spectrograph utilizing Bragg reflection from a crystal was flown in a rocket equipped with a sun-pointing control. Solar emission was recorded in about 12 lines between 12 and 25 Å. The spectrograph used a plane crystal of KAP and a large, thin-window geiger counter to record the reflected intensity. The crystal was slowly rotated so as to scan the X-ray spectrum twice during the high altitude portion of the flight. The sun itself served as the entrance slit for the spectrometer. The angular resolution for monochromatic X-rays was such that a strip of width about 2.5 minutes of arc was seen by the spectrograph at any moment during the spectral scan. With this angular definition it was possible to determine the approximate size of the solar X-ray source and to determine whether a line originated from the solar disk as a whole or from a single strong plage. X-ray lines originating from the entire solar disk had a measured line width about four times that of lines coming from a plage. The wavelengths, relative intensities, and emitting source are listed in the table below. These are preliminary raw data.

The experiment was carried out by R. Blake under the direction of H. Friedman and T. A. Chubb at the E. O. Hulburt Center for Space Research, U. S. Naval Research Laboratory.

TABLE

PRELIMINARY DATA X-RAY SPECTRUM (BLAKE)

λ *Å*	*Intensity counts/ sec*	*Back-ground counts/ sec*	*Source* *P* = *PLAGE* *D* = *DISC*	*Comments*
13.7	200	340	P	Ne IX?
15.0	1 200	240	P	
15.2	680	240	P	
16.0	280	180	P	
16.7	700	140	P	
16.9	240	140	P }	
17.0	850	140	P }	Could be one line
17.05	880	140	P }	
17.65	160	140	P	
17.7	200	140	P	
18.9	960	120	D	O VIII? contains some Plage contributions
20.8	WEAK	120	D	
21.5	1 600	100	D	O VII?
21.7	720	100	D	O VII?
22.0	1 300	100	D	
23.2	180	100	P	
24.8	300	100	P	

L. Houziaux. In Liège, Gabriel and I have attempted to interpret the high dispersion profile of the solar Lyman β hydrogen line, obtained by Garrett, Purcell and Tousey at NRL on 1962 August 22 (exposure No. 10), and kindly communicated to me by Friedman. Taking into account levels 1 and 3 of the hydrogen atom only, one finds that collisional processes dominate radiative processes by a factor of 100 or more. It is quite possible to account for the observed central reversal and the separation of the peaks. A part of the central reversal may be due to the $2p^2 3d^3D - 2p^3\,{}^3P$ transition of O I. However, as Thomas has pointed out to me, neglecting the influence of the second level of the H atom may alter the source function in a significant way.

SESSION III

THE ACTIVE PHOTOSPHERE AND CHROMOSPHERE; SPOTS, PLAGES, FLARES AND PROMINENCES

SPECTRA OF SOLAR ACTIVITY REGIONS

A. SEVERNY
Crimean Astrophysical Observatory, U. S. S. R.

From the brilliant lecture of Professor Minnaert we have learned how poor our knowledge is of the physical state of the undisturbed quiet solar atmosphere. But if we turn our attention to the phenomena of solar activity the position is still worse because only during the two past decades we accumulated more or less systematically important spectroscopic data about the processes in solar active regions, and yet we are very far from a clear understanding of the physical nature of these processes.

This paper represents an attempt to collect the most reliable results relating to some spectroscopic phenomena in active regions. However, the meaning of "active" region is so broad that we are forced to confine ourselves by considering only one class of phenomena which may be described as "explosion like" phenomena (flares, moustaches, active plages) and which have been examined most intensely during the past decade in the Crimean astrophysical observatory. Moreover we will concentrate on the spectroscopic data relating to the macroscopic state of phenomena instead of on microscopic characteristics as excitation, ionization of atoms, etc.

Looking for spectroscopic evidence of corpuscular streams out of active regions we found (1952) well-pronounced *peculiar asymmetry* of K-, H- and H_α-line profiles in the regions of bright Ca II plages. The profiles were obtained for more than 30 plage regions by averaging 3–9 separate photoelectric records obtained at the solar tower (dispersion 0.5 Å/mm).[1] The speed of scanning was 1 record per 4 sec and the mean error of one separate measurement of rest

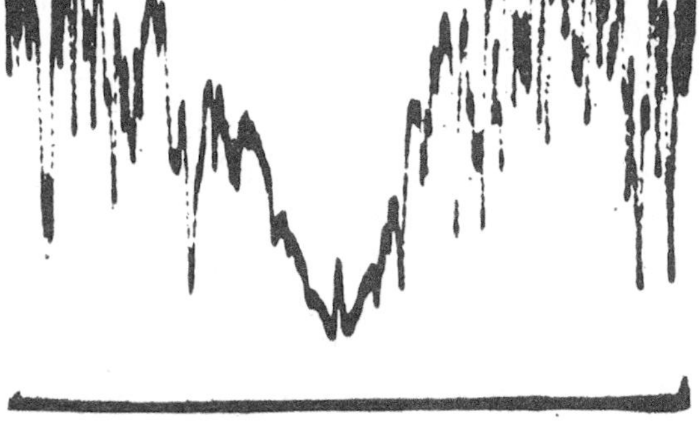

Fig. 1. An example of K-line record in a plage.

C. de Jager (ed.), The Solar Spectrum. All rights reserved intensity was 0.3%. Figure 1 shows the example of K-line record and on the top of Figure 2 the run of excess of the rest intensity in the sense: plage minus undisturbed atmosphere, (Δr_1), is shown. Most interesting, however, is the behaviour of asymmetry of profile in the sense r_ν (blue) — r_ν (red) = Δr_2 (Figure 2, below), showing in most cases a positive blue excess, as if the emission

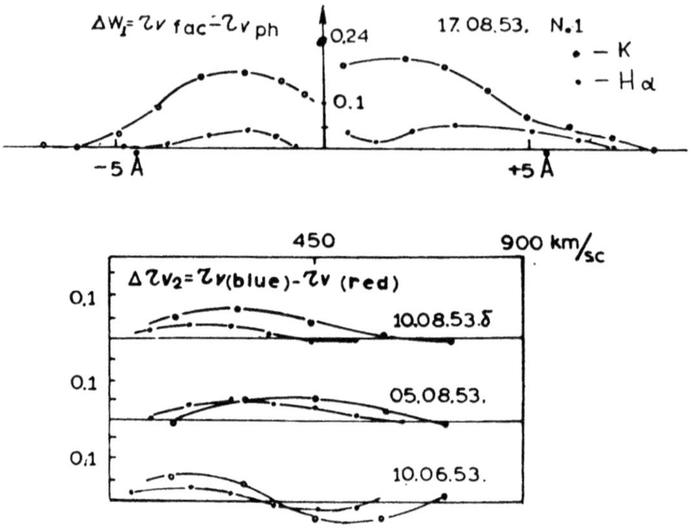

Fig. 2. The runs of excess of the rest intensity in the plage (top) and of the asymmetry (below).

of the blue wing were larger and more extensive than that of the red one. As a rule there is the general agreement in the run of asymmetry Δr_2, for the H-, and K-line on the one side and the H_α-line on the other, although the profiles of emission excess may differ essentially for both lines. That is quite natural, due to the difference in the conditions of their formation in the solar atmosphere. For different plages the position of the maximum of asymmetry changes from several hundred to 1 000–1 500 km/sec, but no dependence center-limb was found for this position. Therefore the phenomenon of asymmetry cannot be ascribed to the process of pure radial outstreaming of corpuscules which process is usually adopted to explain the 27-days recurrent Earth magnetic perturbations. Moreover, if these perturbations are connected with Ca-plages[2], the observed effect does not probably have any relation to plages and we should look for some other phenomena responsible for the asymmetry.

It should be noted here, however, that the well-pronounced depression of the blue wing has been observed on several rare occasions, for instance during the 1^+ flare of July 10, 1952 where the depression was found at a distance of $\simeq$ 10 Å from the center to the red side, as Figure 3 shows. The measured equivalent width of the depression

$$W = 0.2\,\text{Å} = \frac{\pi e^2}{mc^2} Nf \cdot 10^8(\text{Å}) = 1.37 \times 10^{-13} N\,,$$

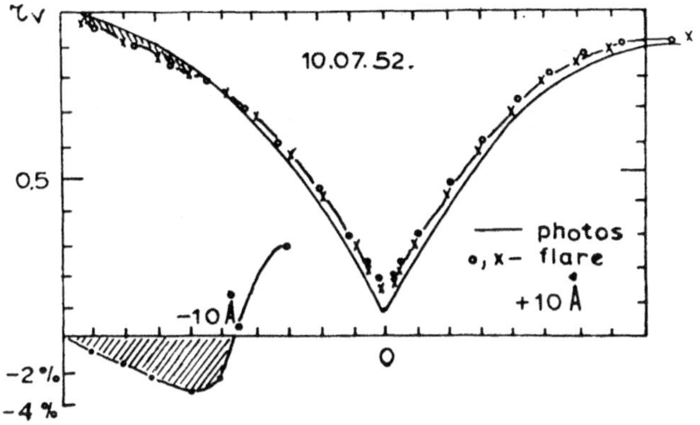

Fig. 3. K-line photoelectric record in the flare 1+ of July 10, 1952.

if we assume the cloud of Ca II particles to be optically thin. It gives $N = 1.5 \times 10^{12}/\text{cm}^2$ and assuming the length of the cloud 10^4 km we get $n = 1.5 \times 10^3\ \text{cm}^{-3}$ near the surface of the Sun. Taking the area emitting particles of $\simeq 10^{-3}$ of the whole disk we obtain

$$n' = \left(\frac{r_\odot}{D}\right)^2 \frac{S}{S_{\odot}}\, n = 2.17 \times 10^{-8} n \simeq 3 \times 10^{-5}$$

Ca atoms per cm³ near the Earth's atmosphere. Taking the relative abundances of a stream the same as for solar atmosphere we get $n_{\text{H}} \simeq 10$ H-atoms/cm³, which corresponds closely to recent measurements of the particle density during flares made by *Space probes*.[3]

The search for the explanation of the asymmetry in plage regions led us in 1954 to the examination of the very peculiar *fine structure* of the emission in active regions and especially to the phenomenon which we called "moustaches".[4] We found that the most characteristic feature of emission in an active region is its fine structure, the concentration of emission of all kinds in very small grain-points, the size of which does not exceed several seconds of arc and is, frequently, comparable with the circle of scattering of our telescope (0.″3). It should be emphasized that the observation of this fine structure is *only* possible at good seeing. Even if we observe the flares at good seeing, they never showed continuous (along the line) emission reversal; on the contrary the emission, in the form of streaks, does not exceed the characteristic size of the fine chromospheric mottling. Figure 4 and 5 show examples of this fine emission structure in the case of flares (H and H_α-lines). Figure 6 illustrates the appearance of typical moustaches on échelle spectrogram.

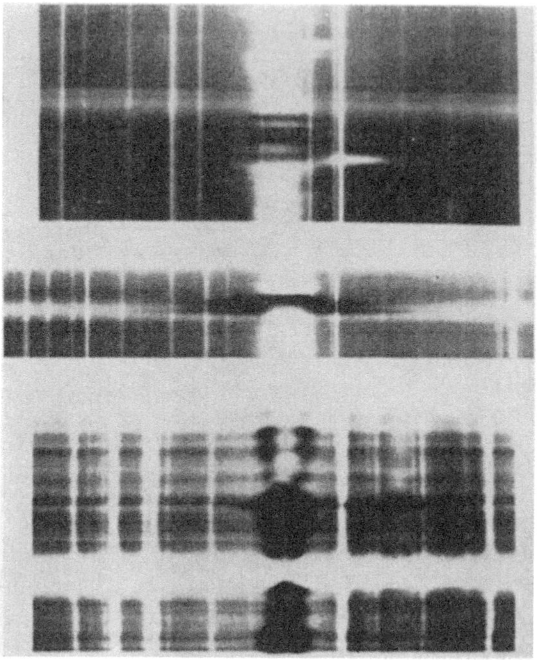

Fig. 4. The examples of the fine structure of emission in flares: the streaks of emission are produced by grains with the size less than 1″.

The following is a short summary on this very interesting phenomenon. By moustaches we mean bright, extensive and very thin (across the dispersion) wings of some strong lines without pronounced emission reversal in the *core* of the line. They are observed usually on the background of the very thin streak of continuous emission.* The size of the region generating these emissions does not exceed several seconds of arc and is very frequently comparable with the circle of scattering (the estimates are very dependent on the quality of seeing). The appearance of moustaches is recorded for all Balmer lines (the extension up to 10 Å), H- and K- lines (bright and extension up to 15 Å), $\lambda\lambda$ 3706; 3737 Ca II, Mg-triplete (5167, 5173, 5184), at some strong iron lines, lines D_3 (near the

* The phenomenon of moustaches was probably first observed by Ellerman who in a very short note[24] pointed out all the important spectroscopic features except the last one (continuous emission).

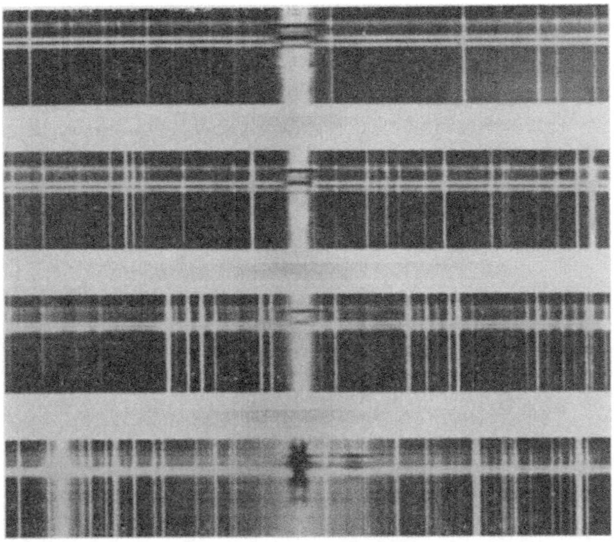

Fig. 5. The fine structure of emission in the Balmer lines and K in the flare 1+, of June 4, 1957 at 3.53 UT.

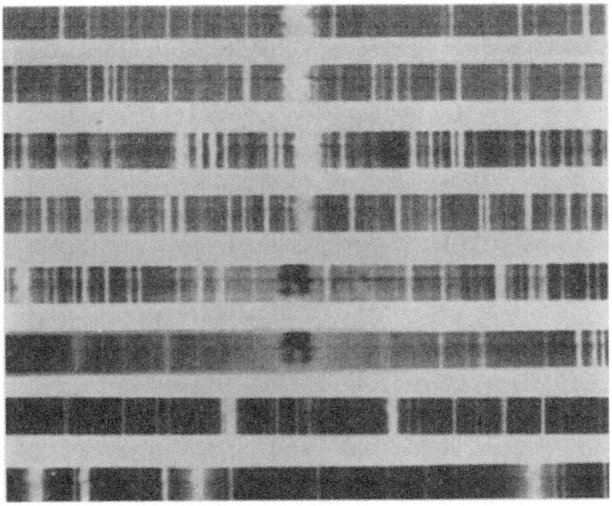

Fig. 6. The spectrum of the active region showing moustaches on July 29, 1959 at 5.07 UT.

limb). The lifetime is from several minutes up to 20^m. The moustaches on H_α-filtrograms are seen as bright points (contrast 1.3) and on K-spectroheliograms their contrast is much higher (up to 3.0) according to Reference 5. The size of these points does not show, apparently, the dependence center-limb and they tend to increase in size when the phenomenon dies out (to be more exact the points become more diffuse). The same effect can be seen from the qualitative investigation of more than 30 spectra. This interesting feature can be seen on the filtrograms taken at the blue side of the H_α-line (the pass-band is 0.5 Å apart from the center). They show almost invariably the appearance of a small dark filament coming from the very point. In Figure 7 we see the appearance of moustaches on H_α-filtrograms taken at different positions of the

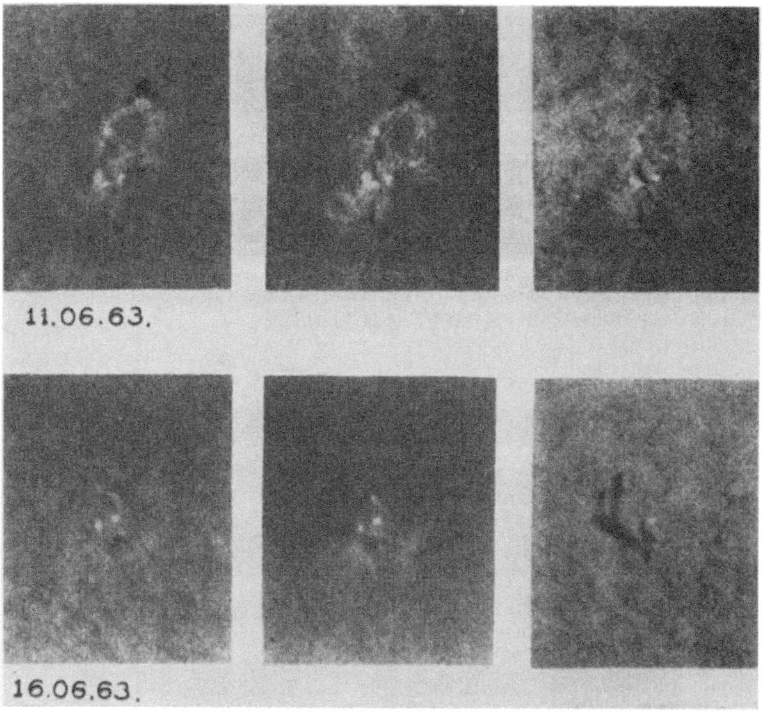

Fig. 7. The appearance of moustaches (bright points) at different positions of the pass-band relative to the center of H_α (+0.5 Å on the left, 0 in the middle, and −0.5 Å on the right).

pass-band. Just like the flares, the moustaches tend to appear near the border of growing small sunspots, at the border of penumbra of large spots and between the p- and f-spots. Besides, they can be observed in plage and faculae regions. They can, like flares, appear recurrently at the same place. They appear frequently in the absence of flares, but flares are frequently accompanied by moustaches. Moreover, the repeated Crimean observations at good-seeing conditions of the broad wings of flares of importance $\geqslant 2$ at flash phase show clearly that extensive wings are resolved into separate thin threads which are identical with moustaches not only in appearance but also in the way of broadening (see below). The emission of moustaches can be seen normally starting from

$\simeq 0.7$ Å apart from the center of the dark H_α-line and leading to the depth of the grain generating moustaches $\tau_0 \leqslant 0.4$. However, the phenomenon can sometimes (but rarely) also be observed in the chromosphere and even above the limb. This means that the phenomenon can appear at different depths in solar atmosphere in the region $0 < \tau_0 \leqslant 0.4$. (See Figure 8 illustrating the appearance of moustaches on the disk and at the limb). The continuous emission background of moustaches forms a subject extensive enough for a second lecture so that we would like to refer to our paper of Reference 4.

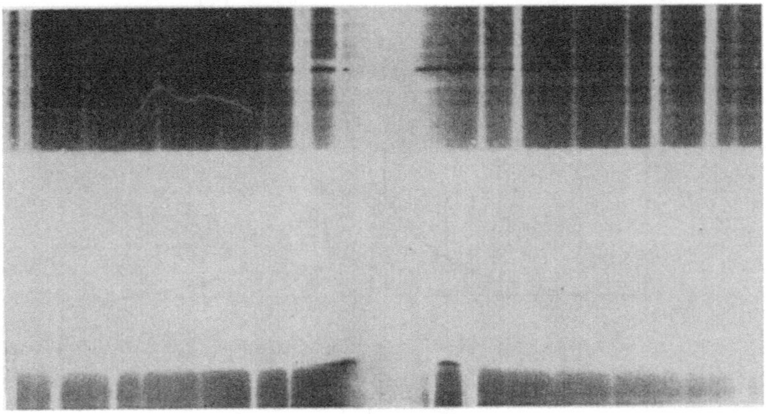

Fig. 8. The appearance of moustaches at the disk (top) and at the limb (below).

The most interesting feature of moustaches is, however, the *asymmetry* of line profiles: in most of the cases the blue wing is more extensive and brighter than the red one irrespective to the position on the disk – the same effect as we observed photo-electrically in bright plages. Using a sufficiently high slit ($\simeq 10''$) for photoelectric records the averaging effect over the big area produce some composite profile with a less pronounced asymmetry in separate moustaches. Furthermore the blue asymmetry of moustaches is frequently accompanied by the shift of absorption line *as a whole* to the *violet* side independent on the position at the disk. Figure 9 shows these displacements clearly. The careful consideration shows also that these displacements for some lines are larger than for the other. Some strong lines even show the absorption features like "flag" or dark ejection from the blue side (for instance Sodium D-lines and Mg-triplet, all Balmer lines) see again Figure 6. This fact is closely connected with the appearance of dark filament from moustaches seen at the blue H_α-filtrogram. Let us consider the blue shifts of the lines, now using Balmer lines. The profiles of these lines are shown on Figure 10 for typical moustaches, we see that the absorption feature (the "hole") in the center of the lines is

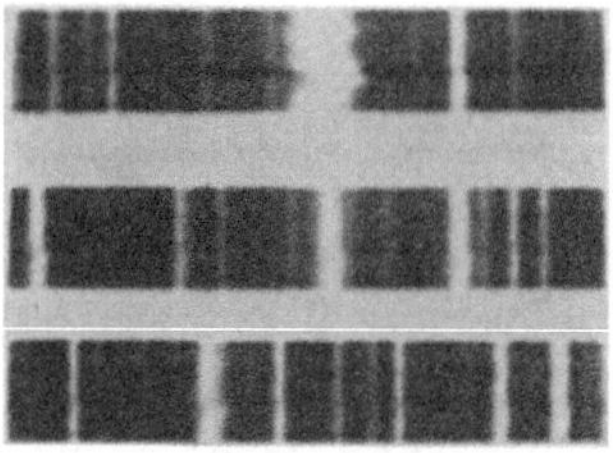

Fig. 9. The difference in the displacements of different lines above the moustaches on July 29, 1959 at 5.07 UT.

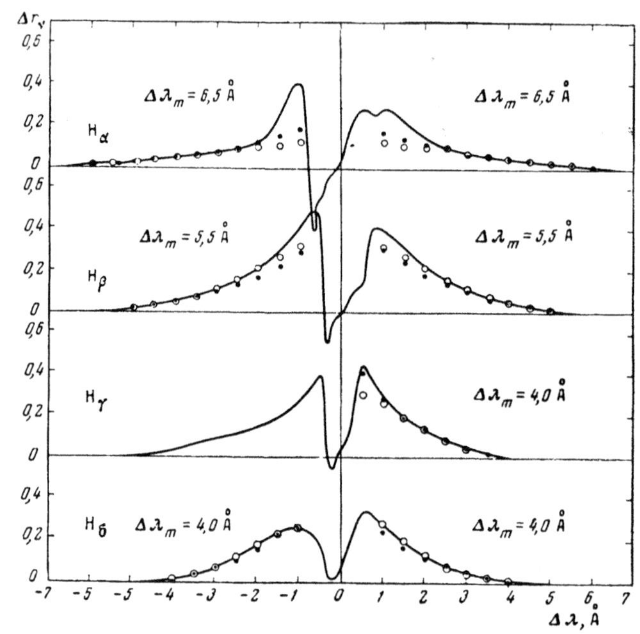

Fig. 10. Line profiles of Balmer lines in the same moustaches (see Figure 9).

shifted gradually towards the violet when we pass from higher to lower members of the B-series. The mean shift for three plates gives the following run of $\Delta\lambda/\lambda$ (see Table 1).

TABLE 1

	$\Delta\lambda$(Å)	$\frac{\Delta\lambda}{\lambda}\cdot 10^3$	τ_0
H_α	−0.77	1.17	0.29
H_β	−0.43	0.88	
H_γ	−0.35	0.81	
H_δ	−0.24	0.59	0.41

i.e. $\Delta\lambda/\lambda$ diminishes almost in two times while it should be constant in case of uniform rise of the reversing layer with a constant velocity. Using the De Jager model of H-lines we may estimate the optical depths to be $\tau_0 = 0.41$ and $\tau_0 = 0.29$ for H_δ and H_α respectively, corresponding to the same absorption feature.[6] This gives a substantial increase of velocity (2 times) outward (see Figure 11, top). The velocity shifts for more than 100 metallic lines confirmed this conclusion for phenomena at different positions at the disk ($\varrho = 0.93$; 0.80; 0.44) as Figure 11 (below) shows.[7] Therefore we are faced with the fact of substantial *outward velocity gradient* in the region of moustaches.

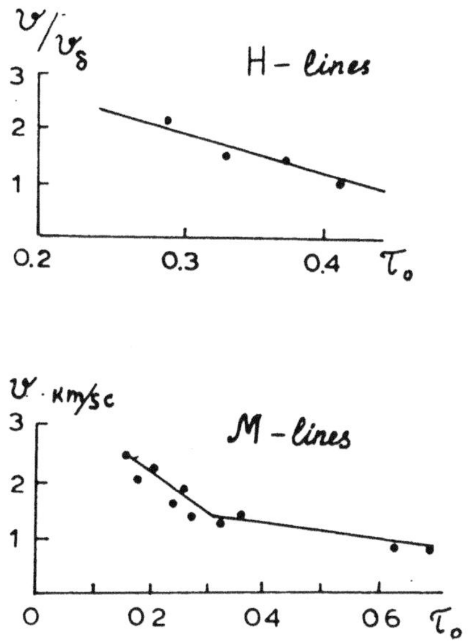

Fig. 11. The dependence of velocities on the depth in moustaches (for Hydrogen lines: top; and for metallic lines: below).

Now, if we consider carefully the kink of the K-line and some other strong lines, we observe a slight absorption feature and a slight kink or jet also towards the *red* side. The position of the kink on the disk is slightly different from the position of moustaches. Moreover in some cases we have observed the different directions in shifts of Hydrogen and metallic lines. Figure 12 shows the *blue* kink of H-lines and the *red* shifts or diffuse red component of metallic lines (as if we have *some explosion* a little above the reversing layer) producing a divergent shock front propagating upwards and downwards.

The character of the motions around moustaches is the problem to be solved. We do not know whether we can describe these motions as a manifestation of contraction (collapse) or expansion, but we must reject definitely the picture of

TABLE 2

BROADENING OF WINGS IN BIG FLARES AND MOUSTACHES

No.	*Date*	*Importance*	*Profiles considered*	*Character of wings*	*Broadening of wings* $(\Delta\lambda_D/\lambda)$ 10^3	*Velocities* ξ_t, *km/sec*
1	04.06.57/12:07	2+	H_α–H_δ	Moustache, $\Delta\lambda_\alpha \simeq 10$ Å	D, $\Delta\lambda D/\lambda$ = const = 0.66	207
2	03.09.57/17:30	3	H_α–H_γ	Asymm. wing, $\Delta\lambda_\alpha > 10$ Å	D, $\Delta\lambda D/\lambda$ = 0.9–0.7; 0.75	238
3	06.09.57/07:25	3 (2+)	H_α–H_{13}	Moustache, $\Delta\lambda_\alpha = 10$ Å	D!, $\Delta\lambda D/\lambda$ = const = 0.79	250
4	21.09.57/08:01	3	H_α–H_ε, H, D_3	Moustache, asym., $\Delta\lambda_\alpha = 9$ Å	D!, $\Delta\lambda D/\lambda$ = const! = 0.74	234
5	24.09.57/08:15	2	H_α–H_ε, H	Limb flare, $\Delta\lambda_\alpha = 5$ Å	D!, $\Delta\lambda D/\lambda$ = 0.3–0.2 = 0.24	78
6	29.07.59/08:06	—	H_α–H_9	moustache, typical, $\Delta\lambda_\alpha = 10$ Å	D, $\Delta\lambda D/\lambda$ = const. = 0.61	190
7	25.07.59/11:41	—	H_α–H_{11}, D_3, K	moustache, wing, $\Delta\lambda_\alpha = 8$ Å	D?, $\Delta\lambda D/\lambda$ = 0.95–0.45 = 0.86	258
8				sym. moustache, $\Delta\lambda_\alpha = 7$ Å	D?, $\Delta\lambda D/\lambda$ = 0.7–0.34 = 0.55	165
9	25.07.59/12:07	—	H_α–H_δ, K	moustache, wing, limb $\Delta\lambda_\alpha = 6$ Å	D, $\Delta\lambda D/\lambda$ = const = 0.53	160
10	25.07.59/12:23	2 (1+)	H_α–H_{11}, K, D_3, λ 4472	limb, sym. flare $\Delta\lambda_\alpha = 6$ Å	D!, $\Delta\lambda D/\lambda$ = const = 0.38	113

pure radial streams directed outwards. We think that the concept of some contraction preceding to expansion, or the coexistence of both these processes in small region of solar atmosphere is the most probable (see below).

More information on the nature of the phenomena of moustaches as well as on the nature of flares may be obtained from the examination of the cause of the broadening in the wings of flares and moustaches. The problem of broadening of emission in flares looks at first glance most controversial, due to the difference in results obtained. On the one side there is reliable spectroscopic evidence of Stark broadening in the wings of flares.[8-10]

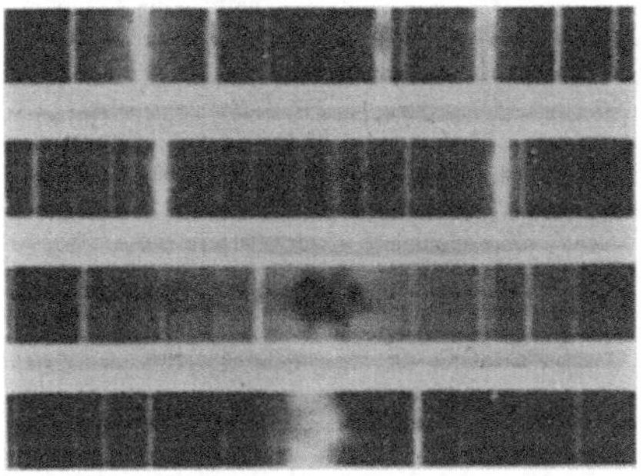

Fig. 12. The opposite kink of Hydrogen and metallic lines in moustaches on June 25, 1959.

On the other side there is even more numerous and reliable evidence that Doppler (or Gaussian) profiles fit emission wings very satisfactorily especially for the extensive wings of flares in flash phase and for moustaches.[6,11-14] Table 2 gives an idea how this representation agrees with the observations. The most reliable are the data relating to the limb phenomena[8,11] which are plotted in Figure 13. Figure 14 illustrates these limb phenomena. Disregarding the scatter of the points for the blue part of spectrum due to blends, we see from Figure 13 that the relation $\Delta\lambda_D/\lambda$ = const holds true not only for H-lines but for Helium and H and K. We would like also to emphasize that the broadening of moustaches is practically the same as that of broad wings of flares in flash phase.

The third possibility was proposed by observers of Sacramento Peak[15] who found for two flares above sunspots the best representation of observed profiles in the form of "non-maxwellian" turbulence $e^{-a\Delta\lambda}$.

At least we may assume the fourth possibility that the broadening in the extensive wings are due to some explosion (or expansion) of a small region in the form, for instance, of a spherical volume with some gradient of velocity inside.* The distributions of intensity inside the line can easily be found in the

* When the characteristic size of the fine structure is comparable with the usual width of the slit, such assumption of spherical symmetry seems to be reasonable.

manner similar to that for the envelopes of Novae and Be-stars.[16] For the spherical shell moving with the velocity v we have intensity:

$$\mathrm{d}I = K\varepsilon \cdot 2\pi r^2 \frac{\mathrm{d}r}{v(r)} .$$

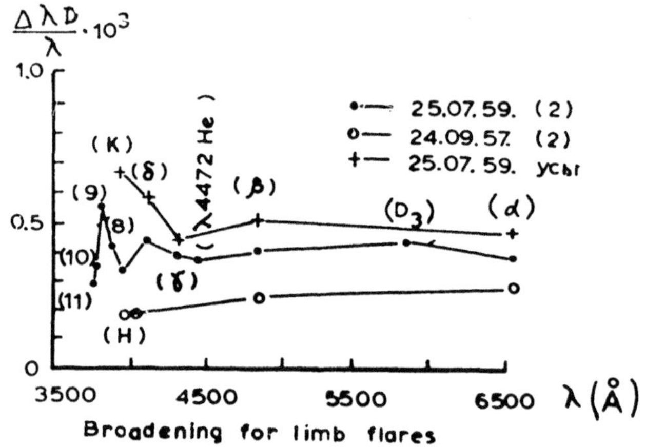

Fig. 13. The run of $\Delta\lambda_D/\lambda$ for the different lines in case of some limb phenomena.

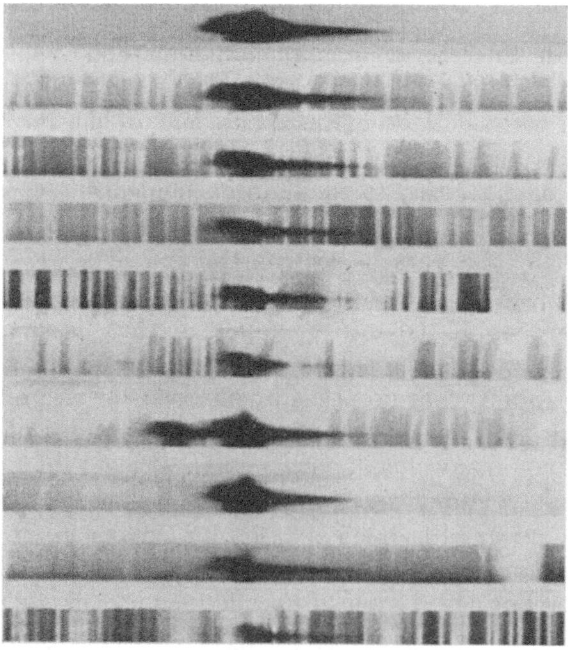

Fig. 14. The Limb flare (1+) on July 25, 1959 at 9.23 UT.

Taking $v \sim r^m$, $E \sim n \sim 1/vr^2$, (E is the emission per cm^3 which we assume to be proportional to the number density of H-atoms) we have

$$I(\Delta\lambda) = \int_{v}^{v_m} v^{(1/m)-3} \mathrm{d}v \sim \ln\frac{v_m}{v} = \ln\frac{\Delta\lambda m}{\Delta\lambda}, \qquad (2)$$

assuming $v \sim \sqrt{r}$, (v_m is maximal velocity). The reason for assuming this law $v \sim \sqrt{r}$ is quite simple: Formula (2) fits observed profiles perfectly, even better than the Doppler effect does.* It can be seen from Figure 15 where all the sources of broadening are compared for the most reliable case of moustaches at the limb where the influence of the underlying background can be neglected. We see that in this case the mechanism of the Stark broadening must be definitely excluded. The law $e^{-a\Delta\lambda}$ is better, but still worse than Doppler, and Doppler is worse than Formula (2).

Let us consider now the possible sources of these discrepancies. Observing the flare at the disk the emission wings may be influenced very much by the choice of the so-called "undisturbed" underlying atmosphere. Recently, we found, for instance that our profiles of H-lines for flares *above Sunspots* are also in good agreement with the law suggested by the American observers. But in the case of a spot the question what should be considered as underlying "undisturbed" *atmosphere* is even more acute than in case of a flare apart from the spot, because we usually do not have observations of the particular spot before and after a flare. (We measured the "undisturbed profile" in some other spots which were similar to the spot in question).

The second source is the lack of good seeing, which is even more serious because of the fine structure of emission. This is the reason why we have measured no more than 10% of all spectra of flares and moustaches that we have. For instance, the overlapping of different moustaches composing the broad wings of a flare may lead to the profile $e^{-\alpha\Delta\lambda}$ due to trembling and bad seeing even if the profile of separate moustaches is of the form $e^{-\alpha\Delta\lambda^2}$.[6]

The third source of differences is the difference in the phase of the phenomena. For instance SVESTKA found[12] for one flare that the profiles were represented by a formula $e^{-a\Delta\lambda^m}$ where m was changing from 1.2 to ~ 2 during the flare ($m \simeq 2$ near maximum). However, in the case of a small solar image, as in Ondřejov, the difference in time may be masked by the difference of places due to trembling and errors of guiding. Even atmospheric dispersion may lead to different profiles as we know from our experience.

But, despite of these critical remarks, we think that all these differences are quite real and the problem is to find the physical cause for them. In this

* The same formula was obtained in Reference 7 for the linear jet with a gradient of velocities $v \sim z$. In this case the continuity equation gives $n \sim 1/v$ and assuming the jet to be transparent we get (2) if we assume that, for a given mean velocity v, a dispersion of turbulent velocities exists in the range of $(-v, +v)$.

respect we would like to draw your attention to some experiments carried out at the Crimean observatory with the linear pinched discharge in Hydrogen.[17] We found the well-pronounced difference in broadening of Balmer lines: they are broadened mostly by linear Stark effect when we look *across* the axis of discharge and by turbulence (profiles of the type $e^{-a\Delta\lambda^2}$) when we look *along* the axis of discharge.

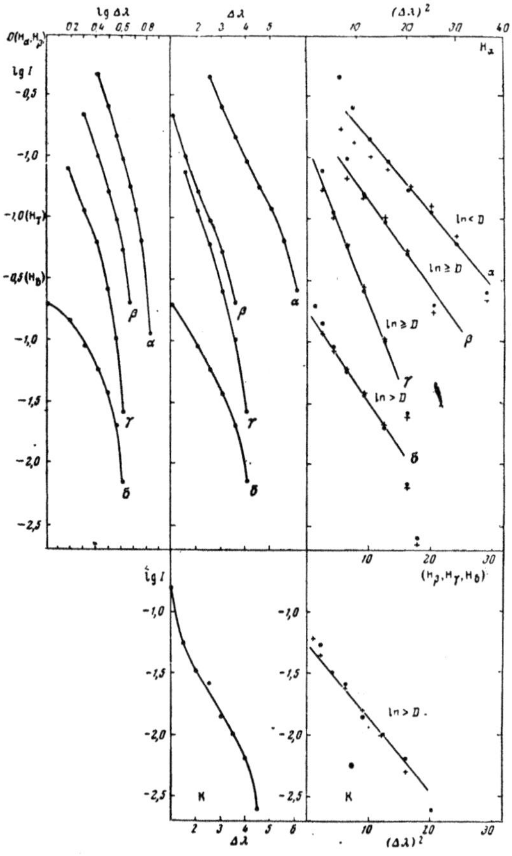

Fig. 15. The comparison of different cause of broadening in case of moustaches at the limb on July 25, 1959 at 9.07 UT.

Furthermore, the recent investigations of N. STEPANYAN (Crimean Observatory)[18], show remarkable difference in height of emission as well as in the time sequence of emission when we compare hydrogen and metallic lines. Figure 16 shows that Hydrogen broad emission lines and some strong metallic lines in the limb flare of class 2 (July 25, 1959), are found considerably higher (7000 km up to 12000 km) in the chromosphere than the comparatively narrow emission of the lines where more than 500 metallic lines in emission recorded on the echelle spectrogram (heights 2500–1000). MICHARD *et al.*[19] come to the same qualitative conclusion from statistical considerations of 26 flares. By a qualitative consideration of many spectra of flares SVESTKA *et al.*[19] have also found that metallic lines appear in emission at the same places where the core of H-lines

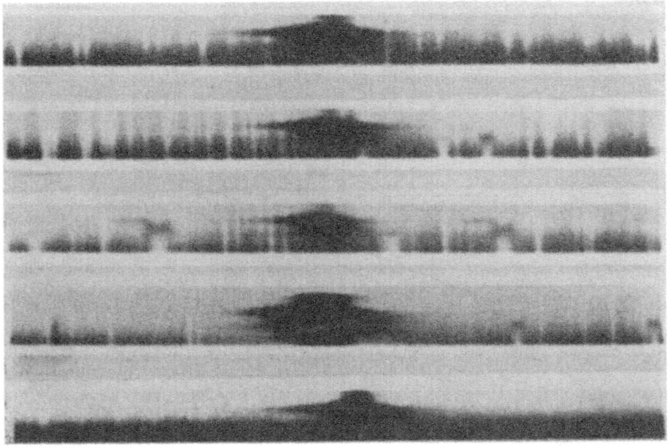

Fig. 16. The difference in the height of emission of Hydrogen broad emission and emission of weak metallic lines for the limb flare (2) on July 25, 1959.

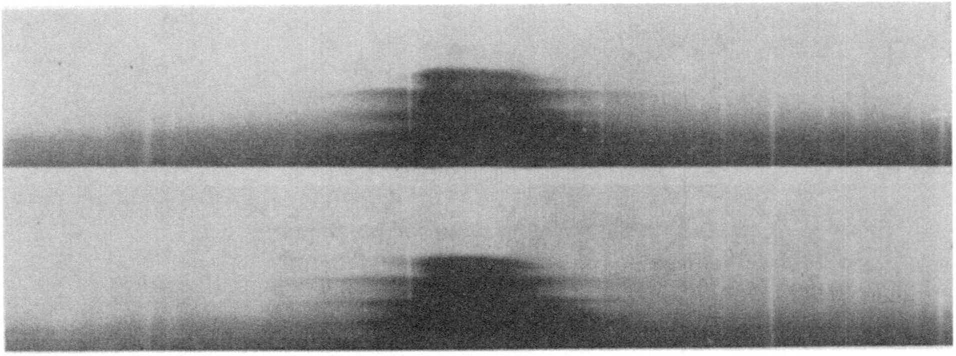

Fig. 17. The "Sandwich structure" in H_α of the flare on June 1, 1957 at 4.05 UT.

originates while the position of broad emission of H-lines is shifted relative to the position of metallic emission. Therefore it is difficult to avoid the conclusion based on all these results, that the position of the flare itself may differ considerably from those parts of the chromosphere which are disturbed by flare (of course this division is rather conventional). This can explain why we sometimes observe in flares with their high temperatures the lines of metals with the low I.P. and E.P. A flare is rather similar to a sandwich of different physically sheets of solar plasma than a uniform bulb, and Figure 17 illustrates this.

We think that we have here also a parallel to the study of active prominences by TANDBERG-HANSEN and ZIRIN.[21] For these authors have found that the active prominence spectra are divided into a "hot" class (T ⩾ 50000°) with intense He I emission and a "cool" class (T ≃ 10000) with weak He and

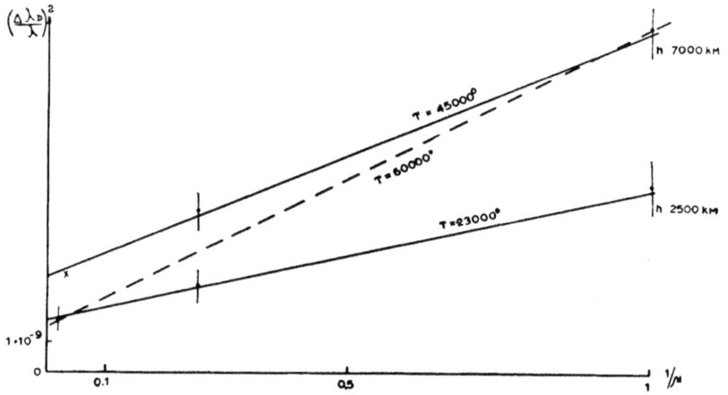

Fig. 18. The dependence of $(\Delta\lambda_D/\lambda)^2$ on $1/\mu$ for the different layers of the limb flare (2) on July 25, 1959.

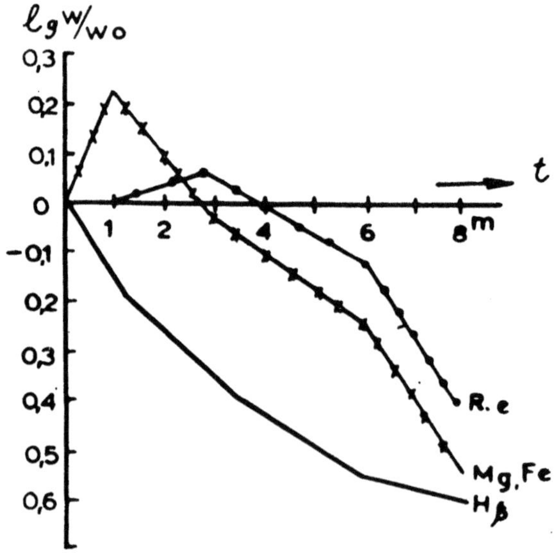

Fig. 19. The dependence of equivalent widths of Hydrogen H_β, metallic lines and Rear Earth lines with time in the limb flare 2 on July 25, 1959 ($t = 0$ corresponds to the maximum H_α-intensity, Wo is the intensity of continuum at the border of the disk).

strong metals; moreover they found that the lines of He II show greater broadening than He I and point out to an origin in a *different region*. If we now plot the graphs $(\Delta\lambda/\lambda)^2$ versus $1/\mu$ for the lower "metallic part" and for the "Hydrogen-Helium" higher part of the flare we get in case of the above mentioned limb flare two straight lines with moderate temperatures (Figure 18). But if we observe the same flare at the disk we could easily adopt the value $\Delta\lambda/\lambda$ for Hydrogen from the "hot part" and for metals from the lower "cool" part. The result would be a dashed line pointing to temperature about 60000°, which is obviously fictive.[14,19] Therefore we should carefully distinguish the place to which we refer before speaking of a particular mechanism of broadening

Fig. 20. The appearance of some limb flares in H_α on June 15, 1956 at 6.39 UT (top), on September 24, 1957 at 8.15 UT (middle), and on December 3, 1958 at 12.54 UT (below).

and try to avoid if possible the overlapping of the different (hot and cool) parts. Moreover, the same investigation[18] shows clearly that first Hydrogen and Helium can reach the maximum of their emission, and then, a few minutes later, the metallic lines become brightest, and afterwards the Rear Earths reach their maximum (see Figure 19). All this observed phenomena may have the simple meaning that some agency like shock front is coming out of the center of the explosion (the "core" of the flare) a little above the chromosphere and when propagating down it meets the chromosphere and produces here a violent excitation of atoms. From the time-delay of the maximum intensity of metals with respect to that of Hydrogen and from the difference of heights we obtain a velocity of shock-front ~ 100 km/sec which is in good agreement with the velocities of contraction of "conical hills" of limb flares measured directly in Crimea on H_α-filtrograms.[22] (See Figure 20.) In this connection the results of Czechoslovakian astronomers should also be mentioned.[20] They found by statistical analysis that before they reach their maximum flares frequently show blue asymmetry and after that they mostly show red asymmetry in central parts of the H-lines. Near the limb this effect almost vanishes. This result is in agreement with the just mentioned Crimean measurements of the dilatations and contractions of the limb flares and it also agrees with the fact that the area of the flare is changing almost synchronically with H_α-intensity.[23] Summarizing all the just-mentioned spectroscopic and filtrographic evidence we come again to the concept of some explosion: a rapid expansion (or dilation) of a small portion of the solar plasma* and after that a comparatively slow contraction. Is there any very rapid contraction before expansion phase (flash phase) or is the first expansion accompanied by a contraction in a different direction (for instance, around vertical axis)? As yet we do not know. There are still so many questions in the whole problem that they present a broad field of investigation to the activity of several generations of solar physicists.

References

1. Nikonov, V. B., Severny, A. B.: 1951, *Dokl. Akad. Nauk SSSR*, **57**, No. 2, 217. Severny, A. B.: 1957, *Publ. of the Conf. on Solar Research*, p. 40.
2. Mustel', E. R.: 1960, *Astron. Ž.* **37**, 403. Mustel', E. R.: 1962, *Astron. Ž.* **38**, No. 28a, 385.
3. *COSPAR Information Bulletin*, 1963, No. 14.
4. Severny, A.: 1956, *Astron. Ž.* **33**, 74. Severny, A.: 1957, *Izv. Krimsk. Astrofiz. Obs.* **17**, 129.
5. McMath, R., Mohler, O., Dodson, H.: 1960, *Proc. Natl. Acad. Sci. U. S.* **46**, 165.
6. Severny, A.; Koval, A.: 1961, *Izv. Krimsk. Astrofiz. Obs.* **26**, 3.
7. Koval, A.: 1962, *Izv. Krimsk. Astrofiz. Obs.* **28**, 241.
8. Mustel', E. R., Severny, A. B.: 1952, *Izv. Krimsk. Astrofiz. Obs.* **8**, 19.
9. Svestka, Z.: 1960, *Bull. Astron. Inst. Czechoslovakia* **11**, 167.
10. Suemoto, Z., Hiei, E.: 1959, *Publ. Astron. Soc. Japan* **11**, 185.
11. Severny, A.: 1958, *Astron. Ž.* **35**, 355. Severny, A.: 1959, *Izv. Krimsk. Astrofiz. Obs.* **21**, 131.
12. Svestka, Z.: 1961, *Bull. Astron. Inst. Czechoslovakia* **12**, 73.
13. Polupan, P.: 1959, *Circ. Astr. Obs. Kiev*, No. 69, 16.
14. Dunn, H., Jefferies, J., Orrall, E.: 1960, *Observatory* **80**, 31.
15. Jefferies, J., Smith, E. V., Smith, H.: 1959, *Astrophys. J.* **129**, 146.

* In Reference 22 we spoke about cumulative shocks, the cumulative effects being produced by cusped field geometry of the magnetic field.

16. GERASIMOVIČ, B. P.: 1933, *Z. Astrophys.* **7**, 335. WILSON, O. C.: 1934, *Astrophys. J.* **80**, 259.
17. BABIN, A., LUKIANOV, S., SEVERNY, A., SIDOROV, G., SINITZIN, V.; STEŠENKO, N.: 1962, *Izv. Krimsk. Astrofiz. Obs.* **27**, 52.
18. STEPANJAN, N.: 1963, *Izv. Krimsk. Astrofiz. Obs.* **30**.
19. MICHARD, R., SERVAJEAN, E., LABORDE, G.: 1959, *Ann. Astrophys.* **22**, 887.
20. SVESTKA, Z., KOPECKY, M., BLAHA, M.: 1961, *Bull. Astron. Inst. Czechoslovakia* **12**, 229. SVESTKA, Z., KOPECKY, M., BLAHA, M.: 1962, *Bull. Astron. Inst. Czechoslovakia* **13**, 37.
21. TANDBERG-HANSEN, E., ZIRIN, H.: 1959, *Astrophys. J.* **129**, 408, 414. TANDBERG-HANSEN, E., ZIRIN, H.: 1960, *Astrophys. J.* **131**, 717.
22. SEVERNY, A., ŠAPOŠNIKOVA, E.: 1960, *Izv. Krimsk. Astrophys. Obs.* **24**, 235.
23. SEVERNY, A., ŠAPOŠNIKOVA, E.: 1954, *Izv. Krimsk. Astrophys. Obs.* **12**, 3.
24. ELLERMAN, F.: 1917, *Astrophys. J.* **46**, 298.

ABOUT THE FLARE PROBLEM

K. O. KIEPENHEUER
Fraunhofer Institut, Freiburg i. Br., Germany

Abstract. The macroscopic and microscopic structures of flares are discussed, their relations to spots, to the photospheric magnetic fields and especially to the filaments (prominences) are reviewed. Flares are macroscopically located in a horizontal magnetic field, mostly occupying the dividing line of polarities or the line of vanishing longitudinal field. Flares often reach into regions of very small fields.

Arguments are presented, which make it unprobable that the flare energy is supplied from outside the Hα-volume. It seems furthermore improbable that the flare phenomenon is a macroscopic explosion or collapse. There is spectroscopic and other evidence for the existence of subtelescopic structures. The close relation between flares and filaments – both show a similar magnetic configuration – lets us suspect that the preflare activation of a filament and the resulting subtelescopic distortions and twistings of the associated magnetic field are the basic processes leading to a flare.

Introduction

The physical interpretation of a solar flare is a very complex problem. The reasons are various: The physical processes involved are very far away from thermal equilibrium; flares occur in a layer of the solar atmosphere, which is very difficult to observe; flares come unpredicted, sometimes after several weeks of waiting, and deliver within a few minutes a large amount of informations. The probability that the occurrence of a flare, good seeing conditions and equipment in good working condition coincide is therefore rather small. The radio records of flares and of associated phenomena, as well as the record of their terrestrial effects are by far more complete and better than the optical documentation (monochromatic pictures, photometric results, high resolution spectroscopy).

In this paper the relation of flares with other phenomena in active regions of the sun will be reviewed. Especially the often observed association of flares with dark filaments (prominences as seen on the disc) will be discussed in some detail, because it might be of crucial importance for the interpretation of the flare mechanism.

Structure and Magnetic Field Configuration of Flares

The combination of limb- and disk-observations has not yet given us a reliable threedimensional picture of a flare. Apart from a few white light flares the photosphere seems not to be involved, not even in those cases where it was looked at systematically. From the inspection of a number of Hα-filtergrams it seems even that the lower parts of the active chromosphere are not affected,

C. de Jager (ed.), The Solar Spectrum. All rights reserved.

while the higher structures (as small filaments) become part of the flare and often obtain flare brightness. The highest parts of flares reach according limb observations up to 20000 to 50000 km. The main luminous body of flares therefore is mostly located in the very inner corona, still reaching the coarser structures of the chromosphere. The flare is mostly surrounded by material of a permanent coronal condensation, $n \sim 10^{10} \ldots 2 \cdot 10^{10}$ cm^{-3}, $T > 10^6$ degree: The density of the flare's surrounding is almost the density of a prominence, its temperature, however, is 100 times higher.

As seen against the disc, flares appear as bright ribbons or threads, sometimes made of bright knots merging together after a little while, sometimes composed of small pieces of thread. As seen on the limb, the first beginning of a flare might look like a little bright cone, later transforming into a system of bright loops, rising slowly and showing at the same time downward streaming material.

A flare represents a characteristic feature of a solar center of activity and cannot be understood without discussing it in a synoptic manner. There are some typical flare-spot-configurations, which are represented in Figure 1 in a

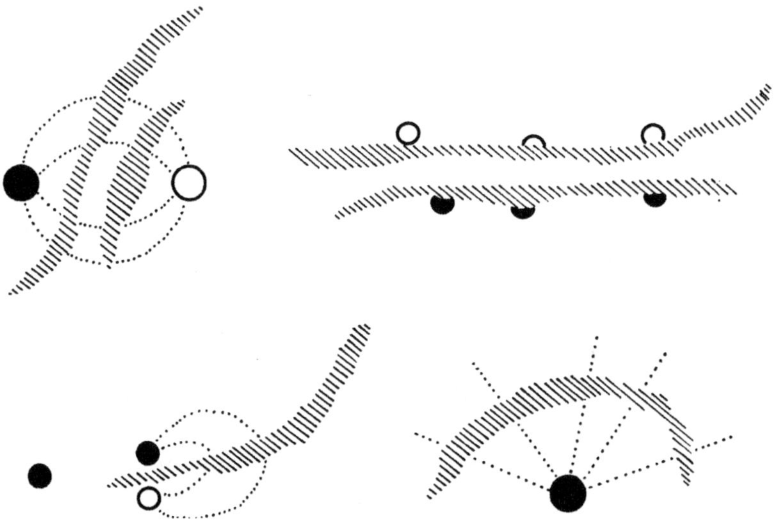

Fig. 1. Typical configurations of flares and spots. Magnetic field lines are indicated by dotted lines, the spot polarities by bold dots and small circles.

simplified form. In addition a great number of maps of the photospheric magnetic fields below flares are available. From the known observations the following results about the magnetic field configurations of a flare can be extracted:

(a) Flares coincide like dark filaments in most cases with the dividing line of magnetic polarities. Along this line the longitudinal field H_L disappears, while the gradient dH_L/dx becomes large.

(b) Flares occur almost only in regions with a horizontal field.

(c) Flares preferably start in regions with great values of dH_L/dx and changing field, sometimes possibly in neutral points of the magnetic field ($|H| = 0$).

(d) The macroscopic structure of a flare extends and propagates across the magnetic field. (The field lines are indicated in Figure 1).

(e) Flares often occur in regions of an obviously homogeneous magnetic field and extend into areas with very small fields (< 10 Gauss).

In general, one can say that the characteristic magnetic configuration of a flare looks much more like that of a prominence (see Figure 2a), as proposed by KIPPENHAHN and SCHLÜTER (1957), than the one presented by SWEET (1956), interpreting a flare as the collision of the coronal fields of two merging spot groups (see Figure 2b).

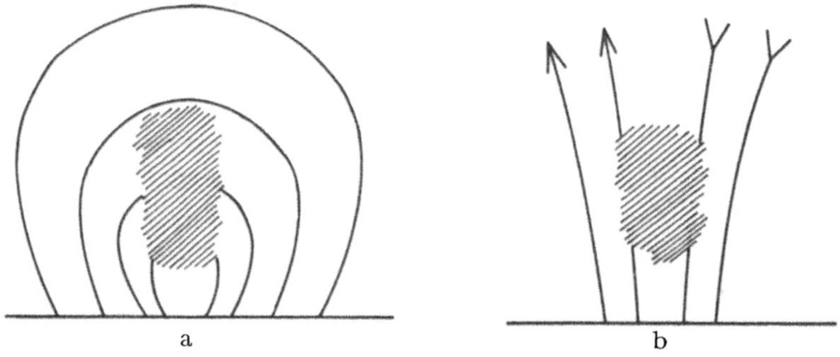

Fig. 2. Magnetic configurations: (a) of a filament according KIPPENHAHN and SCHLÜTER (1957); (b) of a flare according SWEET (1956).

Further conclusions are:

(f) The occurrence of "neutral points" of the magnetic field does not seem to be essential for the shape or the extension of a flare.

(g) Flares often occur almost simultaneously in different parts of an extended center of activity or even in centers a few hundred thousand kilometres apart. The triggering agency obviously travels through the corona with a velocity of the order of 1500 km/sec (BECKER, 1958).

(h) Sometimes flares have the tendency to repeat themselves within a few hours, in the same location, even showing the same time variation of their Hα, Radio and X-Ray emission (DODSON, 1949; HANSEN, GORDON, 1960; FORTINI, 1963), their mechanical output as the Moreton blast (ATHAY, MORETON, 1961) and the flare surges. This implies that the magnetic field configuration around the flare survives the flare unaltered or will be recovered within an hour.

Inspite of the fact that quite many investigations of the magnetic fields

around flares have been undertaken, there is no convincing evidence available yet that the measured photospheric field changes below a flare (recorded in the form of maps of the photospheric longitudinal field) are really significant for the flare mechanism. No measurements of the magnetic field in the luminous volume of the flare itself are available. The relation between the photospheric field and the fields in the very vicinity of a flare on the other hand are quite complex, especially along the line of disappearing longitudinal component. There have even been flares without any field changes in their surroundings. No field change as fast as a flare develops has ever been recorded.
So we might state:

Magnetic field changes around flares might happen quite often, but do not seem to be essential for the occurrence of a flare as such. More field measurements with higher resolution in space and time are needed to get a clear picture of the effects of changing fields.

Association of Flares and Filaments

Next to the sunspots and associated plage areas the dark filaments in an active region of the sun are often closely related to a flare. Sometimes a filament itself or part of it becomes a flare, even without changing its shape. We know for some time now that the magnetic configuration of a filament is very similar to that of a flare. This holds for larger filaments in active regions as well as for grown up quiescent filaments far away from sunspots. In Figure 3 some characteristic types of flare-filament-association are shown, which are summarised in the following section:

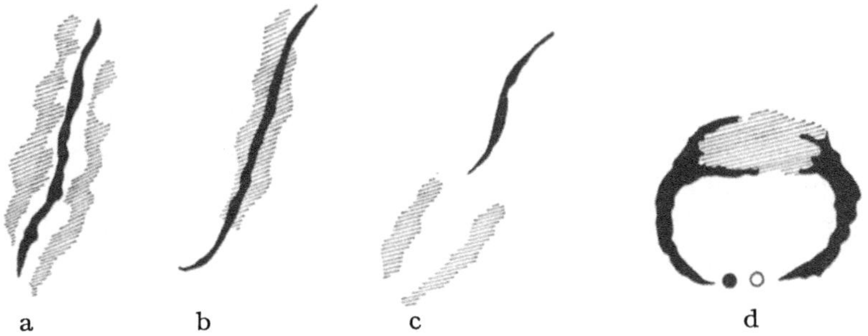

Fig. 3. Characteristic flare-filament-configurations (flares are hatched).

(1) The flare develops or propagates, along an existing filament (Figure 2a), quite often in two lanes, which show an expanding motion away from the filament with velocities ranging from about 2 to 20 km/sec (Bruzek, 1957; Malville and Moreton, 1963). The motion comes to a halt after a while. It

is interesting to note that the same expanding motions are observed for the bright quasi-flare-lanes which develop along a quiescent filament during its activation or disappearence.

(2) The flare coincides with a filament (Figure 2b), part of which might reach flare brightness. Very often the filament or parts of it survive the flare unchanged. Sometimes the flare consists of preexisting coarse structures of the chromosphere, which stick a little higher into the inner corona. Sometimes flares occur in the prolongation of filaments (see Figure 3c), flanking the "chromospheric wake" of a filament, manifested by chromospheric thread structures looking like the wake waters of a ship. There are of course also quite many flares without filaments. In some of such cases, the chromospheric structure underneath the flare looks similar to the underground of a filament.

(3) There are flares without filaments, but even then the chromospheric fine structures underlaying the flare often have the typical appearance observed in the prolongation of filaments.

Quite often (BRUZEK, 1957 and others) the flare filament becomes distinctly darker an hour or more before the onset of the flare. This can be observed best in the blue wing of Hα, less pronounced in the center of the line. An example is given in Figure 3d. This preflare activation of a filament means an increase of internal turbulence from 10 to $\sim$ 30 km/sec, a systematic upward motion of the whole filament and possibly also an increase of the number of hydrogen atoms/cm^3 in the absorbing structures of the filament. The density of the coronal surrounding is already that of a quiet filament ($\gtrsim 10^{10}$ cm^{-3}), its temperature, however, 100 times higher. Assuming pressure equilibrium between filament and corona, a filament density of 10^{12} cm^{-3} is required. Still higher densities in certain parts of the filament are needed, if BECKER's (1959) observations of structures of activated filaments in white light would be interpreted in terms of electron scattering of by H^- absorption.

(5) There is a great number of flares which break out in the midst of an activated or rising filament. The change from Hα-absorption to emission might take only a few minutes. The disappearance of the associated filament might occur in the form of ejection or by raining down into the chromosphere. Many striking examples of this type have been shown by G. MORETON with the 10-sec-Hα-patrolfilms, obtained at Lockheed Solar Observatory, Los Angeles.

All these observations, especially those showing the very close interaction of active filaments and flares make it clear that these two phenomena should be treated together.

Energy and Mass Balance of Flares

The total energy output of a flare exceeds the thermal energy content of the flare's volume by a factor of the order of 1000. The estimates of the radiative output of a flare are uncertain mostly because of the unknown intensity of the emitted continuum, which even is unknown in the visible spectrum. The large amount of energy produced by a flare in the form of accelerated particles or clouds or shockwaves is also difficult to estimate.

The following data represent the generally assumed orders of magnitude:

Optical output. Most of the energy goes into an enhanced line emission and in a continuum of unknown intensity, which in the visible range seems to be less than 1% of the photospheric continuum. Altogether the optical emission is certainly $\lesssim 6 \cdot 10^8$ erg/cm²/sec. Supposing that this emission occurs in a macroscopic depth of $5 \cdot 10^8$ cm and within 1000 seconds on the average, about 1000 ergs/cm³ in the flare's volume will be needed. This figure leads for importance 2 flares ($\sim 10^{18}$ cm²) to a total optical output of $\lesssim 6 \cdot 10^{29}$ ergs, for a 3+ flare ($\sim 5 \cdot 10^{19}$ cm²) to $\lesssim 10^{32}$ ergs.

The "mechanical" output of a flare appears in the form of flare surges, solar blast waves, interplanetary blasts, proton events, and solar cosmic rays.

Flare surges. Their total volume ranges roughly around 10^{27} cm³, their velocity is $\sim 10^7$ cm/sec, their density probably around 10^{11} cm^{-3}. Their total kinetic energy might therefore come up to the order of 10^{28} ergs, their mass to 10^{14} grams.

Solar blast. Discovered by MORETON (*l.c.*), this ejection propagates with a velocity of 1000 to 1500 km/sec across the solar disk. When observed in the wing of Hα it looks like a rather transparent cloud, being shot out by the flare in certain directions and affecting filaments and certain chromospheric structures up to distances of more than a solar radius. The total mass might also be of the order of 10^{14} grams. This would correspond to a kinetic energy of $\gtrsim 10^{30}$ ergs.

Interplanetary blast. This phenomenon is probably the same as the "solar blast". It produces one or two days later a geomagnetic storm. With a particle density of 30 cm^{-3} in the earth orbit (PARKER, 1963), a volume of 0.1 a.u. in diameter and a velocity of 1500 km/sec, a total mass of $\sim 2 \cdot 10^{15}$ grams and a total kinetic energy of $\sim 10^{32}$ ergs is supplied.

Proton event. According PARKER (1963) about 10^{10} grams of protons are accelerated to an energy of about 30 MeV or a velocity of $0.7 \cdot 10^{10}$ cm/sec corresponding to a flux of 10^4 protons/cm²/sec within a volume of 1 (a.u.)³. The total energy involved amounts to about 10^{30} ergs. It might be added here that the solar cosmic ray events certainly do not carry more energy than this.

Summarizing one can say that in most cases the mechanical energy output of a flare exceeds largely the energy radiated by a flare.

The mass flow connected with the above events can amount to more than 10^{15} grams. Most of this mass is contained in the interplanetary blast, which carries probably more material than that contributing to the optical emission of the flare. The ejected mass exceeds even the mass contained in the permanent coronal condensation housing the flare.

Before looking to the powerful energy-sources for the flare mechanism, one should try to make sure, whether the energy is supplied from within the luminous volume of the flare or is transferred from the corona or from the photosphere or chromosphere below the flare.

Supply of energy from the corona is quite unprobable, because the thermal energy of the total corona is less than the output of a great flare. In addition, the structure of the corona as a whole and also the structures in the coronal condensation around the flare seem to survive the flare unchanged.

The transfer of energy to the flare from the underlaying photosphere as suggested by WARWICK (1962) and others requires a very strong flux of particles through the chromosphere, which certainly would be much denser than the solar blast travelling through the corona. If only 1 % of the momentum of this particle stream would be transferred to the chromosphere, the latter would be pushed away.

Also the observed triggering of flares makes it rather improbable that their energy is supplied from the outside. This observation suggests more that the energy is available inside the flare's volume ready to be released within a few minutes.

Under such circumstances it seems that the only source of energy, which would be sufficient to produce a flare, is the magnetic field within the flare's volume. In spite of the fact that we are still not able to understand completely the annihilation of a field of a few hundred gauss within a period of a few hundred seconds, this process is obviously the one we have to look for.

The Flare Mechanism

The close relation between filaments and flares makes us suspect that the physics of flares and filaments are of the same kind.

In the KIPPENHAHN-SCHLÜTER (1947) model of a quiescent filament, the material hangs in the horizontal sections of the magnetic lines of force in the corona. These lines connect usually the two parts of a bipolar magnetic region, the filament coinciding then with the dividing line of the two polarities.

As soon as such a filament becomes activated, no matter whether far away or close to sunspots, the field-lines will become increasingly distorted because of the increased turbulence and density in the filament. For a turbulent velocity of 30 km/sec and a density of $3 \cdot 10^{11}$ cm^{-3}, equality of kinetic and magnetic energy is reached for 10 gauss, with 10^{13} cm^{-3} for 50 gauss! This field perturbation, which probably occurs mostly on a subtelescopic scale, will result in numerous twists of the field-lines and in collisions of regions with different vectors of magnetization. While encounters of regions with fields will be hindered because of the repulsion of the associated current systems, encounters of fields will be favoured.

These (subtelescopic) field collisions will probably act as in the macroscopic interaction described by SWEET (1956), and by GOLD and HOYLE (1960).

It is quite clear, however, that without introducing interpenetration of the colliding fields, no fields will be annihilated and the energy output will not exceed the content of turbulent energy, which is by far not adequate. The field collisions must end in a kind of pinch effect.

There is a number of observations, which favour the concept of a flare being built up by many subtelescopic or micro-instabilities.

Extremely underexposed filtergrams of flares, obtained under excellent seeing conditions and showing no trace of the solar disk sometimes show very fine well-blackened points of about 1″ in diameter or perhaps less. This means that a flare, at least, during the flash phase shows a very strong fluctuation of brightness in dimensions, which probably are "subtelescopic". In addition, one has visually the impression that – during short periods of excellent seeing – there is a fine hairy structure, just as it can be seen sometimes in flare surges and certain prominences.

Furthermore it has been shown by SUEMOTO and HIEI (1959) and others that the line emission of a flare in the Balmer series must occur in a total depth of only 10^6 cm. The macroscopic dimension of a flare being of the order of 10^8 to 10^9 cm, this means that there must exist numerous still finer structures in a flare distributed all over the volume. They come to this conclusion by assuming that the broadening of the higher members of the Balmer series is produced mostly by Stark effect and that the density of the emitting structures must be of the order of 10^{13} cm^{-3}.

There are two more arguments in favour of small scale structures in a flare: If one assumes that the duration of the flash phase of a flare represents somehow the decay time of a magnetic field σa^2 (σ, conductivity, a, dimension of the conductor), then, even under very favourable conditions, subtelescopic values for a will result. Also the rapid preflare-activation resp. densification of filaments is very difficult to understand in terms of a structureless simultaneous increase of density all over a volume of more than 10000 kilometres extension. Even this process looks much more feasible in the form of many subtelescopic globules, condensed out of coronal material without need of supply of material from the outside across distances > 10000 km. It would be worth-while therefore, if this type of micro-instability in a coronal plasma would be investigated by the experts in this field. In addition, highly resolved spectra of activated filaments are urgently needed.

It is a pleasure to record here my sincere thanks to Dr. Bruzek, who assisted me greatly in discussing the numerous observational aspects of flares.

References

ATHAY, R. G. and MORETON, G. E.: 1961, *Astrophys. J.* **133**, 935.
BECKER, U.: 1958, *Z. Astrophys.* **44**, 243.
BECKER, U.: 1959, *Z. Astrophys.* **48**, 189.
BRUZEK, A.: 1957, *Z. Astrophys.* **42**, 76.

Bruzek, A.: 1958, *Z. Astrophys.* **44**, 183.
Bumba, V.: 1958, *Izv. Krimsk. Astrofiz. Obs.* **19**, 105.
Bumba, V.: 1962, *Izv. Krimsk. Astrofiz. Obs.* **23**, 437.
Dodson, Helen W. and Hedeman, E.: 1949, *Astrophys. J.* **110**, 242.
Fokker, A. D.: 1963, *Space Sci. Revs.* **2**, 88.
Fortini, T.: 1963, *Astrophys. J.* **138**, 278.
Gold, T. and Hoyle, F.: 1960, *Mon. Not. Roy. Astron. Soc.* **120**, 89.
Hansen, R. and Gordon, D.: 1960, *PASP.* **72**, 194.
Howard, R. and Severny, A.: 1963, *Astrophys. J.* **137**, 1242.
Howard, R. and Babcock, H. W.: 1960, *Astrophys. J.* **132**, 218.
Jefferies, J. T. and Orrall, F. O.: 1961, *Astrophys. J.* **133**, 963.
Kippenhahn, R. and Schlüter, A.: 1957, *Z. Astrophys.* **43**, 36.
Martres-Tropéz, M. J. and Pick-Gutman, 1962, *Compt. Rend.* **254**, 3975.
Moreton, G. and Malville, J. M.: 1963, *PASP* **443**, 176.
Parker, E. N.: 1963, *Astrophys. J. Suppl.* **7**, 177.
Smith, S. F. and Ramsey, E.: 1964, *Z. Astrophys.*, **60**, 1.
Suemoto, Z. and Hiei, E.: 1959, *Publ. Astron. Soc. Japan* **185**, 11.
Sweet, P. A.: 1956, *IAU Symposium* **6**, paper 14.
Unsöld, A.: 1962, *Physik der Sternenatmosphären*, p. 323.

DISCUSSION ON THE ACTIVE CHROMOSPHERE

J. EVANS
Chairman
Sacramento Peak Observatory, Sunspot, New Mex., U.S.A.

M. RIGUTTI
Secretary
Firenze

W. DE GRAAFF
Secretary
Utrecht

PECKER. On spectra secured by Dr. Michard, Miss Sivirine of the Meudon Observatory, has been measuring several Hα and Hγ profiles in moustaches. These are defined essentially by the measured parameters *a*, *b*, *c* in Figure 1. Note that *a* is positive, contrary to the more extreme cases described by Dr. Severny. She has computed what type of perturbation can account for these results: its depth is well determined by *a*; ($\tau_0 \approx 10^{-5}$ to $5 \cdot 10^{-5}$). The data *b* and *c* give two equations almost identical to each other; the system is thus badly determined: one only can say that $\Delta P_e/P_e = 40\%$ and $\Delta T = 100\,^\circ$K; but an increase of the pressure perturbation, and a decrease of the temperature effect is largely allowed.

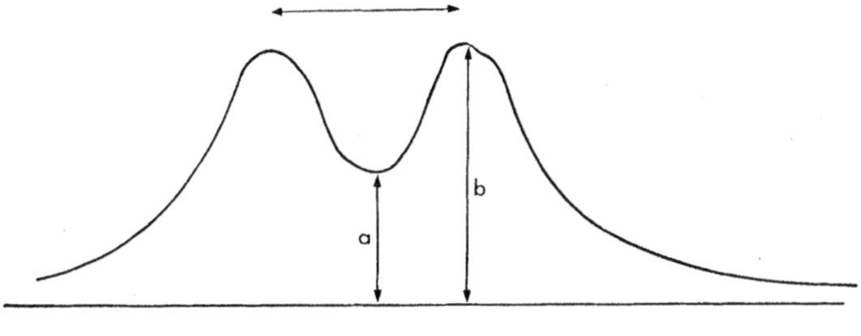

Fig. 1

MORETON. This remark concerns the matter of the asymmetry in the Hα line in flares, mentioned by Dr. Severny and Dr. Kiepenheuer. This line asymmetry has been attributed to an invisible cloud which ascends before the flare maximum and descends after flare maximum. S. H. SMITH and H. E. RAMSEY (*Astrophys. J.*, in press, *Z. Astrophys.* **60**, 1) have recently studied flare-associated filaments, confirming and amplifying the work by Bruzek, and find (1) the motion of filaments, often beginning well before a flare occurs, represents the beginning of the ascent of a filament into the corona; (2) As the flare approaches maximum brightness, the velocity of the absorption filament

C. de Jager (ed.), The Solar Spectrum. All rights reserved.

increases up to about 100 km/sec; (3) After flare maximum, absorption features form around the active region, suggesting decending absorption material; and (4) the study suggests that more than half of all major flares are associated with these gross absorption clouds.

It is further suggested that this feature, commonly observed on filtergrams made at wave lengths displaced 0.5 Å from Hα, is the cause of the Hα line asymmetry in flares.

ZIRIN. We have also observed the effect noted by Moreton in two cases, November 20, 1960 and July 20, 1961. In both cases an active prominence appeared above the limb and was blown off 10 or 20 minutes later by a huge flare.

LÜST (to Kiepenheuer). Is there any recent change in the estimates of the energy output of a flare, particular in the optical emission?

KIEPENHEUER. The energy estimate, I gave are mostly based on PARKER's paper for the blast wave, on own estimates for the optical output. They do not include the "white" flares.

MICHARD. (1) "Mechanical" phenomena such as surges, "blast waves" are observed in only some flares; their energy outputs seems to be difficult to estimate. Therefore no general argument about flares can be derived from their consideration. (2) I believe that the spectroscopic evidence for "microstructure" in flares is not overwhelming! In the analysis of flare spectra various authors have found the wings of Balmer lines could not be explained by the Stark effect, an assumption which, on the contrary, is the basis of Suemoto and Hiei's results.

KIEPENHEUER. Even if Suemoto's result is wrong by a factor of 10, we would still have emitting layers small as compared to the total thickness of the flare.

ATHAY. Dr. Kiepenheuer's remarks on fine structure in flares have a number of points still being unclear: (1) The evidence for the energetic phenomena being confined to a fraction of about 0.1 of the flare volume seems weak and this postulate seems to be unnecessary; (2) Localized regions within flares with densities of the order of 10^{15} cm^{-3} would produce more Balmer continuum radiation than is observed in flares unless the volume of such regions is less than about 10^{-3} of the total flare volume, as it is normally inferred to be; (3) The strong magnetic field energy that is supposed to exist in the vicinity of the prominence, in this picture, is generated by turbulent motion of the prominence itself. Thus, the flare energy derives ultimately from the kinetic energy of the prominence. Would you comment on these points?

KIEPENHEUER. You are certainly right that part of the energy output of the flare must come from the turbulent energy of the associated filament. But as far as I can see it now, the turbulent energy is mostly used to bring together the fields of opposite direction (in a micro scale). As soon as they have reached a certain contact they will interact and annihilate without using up mechanical energy of the filament. To your second question: I think most of the observers agree that at least the surges, probably also the blast wave comes from a volume, which is small as compared to the luminous volume of the flare.

JEFFERIES. Measurements of the electron scattered and Balmer free bound continua made by Orrall and me indicate a consistent model of an active flare like loop, to have an electron density $\approx 2 \cdot 10^{11}$ cm^{-3} spread over a distance $\approx 10^{9}$ cm. A clumped model presents the difficulty that it requires correspondingly higher electron densities which appears to lead to problems in reconciling the observed and estimated overlap of the Stark broadened lines of H. Such a model would be clearly in disagreement with those proposed by Suemoto, Hiei and Hirayama.

It should also be mentioned that line profiles are a notoriously unsatisfactory way of determining kinetic temperatures and turbulence; we have shown that for one flare the observations of Ti II and Balmer line widths interpreted in this way will give temperatures which are indeed negative.

MORETON. I wish to call attention to two observations which may be pertinent to Dr. Kiepenheuer's discussion: (1) William Miller, of RCA Long Island Laboratory has conducted white light cinematographic observations of sunspots. He has reported at least one observation of absorption blobs apparently ejected from a sunspot (a photograph of which is reproduced in Menzel's book *Our Sun*). (2) Robert Coates of NASA once mentioned to me that careful photometric measurements of flare light curves showed that some flares definitely have fluctuations in their Hα intensity during their declining phases.

DE FEITER. When comparing the results of the method of Redman and Suemoto for determining electron densities with and without electron damping, one should naturally start from the same set of data. When doing this, I found only a change of a factor of about two in the deduced electron density, instead of the factor 10 which Jefferies mentioned.

JEFFERIES. Unless my arithmetic is at fault this suggests that there is an inconsistency between the Redman-Suemoto and the modified Inglis-Teller methods for inferring the electron density.

DE JAGER. A question to John Jefferies. You say that the flares studied by you have twice the N_e of prominences. So their N_e must be between 10^{10} and 10^{11} cm^{-3}. You further say that the introduction of Inglis-Teller theory instead of Kolb reduces the "classical" N_e by a factor 10. This would bring N_e back from $10^{13.3}$ to $10^{12.3}$ cm^{-3}.

So there is still a factor 100 between your two densities; what is the source of this discrepancy?

JEFFERIES. Discrepant electron densities in limb and disk flares in Hα have often been reported. I don't believe that this is a difficulty since I think they refer to different regions. The densities in disk flares are estimated from Hα line wings which arise in photospheric regions excited by particle or electromagnetic radiation from the primary phenomena. These regions would of course, be unobservable in the limb Hα emission. In fact, Michard and Smith and Smith have pointed out many such differences which I think, can be reconciled in a similar manner.

ZIRIN. We must remember that flares are not a cool, quiet phenomena, but a hot one. In every case when a flare is seen on the limb, a coronal condensation of yellow line is observed. The continuum intensity gives us the tendency directly from consideration of Thomson scattering. Densities are around 10^{10}–10^{11} and temperatures 3–4 million degrees.

The problem with Jefferies' *Balmer continuum* to *scattering continuum* ratio is that the Balmer continuum is sensitive to the cool parts of the flare, while the scattered continuum comes from all parts. In most cases the continuum for $\lambda > 3700$ Å comes mostly from the corona, so spurious results may be obtained.

The most important point to determine is the relation between the hot coronal phenomenon, where the X-rays decimeter radiation, etc. are produced and the surface Hα phenomenon, which is well observed but only shows cool parts of the flare.

SMERD. The fast-drift bursts in the explosive phase of a radio flare suggest that individual discharges occupy $\lesssim 1$ sec may repeat themselves quasiperiodically for several minutes. The regular repetition seems to imply repeated ejections from the same explosion centre. The ejection of plasma blobs at supersonic speeds are needed to interpret slow-drift Type-II bursts, the microwave Type-IV and the moving metre-wave Type-IV bursts. There is evidence for multiple shock fronts moving in different directions, all probably originating at the time and place of the original phase explosion.

Vast quantities of electrons seem to be the needed to account for Type-IV emission and large plasma streams are required to provide them as well as the energy needed to drag magnetic fields out to the coronal heights at which the moving metre-wave Type-IV is observed. The above radio evidence suggests *fine-structure in time* during the explosive phase of a flare; it does not seem to support the idea of "sub-telescopic" *spatial fine-structure* in the plasma-instability responsible for the flare explosion.

G. E. MORETON

Kinematics of Solar Flares

The occurrence of high velocity motions in flares and the chromosphere constitutes one of the most important problems of solar physics. One method of acquiring data on transverse or horizontal motions in the solar atmosphere is by means of high time-resolution (≈ 10 sec) monochromatic cinematography. Results of a study employing this technique indicate that flares fall into two distinct catagories. Most flares are characterised by gradual brightening to maximum intensity, and slow (~ 10 km/sec) expansion of the flare borders. Flares in this group commonly initiate the flare-surge event, i.e., the ejection of matter in a brush-like shape, with a typical velocity of around 150 km/sec. Flares of the second category are less frequent than the former, but they are characterized by more violent development and usually exhibit a distinct acceleration phase in their rise to maximum intensity and area. The flare expansion is often in a preferred direction with a velocity $\sim$ 100 km/sec. In a significant number of cases, extensive cloud or wave-like disturbances are ejected at this "explosive" phase. These disturbances are observed to propagate over distances as great as a solar radius, with a characteristic velocity $\sim$ 1 000 km/sec.

Explosive phase flares are much more frequently associated with radio bursts than are regular flares: all explosive flares of importance $\geqslant 1^{+}$ have an associated burst at 10 cm, and 80% are associated with Type-III's. About half as many of the explosive 1 and 1^{-} flares are associated than with flares in general. The relevance of the time of the explosive phase, a period of 20 or 30 seconds, is displayed by the close time-association with radio bursts: 75% of the 10 cm and type III's occur in a 90 second interval around the explosive phase time. In contrast, only about 20% occur in the corresponding time intervals centered on the start and maximum of the flare.

Since the explosive class of events are definitely in the minority, it may be significant that *all* eight high energy (> 20 keV) X-ray events measured by rockets and balloon techniques are connected to flares of the explosive type. Perhaps most significant is the fact that in five of these events, where a complete time history of the X-ray flux is available, the hard X-rays are distributed in time about an interval *following* the explosive phase by about two minutes. In contrast, low energy X-rays, inferred from the occurrence of SID's, show little preference for occurrence with the explosive time, indicative of their production by the normal flare process.

The explosive development in a flare is often limited to only 20 to 30% of the total flare area. This, coupled with the above, suggests that the explosive phase marks the relevant time for the initiation of non-thermal energy production in a localized region of an otherwise normal flare. High velocity (1 000 km/sec) disturbances are initiated and Type III and centimetric continuum are enhanced, at this time. Subsequent particle generation, requiring about 2 minutes results in bremsstrahlung X-rays. The two-minute lag has been predicted by E. N. Parker, who employed the Fermi process to accelerate particles to relativistic energies, in his 1958 flare model.

D. H. MENZEL

The Filamentary Structure of Solar Prominences*

The filamentary structure of solar prominences persists for times that are long

* Also on behalf of Lowell Doherty. This research was supported by the U. S. Air Force Cambridge Research Laboratory.

compared with the theoretical radiative life of the prominence material. The structure of a prominence depends on the interplay of a number of complex processes of energy exchange. Prominence material radiates energy in continuous and line emission. It gains energy by absorption of radiation from the corona, the chromosphere and the photosphere. Heat flows into the prominence by conduction from the surrounding hot coronal gas. The rate of heating depends on the magnitude and structure of the shielding magnetic fields. Mechanical processes, such as shock waves or magnetohydrodynamic processes can also affect the energy content of the solar material.

Orrall and Zirker first investigated the fine structure of prominences, on the assumption that heat conduction from the corona exactly balanced the radiative losses of a cylindrical, isobaric filament. Their model was optically thin in Lyman α, with a central temperature of 30000 °K. With the low conductivity imposed by the high magnetic field, the shell from the hot corona to the cooler core became extremely thin. Their study led to filaments only a kilometer or so in diameter.

We also consider idealized cylindrical filaments with heat conduction. We have, however, to include the effects of radiative transfer in the resonance lines of H I, He I and He II. This assumption enables us to construct models having central temperatures lower than that of the model of coronal and chromospheric radiation and the radiative loss from heavy ions as well as from hydrogen and helium. The calculations were carried out on IBM 1090 of the Harvard Computation Center.

We find that prominence material will not cool indefinitely, even in the absence of conductive heating. Absorption of chromospheric radiation in the continua of H and He can balance the radiative loss from the prominence, when the temperature has fallen to some particular value, dependent on the total particle density, N_T, and the internal fields. We estimate the ambient radiation field in the far UV from data given by Hinteregger. For $\phi = N_T T = 2 \times 10^{13}$, 2×10^{14} and 2×10^{15}, the respective critical temperatures are approximately 15600, 12100, and 9600°K.

For our most realistic models the central temperature settles down to about 12000 °K. Our filaments have a very steep temperature gradient near the outer boundary and thus resemble those of Orrall and Zirker. However, since we rely on heat sources other than thermal conduction, the filaments can be much larger. Also, because of the lower temperature, an appreciable amount of neutral hydrogen forms, contributing to the conductivity.

F. Q. Orrall

Some Considerations of Mass and Energy in Loop Prominences

In the course of a number of visual and photographic observations of loop prominences, J. T. Jefferies and I have noted that the Hα line occasionally has very strongly developed wings. It is, of course, well known that the Hα line in loops is broadened and distorted by macroscopic notions, but this broadening usually amounts to a few Ångstroms, while the extended wings observed here extend tens of Ångstrom from the line center. Our best observations of this phenomena were made in the loops of 1962 February 19, in which the wings can be followed $\pm$ 60 Å from the line center. These were observed against the sky with the coronagraph. We are convinced that they are not due to instrumental effects. They can be due to Stark broadening only if the emitting region is of the order of 1 cm thick. This seems unlikely, so we must assume that they are due to the Doppler effect. We cannot distinguish between an ordered streaming of material such as that described by Dr. Severny, and a high velocity tail on the thermal velocity distribution. We observe the Hα radiation from fast neutral atoms, but these presumably come from fast protons – most probabily by charge exchange. We do not know if there are equally energetic electrons.

The following considerations of mass and energy loss from loop prominences also seem important for an understanding of the loops and coronal condensations in which they are imbedded. Material is usually seen flowing downward in the loops and frequently at least ten individual threads can be seen with downward velocities of 100 km sec^{-1}. Since the electron density is 10^{11} cm^{-3} and the radius of the threads is $\approx 10^8$ cm, this flow corresponds to a mass loss from the loops to the photosphere of 7×10^{11} g sec^{-1}. Sporadic condensations occur in permanent coronal condensations. The mass of the model coronal condensation of Waldmeier and Müller is about 10^{15} grams, so that if loop prominences form from coronal material, they will drain all the mass from a Waldmeier-Müller condensation in 1.5×10^3 seconds. Most loop prominences last longer than this. The loops of 1962 February 19 were observed for 1.6×10^4 seconds. This implies that the coronal condensation (or the loops) are continually being supplied from the chromosphere.

This downward flow involves considerable kinetic energy. For the long-lived loop system described above, 3×10^{29} ergs were presumably dissipated in the chromosphere at the foot of the loop system.

Finally, this downflow gives us a good estimate of the total radiative energy loss of the prominence, because we have that 7×10^{11} g sec^{-1} of material which was cooled by radiation from 2×10^6 °K to 2×10^4 °K. This means that 2×10^{26} erg sec^{-1} must have been radiated away. For the loops that lasted for 1.6×10^4 seconds this is 3×10^{30} ergs, the energy estimated by Dr. Severny for a class 2 flare.

O. NAMBA

The Profile of the Infrared He I Line in Solar Faculae*

De Jager and Neven observed the profile of the He I 10830 Å line in the normal disk and then I observed it in faculae, respectively, at the Jungfraujoch. From the profiles observed in 20 faculae, assuming the plages to be a homogeneous layer with a constant excitation temperature T_{ex}, I derived three physical parameters for the plages: the optical thickness τ^* in the line center, the Doppler velocity, and the excitation temperature, which are summarized in the accompanying table. I would like to emphasize that from a *single* tracing of the infrared He I line one can at once derive these *three* fundamental physical quantities.

	Plage	*Normal chromosphere*
τ^* (10830)	1.9 (1.3 in the radial direction)	0.08 in the radial direction
ξ_0	8.9 km/sec	$\approx$ 15 km/sec
T_{ex}	5200 °K (likely $\approx$ 5000° or less)	4000–6000 °K (likely $\approx$ 5000° or less)

For comparison I gave in this table corresponding quantities for the normal chromosphere, which were also derived assuming a homogeneous model for the chromosphere. Especially interesting is the difference in "turbulent" velocity ξ_0 between the disturbed and undisturbed chromospheres. "Turbulence" is stronger in the "quiet" chromosphere than in the "active" chromosphere.

But now, as De Jager has told you, the shallow profile of the 10830 line in the normal region (central depth $\approx$ 5 to 10%) does not mean that the line is formed in an optically thin layer. The carefully measured central-depth ratios $r = R_0(10830)/R_0(10829)$, in the normal region indicate a rather large optical thickness of the normal chromosphere, which is more or less the same as in the plages ($\tau^* \approx$ 1 to 2). If this is true, my interpretation and the above comparison should be partly revised. Namely, the observed profile might be considered as a superposition of profiles from many individual, optically rather thick spicules covered by the spectrograph slit. The difference in turbulence between the disturbed and undisturbed chromospheres might be reduced, but not completely, and the still existing difference might be understood as a difference in the velocity distribution of spicules between the plages and the normal chromosphere.

In many respects it may be very interesting to secure more observations of the profile of the infrared He I line.

* A supplementary note on my paper in *Bull. Astron. Netherlands* **17** (1963) 93.

UNDERHILL. The temperature near 5000° which you have derived and which you call an excitation temperature is not an excitation temperature in the sense that it is the temperature to be used with the Boltzmann law to calculate the population of the 2^3S level of He I. This level is metastable and in many respects can be considered to be a ground level, inspite of its E.P. of about 20 volts. I would rather call the temperature a *brightness* temperature at the wave-length of the centre of the line. It describes the monochromatic brightness of the adopted layer representing the faculae. It is not necessarily the temperature permitting to calculate the population in the 2^3S level of He from the total abundance of helium in the line of sight.

NAMBA. My explanation might be insufficient. But the temperature I derived has no direct concern with the observed central intensity of the line and is not a *brightness* temperature or a *radiation* temperature. It is indeed the *excitation* temperature which was introduced to describe the population *ratio* between the 2^3S and 2^3P levels, appearing in the equation of radiative transfer, and which, inserted in the Planckian formula gives the source function for the plage emission proper.

Please see the equation:

$$I_\lambda(\vartheta) = I_0(0,\vartheta)\,e^{-\tau_\lambda} + B_0(T_{ex})\,(1 - e^{-\tau_\lambda})\,,$$

where $I_\lambda(\vartheta)$ is the emergent intensity in the line, and the first term of the right hand side denotes the photospheric light absorbed by the plage layer, and the second term represents the emission from the plage proper, self-absorption being taken into account. The excitation temperature here derived has nothing to do with the high excitation of neutral helium as a whole, which is one of the most difficult problems in solar physics, unsettled as yet.

F. VAN 'T VEER

The Temperature Gradient of the Sunspots*

We have tried to obtain some new information about the structure of a sunspot using the wings of some strong absorption lines, and we shall show here the results obtained with two of them: Mg I 5184 and Na D. The lines have been studied for a number of spots with areas varying from 50 to 175.10^{-6} H, corresponding to a diameter of 15000 to 35000 km. A characteristic parameter defining the wing of a strong absorption line was introduced by MINNAERT** (1935) describing the profile of the wing by the following formula:

$$I_\lambda = I_0 - I_\lambda \cdot c(\Delta\lambda)^{-2}\,.$$

This formula holds to a depression $r = 40\%$. The parameter c which is called

* See also: VAN 'T VEER, F.: 1963, *Ann. Astrophys.* **26**, 185.
** MINNAERT, M.: 1935, *Z. Astrophys.* **10**, 40.

the wing strength can be easily determined and has been very useful in a number of studies of the center-limb variation of some absorption lines.

The figures 2 and 3 give the quotient $c^*/c^\odot$ of our measurements in the spot and in the solar photosphere as a function of the area of the sunspot.

For Mg I 5184, the quotient is increasing up to A = 80; it reaches a maximum there and decreases when the size of the spot becomes still larger.

For Na D_2, the wing strength rapidly increases with the spot-size in the whole range of observed spots.

Our first conclusion is that spots of different sizes possess different physical

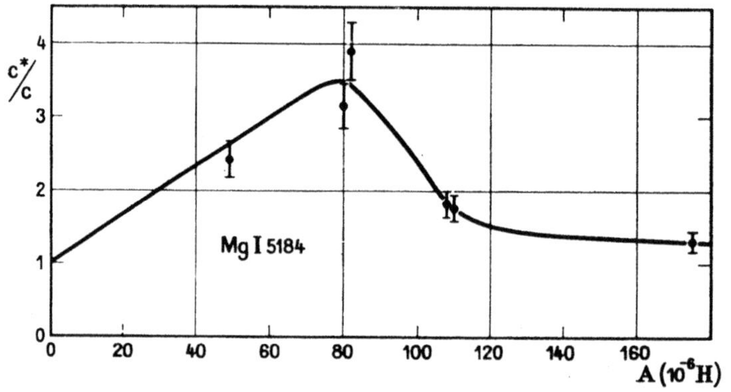

Fig. 2.

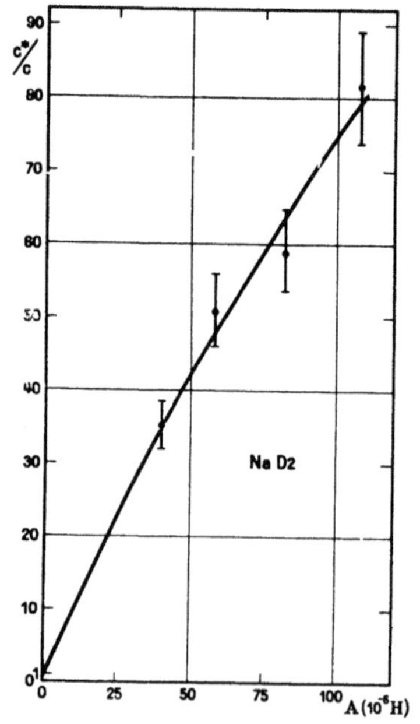

Fig. 3.

parameters. We will now try to answer the following question: What must be the physical structure of the sunspots to explain this empirically determined relation.

PHYSICAL INTERPRETATION

We will turn therefore to the physical interpretation of the parameter c. We must compare two expression for the profile of the wing: (1) The empirical formula which can be written as $r_\lambda = c\,(\Delta\lambda)^{-2}$, for the outer wings of a strong absorption line; (2) A well known theoretical expression using the weighting functions and also valid in the region of the outer wings

$$r_\lambda = \int_0^\infty \frac{\varkappa_\lambda}{\varkappa}\, g(\tau, \mu)\, \mathrm{d}\tau\,.$$

With the aid of the following expression for the absorption coefficient in the line

$$\varkappa_\lambda = K N f \gamma (\Delta\lambda)^{-2}$$

and some algebra we can finally write

$$\frac{c^*}{c^\odot} = \frac{\int_0^1 B^*\left(\frac{y^*}{\mu^*} - z^*\right)\mathrm{d}x^*}{\int_0^1 B\left(\frac{y}{\mu} - z\right)\mathrm{d}x} \cdot \frac{\int_0^1 B\,\mathrm{d}x}{\int_0^1 B^*\,\mathrm{d}x^*}\,,$$

where

$$x = e^{-\tau/\mu}\,; \qquad z = \frac{\zeta\gamma}{\varkappa}\,; \qquad y = \int_0^\tau z\,\mathrm{d}t\,,$$

and ζ is the fraction of atoms in the lower state of the transition.

This ratio is independent of the oscillator strength of the transition and of the abundance of the element in question. This has been made possible by taking the quotient $c^*/c^\odot$. All parameters B, ζ, γ and $\varkappa$ occurring in the last equation can be calculated if the models for spot and photosphere are known. However it is our task to deduce such a model for the sunspot; so we shall introduce some simplifying assumptions in order to describe the model of the sunspot by a limited number of characteristic parameters. After that we shall try to adjust these parameters with the aid of our observations in order to obtain a series of empiric models of the sunspot capable of explaining the behaviour of the Mg and Na lines. We must indeed expect to find a series of models with their parameters varying as a function of the area of the sunspot.

The model of the photosphere which enters also in these computations has always been chosen identical with the second model of Mrs. Böhm-Vitense, extended to the deeper layers.

SIMPLIFYING ASSUMPTIONS

Our simplifying assumptions are the following:

(1) $d\Theta/d \log \tau_0 = -\Psi = \text{constant}$,

(2) $d \log P/d \log \tau_0 = \text{constant}$.

So Θ and $\log P$ are supposed to be linear functions of $\tau_0 = \tau_{5000}$. These two assumptions hold reasonably well for normal solar and stellar atmospheres and they define completely the gradient of the temperature and the pressure in a sunspot. Two supplementary parameters are necessary to define completely the spot model. They are the following: Θ and $\log P$ at an arbitrary depth in the sunspot atmosphere ($\tau_0 = 0.8$)

It is essential for our method to make as few physical hypotheses as possible. So we have rejected the hypothesis of hydrostatic equilibrium which might not be true due to the presence of a magnetic field and also the hypothesis of radiative equilibrium, which is compromised by the presence of the hotter photosphere surrounding the spot.

DETERMINATION OF THE PARAMETERS

We choose our parameters. If the result $\Delta\Theta = \text{constant}$, obtained from the observations of the continuum is correct we must adopt for Ψ the photospheric value $\Psi = 0.25$.

About the pressure gradient we possess very little information. The best thing we can do for the moment is to adopt here also the photospheric value $d \log P/d \log \tau_0 = 0.45$.

With these two gradients given in advance we can now calculate two diagrams giving $c^*/c^\odot$ as a function of $\Theta(\tau_0 = 0.8) \approx \Theta_{eff}$ and $\log P(\tau_0 = 0.8)$. The following step is to determine Θ and $\log P$ with the aid of these diagrams and the empiric values of $c^*/c^\odot$ (Na and Mg) observed in the spots.

The results are given by the first table ($\log P^\odot(\tau_0 = 0.8) = 5.22$)

TABLE 1

A (10^{-6}H)	$\Delta\Theta = \Theta^* - \Theta^\odot$ ($\tau_0 = 0.8$)	$\log P$ ($\tau_0 = 0.8$)
50	0.58	4.02
60	0.57	4.14
70	0.57	4.24
80	0.57	4.31
90	0.60	4.20
100	0.64	4.04
110	0.68	3.79

The agreement is very poor; the temperatures are much lower than we can expect from the measurements of the continuum; besides the gas pressure varies in an unexpected way, with a maximum at A = 80. An improvement of the results can only be obtained by a change of the temperature gradient Ψ; the influence of the pressure gradient being much less important. So we calculated a number of new diagrams giving $c^*/c^\odot$ as a function of Θ and log P for different values of Ψ. The agreement improves when Ψ increases and it becomes very satisfactory for $\Psi = 0.50$ that is twice the photospheric value we adopted in the beginning.

The second table shows the results for Θ and log P obtained with the diagrams based on this new temperature gradient: $\Psi = 0.50$. Now the gas pressure decreases monotonically with increasing spot area and the temperature agrees very well with the values based on the measurements of the continuum.

So we must conclude that the mean temperature gradient in the sunspot is twice the value for the photosphere. Besides it seems evident that the gas-pressure is lower when the spot is larger.

TABLE 2

A 10^{-6}H	$\Delta\Theta = \Theta^* - \Theta^\odot$ ($\tau_0 = 0.8$)	log P ($\tau_c = 0.8$)	$\Delta\Theta_{eff}$ (continuum)
50	0.27	4.26	0.24
60	0.29	4.22	0.28
70	0.31	4.20	0.31
80	0.33	4.17	0.33
90	0.37	3.95	0.36
100	0.41	3.68	0.40
110	0.44	3.44	0.42

The proposed model of the sunspot (A = $80 \cdot 10^{-6}$) is given in Table 3.

TABLE 3

τ_0^*	Θ^*	log P^*	log P_e^*
0.02	1.948	3.38	−2.69
0.05	1.744	3.52	−2.15
0.10	1.596	3.71	−1.70
0.15	1.509	3.79	−1.49
0.30	1.362	3.92	−1.06
0.45	1.274	4.00	−0.80
0.60	1.214	4.06	−0.63
0.80	1.153	4.12	−0.43
1.00	1.105	4.16	−0.31
1.40	1.034	4.23	−0.11
2.00	0.958	4.30	+0.17
3.00	0.873	4.38	+0.72
4.00	0.811	4.44	+1.14
5.00	0.764	4.48	+1.50

PECKER. I have just two questions: (a) what is the physical parameter which makes this enormous difference between behavior of the two lines you analysed (Na D, and Mg I)? (b) You assumed that the influence of the damping constants cancels. This only does if you neglect their depth dependence. If you do not, I am afraid that your results might be questioned since the damping constants are badly known, and are strongly depth dependent. What do you think of this cause of error?

VAN 'T VEER: I made the computations for the ratio of the wing strengths of spot and photosphere: $c^*/c^\odot$. If the damping is due to collisions with the hydrogen atoms as we may assume for photospheres of this temperature and normal pressures, the constant of damping cancels and it has no effect.

UNSÖLD: The dependence of temperature upon optical depth can be checked against center-limb-darkening in the continuum. How is the agreement?

VAN 'T VEER: The earlier models of sunspots were based on the center-limb variations of the continuum. We know now that the effect of scattered light from the photosphere was enormously underestimated for the observations near the limb, making these observations very uncertain. Besides we must not forget that the horizontal optical thickness of the outer layers of a sunspot is very low. By this effect we also observe the light of the hotter material of the penumbra, when we observe the umbra in spots close to the limb.

These two effects will very badly influence the center-limb darkening.

C. ZWAAN

On the Influence of Scattered Light on Sunspot Observations

From a discussion of the measurements in the continua of umbrae of sunspots, one can show that practically all results published as yet are vitiated to a high degree by the stray light of the photosphere. Up to now, nearly all attention has been paid to the *blurring* of the image (chiefly by imperfect seeing). For medium and large spots, however, the *scattering* by the instrument and the atmosphere is much more important in most observational conditions. It is not permitted to assume that the scattered light on the solar disk would be of the same order as outside the solar limb. Assuming plausible distribution functions for the scattered light, one finds that the stray light near the center of the disk is about four times stronger than at 1′ outside the limb, sometimes even more.

Even for the largest spots the observed intensities seem conflicting: MAKITA and MORIMOTO (1960) found intensities much lower than those measured by RICHARDSON (1939) and MICHARD (1953). After having applied corrections for scattered light (using intensities 1′ outside the limb, as measured or estimated by the authors), the values of Richardson and Michard come close to those of Makita and Morimoto, however. (STUMPFF's intensities (1961) for large spots, uncorrected for *scattered* light, are slightly higher).

Generally, from observations by Michard, and more recently by Stumpff, a clear dependence of the intensities on the area of the spot is deduced. This picture is destroyed by the measurements of Makita and Morimoto: they found some of the smaller spots to be as dark as the largest ones. These authors attribute the scatter in the intensities against the date of observation, then "good" and "bad" days clearly appear, the "good" days showing all the spots to be very dark. From the discussion of all these observations we conclude: (a) the intensity hardly depends on the size of the spot; (b) the intrinsic differences are much smaller than suggested by the graphs of Makita and Morimoto.

According to Unsöld, the model of the umbra can be connected with the photospheric model by adopting $\Theta^{u}(\tau) = \Theta^{ph}(\tau) + \Delta\Theta$, since the relative intensities show no significant center-to-limb variations. ($\Theta = 5\,040/T$, u refers to the umbra, ph to the photosphere). From the lowest intensities, measured by the Japanese observers, we derived $\Delta\Theta = 0.48$ (to be increased, if the stray light on the best days was more than 1–2 per cent of the photospheric intensities). This provisional model was completed by $P_g(\tau)$, $P_e(\tau)$, $P_{\mathrm{H}}(\tau)$ and $P_{\mathrm{H}_2}(\tau)$, computed on the assumption that hydrostatic equilibrium prevails in the umbra.

Turning now to the *line spectrum*, we mention the possibility of determining the amount of stray light *from the recorded spectra themselves*. Model computations show the Fe II lines to be absent in the spectrum of the umbra. Certainly they do not show the Zeeman effect in the umbra. Now spectra of a very large spot, taken by Mulders at Mt. Wilson, show clearly visible Fe II lines. If we attribute these lines to photospheric stray light, a provisional estimate near λ 5200 Å shows, that the spectrum contains 60 per cent photospheric and 40 per cent umbral light. (This method had already been applied earlier by Howard). Other spectral features support this high amount of stray light. So the spectrum of the umbra has much greater intensity differences than the first inspection shows: seemingly weak lines turn out to be saturated, sometimes showing wings which have been neglected up to now. Unfortunately, Mulders' best spectra are not photometrically connected with spectra of the adjacent photosphere, so I cannot determine corrected intensities of the umbra.

Let us examine now the very low pressures which are claimed to be found empirically from curves of growth, e.g. by Michard from equivalent widths measured by ten Bruggencate and von Klüber, and more recently by Elsässer. We constructed curves of growth, taking naïvely the observed equivalent widths, uncorrected for stray light, as real, computing the abscissae log X by the method of weighting functions, using our model and the pressures according to hydrostatical equilibrium. We found that the Ti II lines were located at the left of the theoretical curve and the Ti I lines strongly shifted to the right.

Since Ti is practically entirely neutral in umbra:

$$X\,(\mathrm{Ti\,I}) \sim \frac{l(\mathrm{Ti\,I})}{\varkappa(\mathrm{H}^-)} \sim P_e^{-1} \qquad \text{and} \qquad X\,(\mathrm{Ti\,II}) \sim \frac{l(\mathrm{Ti\,II})}{\varkappa(\mathrm{H}^-)} \sim P_e^{-2}$$

($\varkappa$ being the continuous absorption coefficient, l the selective one). Looking at the *relative* positions of the Ti I and Ti II lines only, one would conclude that P_e is much lower than the pressure computed according to hydrostatic equilibrium. Comparing the positions with the theoretical curve, the Ti II lines would indicate somewhat lower pressures, whereas the Ti I lines would point to much higher pressure. In fact, however, the positions are mainly shifted in *vertical* directions, chiefly because of the stray light which has depressed the Ti I lines and has raised the Ti II lines.

In order to test the models and the amount of stray light, we have chosen a representation different from the curve of growth. We restrict ourselves to illustrating the method by one example. Empirically it is found, that the Fe I lines of E.P. = 3.8 V are equally strong in the umbra and in the photosphere. This equality cannot have been vitiated by stray light. Indeed the model predicts this equality, which is quite sensitive to the physical conditions, at about E.P. = 3.8 eV. The ratios of the equivalent widths for other excitation potentials furnish a check on the amount of photospheric stray light.

Although this investigation is not finished yet, I may state that the discussion of the intensities of the continuum and the equivalent widths of Ti I, Ti II, Fe I and Fe II lines points to a *consistent* picture of amounts of stray light and a model, which is characterized by $\Delta\Theta = 0.48$ (perhaps this quantity has to be increased somewhat). This model is in hydrostatical equilibrium and presumably it is in fairly good radiative equilibrium as well.

The model was confirmed not only for large spots ($A_u = 172 \times 10^{-6}$; Mulders; $A_u = 110 \times 10^{-6}$; Žuravlev), but for medium spots as well ($A_u = 41$ to 62 and 97×10^{-6}; ten Bruggencate and von Klüber).

Measurements of magnetic field strengths in umbrae will be complicated by stray light. Measurements of sufficiently weakened lines (e.g., Fe II) will be usefull to correct for the stray light. The method is safe and relatively simple, provided that the stray light is mainly of photospheric origin.

As to van 't Veer's beautiful method, I fear some basic assumptions are uncorrect: (a) there is *no* clear dependence of intensity on area; (b) the influence of the stray light is strongly underestimated. From van 't Veer's model much too high intensities in the continuum are predicted, moreover, this model cannot explain the observed equivalent widths.

VAN 'T VEER: What is the pressure of your model at say $\tau_0 = 1$?

ZWAAN: $\log P = 5$ to 5.5.

VAN 'T VEER: With such a pressure we should predict a wingstrength of the Mg-line, five to six times stronger than we observe in reality. Your model is not capable to explain either the wing of Na D_2 nor the wings of Mg I 5184.

As to the question of scattered light: I did not reduce my observations for scattered light; this being impossible by lack of precise information. If I take a mean scattering parameter the temperature gradient has to be increased with

10 to 20%. So an unexpectedly great amount of scattered light cannot help our models approaching each other. Or the contrary the temperature gradient should become still more steep.

ZWAAN. Your original graphs show, that a correction for scattered light would increase the pressures you derived. I did not check whether my preliminary model can explain the wing strengths.

One may ask whether the mechanisms for line broadening are more complicated in the umbra, e.g. by the hydrogen molecules.

UNNO. Recently, Makita constructed a model of a sunspot with the use of the center-limb observations for continuum and the spectroscopic observations for equivalent widths. In his model, the mechanical balance both in the vertical and horizontal direction is discussed carefully. Reasonable models could be obtained only when the inhomogeneity within the umbra was taken into consideration.

ZWAAN: In the preprint of Dr. Makita's paper the hot component is found to have a radiation temperature of 5700 °K. This is so close to the photospheric radiation temperature that I would prefer to attribute the light from the hot component in Makita's model to the scattered light from the photosphere.

E. H. SCHRÖTER

Some New Observational Results Concerning the Evershed Effect

High resolution sunspot spectra, obtained during good seeing conditions show, that the Evershed-effect line shifts in the penumbrae of sunspots are not real shifts of the whole line, but result from a very strong asymmetry of these lines within the penumbrae. If seeing conditions are extremely good, faint additional lines appear within the penumbra in the red and violet wing of the main line respectively (see Bumba). With the help of photographic equidensitometry these asymmetries (Evershed-shifts) have been measured in various parts of the line profiles. For a metallic line ($g = 0$) the Evershed-effect asymmetries at small depressions amount to a shift of 2–3 km/sec. In the mean part of the line profile one measures 1–1.5 km/sec, the center of the line being practically unshifted ($\lesssim$ 0.3 km/sec).

Stronger lines such as the NaD- and Mg-lines show within the penumbra two faint satellites of approximatively equal intensity, symmetrical to the main line. Although the distance of each satellite-line corresponds to a velocity of 9–15 km/sec, one would measure an Evershed-velocity near zero, in spectra of low resolution (see E. St. John, Evershed, etc.). The comparison of the results for the metallic and Mg-lines connected with estimates of the depth of line formation show that the line asymmetry cannot be due to a depth-dependence of the stream velocity. It seems rather, that the Evershed-effect results from the superposition and smoothing of more or less complex motions within the unresolved fine structure (filaments) of the sunspot penumbra. From an investigation of the difference in the line asymmetries for penumbrae near the sun's center and for others closer to the solar limb in H-type spots, which move across the solar disc I found: the mean, unresolved and smoothed radial motions within the penumbra are inclined to the surface of the sun.

Comparing the obtained inclination angles with observations of the macroscopic configuration of the sunspot magnetic fields (Nicholson, Bumba, Leroy) these overall smoothed (or large scale) motions seem, indeed, to tun along the magnetic lines of forces.

V. Bumba

Some Notes on the Observations of Radial Motions in an Active Region

In my contribution I would like to mention some problems connected with the measurement of radial motions in an active region. These problems arose not only during my study of radial velocities with the aid of an instrument of high resolving power and dispersion at the Crimean Astrophysical Observatory but also during my study of sunspot magnetic fields, development of sunspots and recently during the study of the fine structure of sunspots.

The real resolving power of modern spectral instruments which we call the instruments of high resolving power is still smaller than that of the usual telescopes which allow us to take photographs of the fine structure of granulation, penumbrae and umbrae of sunspots.

The low speed of these instruments means a long exposure time. In the best

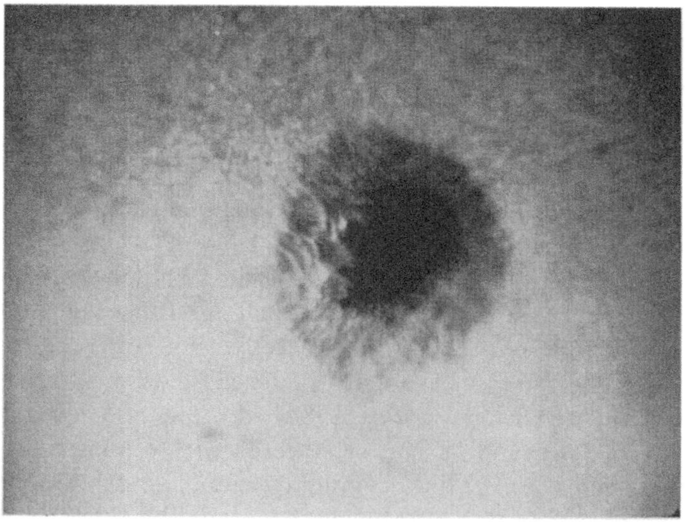

Fig. 1.

case this is some tenths of a second but in most cases it rises to seconds or some tens of seconds. Visually we can observe better with instruments than photographically and only exceptionally is it possible to find on a photographic plate some traces of the fine structure observed visually. This means that the results which are published up to this time, give only the average values of the radial velocities and other parameters

Moreover, the published observational material in the great majority of cases concerns regular spots, and related types of sunspots. These types of spots belong to a late phase of spotgroup development, when the process of disintegration of the local field and the dissipation of its magnetic energy has already taken place*. But for the most interesting phases of development practically no observational material concerning the radial velocities is available so far. This one-sidedness of the observational material may bring one to false conclusions, which occur in various models of sunspots.

The following figures demonstrate some features of the fine structure: the emission grains in the penumbra (figure 1), the specific forms of the light bridges in umbrae (Figure 2) or also in penumbrae with the influence upon the granulation in the neighboring photosphere (Figure 3) and so on.

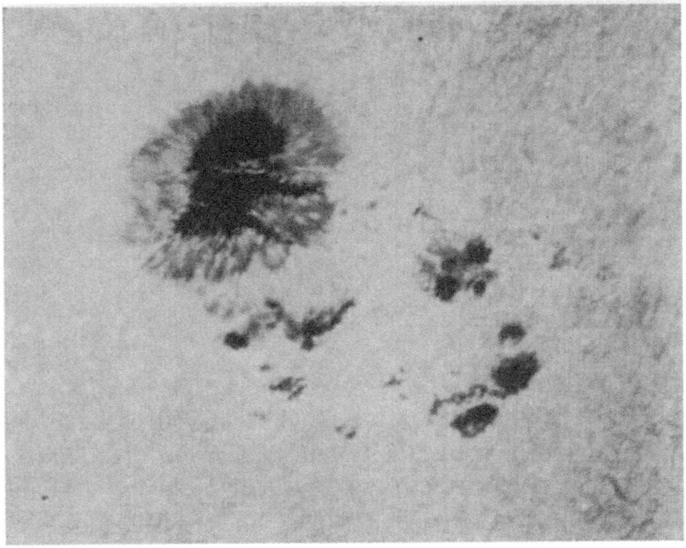

Fig. 2.

What does spectroscopy say about these formations? Did one obtain the spectrograms of these features? Do they show radial motions? There exist in

* Bumba, V.: 1963, *Bull. Astron Inst. Czechoslovakia* **14**, 91.

our observational material some spectra, which suggest how a spectrum can answer these questions.

Several times, we observed visually radial motions of a character somewhat different than the Evershed effect just in the regions of an extensive "penumbra" and small umbrae in the middle regions of bipolar groups. The discontinuity of spectral lines and the absence of "flags" is characteristic in this type of motion. Moreover, in the region of a "penumbra" with a small umbra, these discontinuities of the spectral lines, which represent radial velocities of about 2 km/sec keep on shifting, their life time is short, evidently only a few (2 or 3) minutes.*

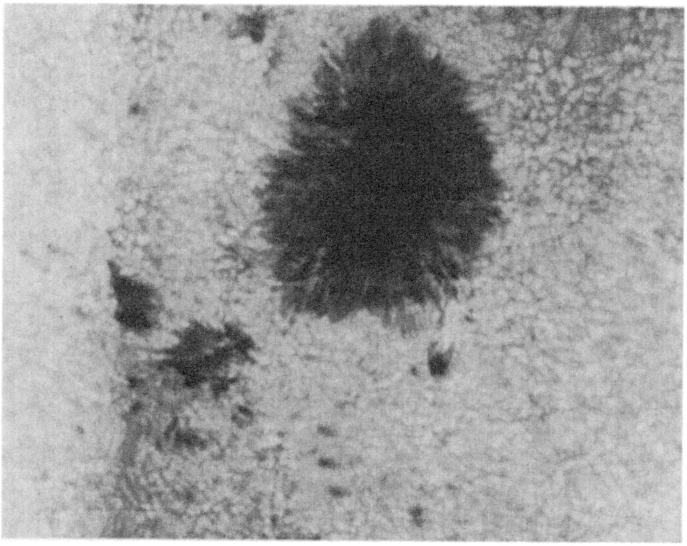

Fig. 3.

Sometimes, we observe irregular motions in the regions of a "shallow penumbra", and small spots in the close neighbourhood of large spots. We can rather suggest than see the strengthening of motion, before all in the dark strips of the continuum of small umbrae. Sometimes, this strengthened motion is again connected with the discontinuity of spectral lines, which means that it is perhaps due to turbulence.

It should be emphasized that with exceptionally good observational conditions the "fine" structure of both the main line and the "flag" can be observed visually in the spectrum of the umbra and mainly in penumbrae spectra. The elements of the fine structure in the "flag" are larger and less dense (more separate from each other) than in the photosphere. Sometimes the "sputnik"

* Bumba, V.: 1963, *Bull. Astron. Inst. Czechoslovakia* **14**, 137.

can be seen also in the spectrum of the umbra, where it is very faint, very difficult to observe and where it attains its maximum Doppler shift.*

During some observations, when the slit of the spectrograph intersects a "gulf" of the penumbra or a foot of the light bridge, the difference between radial velocities in these formations and in the normal Evershed effect was clearly seen.** Indeed this difference might only be due to the different origin of depth of the spectral lines.

In July 1957, at the Crimean Observatory, I obtained a set of spectral cuts of a regular spot (number 130 (Crimea)) during extremely good observational conditions. In some of these spectral cuts we may suspect a fine structure of the lines of a specific and unprecedented character. I obtained these spectrograms six years ago. But I did not risk till this time to publish these results because the character of this fine structure is of such a kind that a sceptic might hesitate to see it and it is impossible to reproduce them. I speak about this uncertain fine structure partly because my remark may be a stimulation for obtaining better observational material which can confirm or disprove my results, partly because this fine structure – if it is real – can show us how the radial motions are connected with the individual formations of the photospheric fine structure.

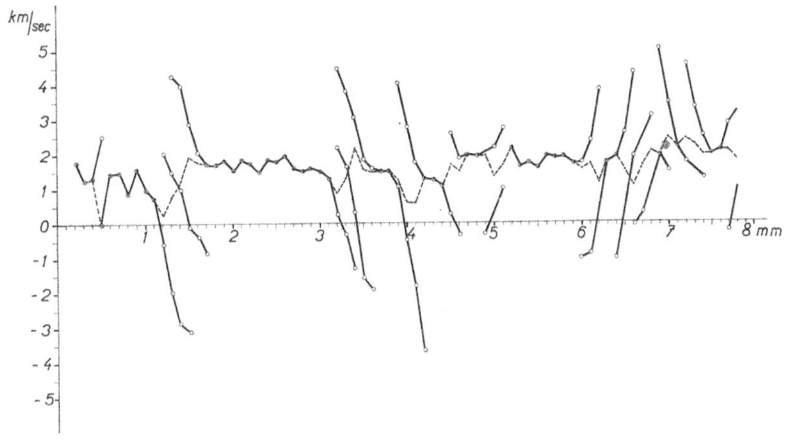

Fig. 4.

Figure 4 demonstrates the results of the measurements of radial velocities in a short section of the spectral line λ 5927.800 Å Fe I in a region of the undisturbed photosphere (the slit was parallel to the solar radius and the values drawn are not corrected for the rotation of the sun). The measurements were made with the aid of a visual spectrocomparator so that the individual details

* Bumba, V.: 1963, *Bull. Astron. Inst. Czechoslovakia* **14**, 137.
** Bumba, V.: 1960, *Izv. Krimsk. Astrofiz. Obs.* **23**, 277.

of the structure mentioned were traced. Just these details were responsible for the fact that the spectral lines with such a fine structure have not their usual zigzag forms, but contain many discontinuities. On the same figure the average values of the radial velocities measured by means of a photoelectric spectro-comparator with two slits (the dashed line) may be seen. (Scale: 0.1 mm correspond to about 420 km at the solar surface, which means that the distance of the most disturbed regions of the line correspond to about $1 \cdot 3 \times 10^4$ km.) In Figure 5 the measurements of the radial velocities in a spectral cut (number 4) in the region of the spot is reproduced (the slit was also parallel to the radius, the scale is the same): above – measured with a visual comparator (not corrected for the rotation) at the bottom – measured with a photoelectric one. The most disturbed parts of the line correspond to the regions of "flags".

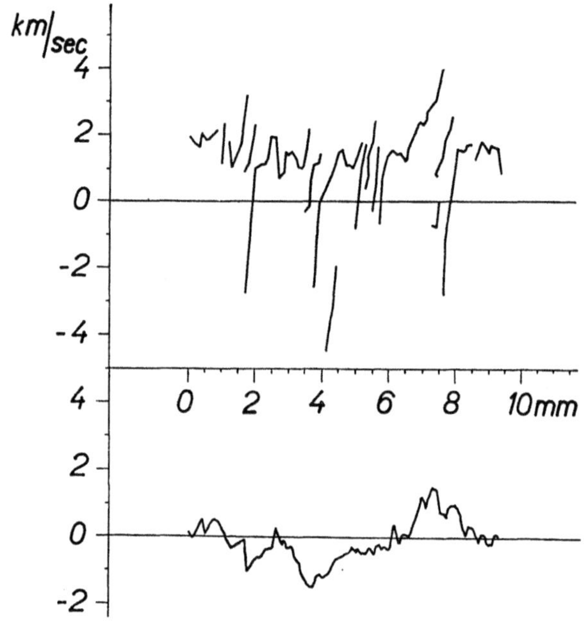

Fig. 5.

With regard to the fact that the reality of the traced details is not guaranteed, I hope you will excuse me for not drawing conclusions from these results. Indeed, if further observational material confirms the reality of the above details, they can help us to solve many interesting questions of the physics of sunspots and photosphere. But I think that for the great majority of the present solar instruments it will be very difficult to solve the problems mentioned unless the speed of these solar instruments is much greater and the exposure time essentially shorter, or unless our instruments are brought outside the atmosphere.

SESSION IV

THE CORONA

THE THEORY OF THE CORONAL SPECTRUM

M. J. SEATON
University College of London, London, England

1. Outline

The corona[1] is an ionized gas at $T \simeq 10^6$ °K. Its base is at a height $h \sim 10^4$ km above the photosphere, corresponding to $\varrho \simeq 1.01$ where $\varrho = (R_\odot + h)/R_\odot$. The electron density N_e varies from $\sim 4 \times 10^8$ cm^{-3} at $\varrho = 1.01$ to $\sim 2 \times 10^6$ cm^{-3} at $\varrho = 2$. The corona is heated by dissipation of mechanical energy originating in the convection zone.[2]

The energy per unit volume scattered by free electrons is $4\pi\sigma N_e D F_\lambda$ where $\sigma = 6.6 \times 10^{25}$ cm^2, $D = (1/2)(1 - (1 - \varrho^{-2})^{\frac{1}{2}})$ and πF_λ is the flux at the surface of the photosphere. Continuous emission is observed in X-ray and radio wavelengths but is masked by electron scattering in the visible. Coronal emission lines are produced by ions having ionization potentials in the range 200–800 eV. All identified lines in the visible are due to forbid transitions in configurations of p electrons; 38 identifications have been made[1,3] the elements represented being Ar, K, Ca, V, Cr, Mn, Fe, Co, and Ni. Ultraviolet lines are due to allowed transitions in positive ions; 482 lines have been listed.[4]

2. Ionization

The equilibrium equation is

$$\frac{N(X^{+m+1})}{N(X^{+m})} = \frac{q_i(X^{+m})}{\alpha(X^{+m})}$$

where q_i is the rate coefficient for collisional ionization and α is the radiative recombination coefficient. The ionization therefore depends on temperature but not on density. Recent work, experimental[5] and theoretical[6], leads to greatly increased confidence in estimates of q_i; α may be calculated with tolerable accuracy using hydrogenic formulae. Calculations[7] show that

$$\frac{N(X^{+m})}{\Sigma_m \; N(X^{+m})}$$

has a maximum value, at $T = T(X^{+m})$, and becomes very small if T differs much from $T(X^{+m})$. The calculated values of $T(X^{+m})$ should be correct to within 20%.

C. de Jager (ed.), The Solar Spectrum. All rights reserved.

3. Temperatures

It is not possible to explain all phenomena in terms of a unique temperature or even, at a given time, a unique temperature variation with height.

When a line has appreciable intensity the electron temperature T_e of the emitting region cannot differ much from $T(X^{+m})$. Different ions give different values of T_e. There is independent evidence that different emissions come from different regions.[8]

From line profiles, temperatures T_D are obtained assuming thermal Doppler broadening. Some typical results are[7,9]:

Spectrum	*Fe X*	*Fe XIV*	*Ca XV*
I, eV (for production of ion)	235	355	820
$10^{-6} T_e$	0.5	1.1	2.5
$10^{-6} T_D$	1.7	2.5	3.7

Both T_e and T_D show the same tendency to increase with increasing I but T_D exceeds T_e by about 1.2×10^6 °K. It is considered that the profiles are affected by the macroscopic motions responsible for coronal heating. This requires that $\sim 1\%$ of the kinetic energy of the corona be macroscopic. The cooling rate is $\sim 10^{-4}$ erg cm^{-3} sec^{-1}. These figures should be considered in connection with the theory of degradation processes.

The absence of Doppler shifts[10] imposes severe limitations on heating mechanisms. The motions must either be on a small scale ($< 10^3$ km) or have a high frequency ($> 10^{-1}$ sec^{-1}). Application of ionization theory to observations for a bright condensation[8] suggests that temperature variations occur on a scale much smaller that 5×10^4 km.

4. Excitation

Excitation is followed by quantum emission and true absorption may be neglected in practically all circumstances.[11,19,21] The probabilities of spontaneous emission, A_{ul}, are known with adequate accuracy, the probability of resonant scattering of photospheric radiation is

$$A_{lu} = (\omega_u/\omega_l)(\lambda^5/2hc^2) D F_\lambda A_{ul}$$

and the probability of collisional excitation is $N_e \alpha_{lu}$. For the ultraviolet lines, excitation is purely collisional and for the visible lines excitation is mainly collisional in the inner corona and radiative further out. In calculating α_{lu} for the forbidden lines, it is necessary to allow for excitation of allowed transitions followed by cascade[12]. Accurate calculations[12,13,14] have been made for Fe XIV. These give:

T, °K	1×10^6	2×10^6
α_{lu}, $cm^3\ sec^{-1}$	3.0×10^{-9}	4.0×10^{-9}

An empirical estimate has been obtained[15] using observations of the polarization* of the Fe XIV radiation; this gives $\alpha_{lu} = 4.8 \times 10^{-9}\ cm^3\ sec^{-1}$.

A general formula[14] should give α_{lu} for allowed transitions correct to within a factor of 2.

5. Composition

Some years ago it appeared that the composition of the corona was the same as that of the photosphere. Recent work has confirmed the earlier coronal abundance estimates but has led to significant changes in the photospheric results[16]. The present position is that the Fe/H abundance ratio for the corona appears to be greater than that for the photosphere by a factor of 20.

I. FORBIDDEN LINES

Expressing intensities in terms of the equivalent width W of the electron scattering continuum,

$$h\nu N_u A_{ul} = 4\pi\sigma N_e D F_\lambda W$$

where N_u is the excited state population. The ratio N_u/N_e is thus directly deducible from observations. ALY, EVANS and ORRALL[17] obtain values of $N(X^*)/N(\mathrm{H})$, where $N(X^*)$ is the total number of ions, of an element X, *observed in excited states*. They find values of $N(\mathrm{Fe}^*)/N(\mathrm{H})$ as great as one half of the photospheric Fe/H abundance ratio and value of $N(\mathrm{V}^*)/N(\mathrm{H})$ exceeding the photospheric V/H ratio by an order of magnitude. Consideration of excitation and ionization processes shows that $N(X)/N(\mathrm{H})$ will be larger than $N(X^*)/N(\mathrm{H})$ by a factor greater than ten.

Detailed abundance analyses, using forbidden lines, have been made by POTTASCH.[18]

II. ULTRA-VIOLET LINES

The emission from the volume element $\mathrm{d}V$ is $h\nu\alpha_{lu}N_e N(X^{+m})\,\mathrm{d}V$. The observed flux at the earth's distance r is

$$\mathscr{F} = \tfrac{1}{2}\frac{h\nu}{4\pi r^2}\int \alpha_{lu} N_e N(X^{+m})\,\mathrm{d}V$$

* Polarization was also observed[15] for the Fe X red line. Dr. Wlérick tells me that P. Charvin has criticized this work on the grounds that the red line, with an upper state having $j = \frac{1}{2}$, should be unpolarized.[29]

allowing for the fact that $\frac{1}{2}$ of the emission is absorbed by the photosphere. Although at the present time we have no height resolution for the ultra-violet lines, from the observations one may obtain chemical compositions[19] and information about the transition region between the chromosphere and corona[20,21].

Assuming a uniform chemical composition, the quantity

$$\theta = \alpha_{lu} N(X^{+m})/N(X)$$

depends on T but not on density. It has a maximum value, $\theta_{\max}$, at a temperature T_i. We may put

$$J = \frac{1}{2} \frac{h\nu}{4\pi r^2} \theta_{\max} \frac{N(X)}{N_e} \int \gamma N_e^2 \mathrm{d}V$$

where

$$\gamma = \theta/\theta_{\max} .$$

From the observations, $(N(X)/N_e) \int \gamma N^2 \mathrm{d}V$ may be deduced and plotted as a function of T_i. Consistent results are obtained assuming that $\int \gamma N_e^2 \mathrm{d}V$, as a function of T_i, is the same for all elements. The relative abundances of the elements giving ultra-violet lines can therefore be obtained. The theory is not applicable to hydrogen lines and Pottasch therefore determines his abundances relative to oxygen.

III. THE RADIO SPECTRUM

The central brightness temperature, T_R, as a function of λ, may be deduced when

$$\int_{h(T)}^{\infty} N_e^2 T_e^{-3/2} \mathrm{d}h$$

is known. Assuming various ratios $N(\mathrm{O})/N(\mathrm{H})$, Pottasch deduces this integral from the ultra-violet results for $(N(\mathrm{O})/N_e) \int \gamma N \mathrm{d}V_e^2$. He obtains good agreement with the observed radio data for $N(\mathrm{O})/N(\mathrm{H}) = 7 \times 10^{-4}$.

IV. DISCUSSION OF ABUNDANCES

Results are collected in Table 1. For the light elements there is good agreement between the corona and photosphere results but for the heavier elements the coronal abundances exceed the photospheric abundances by at least an order of magnitude.

We consider that the apparent composition differences are not due to errors in the coronal analysis. It also seems unlikely that they are due to errors in the photospheric analysis, but some interesting effects have been noted by WARNER.[22] Serious consideration must be given to the possibility of real composition differences. Gravitational diffusive separation will reduce the coronal heavy element concentrations in times of $\sim$ 3 hours. Turbulent mixing is generally assumed to be faster than this. Infall of meteoric matter[19] will affect the composition appreciably in times of $\sim$ 10 days and is therefore unimportant even when compared with gravitational separation. Thermal diffusion may be more important.[17,22] Times of $\sim$ 3 days are required to establish a steady state but this state would involve extremely large composition differences between hot and cold regions. Separation may be appreciable in much smaller times. The time scale of turbulent mixing is the main uncertainty.

TABLE 1

ABUNDANCES IN THE CORONA AND IN THE PHOTOSPHERE; RELATIVE VALUES OF LOG N(X).

X	*Corona*		*Photosphere*[16]
	Visible[18]	*UV and radio*[19]	
H	12.0	12.0	12.0
He	—	11.3	—
C	—	9.2	8.7
N	—	8.2	8.0
O	—	8.8	9.0
Ne	—	7.9	—
Na	—	$\leqslant$ 6.5	6.3
Mg	—	8.0	7.4
Si	—	8.3	7.5
S	—	7.6	7.3
Ar	7.1	—	—
K	5.6	—	4.7
Ca	6.8	$\leqslant$ 6.5	6.2
Cr	6.0	—	5.2
Mn	5.9	—	4.9
Fe	7.9	7.8	6.6
Co	5.3	—	4.6
Ni	6.7	—	5.8

6. X-Rays

Solar X-radiation is highly variable. Pin-hole camera photographs show that the greater part of the emission comes from localised active regions in the lower corona, above Ca plages.[24] In this respect, X-radiation is similar to Fe XIV and Ca XV and to decimetre radio waves. The emission extends to $h \approx 43000$ km.

We shall consider only the quite emission, which may be attributed to thermal processes (non-thermal processes are probably operative during flares). The theory has been developed by ELWERT.[25] Line radiation is produced by collisional excitation, as in the ultra-violet, and continuum radiation is produced by radiative recombination. The total recombination rate is, of course, equal to the collisional ionization rate, but some recombinations will be on excited states and will not give continuum radiation at X-ray wavelengths. Also, rate coefficients for excitation are generally larger than those for ionization. In the range 50–100 Å the calculated emission is mainly due to lines; this appears to be in general agreement with observation.[26]

For $\lambda \leqslant 20$ Å the emission at sun spot minimum is consistent with temperatures not exceeding 1×10^6 °K and perhaps closer to 0.5×10^6 °K. At sun spot maximum, increased emission may be explained by supposing that localised hot regions, with $T = 2 \times 10^6$ or more, cover 1% of the disc.[27]

Measurements in the range 4–14 Å have been obtained from the Ariel satellite[28] with 2 Å resolution. In order to explain the non-flare spectra in terms of thermal processes it is sometimes necessary to assume the existence of regions with $T \simeq 5 \times 10^6$. Temperatures as high as this are consistent with strong Ca XV emission.

7. Conclusion

The problem of interpreting the coronal spectrum cannot be separated from the problems of the structure and dynamics of the corona and the problem of coronal heating mechanisms. Intriguing problems are raised by the apparent composition differences between the material of the corona and of the photosphere.

I am indebted to Dr. S. R. Pottasch for showing me his work in advance of publication.

Post-script (October, 1964). Since the time that this paper was written, in the summer of 1963, Burgess[30] has shown that the process of *dielectronic recombination* gives rate coefficients α which, at coronal temperatures and densities, are at least one order of magnitude greater than the coefficients calculated previously. Allowance for this process[31] increases the calculated ionization temperatures and removes, or at least greatly reduces, the discrepancies between T_e and T_D.

References

1. Van de Hulst, H. C.: 1953, in *The Sun* (ed. by G. Kuiper).
   Allen, C. W.: 1954, *Rep. Prog. Phys.* **17**, 135.
   Evans, J. W.: 1961, *The Solar Corona* (IAU Symposium, 1961).
   Seaton, M. J.: 1964, *Plan. Space Sci.* **12**, 55.
2. Osterbrock, A.: 1961, *Astrophys. J.* **134**, 347.
   Whitaker, E.: 1963, *Astrophys. J.* **137**, 914.
3. Collins, P. D. B : 1963, Thesis, Bristol. (Identifies 5926 with Ar XV, $1s^2 2s\,2p\ ^3P_2 - {}^3P_1$).
4. Ivanov-Holodnyi, G. S. and Nikolskii, G. M., 1962, *Soviet Astron. - A. J.* **5**, 632.
5. Dolder, K. T., Harrison, M. F. A., and Thonemann, P. C.: 1961, *Proc. Roy. Soc.* **A264**, 367 (He^+); 1963, *Proc. Roy. Soc.* **A274**, 546; 1963, *Proc. Phys. Soc.* **82**, 368.

6. Trefftz, E.: 1963, *Proc. Roy. Soc.* **A271**, 379;
Burgess, A. and Rudge, M. R. H.: 1963, *Proc. Roy. Soc.* **A273**, 372.
7. Seaton, M. J.: 1962, Observatory **82**, 111;
Nikolskii, G. M.: 1963, *Geomagnetizm i aeronomija* **3**, 417.
8. Aly, M. K., Evans, J. W., and Orrall, F. Q.: 1963, *Astrophys. J.* **137**, 1313.
9. Billings, D. E. and Lehman, R. C.: 1962, *Astrophys. J.* **136**, 258.
10. Billings, D. E.: 1963, *Astrophys. J.* **137**, 592.
11. Pecker, C. and Thomas, R. N.: 1963, *Astrophys. J.* **137**, 967.
12. Pecker, C.: 1960, *Compt. Rend.* **251**, 1862.
Pecker, C. and Thomas, R. N.: 1962, *Ann. Astrophys.* **25**, 100.
13. Blaha, M.: 1962, *Bull. Astron. Inst. Czechoslovakia* **13**, 81.
Bely, O.: 1962, *Compt. Rend.* **254**, 3167.
Seaton, M. J.: 1964, *Mon. Not. Roy. Astron. Soc.* **127**, 191.
14. Van Regemorter, H.: 1962, *Astrophys. J.* **136**, 906.
15. Mogilevskii, E. I., Nikolskii, G. M. and Nikolskaja, K. I.: 1961, *Soviet Astron. - A. J.* **4**, 225.
16. Goldberg, L., Müller, E. A. and Aller, L. H.: *Astrophys. J. Suppl.* **5**, 45.
17. Aly, M. K., Evans, J. W., and Orrall, F. Q.: 1962, *Astrophys. J.* **36**, 956.
18. Pottasch, S. R.: *Mon. Not. Roy. Astron. Soc.* **125**, 543.
Pottasch, S. R.: 1964, *Mon. Not. Roy. Astron. Soc.* **128**, 73.
19. Pottasch, S. R.: 1963, *Astrophys. J.* **137**, 945.
20. Allen, C. W.: 1961, *Mém. Soc. Roy. Sci. Liège* **4**, 241.
21. Ivanov-Holodnyi, G. S. and Nikolskii, G. M.: 1961, *Soviet Astron. - A. J.* **5**, 31.
Ivanov-Holodnyi, G. S. and Nikolskii, G. M.: 1962, *Soviet Astron. - A. J.* **5**, 632.
Ivanov-Holodnyi, G. S. and Nikolskii, G. M.: 1963, *Soviet Astron. - A. J.* **6**, 609.
22. Warner, B.: 1964, *Observatory* **84**, 14; see also, 1964, *Mon. Not. Roy. Astron. Soc.* **128**, 63.
23. Chapman, S.: 1958, *Proc. Phys. Soc.* **72**, 353.
24. Blake, R. L., Chubb, T. A., Friedman, H., and Unzicker, A. E.: 1963, *Astrophys. J.* 137, 3.
25. Elwert, G.: 1954, *Z. Naturforsch.* **9a**, 632.
Elwert, G.: 1956, *Z. Astrophys.* **41**, 67.
Elwert, G.: 1961, *J. Geophys. Res.* **66**, 391.
26. Elwert, G.: 1963, *Proc. Conf. on the Ionosphere*, Phys. Soc., London.
27. Friedman, H.: *Rep. Prog. Phys.* **25**, 163.
28. Pounds, K. A., Willmore, A. P., Bowen, P. J., Norman, K., and Sanford, P. W.: *Proc. Roy. Soc.* (in press).
29. Charvin, P.: 1964, *Compt. Rend.* **258**, 1155.
30. Burgess, A.: 1964, *Astrophys. J.* **139**, 776.
31. Burgess, A. and Seaton, M. J.: 1964, *Mon. Not. Roy. Astron. Soc.* **127**, 355.

THE CORONAL SPECTRUM

DONALD E. BILLINGS
High Altitude Observatory, Boulder, Colorado, U.S.A.

Abstract. Recent developments in the study of the spectrum of the solar corona are reviewed. These include the X-ray spectrum – both line emission and continuum – the radio spectrum and infrared observations. Recent attempts to interpret the profiles of emission lines in the visible spectrum are discussed in some detail. All spectral studies give evidence of a considerable range of temperatures in the corona.

The Extreme Ultra-Violet Spectrum

The spectrum of the corona is remarkable in that it comprises the shortest and the longest electromagnetic waves that reach us from the sun. Furthermore, only at the extreme ends of the spectrum is the radiation that we receive from the sun strictly coronal in origin. At all other wavelengths the emission from the corona is only a minute part of that from the photosphere and chromosphere, and must be separated out for analysis by very special techniques.

In a sense, the portion of the solar spectrum from about one Ångstrom to a few hundred Ångstroms, now accessible through rocket and satellite techniques, is the "real" spectrum of the corona. In this range the permitted transitions of the emitting coronal ions are found, and the smoothed total coronal emission maximizes. The observation and analysis of this extreme ultra-violet or soft X-ray spectrum is without doubt the most important development in coronal spectroscopy of the past decade.

Earlier observations of the coronal X-ray spectrum (Chubb *et al.*, 1961; Hinteregger, 1960; Mandel'štam *et al.*, 1963) gave measurements of the solar flux within certain wavelength intervals specified by the equipment. Such observations provided a very rough idea of the energy distribution with wavelength. Coronal temperatures were deduced from a comparison of this distribution with a black-body curve. Although a theoretical basis for such a procedure had been given by Elwert (1961), whose computations indicated that integrated emission from the quiet corona would approximate that of a grey body, the method contained great difficulties – both because of the presence of large emission gradients within wavelength intervals used, and because the generalizations concerning the average effect of line emission could not be depended upon to hold for any specific interval in which a strong line might be present. With due recognition of these difficulties, it was concluded that the X-ray data indicated the presence in the corona of two temperature regimes, one below one million degrees, which supplied most of the energy in wavelengths longer than 20 Å, and another at a higher temperature providing emission in the

C. de Jager (ed.), The Solar Spectrum. All rights reserved.

vicinity of 1 Å. Very recently POUNDS, WILLMORE, BOWEN, NORMAN and SANFORD (1963), using the spectral distribution in the range 4–14 Å as given by a proportional counter in the satellite Ariel, have concluded that the shape of the distribution corresponds to a Planck temperature of 1.3 to 2.2 million degrees, but that the total emission is much too great for even the entire sun at 2.2 million degrees. They suggest that the presence of line emission at the longer wavelength end of the range makes the spectrum steeper than Planckian – that the temperature of the emitting regions is about five million degrees – and that if the density in such regions is about 10^{10} cm^{-3}, their dimensions are about 3×10^4 km. In the presence of a flare the shorter waves became enhanced many-fold, requiring temperatures, if the emission be thermal, of one or two orders of magnitude more than normal coronal temperatures.

POUNDS *et al.* have discovered a possible way of separating thermal from non-thermal radiation in a flare, the thermal radiation leading to temperatures $\sim 10^7$ °K.

Between the earlier and more recent X-ray continuum studies, spectrograms obtained by RENSE (1961) and DETWILER, PURCELL and TOUSEY (1961), and spectral scans by Hinteregger, have focused attention on the line spectra. Some forty coronal lines have been identified in these spectra which extend down to 55 Å, through the joint efforts of the various observers, along with Ch. Pecker, Edlén, Zirin and others. Many other lines are present, so closely-spaced that it is difficult to associate them individually with the various transitions that are possible in the coronal ions.

Several important features of the X-ray spectra have been noted. One of the most striking (JEFFERIES, PECKER and THOMAS, 1962; ZIRIN, HALL and HINTEREGGER, 1962) is that lines from high stages of ionization, such as Fe XVI and Si XII, are surprisingly strong. On the other hand, intensities of lines from intermediate stages of ionization are extremely low, indicating to ZIRIN and HINTEREGGER (1962) that the transition from chromosphere to corona is very abrupt. (A conclusion which PAGEL (1963) apparently does not share.) Finally the relative brightness of the lines has lead Zirin and Hinteregger to suggest various revisions in tables of chemical abundance, and POTTASCH (1963) to compute the relative abundance of ten elements in the corona. Pottasch concluded that the abundances of magnesion and iron are considerably greater in the corona than in the photosphere, but are in agreement with abundances in planetary nebula and cosmic rays. He also considered the problem of constructing a model corona from the X-ray line emission data, and arrived at a number of conditions that must hold in such a model; but did not have data to make the model definitive. IVANOV-HOLODNYI and NIKOLSKII (1963), on the other hand, have used the extreme U.V. data to develop a very plausible model for active regions in the corona. They arrive at polytropic models for both active and inactive regions, $Tn_e^k = C$, where C is greater for the active than the unperturbed regions.

The Coronal Radio Spectrum

At the opposite end of the spectrum the corona is a source of radio noise from a few centimeters to several meters in wavelength. The highly transient features of this noise are generally explained by such collective phenomena as plasma oscillations, and thus are not really pertinent to the general subject matter of this symposium. However, the two types of noise designated as quiet sun emission and slowly varying component are generally considered to be thermal in origin, and therefore very properly belong under the topic *Spectrum of the Corona*. Such noise, as a matter of fact, is merely the long wavelength, Rayleigh-Jeans tail of the same grey body emission that maximizes in the X-ray region, and is perceptible only because of the extreme energy-sensitivity of radio receivers.

The steady energy flux from the sun at wavelengths sufficiently long to be attributed entirely to the corona corresponds to a corona of a few hundred thousand degrees, and such temperatures have generally been considered to be those indicated by radio noise measurement. We should note, however, that such radiation originates high in the corona. Shorter wavelengths, of less than half a meter, come in part from the corona and in part from the chromosphere. Thus the brightness temperature for such wavelengths is necessarily intermediate between that of the two regimes in the solar atmosphere. Immediately over sunspots, however, the corona is optically deep in radiation of a few centimeters wavelength. The enhancement in optical depth in such regions is due in part to locally greater density but may also be explained by resonant absorption at various harmonics of the electron gyrofrequency in the sunspot magnetic fields. KAKINUMA and SWARUP (1962), and SWARUP, KAKINUMA, COVINGTON and HARVEY, and MULLALY and ROME (1963) have brought together the results of quite a number of narrow-beam antenna observations for such regions, and point out that the brightness temperatures typically come out in the range 2-4 million degrees. Three explanations of these higher temperatures compared to the longer wavelength quiet sun temperatures are possible: (1) The active region emission is not entirely thermal; (2) the corona is typically much hotter in such regions than outside; or (3) it is only in such optically-dense regions that radio observations may probe close to the chromosphere and detect thereby the actual temperature of the lower corona without diluting the coronal emission with chromospheric emission.

Coronal Infra-Red

A third portion of the invisible coronal spectrum that has recently undergone intensive investigation is the near infra-red. KURT (1962), flying at ten kilometers during the 1962 eclipse, observed the coronal lines $\lambda\,7892$, $\lambda\,10747$ and $\lambda\,10798$. Although he was equipped to observe lines of 1/100 the brightness of $\lambda\,10747$, he did not find the line $\lambda\,8024$ that Lyot had reported. FIROR and ZIRIN

(1962), using image-intensifier tubes on the Climax coronagraph spectrograph, photographed and made photometric studies of the lines λ 10747 and λ 10798 of Fe XIII. They compared these with the brightness of other iron lines in the same regions, and thus succeeded in plotting ion-density distribution curves. Finally they were able to draw some tentative conclusions concerning the excitation processes, the minimum ionization cross-section, and temperature distributions along lines of sight.

The Visible Coronal Spectrum

Compared to the spectacular advances in the study of the widely-spread portions of the invisible coronal spectrum, the work that has been done during the past years in the visible spectrum seems quite pedestrian. It has, in fact, been a continuation, for the most part, of long-term programs and procedures. I will only mention the theoretical work of SEATON (1962), BURGESS (1961), BLAHA (1962), GARSTANG (1962), and Ch. PECKER and THOMAS (1962, 1963) dealing with ionization and excitation processes, since the other paper of the morning involves this subject. The computation of ground-configuration energy levels by Ch. PECKER and RÖHRLICH (1961), directed toward the identification of ultra-violet lines, had led to a check on the identification of several visible lines as a by-product. Specifically they added ten fairly definite identifications of visible lines, and indicated that four identifications probably were not correct.

ALY, EVANS and ORRALL (1963) have recently confirmed from 1952 eclipse plates that the groupings of various coronal lines according to their simultaneous appearance at the solar limb is about as one would expect from the appropriate ionization potentials. Only one line, λ 3643 seemed to behave in an irregular fashion, which throws doubt on its identification. The same study included a detailed analysis of the distribution in brightness of continuum and a number of lines in the intense coronal condensation of that eclipse. The study has yielded emission by several ions versus distance from the center of symmetry in the condensation. The emission distributions with distance suggest a very complex temperature structure in the condensation, and give information on the abundance of various elements. It appears that Fe and V are much more abundant than photospheric observations indicate. Similarly, ZIRIN, from limb flare spectra of the λ 5694 line and continuum, has concluded that calcium abundance is high in the outer solar atmosphere.

Miscellaneous recent studies in the coronal visible spectrum include a detailed determination by Waldmeier of height emission gradients in λ 6374. Waldmeier arrives at the interesting conclusion that λ 6374 emission is proportional to

$$\int_{-\infty}^{+\infty} N_e^2 \, dx \text{ up to } r = 1.09 \,,$$

then becomes proportional to the K-continuum thereafter, indicating that below 1.09 the excitation is collisional, whereas above this height it is radiative. Also we should mention provocative evidence by MOGILEVSKIJ, NIKOLSKIJ and NIKOLSKAJA (1960) that coronal emission lines show a high degree of polarization, indicating that ions excited by radiation maintain their same orientation in space until emission.

The two traditional methods for using the emission lines in the visible spectrum to determine the temperature of the corona have recently come under heavy criticism. The computation of temperature from the ratio of the intensity of lines from two different ions of the same element (usually λ 5303 of Fe XIV and λ 6374 of Fe X) depends on the assumption that emission in the two lines is the same proportion all along the line of sight. This, in turn implies a uniform temperature along the line of sight. Refuting this assumption, DOLLFUS (1962) has studied the λ5303 and λ6374 monochromatic pictures of the corona from Pic du Midi over a period of several years, and has concluded that although the same loops, arches, and other features are frequently bright in both wavelengths, there is always a difference in the appearance of the feature in the two lines. In some cases the λ 5303 emission seems to form a large envelope over a very sharp λ 6374 feature. In other cases, but less often, the situation is reversed. PROKOFIEVA (1963) arrived at a similar conclusion from a close examination of spectrograms from Pulkovo. She also measured the half-widths of green and red line profiles at 33 limb positions, and found no particular correlation between the widths of the two lines. If the emission in the two lines had originated from the same locations in space the temperatures computed from the profiles should have been the same within the limits of accuracy of the measurements, irrespective of whether the source of broadening was entirely thermal, or whether it was a combination of thermal and macroscopic motion. BILLINGS and LEHMAN (1962) made a similar study after first selecting only the spectrograms on which the two lines appeared to have the same intensity variation over a considerable interval on the limb. They found that not only were the λ5303 and λ6374 line width temperatures different, point by point, but that there was a strong tendency for the green line temperature to exceed the red line temperature by about 600000°. Thus, even in regions which look the same in the two wavelengths on the spectrograms, the emission appears to come from different positions in space.

The computation of temperature from coronal line width has been attacked on the ground that one cannot distinguish between thermal and larger-scale motions in the broadening of a line profile, provided the larger-scale motion is of such a nature that it will not disturb the Gaussian shape of the profile appreciably. JEFFERIES and ORRALL (1961), for example, have pointed out that the line profiles from two equally luminous bodies of gas moving in opposite directions along the line of sight could correspond to a temperature two or three times as high as the actual temperature in the gas without being obviously distorted from Gaussian. Of course if the macroscopic velocities were sufficiently great compared to the thermal velocity, the resulting profile would have a dip at its center. WALDMEIER, in the early days of coronagraphy (1940), reported such a dip, and OBAŠEV, very recently (1961), has referred to this report and explained the effect as that of the radial motion of two widely separated jets along the line of sight. However, the early observation of such profile shapes has not been confirmed in the many thousands of higher-dispersion profiles that have been examined during the past twenty years, and should no longer be considered valid.

Along a line of thinking similar to that of OBAŠEV, JALLOP (1961) has raised an interesting point. First, using various temperatures for an isothermal inner corona, he computed the solar wind velocity at various solar radii, according to PARKER's (1960) theory. He then computed the profiles that would be observed at various heights above the solar limb and concluded that the increasing velocity of the solar wind with height would result in a significant increase in profile width with height. Thus for temperatures greater than 1.5×10^6°K, profile widths would increase so strongly with height that the effect would be readily observable at 1.7 $R_{\odot}$. But, he pointed out, the profiles determined by JARRETT and VON KLÜBER (1961), which correspond to temperatures considerably in excess of 1.5 million degrees, did not increase in width up to 1.7 solar radii. Therefore the widths of the profiles could not measure the coronal temperature. To JALLOP's argument, BILLINGS and LILLIEQUIST (1963) replied by pointing out that not only the profiles of JARRETT and VON KLUBER but also profiles from Climax spectrograms became narrower with height. Therefore, if the line widths are to be interpreted as showing a high temperature, they also show a temperature gradient. If such a temperature gradient as well as the temperature is included in the solar wind computation, the resulting solar wind velocity is so low that it has little effect on the line profiles out to 1.7 solar radii.

The most serious objection to the high temperatures given by the coronal lines has been raised by BURGESS (1961) and SEATON (1962) after a careful reconsideration of coronal ionization theory. They have concluded that Fe X ions could not possibly exist at electron temperatures much greater than 700000° K, and that Fe XIV ions would reach their maximum abundance at 1.1 million degrees. Furthermore, LÜST, MEYER, TREFFTZ and BIERMANN (1962) have pointed out that the electron and the ion temperature could not differ by more than a few thousand degrees (although BILLINGS and LEHMAN (1962) and SCHATZMAN (1962) have demonstrated that a greater temperature difference could exist in a temperature gradient). LÜST *et al.* added that the motions which heat the corona could also very well broaden the profiles. In the face of these arguments, it appears imperative to consider the interpretation of the profile half-widths in terms of a combination of thermal and macroscopic motion.

Line Profile Analysis

If coronal line half-widths cannot give us temperature unambiguously, to what use can be profiles be put? Their meaning should not be hopelessly complex, since in an optically thin corona they must indicate motion of some kind. Furthermore, if the corona is heated by a shock phenomenon, there should be a functional relation between the macroscopic and the thermal motion. Consequently, if we possessed sufficient knowledge of the coronal heating mechanism we should be able to separate the thermal from the macroscopic component in the line profiles. Conversely, the study of line profiles may throw some light on the details of the heating mechanism.

Line profiles demonstrate a number of regularities that give hope for their use in analysis of the motion-field in the corona. These are: (1) Significant Doppler shifts in coronal lines are extremely rare. DOLLFUS (1957) obtained an r.m.s. displacement of the λ 6374 coronal line of about 0.01 Å. BILLINGS (1963) has confirmed this result for λ 5303, even in those cases when a major fraction of the emission came from a small feature of dimensions of about 10000 km. Thus we can say that the coronal lines, on the average, do not give evidence of gross velocities greater than 0.01 times the thermal velocity of coronal protons. Larger Doppler displacements do occur, in association with very large line-of-sight velocities in prominences, but these are the exception rather than the rule. (2) There are some generalizations that we can make concerning coronal line widths. In regions bright in both red and green line emission, the width of the green line corresponds closely to 2.5×10^6 °K, that of the red line to about 1.9×10^6. In more active regions in which the green line is bright while the red line is dim, or in which the yellow line, λ 5694 may be seen, the green line width is likely to correspond to three million degrees or more. The yellow line width systematically corresponds to 3.5–4 million degrees (BILLINGS, 1957; ZIRIN, 1959). There is no systematic difference between line widths at the center compared to the edges of bright regions, or in bright equatorial compared to polar regions. There is weak evidence for decreases in line width with height, and for stationary minima in line width versus height (BILLINGS, 1959). (3) Line profiles are very nearly Gaussian. In profiles of the most intense emission, photographed with relatively high dispersion on relatively fine grained film, actual departures from Gaussian may just be distinguished from observational irregularities.

We cannot deduce directly from these observations the nature of the motion that broadens coronal line profiles, but we can place some limitations on it. The smallness of the r.m.s. gross Doppler displacement requires either that the motion be generally symmetric with regard to axes in coronal structures, or that each observation comprises motions that almost balance each other. The latter could be achieved if each line of sight contained a large number of randomly-moving cells, or if during each observation a number of oscillations took place, so that the residual effect on the line position of uncompleted oscillations would be relatively small. From square-root-of-N arguments we can conclude that something over 100 randomly-moving cells would be required along a line of sight to produce the effect that we observe. In the case of well-defined coronal features, of dimensions of about 10000 km, the size of such cells becomes very small. Certainly the large-scale, occasional pulses that are implied by OSTERBROCK (1961) are not observed in the emission corona.

The limitations that profile observations place on motions in the corona have suggested to us certain possible patterns of motion. We have carried out computations to determine the effect on the coronal line profiles of such motion fields. These computations have led to the following conclusions: (1) If the gas is experiencing harmonic oscillation, or a whirling motion, or cylindrically symmetrical inward or outward flow, the profile is distorted so as to make the logarithmic plot of the profile concave upward. See Figure 1. (2) If the line of sight comprises many aggregates of gas, each undergoing a motion of a type included in (1), but with the various axes of the system oriented at random, the resultant profile is very nearly Gaussian. (3) The effect of summing profiles of various widths along a line of sight is to make the logarithmic plot of the result-

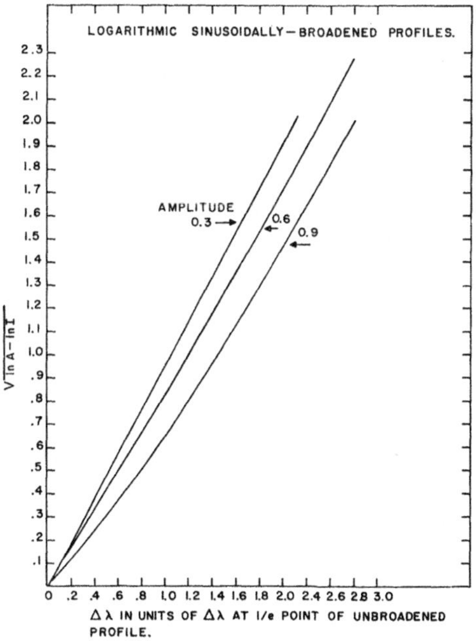

Fig. 1. A logarithmic plot of the profiles

$$I = \int_0^{2\pi} e^{-(x - a\sin\phi)^2} d\phi$$

for various values of amplitude a. A is the maximum value of I.

ant profile concave downward, in a manner almost exactly opposite that produced by the motions described in (1).

Thus if we observe that a profile is Gaussian, and undisplaced, it may mean that the coronal motion is entirely thermal and that the material is isothermal; it may mean that the profile is a combination of thermal and macroscopic motion, the motion being a combination along the line of sight of axially-symmetric motions with the axes randomly oriented; or it may mean that the effects of an axially-symmetric motion and a temperature variation along the line of sight exactly compensate one another.

There is a possibility of separating these effects by comparing the detailed shapes of profiles of sharply-defined coronal features with those from limb positions where no distinct features are present. In the former case we might expect both an axially-symmetric motion and a temperature distribution to be present, the two compensating each other to some extent. In the later case the effect of axially-symmetric motions should be neutralized by summation of various orientations of the axes, leaving the temperature-range effect alone to distort the profile. Thus from the non-distinct-feature case we may evaluate the temperature range, and by applying the result to the distinct-feature case, determine the motion from the profile shape.

We have carried out one analysis of this type, using composite profiles, obtained by adding together many single profiles, in order to minimize observational noise. The distinct-feature composite was obtained by adding together approximately 200 λ5303 profiles from a range of heights and observing times, primarily of three sharply-defined features, (probably parallel loops), lying close together above the limb. The non-distinct-feature composite profile was made up of more than 300 profiles from the same set of observations, but at limb positions where no bright features were present.

The shape of the distinct-feature composite profile agreed with that of a Gaussian distorted by an axially-symmetric motion of amplitude equal to 0.6 the mean thermal velocity. Thus, although the width of the profile corresponded to a temperature of about 2.8×10^6 °K, we concluded that the actual temperature, or more precisely, the most prevalent temperature (T_0) of the Fe XIV ions, lay somewhere along the curve of Figure 2. In this curve the abscissa is the relative line width distribution that might be being compensated in the profile shape by macroscopic motion.

The shape of the non-distinct-feature composite profile agreed with that obtained by summing profiles over a relative width distribution of 0.2. If we assume that in this profile the distortional effects of the motions are smoothed out by random orientations of their axes, we may take 0.2 to measure the actual temperature distribution. Then, if we assume that this distribution also holds for the distinct-feature profile, we conclude from Figure 2 that the greatest number of Fe XIV ions are at a temperature 1.3×10^6 °K. Thus from profile analysis alone, although the widths of the profiles corresponded to about

2.8×10^6 degrees, we have deduced temperatures for the emitting ions acceptable to ionization theory. The validity of the deduction depends primarily on the extent to which the two composite profiles were representative of the situations that they were purported to represent. Not all single bright features have

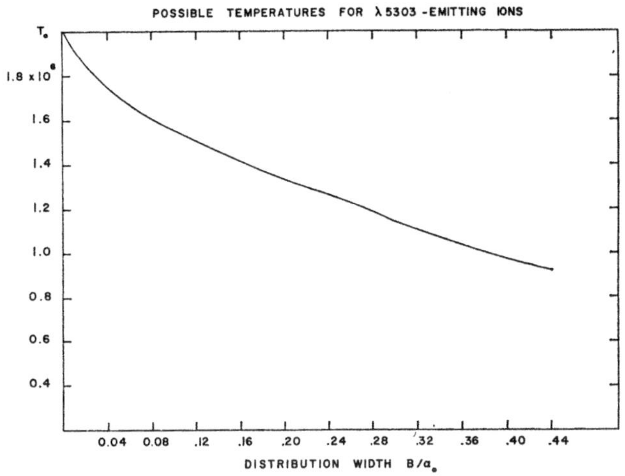

Fig. 2. The possible temperatures of the λ 5303 – emitting ions in the cluster of small bright green-line-emitting features that were studied. The abscissa is the ratio of the breadth B of the line-width distribution to the most prevalent line width α_0, assuming that the effect of such a distribution is canceled by motional distortion.

motion-dominated profiles. With some of the strongest bright features we have studied the converse is the case-suggesting considerable temperature-structure within certain features.

In addition to a temperature value, this method of analysis gives information about the intensity of macroscopic motion and the distribution in temperature. From the data that we have studied, for example, the axially-symmetrical velocity is 30 km/sec. The relative line width distribution, 0.2, describes the temperature distribution for the ions from which the λ 5303 radiation is being emitted. One may deduce from this distribution the distribution of temperature within the entire coronal gas by the use of an ionization theory. Thus, if $f_{5303}(T)$ is the distribution of the λ 5303 emitting ions that we have found from our analysis of the shape of the line profiles, and $\phi_{5303}(T)$ is the theoretical distribution of λ 5303 emitting ions for a flat distribution of all coronal matter with temperature (very nearly the same as the ratio N(XIV)/N(Fe) as a function of temperature, except for a small modification due to excitation effects), the ratio,

$$f_{5303}/\phi_{5303} = F(T)$$

is the distribution of coronal matter with temperature along the line of sight.

Figure 3 shows the $F(T)$ that we have found. The shape of this distribution depends strongly on the precise form that we have assumed for f_{5303}, so we do not consider its detail to be very significant. The minimum, for example, is much too severe to be in agreement with the observations of Firor and Zirin

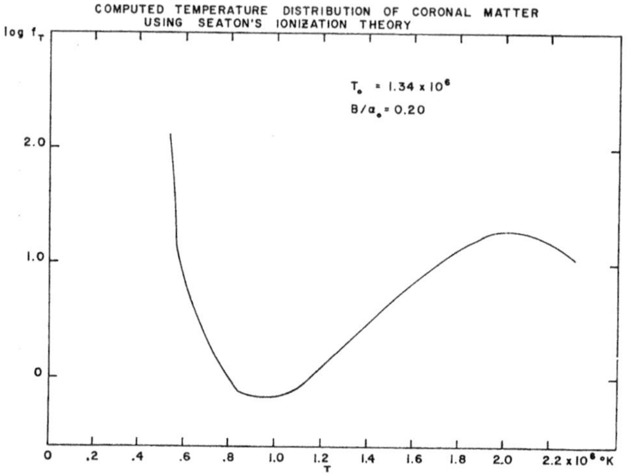

Fig. 3. A distribution of coronal matter with temperature, computed from a comparison of the shapes of profiles in sharp, bright features with those of diffuse features in the corona, and the application of Seaton's computed ratio $N_{Fe\,(XIV)}/N_{Fe}$ versus temperature.

on the simultaneous brightness of several lines of the Fe series. Also the distribution f_{5303} must be cut off very arbitrarily at its lower end to give the observed ratio of red to green line emission. Such a distribution however, agrees with the considerable Fe XVI emission seen in rocket X-ray spectra and with height-gradient observations. We note that by introducing a relatively small modification in our results, i.e., lowering the relative line-width distribution from 0.20 to 0.16 and thereby increasing the temperature T_0 from 1.3 to 1.4×10^6 °K; and at the same time making a small modification in the cross-sections of Seaton which bring his temperature of maximum Fe XIV population from 1.1 to 1.4×10^6 °K we obtain a precisely flat distribution for $F(T)$. We advance the very tentative suggestion that the actual ditribution lies somewhere between that shown in Figure 3 and a flat distribution. Such a distribution may be pertinent only to the day on which the observations were made, and in the particular zone of moderate activity. One should expect considerable change with the solar cycle, and from region to region.

Conclusion

Although the most spectacular advances in coronal science are based on observations at the extreme ends of the coronal spectrum, the understanding of important questions concerning the structure and the heating of the corona depends on the kind of detailed observations that are, at present, possible only in the visible spectrum. Here further interpretations of observations are extremely difficult, but it is possible that we are now making some progress.

References

Aly, M. K., Evans, J. W., and Orrall, F. Q.: 1963, 'A Photometric Study of the Continuum and Seventeen Emission Lines in the Inner Solar Corona', *Astrophys. J.* **136**, 956.

Billings, D. E.: 1957, 'Profile of the Yellow Coronal Line', *Astrophys. J.* **125**, 817.

Billings, D. E.: 1959, 'Velocity Fields in a Coronal Region with a Possible Hydromagnetic Interpretation', *Astrophys. J.* **130**, 215.

Billings, D. E.: 1963, 'Spectroscopic Limitation on Coronal Heating Mechanisms', *Astrophys. J.* **137**, 592.

Billings, D. E. and Lehman, R. C.: 1962, 'Line-Width Temperatures of Distinct Coronal Features', *Astrophys. J.* **136**, 258.

Billings, D. E. and Lilliequist, C. G.: 1963, 'Coronal Temperature Gradient and the Solar Wind', *Astrophys. J.* **137**, 16.

Blaha, M.: 1962, 'Excitation of Fe XIV by Electron Collisions', *Bull. Astron. Inst. Czechoslovakia* **13**, 81.

Burgess, A.: 1961, 'New Results on Coronal Cross-Sections'. *Mém. Soc. Roy. Sci. Liège* **4**, 299.

Chubb, T. A., Friedman, H., Kreplin, R. W., Blake, R. L., and Unzicker, A. E.: 1958, *X-ray and UV Measurements During the Eclipse of October 12.*

Chubb, T., Friedman, H., Kreplin, R. W., Blake, R. L., and Unzicker, A. E.: 1961, 'X-Ray Solar Disk Photograph', *Mém. Soc. Roy. Sci. Liège* **4**, 228.

Detwiler, C. R., Purcell, J. D. and Tousey, R.: 1961, 'The Extreme Ultraviolet Spectrum of the Sun', *Mém. Soc. Roy. Sci. Liège* **4**, 253.

Dollfus, A.: 1957, 'Movements in the Inner Solar Corona', *Compt. Rend.* **244**, 1880.

Dollfus, A.: 1962, 'Properties of Monochromatic Emission of the Solar Corona', *Compt. Rend.* **255**, 3369.

Elwert, G.: 1961, 'Theory of X-Ray Emission of the Sun', *J. Geophys. Res.* **66**, 391.

Firor, J. and Zirin, H.: 1962, 'Observations of Five Ionization Stages of Iron in the Solar Corona', *Astrophys. J.* **135**, 122.

Garstang, R. H.: 1962, 'Transition Probabilities for Permitted and Forbidden Lines of Si X, Fe XIV and Fe X', *Ann. Astrophys.* **25**, 109.

Hinteregger, H. E.: 1960, 'Interplanetary Ionization by Solar Extreme UV Radiation', *Astrophys. J.* **132**, 801.

Ivanov-Holodnyi, G. S. and Nikolskii, G. M.: 1963, 'Extreme UV Radiation and the Structure of the Solar Atmosphere in Active and Undisturbed Regions', *Soviet Astron.-A. J.* **6**, 609.

Jarrett, A. H. and von Kluber, H.: 1961, 'Interferometric Investigation of Emission Lines of the Solar Corona During the Total Eclipse of 12 October 1958', *Mon. Not. Roy. Astron. Soc.* **122**, 223.

Jefferies, J. T. and Orrall, F. Q.: 1961, 'On the Interpretation of Prominence Spectra II', *Astrophys. J.* **133**, 946.

Jefferies, J. T., Pecker, C. W., and Thomas, R. N.: 1962, 'The Structure of Regions of Coronal Line Emission', *Astrophys. J.* **135**, 653.

Kakinuma, T. and Swarup, G.: 1962, 'A Model for the Sources of the Slowly Varying Component of Microwave Solar Radiation', *Astrophys. J.* **136**, 975.

Kurt, V. G.: 1962, 'Coronal Spectrum 7800–12000 Å During Eclipse of 15 February, 1961', *Soviet Astron.-A. J.* **6**, 349.

Lüst, R., Meyer, F., Trefftz, E., and Biermann, L.: 1962, 'Zur Temperatur der Sonnenkorona', *Z. Naturforsch.* **17a**, 259.

Mandel'stam, C. L., Tindo, I. P., Vonon'ko, Ju. K., Šurigin, A. I., and Vasil'en, B. N.: 1963, 'Investigation of the Sun's X-Radiation', *Plan. Space Sci.* **2**, 61.
Mogilevskii, E. I., Nikolskii, G. M., and Nikolskaja, K. I.: 1960, 'The Polarization of Coronal Emission Lines', *Soviet Astron.-A. J.* **4**, 225.
Obašev, S. O.: 1961, 'Some Properties of Coronal Emission Lines in the Neighbourhood of Sunspots', *Izv. Akad. Nauk Kazahskoi Astron. Inst.* **12**, 82.
Pagel, E. J.: 1963, 'Ultraviolet Emission From the Sun', *Plan. Space Sci.* **11**, 333.
Parker, E. N.: 1960, 'Stellar Winds', *Astrophys. J.* **132**, 821.
Pecker, C. and Rohrlich, F.: 1961, 'Forbidden Lines in the Ground Configurations with Special Regard to the Coronal Emission Spectrum', *Mém. Soc. Roy. Sci. Liège* **4**, 190.
Pecker, C. and Thomas, R. N.: 1962, 'Excitation of the Red and Green Coronal Lines', *Ann. Astrophys.* **25**, 100.
Pecker, C. and Thomas, R. N.: 'Some Limits on the Effect of Coronal Self-Emission on the Excitation States of Coronal Ions', *Astrophys. J.* **137**, 967.
Pottasch, S. R.: 1963, 'The Lower Solar Corona: Interpretation of the Ultraviolet Spectrum', *Astrophys. J.* **137**, 945.
Pounds, K. A., Willmore, A. P., Bowen, P. J., Norman, K., and Sanford, P. W.: 'Measurements of the Solar Spectrum in the Wavelength Band 4–14 Å, *Mon. Not. Roy. Astron. Soc.* (in press).
Prokof'eva, I. A.: 1964, 'Solar Corona in Active Regions in the Green and Red Lines', in *The Solar Corona* (ed. by J. W. Evans), Academic Press, New York and London.
Rense, W. A.: 1961, 'Solar Ultraviolet Photospheric and Coronal Lines', *Mém. Soc. Roy. Sci. Liège* **4**, 272.
Schatzman, E.: 1963, 'The Solar Corona', in *Space Age Astronomy* (ed. by A. J. Deutsch and W. B. Klemperer), Academic Press, New York and London.
Seaton, M. J.: 1962, 'The Temperature of the Solar Corona', *Observatory* **82**, 111.
Swarup, G., Kakinuma, T., Covington, A. E., Harvey, G. A., Mullaly, R. F., and Rome, J.: 1963, 'High-Resolution Studies of Ten Solar Active Regions at Wavelengths of 3–21 cm', *Astrophys. J.* **137**, 1251.
Waldmeier, M.: 1940, 'Comparative Observations on the Coronal Lines 5303, 5694, and 6374', *Z. Astrophys.* **20**, 172.
Yallop, B. D.: 1961, 'Coronal Line Widths and Solar Wind', *Observatory* **81**, 235.
Zirin, H.: 1959, 'Physical Conditions in Limb Flares and Active Prominences', *Astrophys. J.* **129**, 414.
Zirin, H., Hall, L. A., and Hinteregger, H. E.: 1962, 'Analysis of the Solar Emission Spectrum from 1300 to 250 Å as Observed in August 1961', *Third International Space Symposium.*

DISCUSSION ON THE CORONA

R. LÜST
Chairman
Max Planck Institut für Astrophysik, München, Germany

D. J. FAULKNER
Secretary
Mt. Stromlo, Canberra, Australia

M. DE GROOT
Secretary
Sterrewacht 'Sonnenborgh', Utrecht, The Netherlands

THOMAS. The equation used by Seaton for the emission in the UV lines actually has implicit in it two assumptions not mentioned by Seaton. The equation assumes that each collisional excitation results in a photon escaping from the atmosphere in that transition. This is true only if: (1) the atom is a two-level atom; i.e. the transition considered may be so represented; (2) either (a) no radiation field builds up in the line, or (b) if a radiation field builds up, the atmosphere is symmetric about its mid-plane so the same emission escapes from both top and bottom. Both these assumptions must be checked in detail. Charlotte PECKER and I had tried to investigate both in some circumstances (*Astrophys. J.* 1963) but the work is only exploratory. However one should realize that he makes these assumptions, when he asks the validity of his results.

POTTASCH. I have attempted to look into the effects of optical depth in some of the resonance lines as well as the effects of the use of a two level atom, on the resultant abundances. The conclusion I came to is that these effects will not introduce an error greater than a factor of two in the abundance. I agree with Dr. Thomas, however, that one cannot be absolutely certain of this conclusion as long as the temperature-density structure of the atmosphere is unknown.

ZIRIN. Dr. Hinteregger has now provided us with data on the XUV spectrum to test Thomas' remark; if there is a large flux in these lines they should be observable in the XUV. There does not, on first examination, appear to be enough flux to produce significant effects on the excitation of coronal lines. I wish to caution on the use of the analysis of Pottasch and of Ivanov-Holodnyi. These methods are keyed to a spherically symmetric atmosphere with smooth temperature variation. Both these calculations fail to explain the missing stages of ionization Ne I–VI, Mg III–VII, Si IV–VIII and Fe III–IX, which indicate the sharp temperature discontinuity, nor do they take into account the fact that the absolute intensity of the lower ionization lines (e.g. O III, C III) are really much weaker than a spherically symmetric model would indicate.

C. de Jager (ed.), The Solar Spectrum. All rights reserved.

SEATON. In interpreting ultra-violet spectra, Pottasch does not make any assumption of spherical symmetry, although a superficial reading of his paper might suggest the contrary.

I am not convinced about the significance of the "missing lines" of Zirin, but would not dare to comment further until I have looked into the matter more carefully.

POTTASCH. I agree with Dr. Seaton that no assumption of spherical symmetry has gone into my calculations. I have looked into the questions of ions missing between Si V and Si XI and Fe X to Fe XIV and would predict that these ions are not observed because their intensities are below the present observable limit. The strength of Fe XV is due to its very high f value.

UNDERHILL. How certain are the coronal motions observed by Dr. Billings?

BILLINGS. Plate dispersion, 2.5 Å/mm. A single displacement is difficult to distinguish from noise, but the average of 16 measurements rises a little above the noise. In any event the measurements give clear evidence of the smallness of the velocities present.

UNSÖLD. If the difference between ionization temperature and temperatures as determined from line widths is due to macroscopic motions, should these not decrease with height? Are there recent observations concerning that?

SEATON. The outer corona is still radiating energy, and for its temperature to be maintained it presumably requires heating by degradation of macroscopic energy. It does not seem to be evident that macroscopic energy is much less further out.

BILLINGS. The motions might be either damped or amplified with distance from the sun.

SCHATZMAN. (1) It is extremely difficult to carry energy in the corona; the refraction and dissipation is very important for longitudinal waves. It is only through magnetohydrodynamic waves, or perhaps through gravity waves, that the corona could be heated.

(2) The corona is filled with a weak magnetic field, and has a very low heat conductivity across the magnetic field. Very large temperature differences can exist for a long time and the corona certainly has a very heterogeneous structure.

MINNAERT. The line-widths of the coronal lines seem to depend clearly on the charge of the ions. This suggests that this line-width could be due to the fluctuating electrostatic fields in the corona, producing motions of the ions. Similar effects have been observed in plasma discharges during fusion experiments.

BIERMANN. The random electric fields in a plasma vary with typical frequencies of the order of the plasma frequency, that is $\approx 10^8$ sec^{-1} for coronal densities. This is much too fast to affect the ions. On the other hand, many experiments indicate that non-thermal conditions easily result in the acceleration of charged particles to energies of many times kT and other similar phenomena. The information derived from these experiments and related theoretical work may prove to be useful for some of the problems we discussed this morning.

ATHAY. Since coronal theories and observations seem to be approaching considerable refinement, perhaps we should pay more attention to the very lowest layers of the corona and the transition region from the corona into the denser, more energetic chromosphere. There is considerable evidence that the corona begins below the 5000 km level and that densities, at least locally, are a factor of 10 greater than those given by Seaton. This is of importance in all problems related to the build-up of radiation in the resonance lines of coronal ions.

A second problem of importance concerns the dynamics of the lower corona. The upward flow of mass in spicules at the 5000 km level is sufficient to replace the entire corona in about 15 minutes. If only 0.1% of the spicule mass escapes into the corona, then the effect of this on the corona may still be as large, or perhaps larger, than the transport effects already mentioned.

SMERD. The measured radio flux density from the whole sun can be expressed as an equivalent temperature, called the apparent disk temperature. The latter is defined as that mean brightness temperature across the circular optical disk which corresponds to the observed flux. The actual brightness at meter waves decreases from the center to well beyond the optical limb; the shape of the radio sun reflects the shape of the corona as observed optically. Even for the central ray the optical depth is only about one and decreases to the longer wavelengths. Apparent disk temperature spectra do not contain sufficient information to derive coronal densities and temperatures; such spectra can be calculated for model atmospheres, and if allowance is made for the observed departures from spherical symmetry they can be compared with the observations.

Such comparisons allow conclusions to be drawn on coronal densities and temperatures; the derived estimates are not very satisfactory since different N and T combinations can account for the same disk temperature. In any case, observed disk temperatures at meter waves should not be identified with a mean kinetic electron temperature in the corona.

THOMAS. This is the next-to-last day of the symposium. I would like to see some exchange of opinion on current assessment of a model of the outer atmosphere. Use the following as a starting point. There is one plateau only, $2.0 \times 10^4 < T < 8 \times 10^4$, and an abrupt rise thereafter. The questions are: (a) reality and extend and the character of the plateau; (b) kind of fluctuations that Seaton implied about the 1.5×10^6 value. (c) departures for spherical symmetry beginning about at 1000–15000 km.

Athay and I have originally proposed a second plateau of about 200000° but work by Charlotte Pecker has shown convincingly that this second plateau cannot exist.

SEATON. Ultra-violet and radio spectra have been interpreted in terms of spherically symmetric models by Allen, by Ivanov-Holodnyi and Nikolskii, by

Oster and by Pottasch. This work seems to indicate that the transition region does not have any very sharp jump in temperature.

POTTASCH. If one assumes spherical symmetry, one may compute a model of the transition region using the ultra-violet data (which is essentially equivalent to the radio data), the continuum observations (λ 4700) as a function of height, and the assumption of hydrostatic equilibrium. The resultant model shows a slow temperature gradient which increases and becomes very steep between 50000° and 150000°K. In fact this temperature range occurs within a height of less than 100 km. This steep gradient occurs at about 3000 km or 4000 km above the limb. The temperature gradient becomes less above this height and the temperature reaches 10^6 °K at about 10000 km. Please notice that such a model cannot explain the Ca II observations which apparently extend above a height of 12000 km. This might argue for a non-symmetric model.

THOMAS. If one uses the observed mean values of the electron density as a function of height, then our calculations of $T_e(h)$ based on hydrostatic equilibrium seem to give too small a $\mathrm{d}T_e/\mathrm{d}h$ relative to the requirements of UV results.

DE JAGER. Anyway it is possible to construct models of the transition layer photosphere-chromosphere that are in hydrostatic equilibrium and that have a steep temperature gradient. The steepness of the gradient depends on the values of the energy input (dissipation) and output (conduction and radiation). The model constructed by Kuperus and me is in hydrostatic equilibrium and has a very steep T-gradient.

BILLINGS. I am commenting on Pottasch's statement that his model corona decreased rather gradually with temperatures towards the photosphere. In such a model there would be a distinct layer of very intense 6374 emission – something that is not observed.

ZIRIN. Our model differs from Thomas' in one respect; there is no high temperature plateau above the cool, spherically symmetric chromosphere, but the temperature rapidly rises to coronal values. Above 1500 km we have only corona and spicules.

There are several observational tests of the presence of such a plateau. Suppose Ne $= 10^{11}$ and $T = 30000°$ there.

(1) Radio data. The central temperature of the sun at 8.6 mm is 8000° and at 3 cm; 13000°. This places a limit on the thickness of this layer, which turns out about 10 km.

(2) Lyman continuum intensity and slope. These both give a temperature around 8000°, placing another limit on the thickness.

(3) 10830 Å line of He I. If the layer were 1000 km thick, we should find it contributing to the 10830 absorption to be concentrated in the spicule bushes up to the limb.

(4) Eclipse measurements of continuum intensity place an upper limit to the electron density at 3000 km around 5×10^{10}, but better data is needed.

(5) Such a plateau would produce 10 or 100 times to great an intensity in all

the lines of low ionization in the XUV, such as O III, C III, etc. If the high temperature plateau exists, why don't we see it?

HINTEREGGER (to Pottasch). Is it correct that your study too suggests a very thin transition region between coronal and lower chromospheric temperatures, say about 100 km?

POTTASCH. Assuming spherical symmetry the ultra-violet data give the integral of the square of the electron density over a given height range, the height range being known by the temperature at each extreme. If we assign an average height to the temperature region we can obtain the electron density for this height from the continuum measurements. Then we obtain a height range which is of the order of 100 km for the region from 50000°K to 150000°K.

ATHAY. Models of the chromosphere derived independently by myself, by Moriyama and by Hiei all have one feature in common, *viz.* the chromospheric line radiation above about 1500 km arises entirely in spicules. This requires that spicule electron densities are about 10^{12} cm^{-3}. Chromospheric continuum data, λ 10830 data and radio data then lead to the conclusion that interspicule electron densities are of the order 10^{10} cm^{-3} or less. My own model predicts that the corona begins, in interspicule regions, at a height of about 3000 km.

THOMAS. Your model gives $T_e \lesssim 50000°$ in spicules. Do I understand then that all ions requiring $T_e \gtrsim 50000°$ to exist will in your model come from the region between spicules? Since spicules occupy only 0.1% of the surface area of the sun, this in effect implies a spherically symmetric atmosphere for lines coming from the interspicular region.

JEFFERIES. Could Athay indicate what fraction of the volume at (say) 1500 km is occupied by spicules? Also could he comment on the fact that his model implies the existence of very great horizontal pressure gradients?

ATHAY. The models referred to give only the relative number of spicules as a function of height and not the height variations of individual spicules.

The models do have a pressure differential from spicule to interspicule regions. This results from the high spicule density and cannot be removed by juggling the temperature structure.

POTTASCH. I would like to clarify my statement that the temperature range from 50000 °K to 150000 °K is contained in a 100 km height. This refers only to a spherically symmetric model. If one uses a model such as described by Dr. Athay, one can no longer use the electron densities obtained from the continuum as before. The density in the interspicule region will now be lower and the height range will be increased, e.g. if the density is lower by a factor of 5, the height range will be increased by a factor of 25.

VAN DE HULST. I should like to raise one question which may be important, but which has not been mentioned at all in the review of the coronal emission spectrum. Is there time enough to establish the ionization equilibrium before the solar wind carries the gas away? I make this suggestion, because Neupert and Lindsay reported at the last COSPAR Symposium (Warsaw, June 1963) that the lack of time to reach equilibrium could explain some features of the observed rocket ultra-violet spectrum. And even before that ALFVÉN (1941) used a crude theory in which there was no recombination at all and the degree of ionization was exclusively determined by the available time.

Qualitatively, this suggestion means that the ionization of gas coming from the sun is less advanced than it would be at the same distance by the equilibrium theory. Hence, the electron temperature derived from ionization ratios in the usual manner must be below the actual electron temperature, which follows e.g. from line width. This difference is in the sense as observed.

Since I do not have Neupert's detailed data, here are a few estimates of my own which may help to assess whether this effect stands a chance of being important. Most observations refer to the range $r = 1.05$ to $r = 1.20$. Electron densities there are of the order of $N_e = 10^8$ cm^{-3}. We assume that the electron temperature is 10^6 °K. Using Šklovskij's older coefficients which do not differ strongly from Seaton's new ones, I find ionization rate from Fe X to Fe XI: $N_e\Phi_{10} = 10^{-2.6}$ (once in 400 sec); and recombination rate from Fe XI to Fe X: $N_e\Phi_{11} = 10^{-3.5}$ (once in 3200 sec), which would give a ratio Fe X to Fe XI of 1 to 8 in equilibrium. Take a solar wind with velocity 400 km/sec, density $N_p =$ 10 cm^{-3}, both measured at the earth. The continuity requires that $N_p r^2 \cdot v =$ constant. So $(1/200^2) \cdot (10^8/10) \cdot (v/400) = 1$, which gives $v = 1.6$ km/sec in the lower corona, so that the distance from $r = 1.05$ to $r = 1.20$, or 100000 km, is travelled in 60000 sec. This looks comfortably long compared with the reaction times quoted, but a number of factors brings the estimates closer together, namely

(1) There have to be 10 or more successive ionizations while the gas moves from chromosphere to corona.

(2) If injection occurs mainly in small streams (above spicules) or at certain time intervals, the injection velocities are larger and the travel times shorter.

I wonder if somebody has given thought to these problems.

BIERMANN. The time dependence of the ionization under coronal conditions has effectively been described in relation to plasma experiments in the same temperature range, first by Knorr in our institute; if I remember correctly his results can be applied without difficulty to the corona. More recent work along the same line has been done, I believe in Harwell.

MAXWELL. Radar observations of the solar corona, made at a frequency of 38 Mc/s at the M.I.T. Field Station at El Campo, Texas, show a Doppler spread on the echoes $> \pm 20$ kc/s, with more energy being returned at positive Doppler frequencies. The appropriate echoing height above the photosphere is approximately 350000 km/sec. The Doppler curve peaks at about $+4$ kc/s, corresponding to a net outward streaming of electrons at velocities of the order of 16 km/sec at a height of 350000 km.

LÜST. Could you also comment on the determination of the coronal temperature by these radar measurements?

MAXWELL. The solar radar observations give good information about echoing heights and the motion of irregularities in the corona, but only limited information on coronal temperatures. Expansion velocities at a height of 350000 km, as predicted by Parker's solution of the hydrodynamic equation and the equation of continuity for an isothermal corona, have been computed for coronal temperatures of 1.0 and 2.0×10^6 °K (Maxwell, Defouw, and Cummings). The predicted velocity for $T = 10^6$ °K is about 4 km/sec, and for $T = 2.0 \times 10^6$ °K about 70 km/sec. The radar observations indicate an outward streaming velocity of about 16 km/sec at height 350000 km and hence suggest coronal temperatures of about 10^6 °K. (It is realized, of course, that the corona is neither isotropic nor isothermal and that such an analysis is, at best, only an approximation.

MINNAERT. Dr. Hinteregger asked about observations which could be made. I should like to call again your attention to the importance of the spectrum between 1800 Å and 3000 Å. There must be numerous emission lines in this region, emitted by the transition regions between the chromosphere and the corona, and some of the missing ions might be found there. In ordinary circumstances these lines are veiled by the underlying continuum; but a rocket during an eclipse might have a good chance.

PECKER. We could fly a coronagraph.

ZIRIN. We have been in an embarassing position in comparing the XUV and visual observations of coronal lines. In the XUV we have identified only the lines of Fe XV and Fe XVI, which are very strong. But in the visual range the strongest lines are those of Fe XIV and Fe XIII. It is possible that the observed line 7059 of Fe XV is abnormally weak because it is in a metastable state. The cross-section to exite such a line might be low. In this case we would have a very broad distribution of ions, from Fe X to Fe XVI, but we still would have the anomaly of absence of the Fe ions below Fe XV.

Hinteregger's new spectra have changed the position somewhat. We have identified some weaker multiplets of Fe X, XI and XIV, using Edlén's measurements. These lines, below 100 Å, are not weaker than those of Fe XV and XVI. We have identified a few moderately strong lines near 200 Å with Fe XIII and XIV, using the known splittings. If the strong lines between 170 and 210 Å turn out to be Fe X to XIV, all will be well.

SEATON. Mr. R. Harman, at University College London, is calculating the energy level structure of ions such as Fe X, XI, XII and XIII using a theory similar to that of Layzer.

Dr. Burgess has made progress in obtaining reasonably simple approximations for the calculation of cross-sections for intercombination.

BIERMANN. In view of the possibility, mentioned this morning, that the oscillation frequency of the propagating waves heating the corona is relatively high ($> 10^{-1}\,sec^{-1}$) I would like to draw your attention to a mechanism proposed by Schlüter and myself at the Cambridge Symposium on Cosmical Gas Dynamics, this is the gyrorelaxation or magnetic pumping.

It has been shown then, that under coronal conditions this mechanism, which I shall not try to describe here, might be quite effective.

In view of the importance of the question whether oscillations in this frequency range, with amplitudes of around 20 km/sec, are present in the corona, it would seem to be desirable to make an effort for observing these oscillations if this is at all possible.

DE JAGER. I have four questions related to the heating of the corona. We know that the chromosphere has an inhomogeneous structure, showing the mottling and the chromospheric network; Leighton has showed (AAS meeting Tucson, April 1963) that the bright Ca II structures are related to a magnetic field ($\approx$ 5 gauss). Apparently the chromospheric heating is strongly dependent on the magnetic field; for the coronal heating above these magnetic parts we have to take into account that the mechanical energy propagates in the form of magnetohydrodynamic waves. Above the quiet parts these waves would rather have the character of shock waves or pressure waves.

(1) What is the mechanism for the heating above the non-magnetic parts of the chromosphere?

(2) If the heating is inhomogeneous, would the inhomogeneous structure be visible in X-ray heliographs? (Thereby, however, taking into account that the scale height for solar X-ray emission is not longer than about 25000 km.)

(3) What would be the influence of the high thermal diffusion in the corona on the problem put in question 2?

(4) Osterbrock showed that in the magnetic parts Alfvén waves are reflected downwards already near an altitude of about 2000 km. Should we hence conclude that there is no efficient mechanism for the heating of the magnetic parts of the corona?

LÜST. In my opinion the influence of the magnetic field goes in the right direction. Of course it is true that transverse waves cannot propagate into the chromosphere and corona. But the normal longitudinal (quasi-acoustic) waves can easily propagate along the lines of force. These waves are even guided by the magnetic fields and therefore more mechanical energy should be transported along this fields into the region where the fields are stronger. In addition the noise production might be even higher in the photosphere if the magnetic field there is strong enough. But the main part of the energy will go into the quasi acoustic waves.

DE JAGER. It seems that we may conclude that we should have shockwave heating in the non-magnetic parts and heating by longitudinal (shockwave-like) waves in the magnetic parts. The mechanical flux is greater in the magnetic parts than in the non-magnetic parts (shown by Kulsrud). Only a part of the flux in the magnetic regions is not reflected; still the flux available for the heating seems larger in these regions than in the non-magnetic parts.

Anyway, since everywhere the waves have the character of pressure waves, a theory for coronal heating should be based on the propagation and dissipation of *pressure waves* rather than *Alfvén waves.*

Indeed Kuperus, basing himself on a coronal heating by pressure waves, could predict in a preliminary theory, that the corona above the magnetic regions should be considerably denser, and slightly hotter than above the non-magnetic regions.

LÜST. The main dissipation mechanism in the transition zone between the chromosphere and corona would be by shock waves. Further out – in the region where the collision frequency is of the same order as the frequency of the waves – the gyro-relaxation effect will contribute as mentioned by Biermann. In addition I should mention that on the laboratory scale it looks that there are even more dissipation mechanisms which we do not understand. But we may hope that further experiments will also help us in this respect for the understanding of the corona.

DE JAGER. What is the relative importance of heating by pressure waves and by gravitational waves?

BIERMANN. Unfortunately I am not yet familiar with Whittaker's work; from what I recall from my own, gravitational waves propagate easily horizontally and should certainly be taken into account in any detailed picture, but they seem to be in a disadvantage as regards vertical propagation.

VAN DE HULST. Is not a gravitational wave traveling in vertical direction identical to a pressure wave?

LÜST. I would not make a sharp distinction between gravitational and sound waves. Below the critical frequency one obtains gravity waves, above it sound waves will propagate. The gravitational waves are propagating in the horizontal direction and they are damped in the vertical direction.

HINTEREGGER. The meaning of the quantity of a so-called "temperature" deduced from the charged particle analyzer measurements conducted on the Venus probe is not necessarily that of an actual temperature in the ordinary sense. Since the analyzer had a highly directional response and since only measurements looking directly along the solar vector were made, one can actually speak only about a certain observed spread in the kinetic energy associated with the radial component of velocity. Only by adding the assumptions of (a) vanishing non-radial components of mass velocities and (b) Maxwellian distribution of velocities in a co-moving reference system, can one rightly speak of temperature. I believe that more defined measurements will be needed before we may be sure about the actual meaning of what is simply called "temperature" of the solar wind.

ZIRIN. We must always remember that the corona is a creature of solar activity; i.e. that most of the corona is produced in regions of magnetic fields (except for the weak corona of sunspot minimum). Thus we must have a mechanism that works in the presence of magnetic fields.

DE JAGER. As a matter of fact also the quiet minimum corona exists (compare e.g. the one of June 30, 1954). Hence even in the absence of all solar activity there must be an outflowing stream of mechanical energy. Even in that case however, the greatest source of the heating may be the (magnetic) chromospheric network. But also the non-magnetic parts lead to a certain heating. Kuperus and I predicted for a corona without any magnetic field a temperature of 6.5×10^5 °K and a density at its basis of $10^{8.8}$ electrons per cm^3.

KUPERUS. About the heating of the corona above the Ca^+ network as has been suggested by de Jager, I would like to draw attention on the very strong dependence of the generation of acoustic noise on the equipartition between turbulent magnetic energy and the energy in the turbulent motions in the hydrogen convection zone as had been shown by Kulsrud (1955). It seems to me that the observed magnetic fields in that network are far too small to expect sufficiently strong turbulent magnetic fieldstrengths for a reasonable mechanical energy increase.

A problem in heating the corona by longitudinal waves is the refraction by the increase in the propagation velocity (Osterbrock, 1961). However one should remind that Osterbrock's calculations are made for a homogeneous magnetic field. If however the field decreases with height, for instance as $H^2(z)/8\pi = N(z)kT(z)$, there would be no marked difference in the propagation of longitudinal "fast" magnetohydrodynamical disturbances and ordinary sound waves. So from Osterbrock's calculations one cannot reject the shockwave heating of the corona.

CHUBB. The following remarks concern the permanence of coronal structure and are related to the question as to whether enough time exists for achievement of thermal equilibrium. Coronal structures have been found to exist for at least one solar rotation by Frank McDonald of NASA. He measured solar protons trapped in plasma clouds as observed in the Explorer XIV satellite. These high energy protons in essence, tagged the coronal structure from which the magnetic storm cloud originated. The interesting point is that 27 days after the first magnetic storm cloud containing high energy protons arrived at the earth, there occurred a second magnetic storm also containing high energy protons. By this time the sunspot group which had originally produced these protons had long since disappeared and solar conditions were quiet. The presumption is that a coronal structure which acted as a source of interplanetary gas and which also served as a reservoir of flare protons had persisted for 27 days and had partaken of the rotation of the sun.

J. C. PERCHE ET J. P. DUMONT

Etude des raies λλ 3388Å, 10747Å et 10798Å dans la couronne solaire

J. Firor et H. Zirin ont déjà publié une étude sur les raies infrarouges du Fe XIII dans certaines conditions. Nous-mêmes avons résolu les équations de l'équilibre statistique dans des conditions différentes en utilisant les paramètres suivants: (1) le facteur de dilution W, (2) la température électronique T_e, (3) la densité électronique N_e. Seuls les cinq premiers niveaux d'énergie du Fe XIII ont été considérés, les autres niveaux n'étant pas encore connus.

Les résultats montrent que pour une distance inférieure à 6′ d'arc ($W > 0.15$ environ) et une densité supérieure à $2.5 \cdot 10^7$ cm^{-3}, le rapport $I(10747)/I(10798)$ varie beaucoup plus avec N_e qu'avec les autres paramètres W et T_e.

Expérimentalement nous avons déterminé la valeur du rapport $I(10747)/I(10798)$ en différents points de la Couronne. Les observations sont faites avec le coronographe de l'Observatoire du Pic du Midi.

Il nous a semblé qu'actuellement le photomultiplicateur est le meilleur moyen pour obtenir le rapport avec une précision suffisante. Nous utilisons donc une installation mise au point par G. Wlérick et nous-mêmes en 1960; le récepteur est un photomultiplicateur Lallemand à 20 étages à couche du type Ag-O-Cs refroidie. Nous avons obtenu la valeur du rapport en fonction de la position héliographique à une distance de 1′ du bord pour une zone active; chaque point est la moyenne de six passages; l'erreur quadratique moyenne vaut environ 0.3. Pour une variation de 6° de l'angle de position, le rapport varie de 3.58 ± 0.35 à $2.13 \pm 0,3$.

Il est donc possible de déterminer la variation de la densité électronique "effective" de la zone considérée. Des calculs pour différents modèles sont en cours, cependant nous pouvons déjà dire que cette méthode semble spatialement plus sélective que la méthode de détermination de la densité électronique à partir de la Couronne K.

La matière de cette communication sera publiée dans les: *Proceedings of the 12th colloquium of Liège 1963* (Infrared Astronomical Spectra).*

WLÉRICK. Rather than to translate the paper of J. C. Perche and J. P. Dumont, I should like to make the following comments:

1. The study of the two infrared lines of Fe XIII is interesting for at least three reasons: (a) The line λ 10747 Å is the strongest coronal line of the wavelength interval λλ 3000–12000 Å; (b) This line can hence be observed up to about 0.5 $R_\odot$; this is in the domain of radio-observations on meter wavelength;

* *Note à la lecture des épreuves.* Ce travail a été complété par des calculs basés sur les résultats théoriques de P. Charvin (communication présentée à la conférence de Nijenrode, Juillet 1963); ces calculs permettent: d'une part, de se rendre compte de l'importance de l'anisotropie du champ de rayonnement sur le rapport d'intensité des raises du Fer XIII et d'autre part d'obtenir les taux de polarisation de ces 3 raises. Ces résultats feront l'objet d'une note aux comptes-rendu de l'Académie des Sciences.

(c) The ratio of the intensities of the two lines, λ 10747 Å and λ 10798 Å is strongly density-dependent as first suggested by Firor and Zirin and as seen in the paper.

2. This density dependence is important for two reasons: (a) It can be checked by comparison of electron density measurements determined by observations of the K-corona such as those made by the K-coronameter or other photoelectric polarimeters; (b) Close to the limb, it is less affected by integration along the line of sight than measurements of the electron corona.

3. The computations of Dumont and Perche were made with the available elements at the present time: they took into account the five lowest energy levels of the ion. Even for these levels, they had to make assumptions concerning the cross-sections for excitation by electron collisions. By comparing their results with those of J. Firor and H. Zirin and with those of V. Kurt, one sees that an exact knowledge of the cross-sections is important to improve the quantitative relation between the electron density and the intensity ratio of the two lines. The excitations by proton collision, introduced recently by Seaton for Fe XIV, were neglected.

Similarly the upper energy levels of the ion were neglected. And, from the work done on Fe XIV by Ch. Pecker-Wimel, R. Thomas and Garstang, one knows that these levels may play an important role. Thus presently, only the trend of the density dependance is known.

4. Experimentally the use of an excellent Lallemand infrared photomultiplier seems the most accurate way to determine the line intensity ratio. It has the drawback to give only one information at a time, but the paper shows that the precision is high enough to give the "effective density" with an already useful accuracy.

In the future, improvements of the image converter technique or the use of an infrared electron camera may permit a high information rate.

5. The fact that experimentally precise measurements are now possible leads me to turn myself toward those that could strongly help us by improving our knowledge of the Fe XIII ion. I think particularly to laboratories like those of Edlén, Garstang, Seaton and van Regemorter.

6. Finally, I would like to mention that an attempt to detect coronal lines at wavelengths between 1.2 and 2.2 microns will be made in the near future. You remember that, in his evening lecture, Professor Minnaert quoted the lack of informations concerning the coronal lines in the near infrared.

ORRALL. Have you also measured λ 3388 which arises from the same configuration? It is perhaps more sensitive to the electron density.

WLÉRICK. Indeed the ratio $I(10747)/I(3388)$ varies more strongly with electron density than the ratio $I(10747)/I(10798)$. However, the measurements have been restricted to the two infrared lines because with a chromatic coronagraph it is quite difficult to compare intensities of lines of so different a wave length.

M. BRETZ. Was this solar region taken for some special reason?

WLÉRICK. The region was chosen because it had a reasonable green line emission. The line 10747 Å can be detected at nearly all position angles, while the line 10798 Å is quite faint, except in condensations. For this reason the measurement of the ratio of intensities was restricted to a region where the 10798 Å line was strong enough.

SEATON. First, I would like to congratulate Mr. Perche on his work. I would also note that somewhat similar work has been done in the USSR by Kurt. The remarkable observational work which can be done using modern techniques adds the incentive needed for the making of good cross-section calculations. Such calculations can be made and now must be made.

R. H. GARSTANG

Comments on Spectral Line Intensities

In his introductory talk Professor Seaton mentioned some calculations on permitted lines of Fe XIV (GARSTANG, 1962a). A few further remarks on this subject would be in order. The calculations were performed to try to answer a question originally posed by Dr. Charlotte Pecker: Why is the oscillator strength of the $3s^2 3p$–$3s 3p^2$ transition array relatively small, if it is the main array from the ground configuration? The answer turned out to be that the array $3s^2 3p$–$3s^2 3d$ is also a strong array. The new feature of the calculations was that configuration interaction between the $3s3p^2\ {}^2D$ and the $3s^2 3d\ {}^2D$ terms were included, as well as the effects of spin-orbit interaction. These effects may make substantial differences to calculated oscillator strengths, often a factor 2, and in a few extreme cases much more. For example the $3s^2 3p\ {}^2P$–$3s3p^2\ {}^2S$ doublet ratio is greatly changed from its LS-coupling ratio of 2 : 1 to 1 : 23.

Although such a large effect is exceptional, caution should be exercised when using intensity data as an aid in the identification of ultra-violet lines. Similar remarks would apply to other coronal ions. For example, in Fe XIII the interacting configurations would be $3s 3p^3$ and $3s^2 3p 3d$.

Professor Menzel mentioned some new tables which are in preparation. These are intended to facilitate elaborate intermediate coupling calculations in complex atoms. Racah techniques are being used to separate the strengths of multiplets when there are several terms of a kind. For example in the array d^3–$d^2 p$ there are two ${}^2D$ terms in the d^3 configuration and 3 ${}^2D$ terms in the $d^2 p$ configuration, and the older tables gave the sum of the strengths of the six multiplets. The new techniques enable six separate line strengths to be obtained. It is important to note however that these separate quantities are *not* the observable quantities and may not normally be used in estimating LS-coupling oscillator strengths. This is because the six quantities are computed in a representation in which the energy is not even approximately diagonal. A numerical example of this was recently given for some forbidden transitions in Fe II

(GARSTANG 1962b, Table II); similar results would be expected for allowed transitions. Factors of 3 or 4 can easily occur between the tabular values in a non-diagonal representation and the true oscillator strengths after diagonalisation of the energy. This difficulty only arises when there are several terms of a kind in a configuration; the effect is of course very similar to that illustrated above for Fe XIV when there are interacting configurations.

Finally, I wish to refer to some recent work at the University of London Observatory. Corliss and Warner have recently measured the oscillator strengths of a large number of faint lines in the Fe I spectrum, and combining their results with older work have compiled a reasonably homogeneous list of 2000 oscillator strengths between 3100 Å and 9900 Å. They have been reduced to the earlier absolute scale of Corliss and Bozman. A recent paper by ALLEN and CORLISS (1963) indicates that the absolute scale is accurate to within a factor of 60 per cent, and may be appreciably better than this. Using this data Warner has rediscussed the solar curve of growth for neutral iron. The spectrum was divided into four wavelength regions. For each region curves of growth were constructed. He has discussed the variation of damping constant and of turbulent velocity with depth in the solar atmosphere. The new oscillator strengths for faint lines enables the curve of growth to be extended on to the faint-line portion and this materially improves the fitting of observed and theoretical curves. This work is of a preliminary nature, and a much more detailed study is needed using refined methods. But one thing seems certain, we should be wary of attributing to unknown physical processes effects which may be due to inadequate oscillator strengths. Warner also finds some evidence that some f-values for Fe I may be as large as 10, and possibly greater. This conclusion needs further checking before it can be accepted without serious reservations.

References

ALLEN, C. W. and CORLISS, C. H.: 1963, *Mon. Not. Roy. Astron. Soc.* **126**, 37.
GARSTANG, R. H.: 1962a, *Ann. Astrophys.* **25**, 109.
GARSTANG, R. H.: 1962b, *Comm. Univ. London Obs.* **57**.

PECKER. What were the exitation potentials of the faint ion lines?

GARSTANG. They range from 1 to 5 volts. Most of those in the blue spectral region are in the 3 to 4 volt range, for longer wavelengths the largest number are in the 4 to 5 volt range.

SEATON. I understand that there are some difficulties in interpreting the work of Corliss and Warner, in particular some of the f-values are very large. I wonder what confidence Dr. Garstang would have in this work.

GARSTANG. The reliability of oscillator strengths of lines between highly excited states is difficult to assess. There is no difficulty in having oscillator strengths in a one-electron spectrum exceeding unity; this happens in hydrogen for the $3d$–$4f$ transition. And on theoretical grounds one would expect the oscillator strengths between states with a given difference of principal quantum numbers to increase with the principal quantum number (e.g. for a transition $nl \rightarrow n + 1, l + 1$ the oscillator strength is proportional to n). Furthermore the total oscillator strength for all transitions between n and $n + \Delta n$, for fixed $\Delta n\,(\neq 0)$, is proportional to n^3, but of course the number of possible transitions increases as n^2, so that individual transitions are not likely to increase more than about proportional to n. The sum-rule is perhaps one of the strongest arguments against very high *f*-values in Fe I; the principal quantum numbers do not exceed four except in a few instances, so that the above arguments also do not favour large *f*-values. Whatever view one takes one should regard the Corliss, Bozman and Warner *f*-values for the higher excitation lines as being more uncertain than their values for lower-excitation lines. This increased uncertainty is due to a systematic effect, rather than to an increase in the random errors of individual lines. It affects lines with upper excitation potentials exceeding about 6 volts.

ZIRIN. Could you estimate the oscillator strength of $3s^2 3p \rightarrow 3s^2 3d$ in Fe XIV, and for $3s^2 3p^2 \rightarrow 3s^3 3p 3d$ in Fe-XIII?

GARSTANG. I must refer you to the published paper for the exact numbers. But the $3p$–$3d$ transitions in Fe XIV certainly have appreciably larger oscillator strengths than the $3s^2 3p$–$3s\,3p^2$ transitions. The position is somewhat analogous to the Wigner-Kirkwood rule for hydrogen, which says that Σf for p–ns transitions is much smaller than Σf for p–nd transitions. Similar remarks of course apply to the other coronal ions.

MÜLLER. I am very happy to learn that at last someone has determined the *f*-values for a number of faint iron lines. The photospheric abundance of iron was always uncertain due to the fact that only the intermediate position of the empirical curve of growth was known. The use of faint lines certainly will facilitate the fitting of the theoretical curve of growth into the empirical one. As to the factor of 10 by which, according to Dr. Garstang, the iron abundance might be raised when employing faint higher excited lines I would be sceptical until someone else has measured independently the *f*-values of the same faint high excitation lines. We have found for some other elements discrepancies in the Corliss and Bozman *f*-values between lower and higher level lines. We must first be quite sure that there are no systematic differences in the *f*-values between lower and higher excitation lines before we interpret the discrepant results in the photospheric Fe abundance by non-LTE or other effects.

GARSTANG. I entirely agree that further checks on the oscillator strengths are needed.

UNSÖLD. In the Institut für Experimentalphysik, Kiel (Prof. Lochte-Holtgreven) extensive measurements of relative *f*-values in Fe I, especially fainter lines, are under way. The lines are checked carefully for blending Fe lines of different excitation.

GARSTANG. Are some often fainter than King's lines?

POTTASCH. In a recent paper in the *Monthly Notices* on the transition probabilities in Fe II you discussed the possibility that several unidentified solar lines might be due to Fe II. You concluded however that with the Goldberg, Müller and Aller iron abundance these lines would be too weak to be observed. What would be the situation with an increased value of the iron abundance?

GARSTANG. When I was in Washington in 1961 Mrs. Sitterly and I examined her solar spectrum ledgers for wavelength coincidences with forbidden Fe II lines. For many lines alternative identifications could be made; for some lines no other identifications could be suggested. My calculations based on faint line theory indicated equivalent widths of the order of 0.1 mÅ, assuming the Goldberg, Müller and Aller value of the iron abundances. Taking both wavelength coincidences and intensities into consideration I felt that forbidden Fe II should be considered absent from the solar spectrum. If the abundance of iron is really 10 or more times the Goldberg, Müller, and Aller value then my predicted equivalent widths would be 1 or 2 mÅ for several of the unidentified lines. These equivalent widths are comparable with the strengths of the lines in the Utrecht Catalogue, and the lines may well be due to forbidden Fe II. It would be premature to conclude that the forbidden lines prove that the increased abundance would certainly lead to contradiction.

F. Q. ORRALL

The Spectrum of Sporadic Coronal Condensations

In their introductory lectures, both Dr. Seaton and Dr. Billings pointed out the difficulties in the interpretation of coronal spectra that arise because we see only the integrated emission over the line-of-sight, so that we may be seeing emission from several very different regions superposed. But in sporadic condensations, the emission comes from relatively small, well defined regions of the corona, so that this difficulty is greatly reduced. It should be possible to use observed intensities to construct a rough model of these condensations.

From the pioneer visual observations of Waldmeier (confirmed by photography through birefringent filters), we know that sporadic condensations form in permanent condensations and often show a loop-like internal structure. Sometimes loop prominences form in sporadic condensations and some individual loops may be seen both in Hα and in the red and green coronal lines. The yellow line is always present in these condensations. J. T. Jefferies and I have collected graded height spectra (λ 5100–6800) in a number of sporadic condensations containing loop prominences and have studied the event of February 19, 1962. All of the individual loops of this event visible in Hα were also visible in λ 5303 Fe XIV, λ 6374 Fe X, λ 5411 He II, λ 6678 He I, D lines and the b group as well as the continuum. There were some loops visible in λ 5303 that showed almost no prominence emission and there were other threads visible only in λ 5694 and λ 5445 of Ca XV.

The permanent condensation had a temperature of 2×10^6 °K judging from the appearance of the yellow line everywhere, and a density of $\approx 2 \times 10^9$ cm^{-3} inferred from the measured electron scattering continuum. The prominence continuum gave an electron density of $\approx 10^{11}$ cm^{-3} for the Hα emitting region. We have interpreted the Balmer continuum of a number of loop prominences by Zanstra's method and always find temperatures $\approx 2 \times 10^4$ °K and we shall assume that this prominence was at this temperature also. Thus in the Hα emitting regions, the product $N_e T_e \approx 2 \times 10^{15}$ °K cm^{-3} while in the permanent condensation this product is $\approx 4 \times 10^{15}$ °K cm^{-3}. This suggests rough pressure equilibrium between the prominence and the rest of the condensation, as Zanstra has observed in ordinary prominences (but rather surprising in loop prominences which are thought to be compressed by the magnetic field). Let us assume that $N_e T_e = 4 \times 10^{15}$ °K cm^{-3} everywhere in the condensation, so that given the temperature we can find the density. But what is the temperature of the regions that emit the red, green and yellow loops? Dr. Seaton's curves of coronal ionization *versus* temperature show rather strong maxima. It is thus reasonable to assume that most of the emission of a given ion comes from regions where the temperature is that of the ionization maximum. This temperature is readily found from Dr. Seaton's calculations and is shown in the table below for the three lines. Having the electron temperature and density we may compute the emission using the excitation calculations of Ch. Pecker and Thomas, the ionization calculations of Seaton and the "coronal values" of the abundance of the elements. The resulting emission ε in erg cm^{-3} sec^{-1} steradian^{-1} is also shown in the table, along with the intensity I (erg cm^{-2} sec^{-1} steradian^{-1}) measured in the brighter loops. Now we may compute a length L for the thickness of the emitting region since $I = \varepsilon L$. This is also given in the Table:

λ	T	ε	I	L
6374	0.45×10^6 °K	120×10^{-8}	2.5×10^2	1.5×10^8 cm
5303	1.0×10^6	65×10^{-8}	10×10^2	15×10^8
5694	2.4×10^6	0.8×10^{-8}	3×10^2	370×10^8

These lengths suggest the following model for the loops of a sporadic condensation: the length L for λ 5303 is about equal to the observed thickness of the green line loops. Imbedded in some of these thick λ 5303 loops are bundles of fine Hα prominence loops having individual diameters of 2×10^8 cm (Such bundles of fine Hα loops are often seen with good seeing). The red line is formed in the transition region between the Hα emitting region and the surrounding λ 5303 region. This transition takes place abruptly as indicated by the small value of L for λ 6374.

The value of L for the yellow line is too large. The assumption of pressure equality between the bright threads of yellow line emission and the surrounding coronal condensation must be wrong (unless the Ca abundance is very much higher). We believe that the yellow line loops are in fact more dense (10^{10} cm^{-3} for the electron density would give these regions their observed width of $\approx 1.5 \times 10^9$ cm), and at a higher pressure than the rest of the sporadic condensation. We suggest that they are the first stage in the process of compression and cooling required to produce an Hα loop from coronal material.

LÜST. What was the geometry in observing this feature? Did you look through only one feature in which the Fe XIV line originated?

ZIRIN. I wonder if by your remark you refer to the time average? In each loop, material is rapidly condensing from the corona and dropping to the surface. Consequently at any time we see a time average for many loops. Those stages which are stable will be bright, and those in which cooling is fast will only be present in a few loops at any one time and thus be weak.

ORRALL. By a rapid transition from the corona to prominence I mean the transition in space. Of course the exact time required for the transition would also be of great interest.

SCHATZMAN. What is the life time of the condensation?

ORRALL. At least four hours, but it may have been longer.

SESSION V

PARTICLE AND RADIO EMMISSION FROM THE SUN

PARTICLE AND RADIO EMISSION FROM THE SUN

E. SCHATZMAN
Institut d'Astrophysique, Paris, France

Introduction

Abstract. The problem of particle and radio emission from the sun is very intricate, and this report is an attempt in organizing the observed facts around the smallest possible number of physical ideas. No attempt has been made for completeness, and there is no doubt that this report has been distorted by the attempts of the author to understand the nature of the high energy phenomena taking place at the surface of the sun.

The report is divided in 3 sections: (1) Effects of high energy particles moving through the matter, (a) Elementary processes, (b) Wave phenomena; (2) Accelerating mechanisms; (3) Description and interpretation of solar phenomena.

The word *high energy* means only that the energy of the particles is large, compared to the thermal energy.

1. Effects of High Energy Particles

A. ELEMENTARY PROCESSES

1. Description of elementary processes

We are interested here not only by the stopping power of the matter for fast particles, but by all phenomena which can be produced by the motion of fast particles through the matter.

Electrons can produce essentially ionization, excitation; they can radiate by bremsstrahlung; they can excite collective phenomena: plasma oscillations, Čerenkov radiation, but these phenomena will be considered in the next section.

Fast protons can ionize, excite, they can also produce collective phenomena; but they can also capture free electrons, and experience with neutral hydrogen atoms charge exchange, the capture being made either on the ground level or on an excited level.

Fast hydrogen atoms, produced by charge exchange, can be excited or ionized.

All these particles can experience elastic collisions, and the cross sections for all these phenomena should be compared to each other.

Finally for energies larger than, 1 MeV, it is necessary to take into account the nuclear cross sections.

2. Stopping power

We consider electrons of energies much larger than the kinetic energy of the thermal electrons, or than the energy of bound electrons in hydrogen atoms. We have to consider, on one hand, the stopping power of the matter for the elec-

C. de Jager (ed.), The Solar Spectrum. All rights reserved.

trons, and, on the other hand, the various phenomena which can take place during the stopping of these electrons.

Let us combine the energy losses due to the inelastic collisions with the atoms or ions, and to elastic collisions with the electrons, ions or neutrals. We shall consider the energy loss $\mathrm{d}E/\mathrm{d}x$ and use the appropriate symbols to distinguish between elastic and inelastic collisions.

For elastic collisions of a particle of mass and charge m, Z, velocity w with field particles of mass and charge m_1, $Z_1 e$, we have the rate of loss (SPITZER, 1962):

$$\left(\frac{\mathrm{d}E}{\mathrm{d}x}\right)_{e.c} = -\,4\pi e^4 n_1 Z^2 Z_1^2 \log \Lambda\left(1 + \frac{m}{m_1}\right)\frac{1}{mw^2} \tag{1}$$

where

$$\Lambda = \left(\frac{kT}{4\pi n_e e^2}\right)^2 \frac{mw^2}{ZZ_1 e^2}. \tag{2}$$

For inelastic collisions, the expression is slightly different for electrons and heavy particles. For electrons (MOTT and MASSEY, 1949)

$$\left(\frac{\mathrm{d}E}{\mathrm{d}x}\right)_{in.c.e} = -\frac{4\pi e^4 N_i Z_i}{m_e w^2}\log\frac{m_e w^2}{I} \tag{3}$$

when I is a mean excitation energy, N the number of atoms of species i and Z_i the number of bound electrons. For heavy particles (MOTT and MASSEY, 1949):

$$\left(\frac{\mathrm{d}E}{\mathrm{d}x}\right)_{in.c.h} = -\frac{4\pi e^4 N_i Z_i}{m_e w^2}\log\frac{2m_e w^2}{I}. \tag{4}$$

For elastic collisions of electrons on ions and electrons, the field particles m_1 are either electrons ($m_1 = m_e$) or protons ($m_1 = m_H$). The dominant quantity is $[1 + (m_e/m_e)]1/m_e = 2/m_e$.

For elastic collisions of protons or ions and electrons, the dominant quantity is:

$$\frac{1}{m_H}\left(1 + \frac{m_H}{m_e}\right) \simeq \frac{1}{m_e}.$$

If we combine both expressions, putting $[(4\pi e^4)/(m_e w^2)]$ in factor, we obtain:

$$-\left(\frac{\mathrm{d}E}{\mathrm{d}x}\right)_e = \frac{4\pi e^4}{m_e w^2}\left[2n_e \log \Lambda_e + \sum_i N_i Z_i \log\frac{m_e w^2}{I_i}\right] \tag{5}$$

and for protons

$$-\left(\frac{\mathrm{d}E}{\mathrm{d}x}\right)_e = \frac{4\pi e^4}{m_e w^2}\left[n_e \log \Lambda_p + \sum_i N_i Z_i \log\frac{2m_e w^2}{I_i}\right]. \tag{6}$$

If E_{re} is the rest energy of the electron, and E_{rp} the rest energy of the proton, we can write for the electrons:

$$-\frac{\mathrm{d}}{\mathrm{d}x}\frac{E}{E_{re}} = 1.14 \cdot 10^{-24}\left(\frac{E}{E_{re}}\right)^{-1}\left[2n_e\left(13.675 + \tfrac{1}{2}\log\frac{T}{n_e} + \log\frac{E}{E_{re}}\right) + n_0\left(4.837 + \log\frac{E}{E_{re}}\right)\right] \quad (7)$$

and for the protons:

$$-\frac{\mathrm{d}}{\mathrm{d}x}\frac{E}{E_{rp}} = 6.22 \cdot 10^{-28}\left(\frac{E}{E_{rp}}\right)^{-1}\left[n_e\left(17.138 + \tfrac{1}{2}\log\frac{T}{n_e} + \log\frac{E}{E_{rp}}\right) + n_0\left(8.401 + \log\frac{E}{E_{rp}}\right)\right]. \quad (8)$$

For a particle of kind A, colliding on all kinds of particles, we can put

$$E/E_{rA} = y \quad (9)$$

and write

$$-\frac{\mathrm{d}y}{\mathrm{d}x} = \frac{A_A}{y} \quad (10)$$

where A_A is roughly a constant.

After integration of equations (7) or (8), we obtain the *range*, and the *time of stopping*:

$$x_A = \frac{y^2}{2A_A} \quad (11)$$

$$t_A = \frac{\sqrt{2}}{3}\frac{y^{\frac{3}{2}}}{A_A c} \quad (12)$$

where c is the light velocity.

It is useful to calculate the *range* and *life time* of electrons and protons at different levels of the photosphere and chromosphere. We have used the density distribution of VAN DE HULST (1953), though better estimates have been given since. We should consider these values as averages. The results are given in the following table, in which can be found as a function of height, the electron density, the hydrogen density, the range and life time for electrons of 51 and 5.1 keV. (Table I), and for protons of 100 MeV and 1 MeV.

These values can be compared to the values given by WARWICK (1963) for the range of 300 MeV protons.

TABLE I

A. ELECTRONS

h	$Log\ N_e$	$Log\ N_0$	*51.1 keV*	*5.11 keV*	*51.1 keV*	*5.11 keV*
			Range (km)		*Life time (sec)*	
−400	14.8	17.2	0.074	0.0021	$7.4 \cdot 10^{-7}$	$6.3 \cdot 10^{-8}$
−300	13.5	17.0	0.112	0.0031	$1.12 \cdot 10^{-6}$	$1 \cdot 10^{-7}$
−200	12.9	16.7	0.235	0.0063	$2.35 \cdot 10^{-6}$	$2 \cdot 10^{-7}$
−100	12.5	16.2	0.74	0.020	$7.4 \cdot 10^{-6}$	$6.3 \cdot 10^{-7}$
0	12.0	15.7	2.35	0.063	$2.35 \cdot 10^{-5}$	$2 \cdot 10^{-6}$
250	11.4	15.2	7.4	0.160	$7.4 \cdot 10^{-5}$	$6.3 \cdot 10^{-6}$
500	11.0	14.5	29.5	0.80	$2.35 \cdot 10^{-4}$	$2 \cdot 10^{-5}$
750	10.8	13.9	148	2.5	$1.48 \cdot 10^{-3}$	$1.26 \cdot 10^{-4}$
1000	10.7	13.3	590	16	$5.9 \cdot 10^{-3}$	$5 \cdot 10^{-4}$
1500	10.5	12.3	5500	140	$5.5 \cdot 10^{-2}$	$4.6 \cdot 10^{-3}$
2000	10.4	11.4	32500	420	$3.25 \cdot 10^{-1}$	$1.32 \cdot 10^{-2}$
3000	10.2	10.6	100000	1100	1	$3.8 \cdot 10^{-2}$
4000	10.0	10.0	200000	2100	2	$6.5 \cdot 10^{-2}$
5000	9.8	9.2	350000	3900	3.5	$1.2 \cdot 10^{-1}$
6000	9.6	8.4	560000	6300	5.6	$2 \cdot 10^{-1}$

B. PROTONS

h	$Log\ N_e$	$Log\ N_0$	*100 MeV*	*1 MeV*	*100 MeV*	*1 MeV*
			Range (10^3 km)		*Life time (sec)*	
−400	14.8	17.2	0.068	$9.3 \cdot 10^{-6}$	$6.8 \cdot 10^{-4}$	$9 \cdot 10^{-7}$
−300	13.5	17.0	0.10	$1.6 \cdot 10^{-5}$	$1 \cdot 10^{-3}$	$1.5 \cdot 10^{-6}$
−200	12.9	16.7	0.21	$3.2 \cdot 10^{-5}$	$2.1 \cdot 10^{-3}$	$3 \cdot 10^{-6}$
−100	12.5	16.2	0.68	$9.3 \cdot 10^{-5}$	$6.8 \cdot 10^{-3}$	$9 \cdot 10^{-6}$
0	12.0	15.7	2.1	$3.2 \cdot 10^{-4}$	$2.1 \cdot 10^{-2}$	$3 \cdot 10^{-5}$
250	11.4	15.2	6.8	$9.3 \cdot 10^{-4}$	$6.8 \cdot 10^{-2}$	$9 \cdot 10^{-5}$
500	11.0	14.5	34	$5 \cdot 10^{-3}$	$3.4 \cdot 10^{-1}$	$4.7 \cdot 10^{-4}$
750	10.8	13.9	135	$1.9 \cdot 10^{-2}$	1.3	$1.8 \cdot 10^{-3}$
1000	10.7	13.3	540	$7.8 \cdot 10^{-2}$	5.4	$7.4 \cdot 10^{-3}$
1500	10.5	12.3	7000	1.1	69	$1.04 \cdot 10^{-1}$
2000	10.4	11.4	37000	5.1	380	$5.2 \cdot 10^{-1}$
3000	10.2	10.6	160000	$2.1 \cdot 10$	1600	2
4000	10.0	10.0	400000	$5.1 \cdot 10$	3900	4.8
5000	9.8	9.2	850000	$1.06 \cdot 10^2$	8400	10
6000	9.6	8.4	1600000	$2.4 \cdot 10^2$	15000	23

3. Ionization and excitation processes

During its motion, the electron or the proton can excite or ionize the atoms, especially the hydrogen atoms. If Q_{A-B} is the cross-section for some specified process experienced by the particle A, when colliding with particles B, the number of processes over distance dx will be

$$dn = Q_{A-B} N_B dx. \tag{13}$$

Using the relation (10) between range and energy, we can write:

$$\mathrm{d}n = -Q_{A-B}\frac{1}{A_A}N_B y\,\mathrm{d}y\,. \tag{14}$$

The total cross section for ionization of hydrogen atoms by electrons is (MOTT and MASSEY, 1949)

$$Q_i = 0.285\,\frac{2\pi e^4}{|E_0|m_e v^2}\log\frac{2m_e v^2}{0.048\,|E_0|} \tag{15}$$

where E_0 is the energy of the ground state.
Writing $E = (1/2)\,m_e v^2$ we can write:

$$Q_i = 0.285\;4\pi a_0^2\,\frac{|E_0|}{E}\log\frac{4E}{0.048\,|E_0|}\,. \tag{15'}$$

The cross section for excitation of hydrogen atoms by electrons to level n is (MOTT and MASSEY, 1949)

$$Q_{0n} = \frac{f_{0n}}{T_n - T_0}\,4\pi a_0^2\,\frac{E_0}{E}\log\frac{4E}{E_n - E_0} \tag{16}$$

where f_{0n} is the oscillator strength for transition $0 \to n$, and $R(T_n - T_0)$ the wave number. For hydrogen, $T_n = 1/n^2$.

The coefficient is 0.55 for excitation of level 2 and, 0.089 for excitation of level 3.

The following table gives the total number of ionizations produced by a fast particle during its life time (Table 2) at different levels in the chromosphere.

TABLE 2

h(km)	*Electrons*		*Protons*	
	50 keV	*5 keV*	*100 MeV*	*1 MeV*
− 400	150	33	70	0.6
− 300	150	33	70	0.6
− 200	150	33	70	0.6
− 100	150	33	70	0.6
0	150	33	70	0.6
250	150	33	70	0.6
500	150	33	70	0.6
750	150	33	70	0.6
1000	150	33	70	0.6
1500	140	30	70	0.6
2000	100	11	60	0.6
3000	50	5	43	0.3
4000	16	2	26	0.2
5000	7	0.6	9	0.07
6000	2	0.16	2.6	0.02

It should be noticed that the number of ionization produced at the same velocity is slightly larger for the electrons than for the protons. As can be seen from Equations (15′) and (16) and from numerical values, the number of excitations to level 3 is about 1/3 of the number of ionizations.

The electrons ejected from the hydrogen atom have an average energy of the order of 35 eV. They are thermalized in a very short time, much before producing any ionization or excitation.

4. Bremsstrahlung

In the non-relativistic limit, when v_0 and v represent the velocities before and after the collisions, and

$$Z\alpha c \ll v < v_0 \ll c$$

(with $\alpha = 1/137$, and Z the charge of the nucleus), it is possible to use the Bethe-Heitler formula for the cross section for Bremsstrahlung (HEITLER, 1944)

$$\phi_k \mathrm{d}k = \bar{\phi}\frac{8}{3}\mathrm{d}k\frac{\mu}{k}\log\frac{[\sqrt{T_0} + \sqrt{T_0 - k}]^2}{k} \tag{17}$$

with the following meaning of the symbols: k is the energy of the photon; μ is the rest energy of the electron; T_0 is the kinetic energy of the primary electron; ϕ is a suitable unit, in which to express the cross-section for bremsstrahlung:

$$\bar{\phi} = \frac{Z^2 r_0^2}{137} = Z^2 \cdot 5.7 \cdot 10^{-28}\,\mathrm{cm}^2$$

where $r_0 = (e^2/m_e c^2)$ is the classical electronic radius. Energy k goes from T_0 (short wavelength limit) to a certain minimum, corresponding to the screening of the positive charges. If λ_D is the Debye length, it is possible to estimate, following Heitler, $k_{\min}$:

$$k_{\min} = \frac{\hbar c\sqrt{2}}{\lambda_D}\sqrt{\frac{T_0}{\mu}}. \tag{18}$$

Taking $(k/\mu) = x$ as a parameter, and using $(T_0/\mu) = y$ as a variable, we can write:

$$x_{\min} = \frac{\sqrt{2}}{2\pi}\frac{\lambda_c}{\lambda_D}\frac{1}{\sqrt{y}} \tag{18′}$$

where λ_c is the Compton wave length, $\lambda_c = (h/m_e c)$. ELWERT (1961), GREENE (1959), have used the following approximation:

$$\phi \mathrm{d}k = \frac{32}{3}\bar{\phi}\frac{1 - e^{-2\pi\alpha Z/\beta}}{\beta^2}\frac{\mathrm{d}(h\nu)}{h\nu} \tag{19}$$

valid at the short wave length limit. The symbol α is the fine structure constant, $\alpha \simeq (1/137)$ and $\beta = v_0/c$.

The power radiated per second per electron by bremsstrahlung on the protons, in the interval $\mathrm{d}(h\nu)$ is approximately:

$$\mathrm{d}I = \frac{32}{3} n_p v \bar{\phi} \frac{1}{\beta^2} \mathrm{d}(h\nu) = \frac{16\sqrt{2}}{3} n_p \bar{\phi} c \frac{1}{y^{\frac{1}{2}}} \mathrm{d}(h\nu) . \tag{20}$$

The energy loss by radiation is (HEITLER, 1944):

$$-\frac{\mathrm{d}T_0}{\mathrm{d}x} = n_p \mu \frac{16}{3} \bar{\phi} . \tag{21}$$

Using the expression (10) for the range, we calculate the total energy loss by radiation:

$$T_{\mathrm{Rad}} = \frac{8}{3} \frac{n_p \mu \bar{\phi} y^2}{A_e} . \tag{22}$$

We must take for n_p the total number of protons (free, or in hydrogen atoms).

We can estimate the total loss by bremsstrahlung:

$$T_{\mathrm{Rad}} \simeq 3.10^{-10} y^2 \text{ ergs/electron or } 200\, y^2 \text{ eV/electron} \tag{22'}$$

$$(T_{\mathrm{Rad}}/T_0) \simeq 3.7 \cdot 10^{-4} y \text{ or in flight } 1.17 \cdot 10^{-3}\, y \text{ photons/electron.} \tag{22''}$$

4.1 Electron-electron bremsstrahlung. There is some doubt concerning the importance of the electron-electron bremsstrahlung, and we can refer to the compilation of cross-sections made by WANDEL, JENSEN and KOFOED-HANSEN (1959). However, the correction, for energies smaller than 0.1 μ, where μ is the rest energy, does not seem to be important.

5. Charge exchange

A proton passing through hydrogen gas can capture an electron either on the ground level, or on an excited level. The hydrogen atom which is formed can either lose an electron by collision or, if it is excited, transfer its excitation to an hydrogen atom.

The cross-section for capture and loss is fairly well known. For energies smaller than 100 keV, we can use the results of ALLISON (1958). At high energies, the result of BRINKMAN and KRAMERS (1930), using the first Born approximation gives for the cross-section for capture into a 1s state from a 1s state:

$$Q_c(1s, p \to p, 1s) = \frac{1}{10} \cdot 2^{18} \pi a_0^2 \frac{\alpha^2}{y_p \left[\frac{2}{\alpha^2} y_p + 4\right]^5} \tag{23}$$

which numerically, gives:

$$Q_c = 1.22 \cdot 10^{-6} y_p^{-1} \left[\frac{2}{\alpha^2} y_p + 4 \right]^{-5}, \tag{23'}$$

and fits to the curve of Allison. The following table (Table 3), summarizes the results and gives also the cross-section Q_l for the losses, as obtained by ALLISON (1958) (column 4) or deduced from the cross-sections of FITE-BRACKMAN (1958) or from the Born approximation (column 5).

TABLE 3

Energy (MeV)	$\frac{1}{y_p}$	*log* Q_c	*log* Q_l	
0.001	900000	$\overline{16}.35$		losses on free protons
0.002	450000	$\overline{16}.50$		
0.005	180000	$\overline{16}.50$	$\overline{17}.6$	
0.010	90000	$\overline{16}.40$	$\overline{17}.6$	
0.020	45000	$\overline{16}.20$	$\overline{17}.85$	$\overline{17}.60$
0.050	18000	$\overline{17}.70$	$\overline{17}.88$	$\overline{17}.55$
0.100	9000	$\overline{17}.20$	$\overline{17}.78$	$\overline{18}.84$*
0.20	4500	$\overline{18}.03$		$\overline{18}.79$*
0.5	1800	$\overline{20}.35$		$\overline{18}.60$*
1	900	$\overline{22}.74$		$\overline{19}.32$**
10	90	$\overline{28}.92$		$\overline{20}.39$**
100	9	$\overline{34}.92$		$\overline{21}.44$**

* FITE-BRACKMAN (1958).
** Cross-section for ionization.

At a given energy, a proton will continuously capture and lose an electron. At a velocity v the average time spent as a proton is given by the life time for capture of an electron on the hydrogen atoms:

$$t_c = \frac{1}{Q_c v n_H}. \tag{24}$$

In the same way the average time spent as a hydrogen atom is given by the life time for loss of an electron by inelastic collision of the electron:

$$t_c = \frac{1}{Q_L v n_e} \tag{25}$$

n_e should be taken as the total number of electrons (free and bound).

During time dt, the particle is a proton during a time $z\mathrm{d}t$, $z < 1$,

$$z = \frac{Q_L v n_e}{Q_c v n_H + Q_l v n_e}. \qquad (26)$$

We can as well call z the effective charge of the proton, during flight. The equality of Q_L and Q_c takes place a 54 keV in neutral hydrogen. It will be noticed that for high energies, Q_L is so much larger than Q_c, that the proton remains free most of the time.

Below 54 keV, that is to say, below a velocity of $3.3 \cdot 10^8$ cm sec^{-1}, the cross-section for the losses becomes smaller than the gains, and the particle spends most of its time as an hydrogen atom.

It is essentially at these comparatively low energies (at the same velocity, an electron has an energy of only 28 eV) that the excitation and transfer of excitation takes place.

Let us consider the different processus which can take place during flight:

(1) Capture on level n, cross-section $Q_c(p, 1s \to nl, p)$, or shortly Q_{cn} (n = upper level);

(2) Radiation, with probability A_{nl} (l = lower level);

(3) Transfer of excitation, with cross-section $Q_t(nl, 1s \to 1s, nl)$, or shortly Q_{tn};

(4) Loss when on level n with cross-section $Q_L(nl, e \to p, e, e)$, or Q_{nl};

(5) Excitation to level n by collision on the free electrons $Q_e(1s, e \to nl, e)$ or Q_{en}, or on the atoms $Q(1s, a \to nl, a')$, or Q_{an}.

We can introduce the frequency of excitations, of loss, of capture (by charge transfer or radiation capture).

$$\begin{aligned} \nu_{en} &= Q_{en} n_e v + Q_{an} n_a v \\ \nu_{nL} &= Q_{nL} n_H v \\ \nu_{cn} &= Q_{cn} n_H v + R_{cn} n_e v \end{aligned} \qquad (27)$$

and we find the following values for the ratios n_n/n_p; assuming that most of the atoms are in the ground state:

$$\begin{aligned} x_1 &= \frac{n_1}{n_p} = \frac{\nu_{c1}}{\nu_{1L}} \\ x_2 &= \frac{n_2}{n_p} = \frac{\nu_{c2} + \nu_{e2}(\nu_{c1}/\nu_{1L})}{\nu_{e2} + \nu_{2L} + A_{21}} \\ x_3 &= \frac{n_3}{n_p} = \frac{\nu_{c3} + \nu_{e3}(\nu_{c1}/\nu_{1L})}{\nu_{e3} + \nu_{3L} + A_{32} + A_{31}} \end{aligned} \qquad (28)$$

The essential phenomena are the capture on the ground level, the loss from the ground level, the excitation transfer and the radiative transitions.

We can thus write:

$$\frac{n_2}{n_p} = \frac{\nu_{e2}(\nu_{c1}/\nu_{1L})}{\nu_{e2} + A_{21}}\,;\ \frac{n_3}{n_p} = \frac{\nu_{e3}(\nu_{c1}/\nu_{1L})}{\nu_{e3} + A_{32} + A_{31}}\,. \tag{29}$$

The fraction of the time of flight spent in state 3 is:

$$\xi_3 = \frac{x_3}{1 + x_1 + x_2 + x_3 + \ldots} \tag{30}$$

Šklovski (1958) has noticed that the cross-section for excitation transfer is very large:

$$Q_{tn} \simeq \frac{2\pi^3 \mu_{1n}^2}{hv_H} \tag{31}$$

where μ_{1n} is the dipole moment for the corresponding optical transition. We have then:

$$\nu_{tn} \simeq \frac{2\pi^3 \mu_{1n}^2 n_{Hn}}{h} \tag{32}$$

from which we derive the following values:

$$\nu_{t3} = 10^{-8} n_H$$

$$\nu_{t2} = 5.25 \cdot 10^{-8} n_H\,. \tag{33}$$

The radiative transition give the following values for the transition probabilities of the excited states:

$$A\,(3p \to ns) = 1.86 \cdot 10^8 \sec^{-1}$$

$$A\,(2p \to ns) = 6.25 \cdot 10^8 \sec^{-1}$$

The excitation transfer will be competitive with the radiative transitions as soon as $n_H > 5 \times 10^{15}$.

We can see from Equation (29) that most of the Hα emission in flight will take place for energies smaller than 54 keV. At high energies, the ratio (ν_{c1}/ν_{1L}) is small, and the particle is, most of the time, a proton. For example, at 1 MeV, $(\nu_{c1}/\nu_{1L}) \simeq 3 \cdot 10^{-3}$.

The number of emissions of Hα will be obtained by considering, on one hand the excitation by collisions on neutral hydrogen atoms, on the other hand the captures on level $n = 3$. The probability of radiative capture is very small. The most important effect is the excitation by collisions.

Bates (1963) give the following cross-sections, for the excitations on neutral hydrogen atoms:

$$Q\,(1s - 2s,\ 1s - \Sigma) = 4.3\,E^{-1} \pi a_0^2$$

$$Q\,(1s - 2p,\ 1s - \Sigma) = 21\ E^{-1} \pi a_0^2$$

$$Q\,(1s - 3l,\ 1s - \Sigma) = 29\ E^{-1} \pi a_0^2$$

$1s - 2s$ represents the transition experienced by the target, and $1s - \Sigma$ represents all the possible transitions of the projectile hydrogen atom; E is the energy of impact in keV.

To give an order of magnitude, an hydrogen atom of 1 MeV is excited 5400 times by the electrons (calculated for $n_e = 4 \cdot 10^9$). If it is hydrogen atom a fraction $3 \cdot 10^{-3}$ of the time, it will produce in flight about 16 photons Hα.

7. *Synchrotron radiation*

The total energy loss by synchrotron radiation is, for relativistic electrons:

$$\frac{\mathrm{d}E}{\mathrm{d}t} = - 4 \cdot 10^{-9} B^2 E^2 \,\mathrm{MeV} \cdot \mathrm{sec}^{-1} \tag{34}$$

where E is in MeV. However the energy loss by synchrotron radiation becomes important only above 1 MeV.

For non relativistic electrons, we can compare with the collisional losses. The energy losses for non relativistic electrons are:

$$\frac{\mathrm{d}E}{\mathrm{d}t} = - 3.15 \cdot 10^{-15} B^2 E \,\mathrm{MeV} \cdot \mathrm{sec}^{-1}. \tag{35}$$

They are extremely small.

Stein and Ney (1963) have considered the energy losses by synchrotron radiation of electrons of about 10 MeV. We shall take into consideration their result, when discussing the X-ray emission of the Sun.

B. COLLECTIVE EFFECTS

7. *Čerenkov radiation*

If a particle is moving in a plasma, at a speed larger than the thermal velocity of the electrons, a Čerenkov radiation of plasma waves is radiated away in the plasma. M. H. Cohen (1961) gives the power radiated, the spectrum and the angular spectrum. If u_0 is the thermal velocity of the particles, the energy radiated by a particle of velocity u, is:

$$\frac{\mathrm{d}w}{\mathrm{d}x} = - \frac{e^2 \omega_p^2}{8\pi\varepsilon_0 u^2} \log \left[\left(\frac{u}{u_0}\right)^2 - 1\right] \tag{36}$$

or

$$\frac{\mathrm{d}w}{\mathrm{d}x} = - 5.1 \cdot 10^{-23} \left(\frac{f_p}{10^8}\right)^2 \left(\frac{c}{u}\right)^2 \log \left[\left(\frac{u}{u_0}\right)^2 - 1\right], \tag{36'}$$

where f_p is the plasma frequency, $f_p = (\omega_p/2\pi)$.

The spectrum is:

$$I_\omega = 1.62 \cdot 10^{-31} \left(\frac{f_p}{10^8}\right) \frac{c^2}{u^2} \frac{\omega/\omega_p}{(\omega/\omega_p)^2 - 1} \,\mathrm{ergs/cps/cm} \tag{37}$$

It has a cut off at low frequency, for

$$\omega_c = \omega_p \Big/ \left[1 - \left(\frac{u_0}{u}\right)^2\right]^{\frac{1}{2}} \tag{38}$$

and at high frequency, for:

$$\omega \sim \sqrt{2}\,\omega_p \tag{39}$$

due to Landau damping.

The angular spectrum is given by:

$$I_{\theta,\varphi} = \frac{e^2 \omega_p^2}{8\pi^2 \varepsilon_0 u^2} \frac{\cos\theta}{\cos^2\theta - (u_0/u)^2} \text{ ergs sr}^{-1}\text{ cm}^{-1}; \tag{40}$$

when $u \gg u_0$, the angular spectrum has a sharp maximum near 90°, and the spectrum has a sharp maximum near the plasma frequency.

We can easily calculate the total energy radiated away during flight. We can write:

$$\frac{\mathrm{d}w_{\check{c}}}{\mathrm{d}x} = -\,2.55 \cdot 10^{-23} \left(\frac{f_p}{10^8}\right)^2 \frac{1}{y} \log\left[\left(\frac{u}{u_0}\right)^2 - 1\right] \tag{36''}$$

with

$$f_p^2 = \frac{1}{\pi} \frac{N_e e^2}{m_e}$$

and obtain

$$w_{\check{c}} = 2.55 \cdot 10^{-23} \cdot 8 \cdot 10^{-9} \frac{N_e}{A_i} \log\left[\left(\frac{u}{u_0}\right)^2 - 1\right].$$

The following table gives the power radiated away in flight per electrons of 51 keV ($y = 0.1$) (Table 4) together with the plasma frequency, and the low frequency cut off.

TABLE 4

h(km)	*log* $W_{\check{c}}$ *(eV)*	*log* f_p	$f_p - f_c$ *(kc sec*$^{-1}$*)*
− 400	1.34	11.35	1220
− 300	1.04	10.70	275
− 200	0.74	10.40	140
− 100	0.84	10.20	88
0	0.84	9.95	49
250	0.84	9.65	25
500	0.94	9.45	15
750	1.44	9.35	12
1000	1.94	9.30	11
1500	2.71	9.20	9
2000	3.38	9.15	
3000	3.67	9.05	
4000	3.76	8.95	
5000	3.81	8.85	
6000	3.82	8.75	

It should be noticed that the total energy radiated by Čerenkov effect is appreciably larger than the estimate of COHEN. The difference is essentially due to a better estimate of the range of the fast particle.

8. *Coherent emission*

When two streams of plasma move with a different velocity, an instability can appear, known as two stream instability.

In such a case, the stream is the place of a plasma oscillation of large amplitude, with a coherent emission of plasma waves by the electrons of the stream.

GINZBURG and ŽELEZNIAKOV (1958) using a result of BOHM and GROSS (1949) have estimated the electric field produced by coherent oscillations:

$$E_0 \simeq \frac{m_e \omega}{2 e v_0} \left[\frac{16}{3} \frac{u_0^3 m_e}{k T_s} \frac{\omega_s^2}{\omega^2} \frac{\nu_s}{\nu} \right]^2 \tag{41}$$

where: m_e is the mass of the electron; ω the frequency of the wave in the plasma; ω_s the frequency of plasma oscillations in the stream; ν_s the frequency of electron collisions in the stream; ν the frequency of electron collisions in the plasma; v_0 the velocity of the stream ($v_0 \gg v_{th}$); v_{th} the thermal velocity of the electrons.

As noticed by GINZBURG and ŽELEZNIAKOV, when the amplitude of the electric field E_0 is sufficiently large, the wave will differ considerably from sine waves. Harmonics will necessary be present.

9. *Production of E.M. waves*

Plasma oscillations can produce E. M. waves, either if they are diffracted by density fluctuations of the plasma or if they are produced in a density gradient.

The flux of energy in plasma waves will be of the order of $E_0^2 v_{th}$. As calculated by TIDMAN (1960) a certain fraction of this energy will be radiated as E. M. waves, with an efficiency

$$R \simeq \theta_1^2 1.1 \cdot 10^{-2} \left(\frac{c}{v_{th}}\right) \left(\frac{2\pi v_{th}}{\omega_p L}\right) \tag{42}$$

θ_1 is the angle of the cone of propagation of the E.M. waves

$$\theta_1 \simeq \frac{v_{th}}{c} \, . \tag{43}$$

We can write the electric field:

$$E_0 \simeq \frac{m_e \omega_p}{2 e v_0} \left[\frac{16}{3} \frac{u_0^3 m_e}{k T_s} \left(\frac{N_s}{N_{pl}} \right)^2 \left(\frac{T_{pl}}{T_s} \right)^{\frac{3}{2}} \right]^2$$

$$\simeq \frac{128}{9} \frac{m_e \omega_p}{e v_0} \frac{u_0^6}{v_{th\,s}^4} \left(\frac{v_{th}}{v_{th\,s}} \right)^6 \left(\frac{N_s}{N_{pl}} \right)^4 . \tag{41'}$$

The E. M. flux is

$$F_{\text{E.M.}} = 1.1 \cdot 10^{-2} \frac{2\pi c}{\omega_p L} v_{th} \left(\frac{v_{th}}{c}\right)^2 \frac{2^{14}}{3^4} \frac{m_e^2 \omega_p^2}{e^2} \frac{u_0^{10}}{v_{ths}^8} \left(\frac{v_{th}}{v_{ths}}\right)^{12} \left(\frac{N_s}{N_{pl}}\right)^8 \tag{44'}$$

From the expression given by GINZBURG and ŽELEZNIAKOV, we can estimate the ratio of the efficiencies of scattering by the gradient and by the fluctuations:

$$\frac{R_{\text{gradient}}}{R_{\text{fluctuations}}} \simeq \frac{1.1 \cdot 10^{-2} \cdot 3c^2 v_0}{L^2 r_0 \omega_p^3}$$

where r_0 is the classical radius of the electron and L the scale of the gradient For $v_0 \simeq 10^{10}$ cm sec^{-1}, $L \simeq 10^7$ cm, the ratio is:

$$\frac{R_{\text{grad}}}{R_{\text{fluct}}} \simeq \frac{10^{28}}{\omega_p^3}$$

For electrons of 30 keV, the production of radio waves is essentially due to the density gradient for $\omega_p < 2 \cdot 10^9$, that is to say for $N_e < 10^9$.

However, such an electronic density is reached only in the corona, where the temperature, and the scale height, are much larger. It is easy to find that for $v_0 \simeq 10^{10}$, the ratio $(R_{\text{grad}}/R_{\text{fluct}})$ is always smaller than one.

The scattering by the small irregularities of the plasma is the most efficient process of producing the E.M. waves.

The diffracted waves are emitted in a cone of angle $\theta_1 \simeq (v_{th}/v_0)$. It is possible to estimate the brightness temperature of the radiation. If the radiation is emitted in a layer of thickness L, it is possible, following Ginzburg and Żelezniakov, to calculate the brightness temperature: ·

$$T_B = e^{-\tau} \frac{\pi \cdot 2^{13.5}}{3^5} \frac{e^2}{kc} \frac{N_e L}{\omega_p} \frac{v_0^{11} v_{th}^{11}}{v_{ths}^{20}} \left(\frac{N_s}{N_{pl}}\right)^8$$

where τ is the optical depth of the emitting layer. The frequency is approximately the plasma frequency.

Numerically:

$$T_B = 10^{-12.18} N_e^{\frac{1}{2}} L \frac{v_0^{11} v_{th}^{11}}{v_{ths}^{20}} \left[\frac{N_s}{N_{pl}}\right]^8 e^{-\tau}.$$

The coherent emission is important only if $v_{th} s \ll v_0$. In the corona, for $v_{th} \simeq v_{ths} \simeq 10^{8.5}$; $v_0 \simeq 10^{10}$; $L \simeq 10^9$; $\tau = 4$; $N_e = 10^9$, we calculate, following GINZBURG and ŽELEZNIAKOV:

$$T_B \simeq 10^{31.3} \left(\frac{N_s}{N_{pl}}\right)^8.$$

As the observed value is $T_B \simeq 2.5 \cdot 10^{11}$, we find

$$\left(\frac{N_s}{N_{pl}}\right) \simeq 10^{-2.5}$$
$$N_s \simeq 10^{6.5} .$$

The question of the second harmonics, observed in Type II flares remains to be explained. LARENTZ (1963) has considered the interaction of E.M. waves, and shown that it produces E.M. waves having twice the plasma frequency.

C. TRAPPING

10. Drift of particles

Particles in a magnetic field can drift through the magnetic field. For a particle of mass m, and charge q, with a velocity $w_{\parallel}$, parallel to the magnetic field, and $w_{\perp}$ perpendicular to the magnetic field, there is a drift velocity.

$$w_D = w_D' + w_D''$$

with

$$w_D' = \frac{mw_{||}^2}{9} \frac{\mathbf{B} \times (\mathbf{B} \cdot \nabla)\mathbf{B}}{B^4}$$

$$w_D'' = \frac{mw_{\perp}^2}{49} \frac{\mathbf{B} \times \nabla B^2}{B^2}$$

w_D' is the drift to the curvature of lines of force, w_D'' is the drift due to the gradient of the magnetic field.

The tim? variation of the magnetic field (along the unit vector **a**) produces another drift:

$$w_D''' = \frac{B}{B^2} \times \frac{\partial \mathbf{e}_3}{\partial t} \frac{m}{q} w_{||}$$

Due to these drifts, a charged particle, trapped in a magnetic field can leave a magnetic bottle either by going through a neutral point, or by reaching a line of force which goes to infinity.

As an order of magnitude, the time of escape will be of the order of (L/w_D) (see for example STEPANYAN and VLADIMIRSKIJ (1961)), and w_D is of the order of:

$$w_D \simeq \frac{2T_0}{qLB\sqrt{1-\beta^2}}$$

(with the relativistic correction).

Therefore

$$t_{\text{esc}} \simeq \frac{qBL^2\sqrt{1-\beta^2}}{2T_0}$$

For particles of 1 MeV, $B \simeq 100$ gauss, $L \simeq 10^9$ cm, we find

$$t_{esc} \simeq 5 \cdot 10^5 \text{ sec} .$$

It should be noticed that the velocity of escape, in this picture, depends only on the energy of the particle, given the geometry of the magnetic field.

For non relativistic energies, a proton and an electron of the same velocity, the electron will stay much longer in the trap than the proton.

11. Motion near a zero field point (Norton, 1961)

A particle which comes sufficiently close to a neutral line, escapes very quickly, the time of escape being then of the order of $(L/w_{\parallel})$.

If L is the scale of the gradients of the field, near the zero field point, the radius of gyration has to be of the order of L.

If B_0 is a characteristic value of the field ($B = B_0$ at a distance L from the zero field point), we find the distance x

$$x = \frac{E_0 \beta L}{e B_0 x}$$

For E_0 in electron volts,

$$x \simeq \left(\frac{\beta}{300 \; B_0} \; E_{0\text{eV}} L\right)$$

For a velocity $c/3$, a proton has an energy of about 100 MeV, and an electron, an energy of about 50 keV. With $B_0 = 100$ gauss $L \simeq 10^9$ cm, we find:

$$x_{el} \simeq 2 \cdot 10^4 \text{ cm} \; ; \quad x_p \simeq 10^5 \text{ cm} .$$

The particle has to come very close to a zero field point to escape quickly the trap. This happens when the drift brings the particle on a line of force which comes very close to the neutral line.

2. Accelerating Mechanisms

12. Review of accelerating mechanisms

As mentioned by PARKER (1958) in his discussion, the accelerating mechanisms can be classified in two classes: the betatron mechanism and the Fermi mechanism.

In the betatron mechanism, a particle turns in a growing magnetic field. The electric field, induced by the growth of the magnetic field accelerates the particle. If the gyration period is small, compared to the characteristic time of the growth of the magnetic field, the energy of the particle grows like the magnetic field, as a result of the conservation of the magnetic moment.

In the Fermi mechanism, the magnetic field is turbulent, and a charged particle, trapped in the magnetic field, is reflected on the bottle necks of the field. If two bottle necks are approaching each other, the particle reflected back and forth, increases its energy at each collision with the magnetic field in the ratio (V/c) of the velocity of the field to light velocity.

If the velocities of the clumps of the magnetic field are distributed at random, the field grows like $(V/c)^2$, on the average at each collision.

Many mechanisms have been derived from these two mechanisms:

(a) DUNGEY (1958) considers the growth of the energy of a particle as a consequence of a stochastic betatron mechanism. The rate of growth of the energy is proportional to $(V/c)^2$, but is several hundred times larger than in the Fermi mechanism.

(b) The question has been raised by several authors of increasing the efficiency of the Fermi mechanism from $(V/c)^2$ to (V/c). PARKER (1958) discusses the different suggestions which have been made. A system of magnetohydrodynamic waves moving in opposite direction can accelerate. However, when the particle is lost by the magnetic trap, an energy loss takes place, and the Fermi rate is found again.

(c) WENTZEL (1963) has considered the effect of a shock wave moving in a non uniform field. He has considered two possibilities: a shock moving in the direction of the radius of curvature of the lines of force outwards; a shock moving in the direction of the lines of force.

In both cases, the refraction of the lines of force across the shock front plays an important role. In the first case a line of force meets twice the shock front, and is refracted in two places. These two places of refraction are also the place where the particles are reflected like on two bottle necks, until the line of force is overtaken entirely by the shock front.

In the second case the shock moves from a region of weak field towards a region of strong field. One of the bottle necks is stationary, the other one is moving with the shock front.

(d) SWEET (1956), DUNGEY (1953), after GIOVANELLI (1948), have considered the instability of a zero field point, and the behaviour of particles as a consequence of this instability.

Sweet has shown that a magnetic configuration with a zero field point becomes unstable, when a certain characteristic field B_0 becomes larger than a critical value given by $(B_0^2 / 8\pi) \simeq P_G$. The acceleration of a particle near the zero field point is very similar to the betatron mechanism, but, this time, the increase of the magnetic field is very large, and can produce a considerable increase of the energy of the particle.

13. Critics

A variety of critics can be made to these mechanisms:

(a) The betatron mechanism cannot build large energies, as the magnetic field does not change so quickly.

(b) The Fermi mechanism is not very efficient, being only in $(V/c)^2$. The energy losses by collisions are very high in between two magnetic collisions, and the initial velocity must be very high, so that the collisional losses are smaller than the gains: this implies a high energy of injection of the articles to be accelerated.

(c) The Wentzel mechanism, which is a variety of the Fermi-Parker mechanism meets the same criticism. Furthermore, if the velocity of the particle is not much larger than the velocity of the shock front, the number of encounters with the shock front is small, and the total increase of energy with only one shock wave is not large.

(d) In the neutral point mechanism, the effect of the conductivity of the medium is to reduce the efficiency of the mechanism. PARKER (1963) considering the orders of magnitude, disagrees with GOLD and HOYLE, on the speed of penetration of the fields.

To be more precise, we can estimate the velocity of injection by comparing the losses and the gains. For the statistical Fermi mechanism, the ratio y of the kinetic energy to the rest energy should be larger than a certain value:

$$y > y_{cW} = \left(\frac{A_A \Delta x}{4}\right)^{\frac{1}{2}} \frac{1}{\beta}$$

where β is the ratio of the velocity of the magnetic clumps to light velocity, and Δx the size of he magnetic clumps. A_A has been defined in Section 3, and has the value A_p for the protons.

For the Fermi-Wentzel mechanism, the energy of injection would be given by:

$$y > y_{cW} = \left(\frac{A_A \Delta x}{\beta}\right)^{\frac{1}{2}}$$

For $\beta = (10^7/c) \simeq 3 \cdot 10^{-4}$ we calculate the following values of y:

TABLE 5

ENERGY OF INJECTION

$h(km)$	$y_{cF}\,(MeV)$	$y_{cW}\,(MeV)$
0	—	—
1000	750	500
2000	170	60
4000	50	18
6000	26	9

We have taken $\Delta x \simeq 10^7$, for the statistical Fermi mechanism, and $\Delta x \simeq 10^9$ for the Wentzel-Fermi mechanism. We see that the problem of the injection mechanism remains the important problem of the acceleration mechanism.

14. Shock wave structure

The implications of the shock phenomena for the acceleration mechanism are so important, that it is necessary to give a review of the shock wave properties in ionized matter.

(a) *Shock in a perfect fluid.* In a fluid with no viscosity, no thermal conductivity, and no resistivity, a shock front is a perfect discontinuity.

When no magnetic field is present, the relations between the physical conditions ahead and behind the shock front are well known; they express the conservation of mass, momentum and energy.

In the presence of a magnetic field, the situation is much more complicated, and the exhaustive classification of all kinds of shock waves have been given by BAZER and ERICSON (1959).

For our purpose it is sufficient here to recall the properties of the parallel shock and of the perpendicular shock.

In the parallel shock, the direction of propagation is parallel to the lines of force, and the properties of the shock are unaffected by the presence of the magnetic field. In the perpendicular shock the direction of propagation is perpendicular to the lines of force, and the properties of the shock are affected by the presence of the magnetic field. The relations, at the shock front are the following:
Density:

$$N_2 = (1 + \bar{\eta}) N_1$$

Magnetic field:

$$B_2 = (1 + \bar{\eta}) B_1$$

Temperature:

$$T_2 = T_1 \frac{(3 + 4\bar{\eta}) + \frac{5}{6} s_1^{-1} \bar{\eta}^3}{(3 - \bar{\eta})(1 + \bar{\eta})}$$

with

$$s_1 = \frac{4\pi\gamma P_1}{B_1^2}$$

and

$$\gamma = \frac{5}{3}.$$

Velocity of the shock front with respect to the matter ahead of the shock:

$$u_1^2 = \frac{\bar{\eta} + 1}{3 - \bar{\eta}} \frac{5P_1}{\varrho_1} \left[1 + \frac{6 + \bar{\eta}}{6s_1}\right].$$

Ratio of the velocity behind the shock front to the sound velocity behind the shock:

$$\frac{v_2^2}{\gamma R T_2} = \frac{1 + s_1 + \frac{1}{6}\bar{\eta}}{s_1 + \frac{4}{3}\bar{\eta} s_1 + \frac{5}{18}\bar{\eta}^3} .$$

As it is well known, T_2, v_1, increase to infinity when (N_2/N_1) becomes close to 4.

To give an idea of the relation between shock strength and flux of mechanical energy, we consider a shock with no magnetic field. If t_0 is the characteristic time for the decay of energy behind the shock front, f the frequency of the shocks, the flux of mechanical energy is of the order of:

$$F_{\text{mech}} = f t_0 \tfrac{1}{3} \varrho_0 a_0^3 \left(\frac{\bar{\eta}}{1 + \bar{\eta}}\right)^2 \left(\frac{3 (1 + \bar{\eta})}{3 - \bar{\eta}}\right)^{\frac{3}{2}}$$

where a_0 is the sound velocity. For $\varrho_0 \simeq 10^{-10}$ g cm^{-3}, we calculate the following values of $\bar{\eta}$, for $f t_0 \simeq 1$:

TABLE 6

SHOCK WAVE CHARACTERISTICS

F_{mech}	$3 - \bar{\eta}$	v_2 *(km/sec)*
10^7	0.34	33.9
10^8	0.078	74.5
10^9	0.0177	159

It can be noticed that the corresponding shock strength are fairly high, v_2, has been calculated for a sound velocity 7 km/sec, and is also fairly large.

(b) *Shock in a real fluid.* It is also well known that in a real fluid the region (1) ahead of the shock is separated from region (2) behind the shock by a transition region, where the change from ϱ_1, B_1, T_1, etc . . . to ϱ_2, B_2, T_2, etc . . . takes smoothly place. The relations between quantities ahead and behind the shock are still valid.

(c) *Shock in a real gas.* The situation becomes much more complicated when radiation is taken into account. According to SKALAFURIS and WHITNEY (1963) and to LEDOUX and WHITNEY (1961), the structure of the shock is the following.

The increase of temperature in the shock front produces a strong emission of radiation. Ahead the shock region exists a region where the precursor radiation ionizes the gas. Behind the shock front it is possible to find three regions: first the region where collisional relaxation between the external degrees of freedom occurs; second, the region where the internal relaxation (ionization and excitation) occurs, finally, the region of radiative relaxation, where the gas cools, principally by recombination radiation.

The main effect of the precursor radiation is to raise the kinetic temperature in the region of collisional relaxation.

(d) *Structure of the shock front.* For the time being, the gas has been considered as continuous medium. However, when considering the gas as made of ions and electrons, new features appear. When the Larmor radius of the electrons is much smaller than the mean free path, oscillations of the magnetic field can take place from region 1 ahead of the shock, to region 2 behind the shock (DAVIS, LÜST, SCHLÜTER, (1958)).

Between regions (1) and (2) the shock relations are still valid, but, over a distance which is of the order of the mean free path. In a phenomenological treatment, GEIGER, KAPPELER and MAYSER (1960) have shown how the oscillations of the magnetic field take place. After a sudden raise, the field oscillates, with a decreasing amplitude, towards the value behind the shock.

CAMAC *et al.* (1961) following PETSCHEK and KANTROWITZ (1960) have considered, in similar conditions, the presence of a stochastic distribution of the magnetic field in the transition region, the main effect being the irreversible transformation resulting from the interaction of magnetic waves in the shock front. In this last picture, the shock thickness is 4 to 10 times the characteristic length (c/ω_p), whereas the shock thickness in Davis, Lüst, Schlüter's picture is rather of the order of the mean free path. This depends naturally of the importance of the dissipative mechanism.

As can be seen from the discussion during the Varenna meeting (1960), there is no universal agreement on these theories. The nature of the dissipative mechanism which has to be introduced is not clear. In the picture given by Davis, Lüst and Schlüter, it is assumed that collisions between particles is the dissipative mechanism, and this leads to a shock thickness of the order of the mean free path.

In the picture given by Petschek and Kantrowitz, the non-linear interaction of waves produces the dissipation. The question is important. If the mean free path is a characteristic of the shock thickness, we can hardly speak of shock waves in the corona. Beyond the point where $n_e \simeq 10^6$, $T \simeq 10^6$, $R \simeq 140000$ km, we cannot speak of shock waves.

If the Larmor radius of Alfvén velocity characterizes the shock thickness, the shock thickness is of the order of $3 \cdot 10^7 N_e^{-1/2}$, and it is possible to speak of a shock, charge separation becomes important and the shock becomes an electrostatic shock.

15. Shock propagation

The propagation of a shock wave in a non uniform medium in the most general case is a difficult problem. It should take into account the divergence of the wave and its dissipation. For weak shocks, the Eikonal method can be applied and

TABLE 7

LARMOR RADIUS

Electrons

Energy	*Larmor radius (cm)*
50 keV	$840/B$
5 keV	$243/B$

Protons

Energy	*Larmor radius (cm)*
100 MeV	$1.57 \cdot 10^6/B$
1 MeV	$4.48 \cdot 10^5/B$

gives the possibility of calculating the energy repartition over the wave surface; and it is possible to use an approximate expression for the energy dissipation.

The situation is better for plane shocks, in a one dimensional medium. The interesting phenomenon is the very high energy dissipation, due to the irreversible processes taking place in the shock front. Even a very strong shock does not increase much its strength in the decreasing density field.

To give two examples, let us consider first the propagation of a moderately strong shock in a constant magnetic field. The shock strength varies with height in the following way:

$$\bar{\eta} = (4\pi\varrho)^{\frac{1}{2}} B^{-1} \left[\int_{z_0}^{z} \frac{2\pi K}{t_0 B^2} \varrho^{\frac{3}{4}} \mathrm{d}z \right]^{-1}$$

the equation of conservations of energy having been written:

$$\frac{\mathrm{d}}{\mathrm{d}z} \varrho v^2 V = - \frac{K \varrho v^3}{V t_0}$$

and V being the Alfven velocity:

$$V^2 = \frac{B^2}{4\pi\varrho}$$

The shock strength decrease roughly as $\varrho^{-1/4}$.

The other example is obtained by UCHIDA (1961) after elaborate calculation of the propagation of a shock wave in a chromosphere with the decreasing magnetic field of a dipole. The velocity of propagation is every where very near the Alfvén velocity, and is of several thousand kilometers per second.

TABLE 8

PROPAGATION OF A SHOCK WAVE IN A CHROMOSPHERE
(UCHIDA, 1961)

h (10^3 km)	V(km sec^{-1})	$\bar{\eta}$	n_p
8	40000	$1.8 \cdot 10^{-3}$	10^9
40	14000	$4 \cdot 10^{-2}$	$3.2 \cdot 10^8$
69.5	6500	$2.7 \cdot 10^{-2}$	$2.2 \cdot 10^8$
139	2100	$8.7 \cdot 10^{-2}$	$1 \cdot 10^8$
208	1100	$1.5 \cdot 10^{-1}$	$5.4 \cdot 10^7$

16. Acceleration in a shock front (SCHATZMAN, 1963)

The accelerating process is based on the fact that a particle crossing a perpendicular magnetohydrodynamic shock is accelerated. If the magnetic fields ahead the shock and behind the shock are uniform, the shock takes over particle, accelerate it, and leaves it behind, the energy of the particle having increased in the ratio of the magnetic fields. If the magnetic field is not uniform, but presents a clumpiness, the deflection of the charged particles by the clumps provides a statistical mechanisms which sends back at each turn a certain proportion of the particles in the acceleration process. For a particle which keeps going with a shock front of strength $\bar{\eta}$ over a distance x, the energy which can be obtained is:

$$E_{\text{ev}} = \left[E_R^2 + \left(300\,\bar{\eta}\,\frac{Bx}{2\pi}\right)^2\right]^{\frac{1}{2}}$$

where E_R is the rest energy and B the magnetic field.

A particle which has travelled over a great distance has obtained a large energy. At a given point a large number of particles are collected which have travelled over a small distance and a small number of particles which have travelled over a large distance.

The energy spectrum reflects the properties of the statistical mechanism. A heavy particle has experienced many deflections during one turn, and the probability of an appropriate deflection for continuation of the acceleration is not too small. An electron cannot experience many deflections during one turn, and, therefore, most of the electrons are immediately lost for the accelerating process. Fast electrons are likely to be secundaries from the protons of high energy.

The total number of fast particles depends on the injection mechanism. As the radius of gyration is not large, the energy losess are relatively small during one turn, and the conditions of injection are more favorable than in the Fermi mechanism. The condition that the gains are larger than the losses is equivalent to a condition on the density. If the density is smaller than a certain limit which depends on the temperature, the magnetic field and the shock strength, the acceleration can take place. The stronger the shock, the larger the number of particles behind the shock front with a velocity which is large enough for injection in the accelerating process.

It has also to be considered that the injection condition is more favorable for a strong shock, and that a strong shock can accelerate at a much higher density than a weak shock. If the shock is moving in an atmosphere, the total number of accelerated particles is naturally very sensitive to the level at which the acceleration begins. For a strong shock, we obtain a total number of accelerated particles, per cm^{-2} of the order of 10^{16}; but, in such a case, the acceleration begins at a very low level, where the density is of the order of 10^{17} particles cm^{-3}. The acceleration takes place in a region of about 1000 km, and is produced in a few seconds. If a sequence of shock waves travels in the atmosphere, each wave will produce an acceleration.

3. Solar Phenomena

17. Summary of observations

X rays are emitted during flares. De Jager predicted a close association between Type III and X rays (Nice, 1960). The very clear observations of ANDERSON and WINCKLER (1962) of the importance 3 flare (disc flare) of September 28, 1961 show the following sequence of events.

The enhanced Hα emission due to the flare was first detected at 22.02 UT, the maximum Hα emission occured at 22.24 UT.

The X rays appeared first at 22.17. The energy flux of X rays at the peak was $7 \cdot 10^{-7}$ erg cm^{-2} above 20 keV.

Type III bursts are associated with the flare. In the observations of Anderson and Winckler, the profile of the 3.16 cm emission coincides exactly with the profile of the X-ray emission, in shape and in time.

On a dynamic spectrum, several cluster of Type III, or fast frequency drifts, are observed. The first of these begins at 22.16–15 UT and corresponds very closely in time with the main rise of the X-ray burst.

The association of Type III and X rays occurs in 20% of the cases (KUNDU, 1962). Let us first discuss these phenomena. Anderson and Winckler discuss the X-ray emission with the help of an expression due to ELWERT (1961)

$$F(\text{earth}) = 7.10^{-52} n_i N_e \text{ erg cm}^{-2} \text{ sec}^{-1} ,$$

where n_i is the number of ions per cubic centimeter and N_e the total number of electrons above 20 keV.

For an ion density of 10^{12}, this leads to a number of electrons above 20 keV of 10^{33}.

If we consider bremsstrahlung, the numerical values given above give the following results:

Each electron gives about $\frac{1}{2}$ 10^{-8} ergs in X rays; the total X-ray energy emitted is of the order of $2 \cdot 10^{25}$ ergs, corresponding to the bremsstrahlung of $4 \cdot 10^{33}$ electrons. If the life time for the emission is of the order of 10^2 seconds, this corresponds to a density of about $10^{8.5}$ ions per cubic centimeter.

Consider now the power radiated at 3.16 cm. The radio emission amounts to $700 \cdot 10^{-22}$ watt m^{-2} cps. The total power radiated at wavelength shorter than 3.16 cm is of the order of 2.10^{14} watts. The width of the Type III is of the order of 3 seconds, and the energy radiated is of the order of 10^{15} joules, or 10^{22} ergs.

The plasma frequency corresponds to a density of 10^{10}, for which the life time at 50 keV is only of 1 second. Coherent emission over 10^{17} cm^2 could be due to 10^{33} electrons; however, it should be noticed that the total Hα flare area is $970 \cdot 10^{-6}$, or 10^{19} cm^2. The width of the Type III corresponds roughly to the life time of an electron of 50 keV at 10^{10} cm^{-3}.

It seems difficult to accept the hypothesis of coherent Čerenkov emission at 3.16 cm. It seems more likely that this emission is due to synchroton radiation. In a magnetic field of a few hundred gauss, 10^{33} electrons can easily emit the observed power.

The Type III drifts would take place at much higher altitude.

The interpretation of the X-ray spectrum by 3 X-ray pulses, separated by about 40 seconds, and accompanied by Type III bursts suggests clearly that the electrons have been trapped in a magnetic field, and that the X-ray burst represents three separate phases of the same electron bunch along lines of force. The regular decrease in amplitude of the X-ray pulse is then interpreted as due to loss of electrons with small pitch angles entering the solar atmosphere so deeply that they stop mostly by collision loss.

The electrons should spent most of their time in low density regions, and the fast drifts (Type III) should be produced by the ascending motion of the electrons. 10 MHz correspond to a density of 10^6 electrons cm^{-3}, and there is some contradiction between the trapping of the electrons, and the loss of the electrons producing Type III. These conflicting requirements can be met with a convenient structure of the magnetic field, in which the electrons producing the Type III are lost from the trap at each passage of the bunch of electrons.

18. The properties of solar cosmic rays. A summary.

The emission of cosmic rays is most often associated with a strong flare (3 or 3+) and is also associated with a Type IV burst.

The cosmic rays reach the earth some time after the onset of the Hα flare. The time of delay can be very small (a few minutes) and as big as one hour; this is usually interpreted as the time of delay from the sun. Over a distance of $3 \cdot 10^{13}$ cm (twice the Sun Earth distance), the travel time, as a function of the energy, is the following:

100 MeV	70 minutes
200 MeV	30 minutes
400 MeV	23 minutes
800 MeV	20 minutes.

The cosmic ray emission lasts a few hours. The protons spectrum can roughly be described by a power law

$$\frac{dN}{dE} = c^{te} E^{-\alpha}$$

where α ranges from 3 to 6. The protons are accompanied by alpha particles and heavier nuclei. The proportion of alpha particles and medium nuclei varies from flare to flare. Solar alpha particles are observed with an abundance of 3 to 7% of the solar protons of the same rigidity, and even larger. A comparison of charge composition of solar cosmic rays, galactic rays and solar abundance gives the following result (FICHTEL and GUSS, 1962):

TABLE 9

COSMIC RAYS COMPOSITION

	Protons	*Be B*	*C*	*N*	*O*
Galactic cosmic rays (same rigidity)	$2.6 \cdot 10^3$	5	10	5	6
Sun	$20 \cdot 10^3$	10^{-5}	10	2	18
Solar cosmic rays (same rigidity)	2.6 to $50 \cdot 10^3$	0.5	10	6	19

The energy of the particles is in the 100–400 MeV range. The number of particles of higher energy is very small. Estimates of the total energy carried by the particles is difficult to estimate. It may be as high as the total energy released in Hα, 10^{31} to 10^{32} ergs.

The delay in arrival of the particles is due to trapping. Assuming a delay of the order of 10^4 seconds, the particles should be trapped in a region of density smaller than 10^9. The time of trapping for 100 MeV, in a region of 10^9 cm is of the order of 10^3 seconds. In other words, we should have:

$$BL^2 \gtrsim 10^{21}.$$

The trapping should take place either in a larger region, or in a larger magnetic field. Furthermore, the magnetic energy of the trapping region should be much larger than the energy of the trapped particles; therefore, we should have:

$$B^2 L^3 \gg 10^{33}.$$

This implies that the region of trapping is very large: it is suggested to relate this condition to the problem of Type-IV flares.

19. Type IV flares (Boischot-Pick)

The first part of Type IV radiobursts consists of a *continuum radiation*. The metric emission comes from very large sources, 6 to 10 minutes or arc, which can move outward with speeds of the order of 1 000 km/sec.

It seems that the flares can be divided in two categories. Those having a heliographic longitude larger than 60° will give a non polarized emission, and those having a smaller longitude will in general give partly polarized bursts. If we consider the magnetic field of the leading spot of the active region, the sense of rotation seems to correspond to the extraordinary mode of the magneto-ionic theory.

This emission consists of synchrotron emission. Relativistic electrons in the coronal magnetic field will give an emission, mainly on the extraordinary mode, which will propagate through a decreasing density medium.

The flux is of the order of 10^{-19} watts m^{-2} Hz^{-1} for a frequency of $3 \cdot 10^9$. As an order of magnitude, 10^{37} electrons in a magnetic field of 1 gauss can produce such an emission.

It is reasonable to suppose that a magnetic bottle is ejected, and keeps both the electron which produce the Type IV, and the cosmic particles which diffuse ultimately out of the magnetic ring.

20. Structure of the magnetic field

Many observations have been made of the magnetic field at the photospheric level. However, little is known about the magnetic field in the chromosphere. Only guesses can be made from the aspect of the corona: coronal condensations, streamers, prominences.

The lines of force can be guessed from the polarity of the longitudinal field. However, the best informations are obtained from the measurement of the transverse field.

The observations show a line of zero longitudinal field. Along this line, the lines of force are parallel to the solar surface. The lines of force have the shape of arches, going either from one spot to the other, or from one spot to the photosphere. The line of zero longitudinal field seems to play an important role in flares; very often, the flare develops along the line of zero longitudinal field.

It is well known that the calcium plages coincide exactly with the regions of higher magnetic field (Leighton) whereas the region of higher transversal field coincides with the brightest part of Hα facula.

The region of higher magnetic field is also the region of greater mechanical flux (Osterbrock). To the heating of the plage region is certainly associated a solar wind of greater intensity than the normal solar wind.

The loss of mass due to solar wind modifies certainly the structure of the magnetic field. If the lines of force are carried by the matter, a very peculiar structure should develop, which, for the time being can only be sketched (Figure 1).

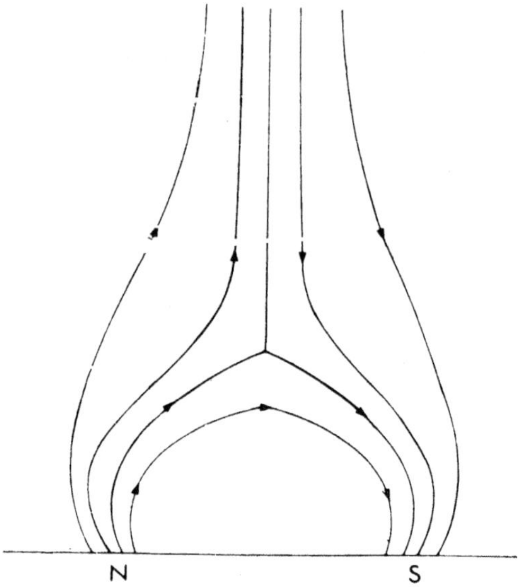

Fig. 1.

A region of weak field (nearly neutral surface) develops, and the whole region looks very much like the streamers. The peculiar magnetic surface from which stems the zero field surface is placed at a certain level, which changes with time. Due to the solar plage wind, the level of the peculiar magnetic surface decreases with time. The following suggestion is made now, in order to interpret the observations of X rays and Type III. The fast electrons which are produced by the accelerating process, can either be trapped all time in the magnetic field, or can reach one of the lines of force which go to infinity.

References

Allison, S. K.: 1958, *Rev. Mod. Phys.* **30**, 1137.
Anderson, K. A. and Winckler, J. R.: 1962, *J. Geophys. Res.* **67**, 4103.
Bates, D. R.: *Atomic and Molecular Processes*, Academic Press, New York, London, p. 549.
Bazer, T. and Ericson, W. B.: 1959, *Astrophys. J.* **129**, 758.
Bohm, D. and Gross, E. P.: 1949, *Phys. Rev.* **75**, 1864.
Boischot, A.: 1958, *Ann. Astrophys.* **21**, 273.
Brinkmann, H. C. and Kramers, H. A.: 1930, *Proc. Acad. Amsterdam* **33**, 973.
Camac, M., Kantrowitz, A. R., Litvak, M. M., Patrick, R. M., Petschek, H. E.: 1961, Int. Atomic Energy Meeting.
Cohen, M. H.: 1961, *Phys. Rev.* **123**, 711.
Davis, L., Lüst, R., Schlüter, A. von: 1958, *Z. Naturforsch.* **13**a, 916.
Dungey, J. W.: 1958, *Cosmical Electrodynamics.*
Dungey, J. W.: 1953, *Phil. Mag.* **44**, 725.
Elwert, G.: 1961, *J. Geophys. Res.* **66**, 391.
Fichtel, C. E. and Guss, T. E.: 1962, *Proc. Int. Conf. on Cosmic Rays* **1961**, **2**, 321.
Fite-Brackman, R. T.: 1958, *Phys. Rev.* **112**, 1141, 1151.
Geiger, W., Kappeler, H. J., and Mayser, B.: 1961, *Int. Atomic Energy Meeting.*
Ginzburg, V. L. and Železniakov, V. V.: 1958, *Astron. Ž.* 35, 694; *Soviet Astron.-A. J.* **2**, 653.

GIOVANELLI, R. G.: 1948, *Mon. Not. Roy. Astron. Soc.* **108**, 163.
GREENE, J.: 1959, *Astrophys. J.* **130**, 693.
HEITLER, W.: 1944, *Quantum Theory of Radiation*, 2nd ed.
DE JAGER, C.: 1961, *Space Research* **1**, 628.
KUNDU, M. R.: 1962, *Proc. Int. Conf. on Cosmic Rays and the Earth Storm* (Kyoto, sept. 4–15, 1961) **2**, 259.
LARENTZ, R. W.: 1963, Symposium on Electromagnetic Theory and Antennas.
LEDOUX, P. and WHITNEY, C. A.: 1961, *Nuovo Cimento* **22**, ser, X, suppl. 1, p. 131.
MOTT, N. F. and MASSEY, H. S. W.: 1949, *The Theory of Atomic Collisions*, 2nd ed., p. 252.
NORTON, R. VAN: 1961, N.Y. O. 9495, Magnetofluid Dynamics Division, Institute of Mathematical Sciences, New York University.
PARKER, E. N.: 1958, *Phys. Rev.* **109**, 1328.
PARKER, E. N.: 1963, *Astrophys. J., Suppl.* **8**, 177.
PETSCHEK, H. E., KANTROWITZ, A. R., FRISHMAN, F. J.: 1960, *Rev. Mod. Phys.* **32**, 959.
PICK, M.: 1961, *Ann. Astrophys.* **26**, 234.
SCHATZMAN, E.: 1963, *Ann. Astrophys.* **26**, 234.
SKLOVSKIJ, I. S.: 1958, *Astron. Z.* **35**, 838; *Soviet Astron.-A. J.* **2**, 786.
SKALAFURIS, A. and WHITNEY, C.: 1963, *Astrophys. J.* **138**, 200.
SPITZER, L., JR: 1962, *Physics of Fully Ionized Gases*, 2nd ed.
STEIN, W. A. and NEY, E. P.: 1963, *J. Geophys. Res.* 68, 65.
STEPANJAN, A. A. and WLADIMIRSKIJ, B. M.: 1961, *Astron. Ž.* **38**, 439; *Soviet Astron.-A. J.* **5**, 326.
SWEET, P. A.: 1956, *Trans. Int. Symp. IAU*, Stockholm, **6**.
TIDMAN, D. A.: 1960, *Phys. Rev.* **117**, 366.
UCHIDA, Y.: 1961, *Publ. Astron. Soc. Japan* **13**, 321.
VAN DE HULST, H. C.: 1953, in *The Sun* (ed. by P. KUIPER).
WANDEL, C. F., JENSEN, T. H., KOFOED, H. O.: 1959, *Nuclear Instruments and Methods* **4**, 249.
WARWICK, J.W.: 1962, *P.A.S.P.* **74**, 302.
WENTZEL, D. G.: 1963, *Astrophys. J.* **137**, 135.

THE VARIABLE SOLAR RADIO SPECTRUM

A. MAXWELL
Harvard Radio Astronomy Station, Fort Davis, Texas, U.S.A.

Abstract. This paper reviews recent experimental and theoretical work concerning the slowly-varying component of solar radio emission, microwave bursts, and the four main spectral types of bursts observed in the meter and decimeter bands. Observational techniques are briefly summarized.

The slowly-varying component is believed to be thermal in origin and to come from overdense regions above plages and sunspots. The physical characteristics of this component are tabulated and recent theories of emission processes are discussed. Microwave bursts, as observed by both fixed-frequency and sweep-frequency receivers, are categorized and emission mechanisms for each category are considered. The characteristics and possible origins of Type-I (noise storm) bursts, which originate high in the corona above active sunspots, are reviewed.

Type-II (slow-drift) bursts are believed to be generated by disturbances propagating outward through the corona, sometimes non-radially, at velocities of about 1000–1500 km/sec, in regions where the electron density is about 10 times that given by the Baumbach-Allen distribution. Measurements at meter wavelengths of the brightness distribution across the Type-II disturbances indicate that the radiation originates in a bright central emitting core, surrounded by an extended halo of scattered radiation. Observations with a swept-frequency interferometer show that the radiation generated in the second harmonic comes from directions corresponding to much lower heights in the solar atmosphere than the fundamental. It has been suggested that radiation in the second harmonic is preferentially back-scattered from the point of generation; the second harmonic is then observed after reflection at much lower coronal heights.

Type-III (fast-drift) bursts, generated by an exciter moving at approximately 0.5 c, are also believed to be generated in density condensations and to have a core and halo brightness distribution similar to that of Type-II bursts. Examination of the burst profiles gives information on the dispersion of the exciter as it moves outward through the corona.

The complex properties of Type-IV (continuum) bursts, as observed throughout the complete radio spectrum (intensities, polarization, diameters, spectral characteristics, motions), are summarized and discussed in terms of existing models based on synchrotron and Cerenkov plasma wave emission processes.

Finally, the relation of the radio bursts to accompanying X-ray and high-energy nucleon emission is briefly examined in terms of existing theories of electromagnetic emission and particle acceleration in the solar atmosphere.

1. Introduction

The radio emission from the sun consists of three main components.

(i) A background thermal emission, which results from free-free transitions of electrons in the field of ions (mainly protons). The thermal emission processes are fairly well understood and the present radio models for the background radiation are in good accord with the observed data.

(ii) A slowly-varying component, which is related to the total area of sunspots visible on the sun, and is most prominent at centimeter and decimeter

 All rights reserved.

wavelengths. This component is difficult to distinguish at frequencies lower than about 200 Mc/s. It is believed to be thermal in origin and to be emitted from regions of increased electron density overlying sunspot and plage areas.

(iii) Transient disturbances, sometimes of great intensity, which originate in localized active areas. These burst radiations are most intense at meter and decameter wavelengths, but are visible at times throughout the whole spectrum. Some of the bursts are believed to result from plasma oscillations, some possibly from bremsstrahlung or amplified gyro-radiation, and others from synchrotron radiation.

This review paper is concerned with the variable solar radio spectrum: namely the slowly-varying component and the transient bursts. We shall not deal with the background thermal emission or its long term variations during the solar cycle. Section 2 summarizes the great variety of observational techniques that have been used to study the various components of solar radio emission. Section 3 lists certain parameters that apply to the solar plasma and surveys the emission mechanisms, both thermal and non-thermal, that have been evoked to account for various types of solar radio bursts. Section 4 discusses the physical characteristics of the slowly-varying component, and current theoretical interpretations of the spectrum, emission height, and angular diameter of the component. Section 5 summarizes observational data concerning microwave bursts, obtained with fixed- and sweep-frequency receivers, and discusses emission mechanisms for the various types of microwave bursts. Sections 6, 7, 8, and 9 deal in turn with bursts in the meter and decimeter wavelength bands known as spectral Type I (noise storm bursts), Type II (slow-drift bursts), Type III (fast-drift bursts), and Type IV (continuum bursts). The physical characteristics of these bursts and their association with flares and sunspots are described, and theories of their generation in the solar corona are discussed. Section 10 examines the relation of radio bursts generated by great flares, to accompanying X-ray and high-energy nucleon emission.

2. Observational Techniques

2.1 *Fixed-Frequency Receivers; Polarization; Flux Density*

Many observations of solar radio bursts have been made with fixed-frequency receivers, attached to antennas of relatively low gain, at numerous frequencies from about 10 Mc/s through 70000 Mc/s. Such equipment has been used to examine the intensity-time characteristics of solar radio bursts and to determine the polarization of the bursts. De Groot (1959) and Vitkevič and Gorelova (1960) have used fixed-frequency receivers at adjacent frequencies to obtain multi-channel information, over a band of about 20 Mc/s, of the characteristics of the burst emission.

The polarization of solar radio bursts is generally determined from antennas of accepting either linear or circular polarization. The polarization parameters are determined by electronic or mechanical switching, with appropriate phase insertions, between two arrays of crossed polarization.

The flux density from the sun is given by

$$F = \int^{\mathrm{SUN}} B(T)\mathrm{d}\omega$$

where $B(T)$ is the brightness. From the Rayleigh-Jeans formula

$$B(T) = 2kT/\lambda^2$$

where k = Boltzmann constant and λ = wavelength. Thus in MKS units

$$F = 2kT\,\Delta\omega/\lambda^2 \qquad \text{watts m}^{-2}\ (\text{c/s})^{-1}$$

$$= 2.09 \times 10^{-32} f^2 T \quad \text{watts m}^{-2}\ (\text{c/s})^{-1}$$

where f is in Mc/s and $\Delta\omega = 6.8 \times 10^{-5}$ steradians for a $^1/_2{}^\circ$ disk. In solar radio astronomy, flux densities are usually expressed as multiples of 10^{-22} MKS units.

2.2 *Sweep-Frequency Receivers*

The spectral characteristics of solar radio bursts may be observed with sweep-frequency receivers (Wild and McCready, 1950). Such equipment uses narrow-band tunable receivers which are repeatedly swept across a wide frequency band. Until the present time, the range of each such receiver has been limited to one octave and most observations have been made with sweep rates of 2 or 3 per second. The equipment at the Harvard Station, from the records of which many of the illustrations in this paper have been taken, covers the band 25–580 and 2100–3900 Mc/s (Thompson, 1961). Equipment of narrower frequency range, 0.13 octave, has been used at a sweep rate of 50 per second (Elgaroy, 1961). The display systems for spectral equipment normally consist of intensity-modulated cathode ray tubes, photographed with a continuous motion camera. The equipments thus record the intensity of the burst emissions as a function of frequency and time.

2.3 *Interferometers*

Historically, two-element interferometers, in which an interference pattern is formed when a discrete source of radiation drifts through the interferometer lobes, were used to resolve and identify regions of intense non-thermal radio emission overlying active sunspot areas on the solar disk. They are unsuitable for the study of individual solar bursts of a few seconds' duration. A swept-lobe interferometer, in which the lobe pattern was swept rapidly over the radio source under investigation, was introduced by Little and Payne-Scott (1951).

A *one-dimensional* radio brightness distribution across the solar disk may be obtained as follows. (i) By Fourier synthesis of information obtained from a two-element interferometer, in which the element spacing is varied along a given azimuth (generally East-West). This method was suggested by McCready, Pawsey, and Payne-Scott (1947) and was first applied by Stanier (1950) who determined a one-dimensional brightness distribution across the solar disk, at sunspot minimum, at a frequency of 500 Mc/s. (ii) By multiple-element interferometers, in which antennas are distributed along one azimuth (the radio configuration being analogous to an optical diffraction grating). This technique was developed by Christiansen and Warburton (1953) with a 32-element interferometer, along an East-West line of 1000 wavelengths. (iii) At ultra-high frequencies by a single, high-resolution fan-beam such as that of the Pulkov radio telescope, which has a beamwidth of 1 arc min $\times$ 1 arc degree at an operating frequency of 9400 Mc/s (Ihsanova, 1959).

A *two-dimensional* distribution of radio brightness across the solar disk may be obtained from a series of one-dimensional strip-scans, taken with a two-element interferometer, in which the elements are arranged at various spacings along a fixed azimuth; further strip-scans are then taken at different orientations. This method was used by O'Brien (1953), but the analysis is complex and considerable time is required to make the observations.

The final aim of interferometric methods is to obtain a two-dimensional radio map of the sun in as short a time as possible. The first instrument capable of obtaining a true two-dimensional radio map was put into operation in 1957 at a frequency of 1420 Mc/s (Christiansen and Mathewson, 1958). It combined the principles of the multi-element interferometer and the Mills' Cross, and consisted of two grating interferometers, arranged in two rows, North-South and East-West, in the form of a cross. The antenna had a directional diagram consisting of a grid of points in the sky. As the earth rotated, the various points in the grid scanned different strips across the sun. The lobe system could also be moved by phase adjustment of the North-South arm, and in this way, a picture of the sun could be built up in a period of about half an hour. The radio heliograms at 1420 Mc/s gave an indication of the brightness distribution of the solar disk corresponding to a height of approximately 40000 km above the photosphere. Similar systems have now been put into operation in other places at different frequencies: for example, at Stanford University at 3200 Mc/s. Multi-element crossed interferometers have been used mainly to study the microwave radio emission from centers of slowly-varying activity, and the longer lived radio outbursts initiated by flares. At present, no instrument can compile a radio map of the sun in less than about 30 minutes.

2.4 *Sweep-Frequency Interferometer*

The sweep-frequency receivers described in Section 2.2 measure the intensity of solar bursts as a function of frequency and time. It is obviously desirable to supplement the spectral data with information concerning the position of the bursts on the solar disk. This has been done with a sweep-frequency interferometer, which has been designed to measure one coordinate of solar radio bursts, nominally hour angle, as a function of frequency and time. So far, only one such interferometer has been constructed (WILD, SHERIDAN, and NEYLAN, 1959) and it is located in Australia. It covers the range 40–70 Mc/s and its sweep rate is twice per second. The relative length of the transmission lines to the antennas can be varied extensively, and in this way the interferometer lobe system is kept on the sun for approximately ± 2 hours about transit. The interferometer uses two antennas separated by 1 km on an East-West base line. A second interferometer system, at a spacing of 0.25 km, resolves the lobe ambiguities associated with two antenna interferometry. The equipment has automatic lobe switching and phase calibration, and the output is displayed as a single intensity-modulated line on the record of a facsimile recorder. The instrument measures the centroid of the radio source, instantaneously at many frequencies, to an accuracy of about ± 1 arc min.

3. Emission Mechanisms

In this section we discuss the emission mechanisms that have been invoked to account for the variable components of solar radio emission. For the most part, the theories of emission will be given by reference and only the main conclusions will be summarized.

3.1 *Escape Conditions; Plasma Parameters*

Radio waves from the sun at any given frequency f can escape only from outside the boundary at which the electron density N in the ionized solar atmosphere reduces the refractive index n to zero. In the absence of a magnetic field, $n = \sqrt{(1 - f_0^2/f^2)}$, where $f_0 = \sqrt{(Ne^2/\pi m)}$ = the plasma frequency, and e and m are the charge and mass of an electron. The electron density decreases with increasing height, and thus radiation at centimeter wavelengths can escape from the low levels of the solar atmosphere, the chromosphere, whereas at meter wavelengths emission is restricted to the outer levels, the corona. Ray trajectories are also governed by the refracting properties of the medium and give rise to directivity in the emission.

Table 1 lists numerical values of various parameters of the solar plasma at sunspot maximum, for two distinct levels; namely the upper chromosphere at a height of 20000 km, and the corona at a height of 200000 km. Two sets of data are given for these levels. One set corresponds to conditions in the quiet atmosphere, with values of electron density corresponding to those given by ALLEN (1963). The other set refers to conditions above an active sunspot. For the latter case the sunspot is assumed (i) to be bipolar, the photospheric field strength of each spot being 1000 gauss and the spot separation 35000 km (UCHIDA, 1960); (ii) to have a dipole field extending outward into the solar atmosphere; (iii) to form the base of a coronal streamer, which extends outward approximately radially for several solar radii, and which has an

TABLE 1

NUMERICAL VALUES FOR SOME PARAMETERS OF THE SOLAR PLASMA AT SUNSPOT MAXIMUM

	Undisturbed Plasma		*Model Streamer Above Bipolar Sunspot*		*Reference*
	Upper Chromosphere 20000 km	*Corona 200000 km*	*Upper Chromoshere 20000 km*	*Corona 200000 km*	
Electron density N (cm^{-3})	3×10^8	5×10^7	3.2×10^9	3.8×10^8	Undisturbed N: ALLEN (1963) Streamer N: $10 \times$ Baumbach-Allen
Temperature T (°K)	10^6	10^6	2×10^6	2×10^6	Undisturbed T: ALLEN (1963) Streamer T: assumed value
Velocity sound v_{sound} (km/sec)	120	120	170	170	$v_{sound} = \sqrt{\gamma RT} \times 10^{-5}$ km/sec
Thermal velocity of electrons v_{te} (km/sec)	6750	6750	9550	9550	$v_{te} = \sqrt{3kT/m} \times 10^{-5}$ km/sec
Collision frequency (protons-electrons) ν (c/s)	13	2	48	6	$\nu = 42 \times NT^{-\frac{3}{2}}$ c/s
Plasma frequency f_0 (Mc/s)	156	64	508	175	$f_0 = \sqrt{Ne^2/\pi m} \times 10^{-6}$ Mc/s
Model dipole field H (gauss)	—	—	250	3	UCHIDA (1960)
Gyro frequency f_H (Mc/s)	—	—	700	8.4	$f_H = 2.8$ Mc/s per gauss
Alfvén velocity v_A (km/sec)	—	—	9640	340	$v_A = \sqrt{H^2/4\pi\varrho_0} \times 10^{-5}$ km/sec

electron density distribution 10 times the normal Baumbach-Allen distribution (refer Section 7.6).

The table shows that in the corona the gyro-frequency is generally less than the local plasma frequency, whereas in the chromosphere the gyro-frequency may be greater than the local plasma frequency. In both the chromosphere and the corona the collisional frequency is much less than either the gyro-frequency or the plasma frequency.

We now summarize various emission mechanisms in turn.

3.2 *Bremsstrahlung*

Bremsstrahlung or "braking radiation" is emitted when electrons are deflected by other charged particles. It corresponds to the free-free transitions of quantum theory. SMERD (1963) has shown that bremsstrahlung from thermal electrons may be important in the generation of bursts if there is a sudden localized heating in the solar plasma. In the case of centimeter wavelength radiation, from the chromosphere, the optical depth is much greater than unity and the brightness temperature increases linearly with T. In the case of meter wavelength radiation from the corona, the optical depth is small and the brightness temperature is proportional to $T^{-1/2}$, so that localized heating will cause a decrease in brightness temperature. Smerd concludes that bremsstrahlung can generate solar bursts at centimeter wavelengths but not at meter wavelengths. Non-thermal electrons may also generate bremsstrahlung, but it will not be of great brightness.

3.3 *Gyro-Radiation*

Weakly relativistic electrons in a plasma permeated by a magnetic field, of strength H gauss, will gyrate and will radiate emission at the fundamental gyro-frequency, $f_H = eH/2\pi mc$, and at a few harmonics of that frequency. The application of this mechanism in the solar plasma was first considered by GIOVANELLI (1948). The mechanism has been re-examined by TAKAKURA (1956), TWISS and ROBERTS (1958), and by SMERD (1963). Twiss and Roberts point out that gyro-radiation is generated mainly in the *extraordinary* mode and that, as the fundamental can not escape in the extraordinary mode, energy should be radiated mainly in harmonics of the extraordinary mode. However, as regards noise storm radiation, the observations (Section 6.3) strongly suggest that the radiation is in the *ordinary* mode, which is in direct conflict with the theoretical predictions. Twiss and Roberts also find it difficult to account for the observed intensity of noise storms in terms of a gyro-model. On the other hand, Smerd believes that the polarization measurements are not conclusive and that objections to the idea of "bunching" of electrons, which would generate coherent emission and thus explain the high intensity of the bursts, are invalid in the presence of a magnetic field. Smerd therefore considers that mechanisms of gyro-radiation can no longer be rejected. He finds that, for a positive gradient in electron energy distribution, and in the presence of Doppler and magnetic spread, "resonance" amplification can occur in the gyro case. Smerd therefore believes that gyro-amplification can account for Type I and other bursts superimposed on a high level continuum background. The origin of these bursts would be electron streams in localized regions, over which the magnetic field is effectively constant.

3.4 *Synchrotron Radiation*

Radiation by the synchrotron mechanism, from electrons of relativistic or near-relativistic velocities trapped in a magnetic field, has been dealt with extensively in the literature (SCHOTT, 1912; VLADIMIRSKIJ, 1948; SCHWINGER, 1949; and others). This emission, in which many harmonics are emitted, is the relativistic form of gyro-radiation. A single relativistic electron of energy E orbiting in a magnetic field of H gauss radiates a series of harmonics of fundamental frequency

$$f_1 = f_H(mc^2/E) = f_H\sqrt{1 - \beta^2}$$

where β is the velocity of the electron relative to c, the velocity of light. The radiation is concentrated in the plane of motion of the electron. The harmonics lie close together and in most cases E and H are not single valued, so that the radiation appears as a continuum. The maximum energy is radiated at a frequency

$$f_{\max} = f_H(E/mc^2)^2$$

and this frequency also gives a rough upper limit to the emission spectrum. The maximum frequency at which the radiation appears is thus proportional to H and to E^2.

The possibility that synchrotron radiation might account for Type IV bursts was suggested by BOISCHOT and DENISSE (1957). Synchrotron emission processes in the solar plasma have been examined in detail by TAKAKURA (1960) and by GINZBURG and ŽELEZNIAKOV (1961). The polarization of the radiation is generally elliptical and the extraordinary mode predominates, owing to the sense of rotation of the electron. However, TAKAKURA (1960) has shown that for steep pitch angles there may be appreciable radiation in the ordinary mode. Takakura has also shown that under certain conditions the escape conditions may favor the ordinary mode and has suggested that such conditions account for the observed change of polarization of Type-IV bursts between the centimeter and the decimeter bands (Section 9.4).

3.5 *Plasma Oscillations*

We shall not review here the extensive literature on plasma oscillations in the solar atmosphere. For detailed discussion of these topics the reader is referred to BÖHM and GROSS (1949, 1950), DENISSE and DELCROIX (1961), GINZBURG and ŽELEZNIAKOV (1958, 1959, 1961). We shall, however, briefly summarize the critical conditions required for setting up plasma waves by superthermal electron streams or other disturbances, and shall discuss subsequent scattering mechanisms. It is generally believed that, in considering intense radio bursts from the solar plasma, we are dealing with coherent emission, which is inherently orders of magnitude stronger than incoherent emission.

GINZBURG and ŽELEZNIAKOV, in considering radio emission from a magneto-active plasma, show that there are three refractive indices to be considered. The first two, n_1 and n_2, correspond to the refractive indices for the ordinary and extraordinary modes of transverse electromagnetic wave propagation. The third refractive index, n_3, corresponds to that of a longitudinal plasma wave, and is given by

$$\pi_3^2 = \frac{1 - \omega_0^2/\omega^2}{v_{te}^2/c^2}$$

where $\omega_0 = 2\pi f_0$, $\omega = 2\pi f$ and v_{te} = r.m.s. velocity of the thermal electrons $= \sqrt{(3kT/m)}$. T is the electron temperature and k is the Boltzmann constant. The dispersion relation for these waves is given by

$$\omega = \sqrt{\omega_0^2 + \kappa^2 v_{te}^2}$$

where $\kappa = \omega_0/v_{\mathrm{ph}}$ = wave number, and v_{ph} = plasma wave phase-velocity. For plasma waves to be excited by a stream of electrons of velocity $\mathbf{v}$, the Čerenkov condition requires

$$\beta \pi_3 \cos\theta \geqslant 1$$

where $\beta = \mathbf{v}/c$, and θ is the angle between the electron stream and the direction of propagation of the plasma waves. The latter equation corresponds to the condition

$$\mathbf{v} > v_{te} .$$

That is, for a stream of electrons to excite longitudinal Čerenkov plasma waves, the stream electrons must have a velocity greater than the thermal velocity of the electrons in the plasma. In the corona, this implies that the stream particles must have velocities greater than about 7000 km/sec.

Ginzburg and Železniakov further show that longitudinal plasma waves couple energy into transverse electromagnetic waves by scattering on fluctuations δN in the electron concentration, where

$$\delta N = \delta N' + \delta N''$$

Here, $\delta N'$ corresponds to *Rayleigh scattering* on fluctuations of density in the plasma and $\delta N''$ represents *combination scattering* of plasma waves on plasma waves. Rayleigh scattering is not accompanied by a change of frequency, but combination scattering preferentially radiates energy at the second harmonic of the local plasma frequency. The efficiency of the energy transfer from kinetic to electromagnetic modes for these scattering processes in the corona is believed to be about 10^{-6}. The angular distribution of the scattered electromagnetic radiation has been investigated by GINZBURG and ŽELEZNIAKOV (1958, 1959, 1961) and also by SMERD, WILD and SHERIDAN (1962). As would be expected, they find the electromagnetic radiation to be beamed strongly in directions orthogonal to the direction of propagation of the plasma wave.

Plasma waves may, of course, be set up in other ways. For example, at the time of a flare a large amplitude disturbance may be propagated through the corona as an acoustic shock wave. Irregularities in the magnetic field or diverging field lines would locally generate transverse modes that could eventually escape as radio frequency emission. If energy from the disturbance is started in transverse (Alfvén) modes, the traveling waves may also incorporate components that may eventually escape from the atmosphere as Type-II radio-frequency waves. Energy may also be propagated in high-velocity transverse modes, such as an electron plasma shock. Energy in these modes will be propagated along a strong field region with a velocity intermediate between the thermal velocity of electrons and the velocity of light. These transverse shock waves will also excite the plasma as they move outwards and may generate Type-III bursts.

4. The Slowly-Varying Component

4.1 *Observational Data*

The slowly-varying, or *S*-component, of solar radio emission was recognized by COVINGTON (1948) at 2800 Mc/s and by LEHANY and YABSLEY (1949) at 600 and 1200 Mc/s. The source of the radiation, above sunspots and chromospheric plage areas, was established from eclipse observations. The intensity of the radiation changes slowly from day to day and is roughly proportional to the total sunspot area on the sun. The component also continues to be radiated from plage areas, which may remain for several rotations in areas where sun-

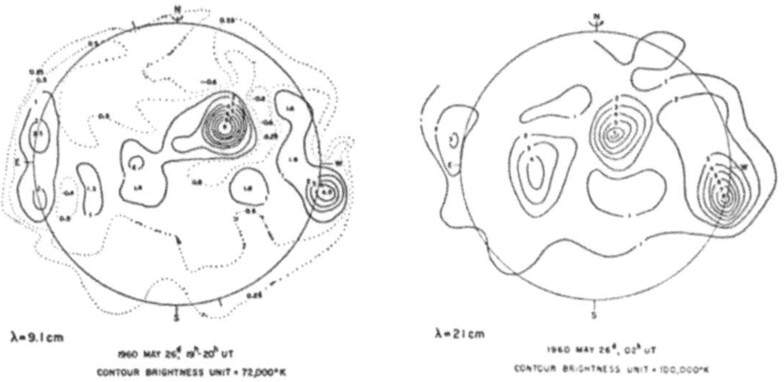

Fig. 1. Radio heliograms at wavelengths of 9.1 cm (Stanford) and 21 cm (CSIRO). Reproduced from SWARUP *et al.* (1963).

spots have previously been observed. The slowly-varying component is most marked at centimeter and decimeter wavelengths. It has been observed at a wavelength of 176 cm (BOISCHOT, 1958) but generally cannot be resolved at longer wavelengths. Radio heliograms, showing bright areas responsible for the slowly-varying component, are presented in Figure 1.

Because of the refracting properties of the solar atmosphere, the radiation shows an approximately cosine variation in intensity as the bright area moves from the center of the solar disk to the limb. The height of the emitting region as a function of wavelength has been determined (i) from direct observations at the limb (normally difficult because of refraction effects) and (ii) from the rate of travel of the radio emission regions across the disk (compared with the rate of travel of the associated optical region on the photosphere). Some general statistical features of the slowly-varying component are listed in Table 2. The radiation is generally partially circularly polarized at wavelengths shorter than 10 cm. The sense of polarization is opposite for the two components of a bipolar sunspot group and each corresponds to the extraordinary wave of magneto-ionic theory. At longer wavelengths the radiation is virtually unpolarized, indicating that strong magnetic field effects from plage regions extend only to heights of approximately 20000 km in the solar atmosphere. The intensity of the radiation reaches a maximum at a wavelength of about 10 cm.

TABLE

CHARACTERISTICS OF THE

Frequency (Mc/s)	*Wavelength λ (cm)*	*Average Source Size at CMP (arc min)*	*Brightness Temperature T_b (× 10^6 °K)*	*Average Flux for Sunspot Area 10^{-3} (× 10^{-22} MKS units)*	*Average Emissio[n] Height (on disk[)] (× 10^3 km)*
9500	3.2	2.8	—	20	15
		1.5–10	0.5–0.1	—	—
		1–2	0.3–1	—	50
4000	7.5	3–5	0.7	25	15
3300	9.1	2.8	0.9	22	18
2800	10.7	2.4	2.7	24	15
1420	21	4–8	1.5	—	40
		3.6	—	13	63
500	60	<3–6	5	4	35
340	88	—	1–10	—	130
169	176	>10	<2	—	200

A study of an active region using combined optical and radio techniques was made by CHRISTIANSEN *et al.* (1960) for an active region which crossed the central meridian of the sun of December 3–4, 1957, at 18° S. The radio observations were made by means of highly-directional grating-type arrays operating at wavelengths of 7.5, 21, 88, and 176 cm. The results showed that the size and shape of the radio bright region conformed very closely to the size and shape of the underlying plage region. Brightness temperatures were measured from the flux, and electron temperatures and densities for the radiating region were then deduced.

SWARUP *et al.* (1963) studied ten strong sources of the slowly-varying component with high resolution interferometers at wavelengths of 3.2, 7.5, 9.1, 10.7, and 21 cm. The ten radio sources were all associated with sunspot groups of area > 500 millionths of the solar disk. These observers confirmed that the flux density of the slowly-varying component increased with decreasing wavelength to about $\lambda = 10$ cm, and then decreased at shorter wavelengths. At $\lambda = 10.7$ cm, they found that the brightness temperatures of the ten sources ranged from 1.6 to 3.8×10^{6}° K, assuming that the north-south widths of the sources equalled the observed east-west widths. The brightness temperatures showed good correlation with both sunspot areas and magnetic field strengths.

2

SLOWLY-VARYING COMPONENT

Polarization (per cent circular)	Antenna: Beam (arc min)		Antenna: Place	Reference
30	Fan	2.2	Nagoya	SWARUP *et al.* (1963)
—	Fan	1.8	Nançay	KUNDU (1959)
—	Fan	1	Pulkovo	IHSANOVA (1959)
10–15	Fan	4.5	Nagoya	KAKINUMA (1956); SWARUP *et al.* (1963)
10	Pencil	3.1		
	Fan	2.3	Stanford	SWARUP *et al.* (1963)
10	Fan	1.2	Ottawa	SWARUP *et al.* (1963)
<2	Pencil	3.0	Sydney	CHRISTIANSEN *et al.* (1960)
<2	Fan	2.0		SWARUP *et al.* (1963)
—	Fan	8.7	Sydney	SWARUP and PARTHASARATHY (1958)
—	Fan	4.8	Maryland	FIROR (1959); CHRISTIANSEN *et al.* (1960)
—	Fan	3.8	Nançay	BOISCHOT and SIMON (1959); CHRISTIANSEN *et al.* (1960)

4.2 *Interpretation*

WALDMEIER and MÜLLER (1950) and WALDMEIER (1956) proposed a model "coronal condensation" to account for the slowly-varying component. This model envisages an approximately hemispherical condensation of diameter approximately 125000 km (2.7 arc min) with densities approximately 20 times normal at heights of 20000 km, decreasing to approximately normal values at 80000 km. The model apparently gave best agreement with the radio observations for temperatures of the order of 6×10^6 °K.

In their combined optical and radio study of a particular active region CHRISTIANSEN *et al.* (1960) derived a detailed model for the electron density and temperature distribution as a function of height above the optical bright area. The results of the analysis suggested that the radiation originated from regions in the corona in which the density was considerably higher than in the surrounding undisturbed corona. For this particular set of observations the density was apparently three times that of the Baumbach-Allen model. By contrast the electron temperatures in the region of increased electron density were approximately normal. The radiation at $\lambda = 7.5$ cm was 20 per cent circularly polarized, and its sense of polarization corresponded to the extraordinary component from the predominantly south magnetic polarity of the region. To account for the polarization at $\lambda = 7.5$ cm, magnetic fields of approximately 1000 gauss would have to extend over an area of diameter 1.5 arc min at heights of approximately 18000 km. Optical measurements of the magnetic field strength gave 1000–2000 gauss at photospheric levels. From the combined observations it was therefore concluded that the slowly-varying component of solar radio emission is thermal in origin and has its source in dense regions which overlie plage regions and extend radially outwards.

NEWKIRK (1961) compared optical observations of high density regions in the corona, made with a K-coronameter with the radio data. The optical equipment gave measurements up to 1 radius from the sun, and had a resolution of 1 arc min. Newkirk suggested that coronal condensations of the type envisaged by Waldmeier and Muller give rise to a small "button" of radio emission, which lasts for only a few days. This short-lived region, at times of maximum development, may be associated with emission of the coronal yellow line, at $\lambda = 5694$ Å, and temperatures of about $2–4 \times 10^6$ °K. Newkirk further suggested that the source of the slowly-varying component was a coronal streamer, similar to that proposed by Christiansen et al.

ŽELEZNIAKOV (1962) suggests that gyro radio emission by electrons in resonance layers $\omega = 2\omega_H, 3\omega_H$ above sunspots, and bremsstrahlung away from sunspots, can account for the effective temperature, brightness distribution, directivity, sense of polarization, and degree of polarization of the slowly-varying component, as a function of frequency. Under some conditions at longer wavelengths, with high values of electron density, bremsstrahlung emission in higher levels may predominate.

The idea of resonance absorption has been further examined by KAKINUMA and SWARUP (1962), who show that the decrease in flux density of the slowly-varying component at wavelengths < 10 cm and the observed degree of polarization are not consistent with simple magneto-ionic theory, in which only the effect of electron-ion collisional absorption is considered. In order to explain observed values of brightness temperature, $2\text{–}4 \times 10^6$ °K, near wavelengths of 10 cm by simple magneto-ionic theory, they find it necessary to assume values of electron density of 20–40 times the normal values. Nor can the decrease in flux density with frequency be explained merely by assuming a greater optical thickness for the extraordinary wave in the corona, since the observed degree of polarization at $\lambda = 10$ cm is much lower than that at 3 cm.

Whilst retaining a thermal hypothesis for the origin of the radiation, to account for the steadiness of the radio emission over periods of days, Kakinuma and Swarup suggest that radiation at the gyro-frequency and its harmonics is emitted by thermal electrons in regions of greater density over a sunspot group, should be taken into account to explain the observed spectrum. The resonance absorption mechanism requires magnetic fields of 600 gauss at height 20000 km, and 250 gauss at 40000 km. The analysis of Kakinuma and Swarup shows that on gyro-theory, densities of 5–10 times normal above a sunspot area are sufficient to explain simultaneously the observed brightness temperatures, flux densities, and polarizations. The theory also explains the small size of the source in the wavelength range 3–10 cm. Electron temperatures in the region of the inner solar corona above a large sunspot group are estimated at about $2\text{–}4 \times 10^6$ °K.

Estimates of electron temperature of $2\text{–}4 \times 10^6$ °K are possibly somewhat high, when compared, for example, with estimates of coronal electron temperatures made from ionization theory (SEATON, 1963), or from Doppler broadening of the coronal green line, $\lambda = 5303$ (BILLINGS, 1963). The appearance of the coronal yellow line, $\lambda = 5694$ Å, indicative of coronal temperatures of $2\text{–}4 \times 10^6$ °K, is generally limited to a few hours and therefore can not be invoked to account for high temperatures lasting for many days above active regions. If non-thermal processes are required, one may consider the possibility of a quasi-stable regime of relatively low-energy magnetohydrodynamic shock waves, generated continuously above the plage region, which may cause FERMI (1949, 1954) acceleration of electrons in the vicinity of the plage.

5. Microwave Bursts

5.1 *Observational Data*

Microwave bursts have been investigated mainly with fixed-frequency equipment. Relatively few spectral observations have been made in the microwave band, so that it is at present not entirely possible to unify the microwave bursts with the spectral categories used in the meter wavelength range. It is believed that there are three main categories of microwave bursts.

(i) *Impulsive bursts*, lasting about 1–5 minutes, brightness temperatures up to 10^9 °K, intensity often up to 1.00×10^{-22} MKS units, angular diameter about 1 arc min, with rapid intensity variations, rapid rise and fall (COVINGTON and HARVEY, 1958; COVINGTON, 1959; KUNDU and HADDOCK, 1961), and polarization generally $<$ 50 per cent circular in the extraordinary mode (TANAKA and KAKINUMA, 1959).

(ii) *Type-IV microwave bursts*, usually lasting 5–40 minutes, brightness temperatures up to 10^9 °K, intensity often up to 2000×10^{-22} MKS units, angular diameter 2–4 arc min, of moderate variability, fairly rapid rise and relatively slow decay, polarization generally $<$ 50 per cent circular in the extraordinary mode (for references see Section 9.1, Table 4). These bursts are often preceded by an impulsive burst and are thought to be part of the generalized Type-IV burst.

(iii) *Long-enduring bursts*, often lasting 1–3 hours, brightness temperature about 0.5–1 $\times 10^6$ °K, intensity about 5–40 $\times 10^{-22}$ MKS units, angular diameter about 1–2 arc min, with no rapid variations in intensity, relatively slow rise and fall (COVINGTON and HARVEY, 1958; COVINGTON, 1959), and polarization corresponding to that of the background slowly-varying component appropriate to region (refer Section 4.1).

Intensity-time profiles of the three main categories of microwave bursts, as observed with a fixed-frequency receiver at a wavelength of 10.7 cm (2800 Mc/s) are shown in Figure 2 a, b, c, d. Type-IV microwave bursts were not listed as a separate category by COVINGTON and HARVEY (1958), who distinguished only two main categories of bursts: impulsive and long-enduring. In this two-fold division, the Type IV microwave burst would appear as an intense, variable, impulsive burst of duration 30–60 min, or a complex group of impulsive bursts of similar duration. Measurements of the diameter of the microwave bursts, their precise emission height, and any motion of the emitting source, have generally been limited by the angular resolution of observers' equipment. In most cases the resolution is not better than 1 arc min, corresponding to diameters of 50000 km in the solar atmosphere.

COVINGTON and HARVEY (1958) believe that all bursts in the three categories listed above are always associated with flares. The reverse correlation does not hold. They have also found (1961) that the larger impulsive microwave bursts often coincide with, or just precede, the explosive or flash phase of a flare.

Spectral characteristics of microwave bursts, as recorded at the Harvard Station with a sweep frequency receiver covering the band 7.5–15 cm (2100–3900 Mc/s), are shown in Figure 2 e, f. The spectral equipment had a sensitivity limit of 20×10^{-22} MKS units, which generally precluded the recording of the long-enduring bursts and the less intense impulsive microwave bursts. Figure 3 shows bursts in the microwave band which have the spectral characteristics of generalized Type-III bursts. These bursts may correspond to the "variable" microwave bursts sometimes seen with fixed-frequency receivers. If the bursts

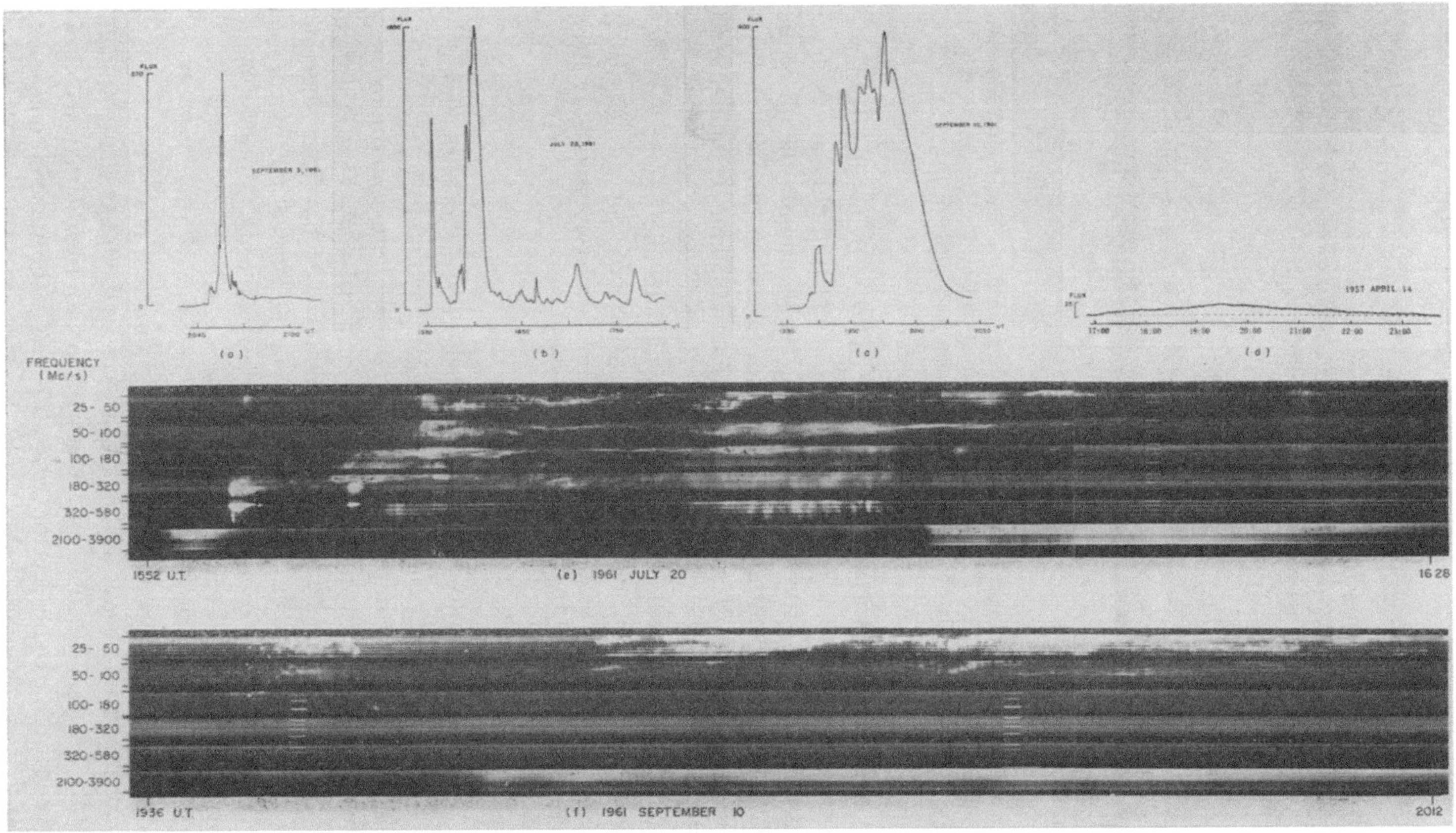

Fig. 2. Microwave bursts. (a) Short-lived impulsive burst. (b) Impulsive burst followed by Type-IV microwave burst. (c) Impulsive burst followed by Type-IV microwave burst. (d) Long-enduring burst. The intensity-time profiles were taken with a fixed-frequency receiver at 2800 Mc/s (Covington). (e) Spectral record showing microwave bursts and associated decimeter and meter wavelength activity. (f) Spectral record showing microwave bursts and accompanying decimeter and meter wavelength activity. The spectral records were made at the Harvard Station.

The intensity-time profiles of bursts (b) and (c) may be compared directly with the intensity-time profiles at frequency 2800 Mc/s on the spectral records (e) and (f).

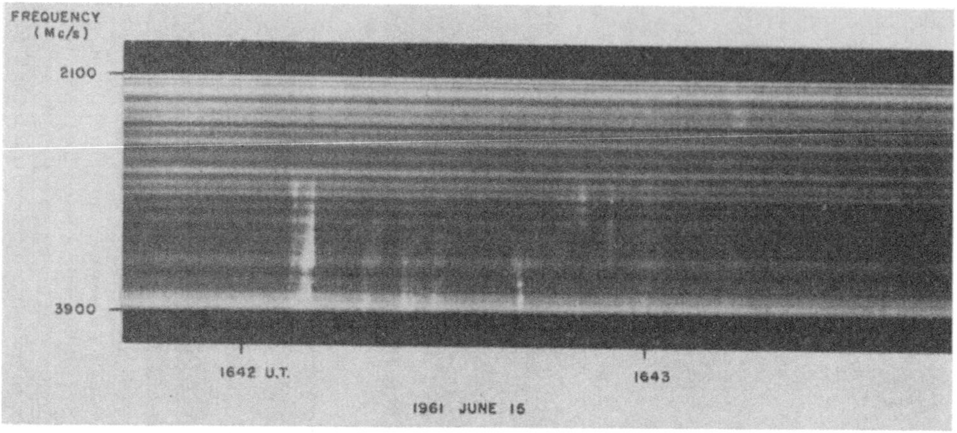

Fig. 3. Spectral record of short-duration, sharp-profile bursts in the microwave band. These may be generalized Type-III bursts. The complete spectral characteristics of the burst, over the band 25-3900 Mc/s, are shown in Figure 15c.

are indeed microwave Type-III bursts and are generated by plasma oscillations in the upper chromosphere, it is probable that the escaping radiation is that of the second harmonic. HUGHES and HARKNESS (1963) estimate the absorption suffered by the second harmonic radiation in escaping through the solar atmosphere to be about 10 db, compared with about 30 db for the fundamental.

COVINGTON and DODSON (1953) and TANAKA and KAKINUMA (1960) have reported a temporary decrease in microwave emission following an impulsive burst. It is believed that this decrease in flux density corresponds to an absorption of the background S-component of solar emission by surge-like matter of low temperature ejected at the time of a solar flare.

5.2 *Emission Mechanisms*

Impulsive bursts. The intensity, variability, and short duration of the impulsive bursts strongly suggest that they are non-thermal. HACHENBERG and WALLIS (1961) have suggested that impulsive bursts might be thermal in origin but as the brightness temperatures are often of the order of 10^7–10^9 °K, it is difficult to see how a thermal mechanism can be sustained. It has been suggested alternatively that the impulsive bursts may be generated by synchrotron radiation (TAKAKURA, 1959) or by amplified gyro-radiation (SMERD, 1963) or possibly by a combination of both mechanisms.

Type IV microwave bursts. Bursts such as those shown in Figure 2 b, c, e, f are generally ascribed to synchrotron radiation, and their characteristics have been examined by a number of authors. TAKAKURA (1960) has shown that the duration of the bursts, 5–40 minutes, is compatible with the theoretical collision damping time for relativistic electrons, $\beta = 0.7$–0.9, in the upper chromosphere, where the density is about 10^9 particles cm^{-3} and the magnetic field strength about 500–2000 gauss. TAKAKURA (1960, 1961) is also of the opinion that the observed polarization characteristics, 10–40 per cent circular polarization in the ordinary mode at frequencies higher than 2000 Mc/s and 100 per cent circular polarization in the extraordinary mode in the frequency band about 500–2000 Mc/s, are in accord with a synchrotron emission model. KAKINUMA (1958) and TAKAKURA (1961) attribute the reversal in polarization to differential absorption of magneto-ionic components, the absorption being a function of frequency. On the other hand, COHEN (1961) suggests that the reversal results from a propagation phenomenon, caused by magneto-ionic mode coupling in a quasi-transverse region, high in the corona, where the field strength is of the order of 2–6 gauss.

Long-enduring bursts. These bursts are characterized by a very gradual rise and fall, and often last several hours. They generally start at the starting time of an optical flare, although the peak emission and the termination of the radio burst are often considerably delayed with respect to the corresponding features of the optical brightening. The energy spectrum decreases sharply at frequencies lower than 1000 Mc/s and implies a thermal origin for the radiation. It is believed that the long-enduring bursts originate from thermal emission (bremsstrahlung) from a dense region of relatively small dimensions, and of somewhat higher than normal temperature, at the top of the chromosphere. The dimensions correspond approximately to those of the optical flare region.

6. Noise Storm Bursts (Spectral Type I)

6.1 *General Characteristics*

Noise storms comprise a series of short-lived bursts at meter wavelengths, more or less randomly distributed in time. Examples of the bursts are shown in Figure 4. They are generally superimposed on a background "continuum" radiation, which is enhanced above the quiet-sun level. During periods of intense activity hundreds of these bursts occur every hour, and they may continue for a period of several days. Noise storm activity is often concentrated in one particular band of the spectrum, for example 40–125 Mc/s (Figure 4a), and may gradually drift into a slightly different band during the course of a day. The bursts are rarely seen at frequencies above 250 Mc/s, although isolated bursts occasionally occur at higher frequencies. During sunspot maximum, noise storm bursts were observable on the Harvard spectrograph records, sensitivity limit 5×10^{-22} MKS units at 100 Mc/s, for approximately 13 per cent of the time (MAXWELL, HOWARD, and GARMIRE, 1960).

6.2 *Flux Densities ; Polarization*

The flux densities of the radiation range from the barely perceptible up to values of 100–1000 times the background thermal radiation, which is of the order 5×10^{-22} watt m^{-2} (cps)$^{-1}$ at 100 Mc/s. The radiation is usually almost

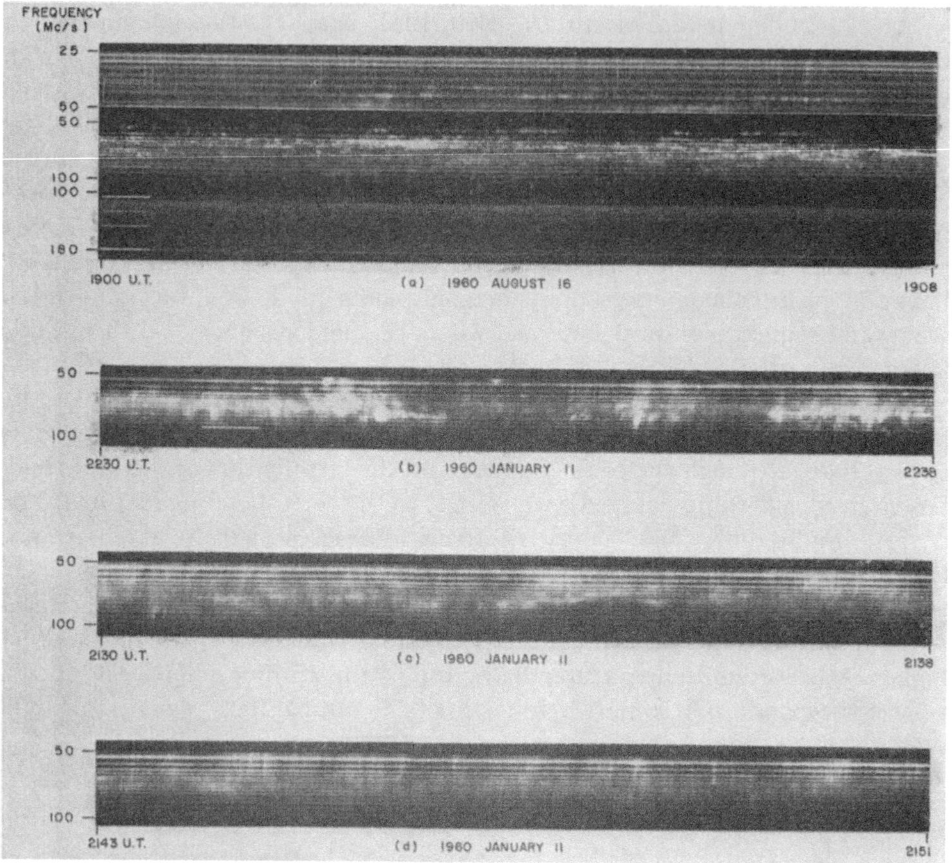

Fig. 4. Type I (noise storm) bursts. (a) Record showing Type-I bursts confined to low-frequency band. (b) A group of narrow-band bursts with wide-band background structure. (c) and (d) Wide-band Type-I bursts; at frequencies < 100 Mc/s the wide-band Type-I bursts often have the characteristics of a storm of faint Type-III bursts.

completely circularly polarized (PAYNE-SCOTT and LITTLE, 1951; FOKKER, 1960), the sense of rotation being determined essentially by the polarity of the largest emitting sunspot: right-hand circularly polarized waves are associated with a spot of negative polarity, and left-hand circularly polarized waves with a spot of positive polarity. This radiation corresponds to the ordinary component of magneto-ionic theory (PAYNE-SCOTT and LITTLE, 1951; FOKKER, 1960). When the emitting source nears the solar limb the radiation is often depolarized (SUZUKI, 1961; KAI, 1962).

6.3 *Fine Structure*

The fine structure of noise storm bursts is complex and varied. At least two distinct types may be distinguished on the Harvard spectral records (MAXWELL,

Swarup, and Thompson, 1958): narrow-band bursts, with bandwidths of a few megacycles and average duration a fraction of a second, sometimes coalescing to give durations of almost a minute (Figure 4a, b); and wide-band bursts, with bandwidths up to 30 Mc/s and lifetimes less than a second (Figure 4c, d). These two categories have also been observed by Vitkevič and Gorelova

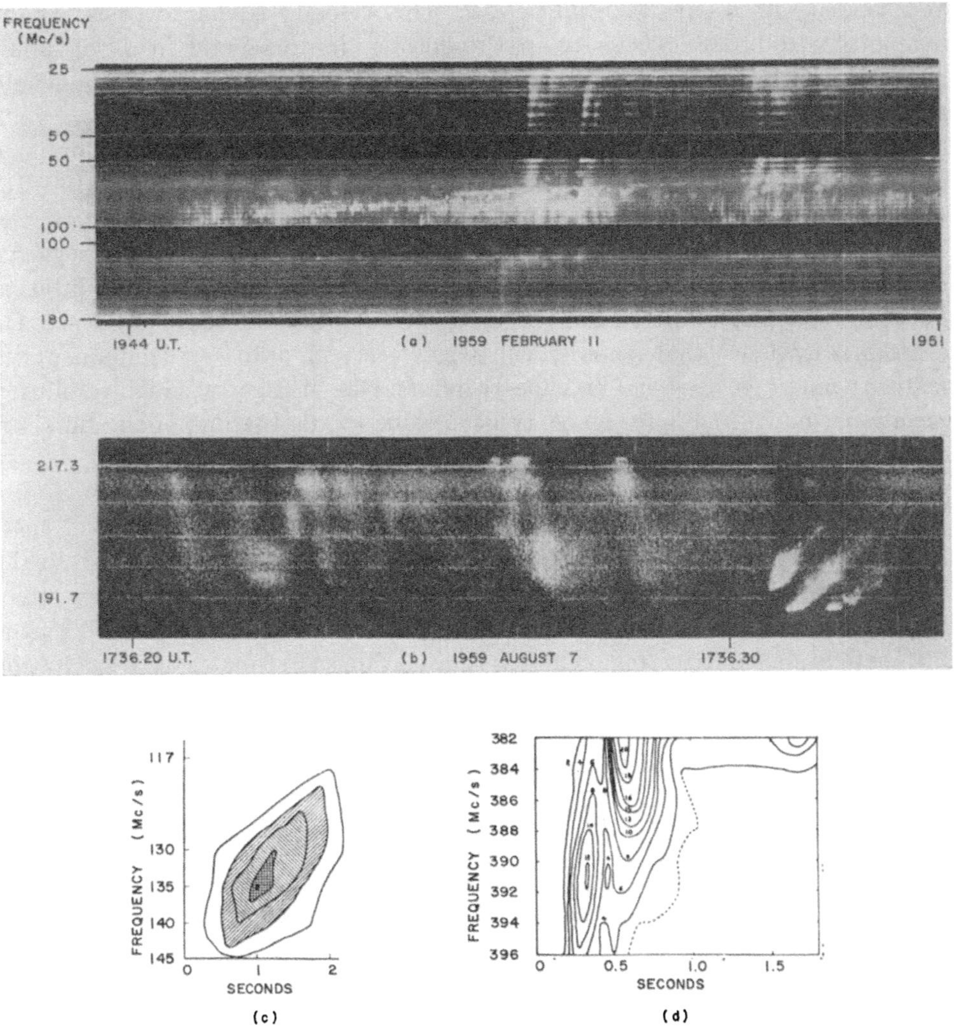

Fig. 5. Type I (noise storm) bursts. (a) Drifting structure, with drift rate comparable to that of Type-II bursts. (b) Fine structure of noise storm bursts in the band 190–215 Mc/s, as observed with a receiver of sweep rate 50 c/s (Elgaroy, 1961; note inverted frequency coordinate). (c) Contours of a wide-band noise storm burst, synthesized by Vitkevič and Gorelova (1960) from fixed-frequency receivers. (d) Fine structure of noise storm burst, with possible Type-III drift, synthesized by de Groot (1959) from fixed-frequency receivers.

(1960). The contours of a wide-band burst, synthesized by these authors from fixed-frequency receivers, are shown in Figure 5c. Similar synthesis of observations with fixed-frequency receivers has been carried out by DE GROOT (1959). An example from his records is shown in Figure 5d.

On occasions a drifting structure may be seen in noise storm bursts. This is illustrated in Figure 5a. The drift rate is comparable to that of the Type-II bursts and may be caused by a small disturbance moving through the corona at a velocity of about 1000 km per sec. This velocity is characteristic of the magnetohydrodynamic velocity in the corona, for magnetic fields of a few gauss. It is to be expected that any localized disturbance in such a region would tend to propagate at this velocity.

ELGAROY (1961) has studied the fine structure of noise storm bursts in greater detail with a sweep frequency receiver operating in the band 190–215 Mc/s at a sweep rate of 50 times per sec. He finds that the largest number of bursts have no frequency drift, but bursts with both positive and negative frequency drift also occur. Some storm bursts have echoes, the time interval between recording of direct and reflected rays being of the order of 0.5 sec. He also finds evidence that some storm bursts are split into two components, a feature sometimes observed in Type-II bursts. One of Elgaroy's high resolution records is shown in Figure 5b. A typical value of the lifetime of the bursts is about 0.4 sec with an apparent lower limit of approximately 0.2 sec.

6.4 *Emission Height; Angular Diameter*

The frequency range of noise storms indicates that they originate high in the solar corona, and this supposition has been amply confirmed by many direct interferometric measurements. The radiating height may be determined by the drift rate of the emitting region across the solar disk. Estimates of the radiating height as a function of frequency are summarized in Table 3. The emission height may vary considerably from case to case, but generally remains fairly stable over a period of days and approximates that of the appropriate plasma frequency

TABLE 3

EMISSION HEIGHT AND ANGULAR DIAMETER OF NOISE STORMS

Frequency (Mc/s)	*Emission Height above Photosphere (in units of $R_0 = 696000$ km)*		*Angular Diameter (arc min)*	*Reference*
	Range	*Average*		
19.7		1.9		SHAIN and HIGGINS (1959)
100	0.3–1.0	0.5		PAYNE-SCOTT and LITTLE (1951)
169	0.3–1.0	0.35	< 10	BOISCHOT (1958)
200	0.2–0.3	0.3	1–6	SUZUKI (1961); KAI (1962)
255	0.26–0.44	0.33	4–7	FOKKER (1960)

level. The average height increases from center to limb, as would be expected from a consideration of escape levels calculated from refracted ray trajectories.

Data on the average diameters of the emitting sources are also summarized in Table 3. The diameters of individual bursts at 100 Mc/s are often < 1.5 arc min (GOLDSTEIN, 1959), and the positions of the bursts may be scattered about the over-all radiating area. The diameter of the enhanced background continuum radiation is usually about the same as the over-all radiating area and is generally about 1–6 arc min (KAI, 1962).

6.5 *Association with Sunspots*

MCCREADY, PAWSEY, and PAYNE-SCOTT (1947) showed that noise storm radiation from the sun originated in a small area, immediately above a sunspot. Their results were subsequently confirmed and extended by many observers. Although noise storms are always associated with sunspot groups, the fact that many large groups have no associated noise storm suggests that there is some sort of threshhold effect. PAYNE-SCOTT and LITTLE (1951) showed that sunspot groups containing no spots of area > 400 millionths of a solar disk are seldom accompanied by recognizable storms, but that when a group contains a larger spot, a noise storm is observed in two out of three cases. Changes in the intensity of noise storms have no statistical association with flares (SWARUP, STONE, and MAXWELL, 1960; SMITH and MCINTOSH, 1962), although on occasions a large flare may initiate or increase noise storm activity (DODSON, 1958), and on rare occasions may decrease noise storm activity.

6.6 *Emission Mechanisms*

There is no agreement concerning the actual mechanism in the corona that gives rise to noise storm bursts, beyond the fact that it is certainly non-thermal: the very high intensities and rapid changes make a thermal origin impossible. Theories of phase coherent mechanisms (plasma oscillations) have been discussed by SHKLOVSKY (1946), BOHM and GROSS (1949, 1950), and by GINZBURG and ŽELEZNIAKOV (1958, 1961). The theories examine the setting up and maintenance of plasma oscillations, damping or amplification, and the manner in which electromagnetic radiation can escape. DENISSE (1960) has suggested that noise radiation may be generated by Čerenkov plasma oscillations excited by particles initially ejected during a flare, and subsequently remaining trapped for several days in magnetic configurations in the corona. This mechanism is discussed further in Section 9.7, in connection with continuum storms. The characteristic life-times of Type-I bursts, about 0.2–2 sec, are of the order of the collisional damping time for a delta-function transient at the appropriate emission level in the corona.

Theories of gyro-radiation as a mechanism for noise storm radiation have been discussed in Section 3.3. TWISS and ROBERTS (1958) consider it difficult to account for noise storm radiation in this way; SMERD (1963) believes that under certain conditions gyro-radiation can be amplified and emitted in appreciable quantities.

It has been suggested that the enhanced background radiation accompanying noise storm bursts may consist of a large number of overlapping bursts (for example, TAKAKURA, 1959) but spectral records, on which the background radiation appears to be essentially continuous in time and frequency, with noise storm bursts clearly superimposed on the background, essentially contradict this suggestion. Alternatively, the background radiation could result from synchroton radiation.

7. Slow-Drift Bursts (Spectral Type II)

7.1 *General Characteristics*

The regularly defining characteristic of the slow-drift burst is a narrow band of intense radiation which drifts slowly, and often irregularly, towards lower frequencies. Examples are shown in Figures 6 and 7. The magnitude of the drift rate is a function of frequency and at 100 Mc/s is about -0.3 Mc/s per sec. The bursts have a total duration of the order of 10–20 min and generally consist of emission at a fundamental frequency, with a strong second harmonic and no higher harmonics. Most of the bursts are of high intensity, $> 10^{-19}$ MKS units at 100 Mc/s. During the sunspot maximum there were about 70 bursts per year, reckoned on a continuous observing basis.

About 80 per cent of Type-II bursts are accompanied or preceded by fast-drift bursts (spectral Type III). The relation in time of Type-III bursts to Type-II bursts (Figure 6b) suggests that both result from a single large disturbance at approximately photospheric level (WILD, MURRAY, and ROWE, 1954). It is presumed that the low level disturbance gives rise to two distinct disturbances traveling at markedly different velocities: one at about 1000 km per sec giving rise to the Type-II burst, and the other at 100000 km per sec causing fast-drift bursts. In approximately 30 per cent of cases the slow-drift bursts are accompanied by continuum emission in the form known as spectral Type IV. The combination of Type-II, Type-III, and Type-IV bursts, in association with a great solar flare, constitutes the spectral characteristics of the "outburst" described in the early literature of solar radio astronomy. The total radio energy emitted in such an outburst, in the band 10 Mc/s-50 Gc/s observable on the earth, is probably about 10^{24} ergs. By contrast, the total energy radiated by a great flare probably exceeds 10^{32} ergs (PARKER, 1957).

7.2 *Polarization; Brightness Distribution*

Observations of polarization are meager, but existing evidence suggests that the bursts are randomly polarized to a high order, perhaps 95 per cent (KOMESAROFF, 1958).

Measurements on the brightness distribution across the emitting centers of

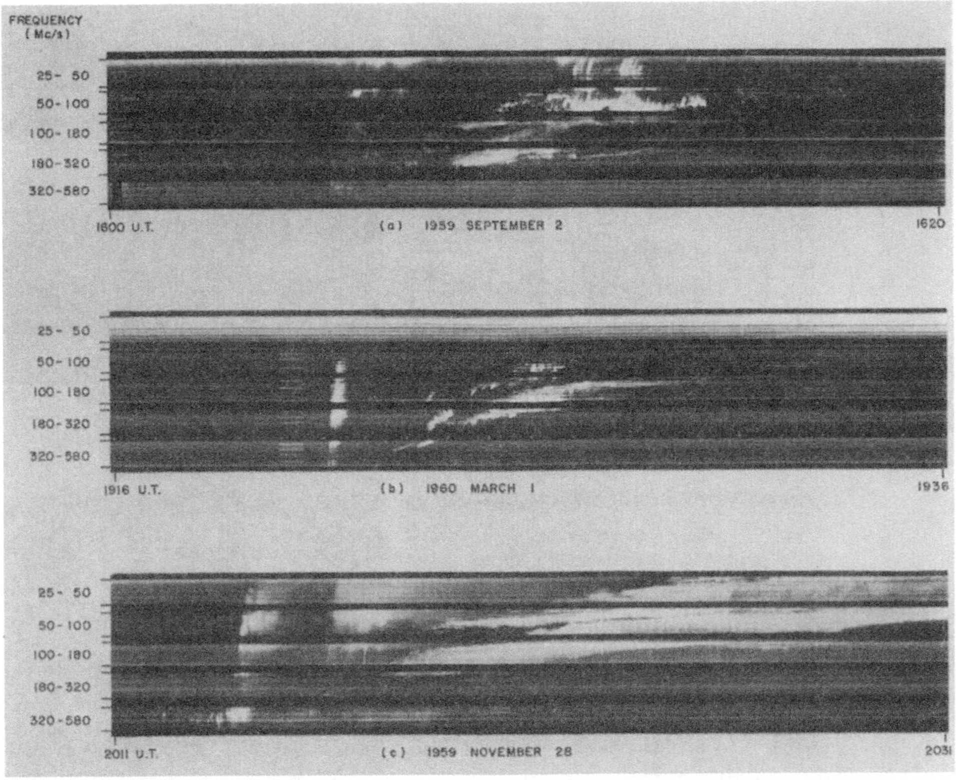

Fig. 6. Type-II (slow-drift) bursts. (a) Burst showing fundamental and second harmonic emission, with a group of Type-III bursts issuing out of the tail of the drifting Type-II bands. (b) Burst indicating a common origin for the exciters of the Type-II burst and the preceding Type-III bursts. (c) Burst showing approximately 1 minute delay between onset of second harmonic and fundamental. The Type-II burst is preceded by inverted U bursts and Type-V continuum radiation.

Type-II bursts have been made by WEISS and SHERIDAN (1962) with an interferometer operated at 40 and 60 Mc/s and with two fixed antenna spacings of 1 km and 0.25 km. Assuming the emission source to be symmetrical, the observers find the measurements to be consistent with a brightness distribution at 40 Mc/s the same as at 60 Mc/s. Then, by combining the observations at two frequencies to obtain five points on a fringe visibility curve, the authors deduced a brightness distribution consisting of a central condensation, or core, superimposed on a diffuse background or halo. This core and halo distribution is shown in Figure 8.

7.3 *Band Splitting*

The spectral characteristics of Type-II bursts are often complex. In general, no two bursts are exactly alike, but certain features that recur from time to

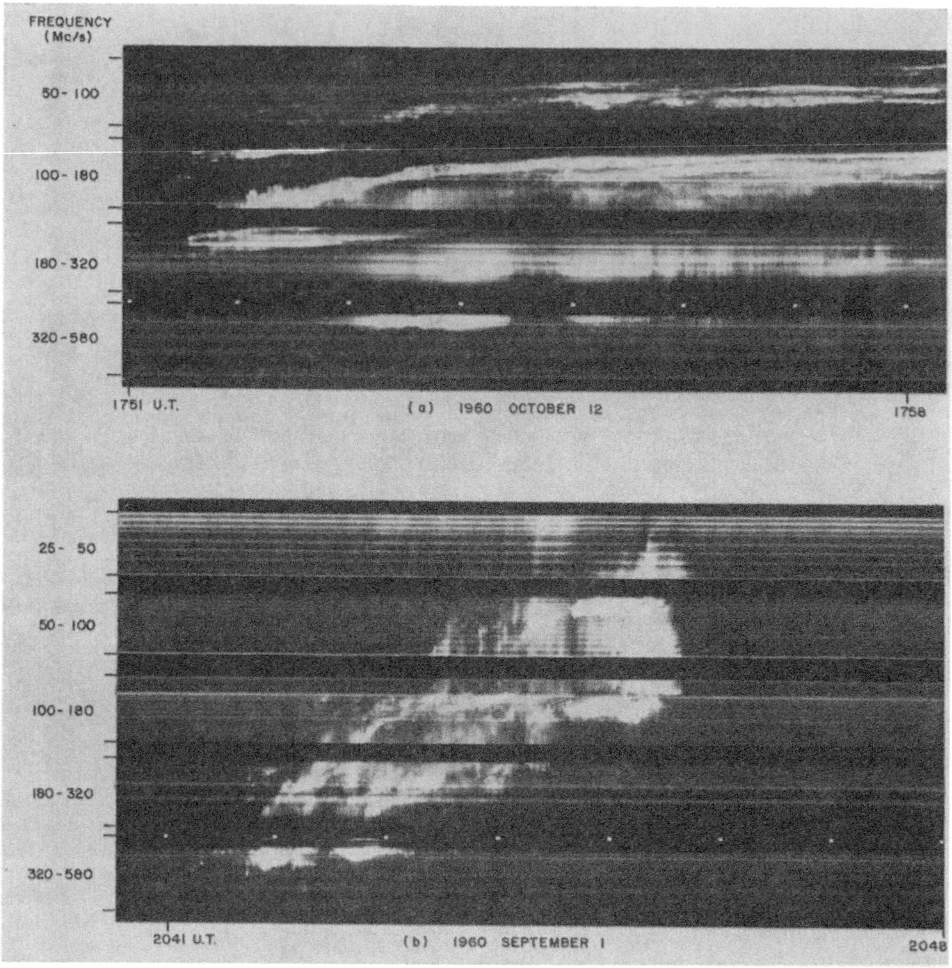

Fig. 7. Type-II (slow-drift) bursts. (a) Band-splitting in the fundamental and second harmonic emission. (b) Generation of Type-III structure in the Type-II disturbance.

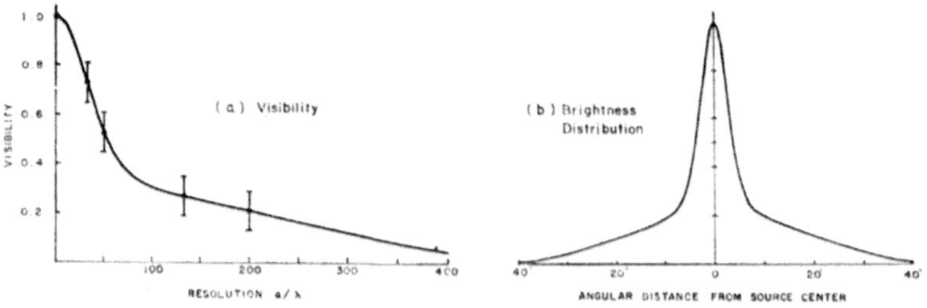

Fig. 8. (a) Average fringe visibility curve for 8 Type-II bursts. (b) Deduced brightness distribution, showing central emitting core and scattering halo. (WEISS and SHERIDAN 1962).

time warrant some attention. Of these, a prominent characteristic is the splitting of each main band of emission into two subbands (Figure 7a). The splitting might be accounted for in terms of an outward moving disturbance which consists of a double shock front. Alternatively the effect may result from magnetic splitting, analogous to the Zeeman effect. The latter explanation, however, meets with difficulties in explaining the observed random polarization; nor is it consistent with the number of components predicted by Zeeman theory, which would suggest triple splitting in the harmonic (ROBERTS, 1959). STURROCK (1961) has suggested that the two ridges may result from preferential excitation, in a magnetic field of approximately 20 gauss, of two frequencies at the extremities of the band of plasma waves. These two frequencies, he believes, would propagate at different inclinations to the magnetic field. If it should transpire that the band splitting of Type-II bursts is indeed magnetic in origin, the radio data will provide extremely important information about magnetic field strengths in the corona.

7.4 *Absence of Third and Higher Harmonics*

There is no quantitative explanation for the absence of third and higher harmonics in the spectrum of Type-II bursts. ROBERTS (1959) suggests that the observed intensity ratios may be explained most directly by assuming that the bursts are generated by non-linear oscillations of the plasma, with the limitation that the oscillations are not sufficiently non-linear to radiate appreciable third harmonic; the radiation efficiency in the fundamental mode is assumed to be considerably less than in the second harmonic. Krook (unpublished) suggests that in a large amplitude shock, energy is transferred through the sequence of harmonics. In this case, since the second harmonic is well above the local plasma frequency, it may readily escape, and its amplitude in the travelling disturbance will remain small. Consequently, little energy will be transferred to the third and higher harmonics, which cannot appear, except as transients.

GINZBURG and ŽELEZNIAKOV (1958) point out that, if a combination scattering process is operative, energy transferred into electromagnetic modes is preferentially radiated in the second harmonic; Rayleigh scattering on density irregularities will transfer energy into the fundamental.

7.5 *Flare Association*

Type II bursts are generally associated with flares and the distribution of these flares across the solar disk is almost uniform in longitude (MAXWELL and THOMPSON, 1962). This distribution indicates that the radiation from the bursts is emitted in a very broad cone (in contrast to Type I and Type IV bursts, both of which give rise to more intense radiation when the active region is near the central meridian). The average delay between the commencement of a flare and the start of the Type-II bursts is of the order of 10 minutes, which is consistent with an interpretation of these bursts in terms of a disturbance, initiated in the chromosphere at the time of flare, moving outwards at a velocity of approximately 1 000 km per sec.

Presumably the chromospheric flare and the radio bursts have a common origin. It is questionable, however, whether ejective prominences seen in the optical spectrum cause the radio bursts, as suggested by GIOVANELLI and ROBERTS (1958). For example, in Figure 9, the Type-II burst apparently has no direct physical relation, as regards height and velocity of propagation, with any of the optical surges. However, when extrapolated back, with the assumption of constant velocity, the time and height of origin of the Type-II disturbance are the same as that of the flare.

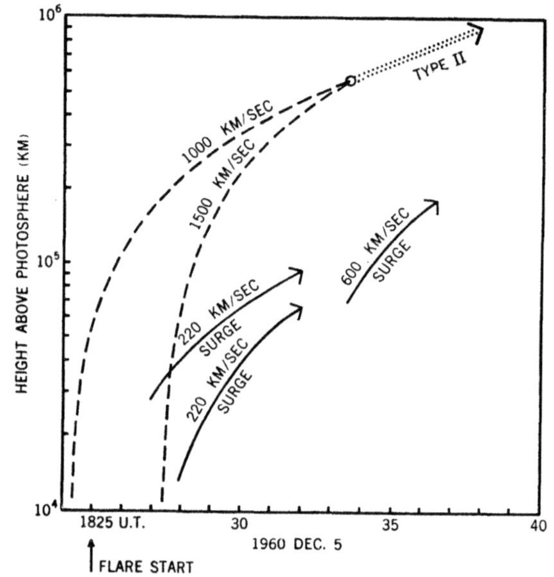

Fig. 9. Optical surges and Type-II burst associated with limb flare. The trajectory for the Type-II burst was calculated on the assumption that the Type-II disturbance was propagated through a coronal streamer of density 10 times the Baumbach-Allen distribution. The dashed lines give possible extrapolated trajectories, at constant velocity, for the Type-II burst. Optical data from G. E. Moreton, Lockheed Solar Observatory, and radio data from Harvard Station. (MAXWELL and THOMPSON, 1962.)

7.6 *Plasma Oscillations; Coronal Streamer Model*

It is generally agreed that Type II bursts originate from plasma oscillations generated by a large amplitude shock front moving outward through the solar plasma. A plasma oscillation mechanism is in accord with the relatively narrow bandwidth of the bursts, the presence of a fundamental and a harmonic, and with position measurements which show that the radiation originates near the appropriate plasma level. We may determine the velocity of the outward moving disturbance from the spectral characteristics of the bursts, by transforming the frequency of the bursts into an equivalent emission height, and then by transforming $-df/dt$ (Figure 11a) into dh/dt. However, we must first consider what model should be taken for the distribution of electron density with height. The

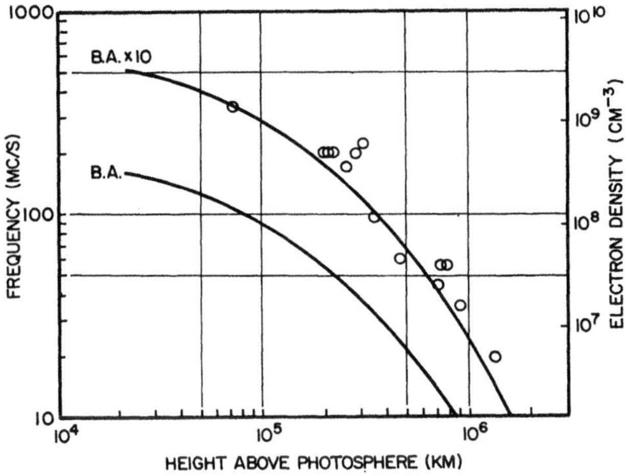

Frequency (Mc/s)	*Height (× 10³ km)*	*Spectral Type*	*Reference*
340	70	I	FIROR (private communication)
255	230	I	FOKKER (1960)
200	210	I	SUZUKI (1961)
200	210	III	MORIMOTO (1961)
200	280	IV	MORIMOTO (1961)
169	250	I	BOISCHOT (1958)
100	350	I	PAYNE-SCOTT and LITTLE (1951)
60	450	III	WILD, SHERIDAN and NEYLAN (1959)
55	700	II	WEISS (1963b)
55	770	IV	WEISS (1963b)
45	700	III	WILD, SHERIDAN and NEYLAN (1959)
26	880	III	ERICKSON (1963)
19	1330	I	SHAIN and HIGGINS (1959)

Fig. 10. Electron density and equivalent plasma frequencies plotted as functions of height in the solar corona. The curve labeled B.A. gives the Baumbach-Allen distribution for electron density. The curve B.A. × 10 refers to the same model with a multiplying factor of 10. References for the data points are tabulated. (Reproduced from MAXWELL and THOMPSON, 1962, with additional data points.)

bursts are strongly flare-associated and probably originate in regions where the electron density is greater than that of the undisturbed solar atmosphere. An analysis of the available optical and radio data (MAXWELL, HOWARD, and GARMIRE, 1959; MAXWELL and THOMPSON, 1962) suggests that the primary disturbance is probably propagated outward along the coronal streamer, in which the electron density is about 10 times the Baumbach-Allen values. Radio data, concerning the height of radio emission centers above active areas on the

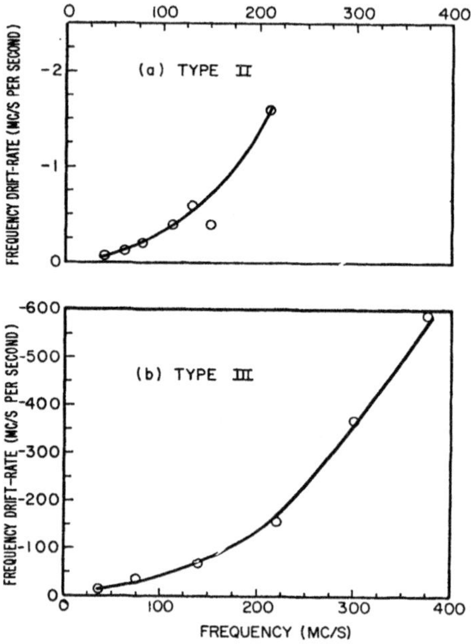

Fig. 11. (a) Average frequency drift-rate for Type-II bursts as a function of frequency. (b) Average frequency drift-rate for Type-II bursts as a function of frequency. The drift-rate for the Type-III bursts is approximately 100 times that of the Type-II bursts at any given frequency.

solar photosphere, are shown in Figure 10. These data are essentially statistical in nature and establish *average* heights of radio emitting centers for a number of frequencies. A similar model has also been adopted by Weiss (1963a). The Baumbach-Allen model (Allen, 1947) is used mainly as a reference. For example, electron densities in the quiet corona during the past sunspot maximum may have been approximately twice those of the Baumbach-Allen model (Newkirk, 1961). This would imply densities in the model coronal streamer only five times that of the surrounding corona.

7.7 *Velocity of Primary Disturbance*

Assuming such a model coronal streamer, we may determine analytically the radial component of the velocity of the outward moving disturbance. The velocity turns out to be approximately 1000–1500 km/sec (Maxwell and Thompson, 1962). This figure is in accord with a few direct measurements made with a sweep frequency interferometer by Wild, Sheridan, and Trent (1959) and WEISS (1963a) but is 3 times the value deduced by ROBERTS (1959), for which the unmodified Baumbach-Allen formula was used. The data in Figure 12a actually suggest a decrease in the exciter velocity, from 1750 km/sec at a height of 150000 km, to 750 km/sec at height 650000 km. While it is possible that the generating shock front suffers a true deceleration as it moves outward,

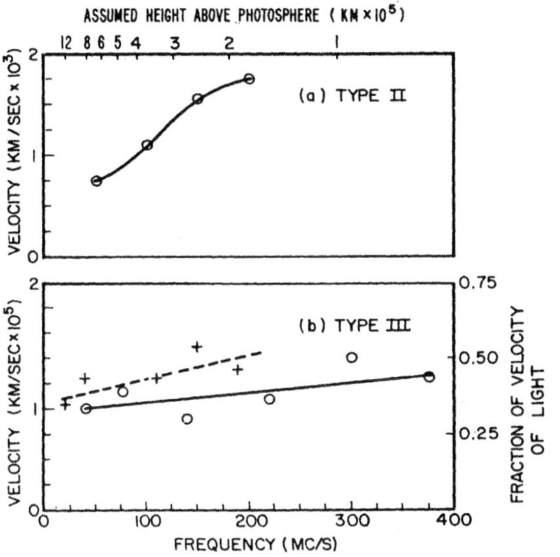

Fig. 12. (a) Average radial velocity component of the exciters of Type-II bursts derived from the drift-rate given in Figure 11a. The exciters were assumed to propagate through a coronal streamer of density 10 times the Baumbach-Allen model (MAXWELL and THOMPSON, 1962). (b) Average radial velocity component of the exciters of Type-III bursts derived from the drift-rate given in Figure 11b. The solid line and points ○ were plotted on the assumption that all bursts were fundamentals. The dashed line and points + were plotted on the assumption that all bursts were second harmonics. The exciters were assumed to propagate through a coronal streamer of density 10 times the Baumbach-Allen model (HUGHES and HARKNESS, 1963).

it must also be remarked that the velocity and deceleration figures are critically dependent on the electron density model assumed. Moreover, deflections from a radial path would also give an apparent deceleration. WEISS (1963a) finds that the trajectories of the exciting shock fronts are indeed often non-radial. Moreover, many Type-II bursts show temporal changes of position at a single frequency, indicating tangential movements in the solar atmosphere with speeds of the order of 1000–2000 km/sec. The spectra of these bursts, with tangential movement, are characterized by little frequency-drift and by broad bandwidth.

The primary disturbance causing Type-II bursts may propagate outwards in a *longitudinal* mode as an acoustic shock wave at a velocity

$$v = M\sqrt{\gamma RT}$$

where $\gamma = 1.7$, R = gas constant, T = temperature of medium, and M = Mach number. The temperature in the corona is approximately constant and is of the order of 10^6 °K. The velocity of sound is about 120 km/sec and the disturbance would therefore require to be traveling at about Mach 10. Purely longitudinal oscillations of the plasma will not give rise to sufficient electromagnetic radiation. However, irregularities in the magnetic field or diverging field lines would locally generate transverse modes that could eventually escape as radio frequency emission.

In the presence of a magnetic field, however, it is more likely that the disturbance will be initiated mainly in the *transverse* modes and will propagate as a magnetohydrodynamic wave. If the magnetic field is quasi-radial it will travel outwards at the Alfvén velocity:

$$v_A = \sqrt{H^2/4\pi\varrho_0}$$

where H = local magnetic field strength and ϱ_0 = density of medium. Inserting values for ϱ_0 and plausible values for H we find the velocity is approximately constant with height and has a value of about 1 000 km/sec (UCHIDA, 1960). If the magnetic field configuration were non-radial, the disturbance would be propagated transverse to the magnetic field and would carry field and material out into the corona. The velocity of a large amplitude disturbance propagated in this manner is given by the transverse magnetohydrodynamic velocity

$$v_T = k'\sqrt{\gamma RT + H^2/4\pi\varrho_0}$$

where k' is a small multiple. Again the predicted velocity is of the order of 1 000 km/sec.

The maximum (starting) frequency of the fundamentals of Type II bursts rarely exceeds 175 Mc/s and the modal values are 75–100 Mc/s. This high-frequency cut-off may result from the fact that, in the chromosphere and lower corona, the main mode of dissipation of a large-amplitude disturbance is by collision and local heating of the medium. In higher regions, where the collision frequency is small, the main dissipation mechanism is the emission of electromagnetic radiation. We may therefore expect substantial transfer of energy into electromagnetic modes only at heights above about 100000 km (MAXWELL and THOMPSON, 1962).

7.8 *Relative Observed Positions of Fundamental and Harmonic Emission*

SMERD, WILD, and SHERIDAN (1962), using the Australian sweep frequency interferometer over the band 40–70 Mc/s, observed the positions relative to the sun's disk of the emission of the fundamental and second harmonic components of Type-II bursts. Contrary to simple theory, they found that the harmonic emission in Type-II bursts appeared to arrive from directions corresponding to much lower heights in the solar atmosphere than for the fundamental. The results of their observations are illustrated in Figure 13. To explain this somewhat puzzling phenomenon, they suggest that radiation in the second harmonic is preferentially back-scattered from the point of generation and is

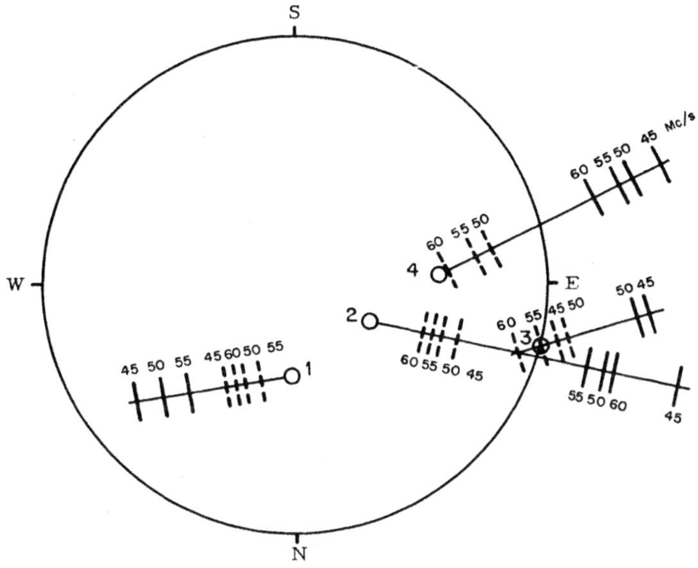

Fig. 13. Mean position for fundamental and second harmonic emission for four Type-II bursts. Positions of associated optical flares are indicated by circles. The radio positions are shown as short lines, perpendicular to lines drawn through the optical flare parallel to the scanning direction. Full lines denote fundamentals and dashed lines second harmonics (SMERD, WILD, and SHERIDAN, 1962).

then observed after reflection at much lower coronal heights. The authors believe that such preferential back-scatter is a natural consequence of the mechanism of burst generation, proposed by GINZBURG and ŽELEZNIAKOV (1958), in which the second harmonic emission is believed to arise from combination scattering of longitudinal plasma waves in a thermal plasma. However, since the velocity of the Type-II disturbance is only about 1000 km/sec, which is much less than the thermal velocity of electrons in the plasma, particles of this velocity in the Type-II disturbance could not of themselves set up Čerenkov plasma waves. The authors suggest that this difficulty may be overcome if it is supposed that the slow disturbance is the seat of continuous ejections of very high energy electrons, which locally generate Čerenkov plasma waves. In support of this proposal they note that the spectral characteristics of Type-II bursts sometimes show principal drifting bands which appear to emit a continuous succession of Type-III bursts (cf. Figures 6a and 7b). The authors examined the theoretical angular power spectrum for second harmonic radiation and showed that, with increasing velocity of the plasma wave, more energy is preferentially scattered backward. One would then expect back-scattered energy to travel to the plasma frequency level appropriate to the second harmonic, deeper in the solar corona, before being reflected outward.

Whilst the theory to some extent accounts for the observations, it also encounters a number of difficulties, which may be summarized as follows. (i) The method of observation did not permit simultaneous measurements of the apparent radiating position of the fundamental and second harmonic. A typical delay time between observation of the fundamental and the second harmonic was about 10 min. By this time, the shock front of the disturbance would travel a further 600000 km, and the fundamental would be generated in a vastly different place. It is highly desirable that the measurements eventually be repeated, with simultaneous observations of the radiating positions of both fundamental and second harmonic. (ii) Second harmonic emission should sometimes appear on the sun's disk as a double source, since the amount of energy back-scattered is theoretically only one or two orders of magnitude greater than that scattered forward. So far, a double source has not been observed. (iii) As the degree of back-scatter of the second harmonic radiation is a function of the velocity of the Čerenkov plasma waves, the whole effect should be much more pronounced for Type-III bursts. Data are only available for six cases of Type-III bursts but these data do not show such a trend. (iv) There is the question of the origin of superthermal electrons in the shock front of the primary disturbance. The observational evidence (Figures 6a and 7b) does, however, suggest that electrons may sometimes be accelerated to relativistic energies in the shock front, and a new mechanism has been suggested by SCHATZMANN (1963) which may account for such acceleration.

8. Fast-Drift Bursts (Spectral Type III)

8.1 *General Characteristics*

Fast-drift bursts are distinguished from other solar radio emission by their rapid drift from high to low frequencies (Figure 14). The drift rate is approximately 100 times that of slow-drift bursts and is of the order of -30 Mc/s per sec at 100 Mc/s. The bursts are of short duration, of the order of 1 or 2 sec at 100 Mc/s, and their duration increases with decreasing frequency. They regularly occur in groups and groups with more than 10 individual bursts tend to be more intense. The average number of bursts per group is about 4. At times, storms of fast-drift bursts may occur, and sections of spectral records may then be covered by hundreds of faint bursts. Reckoned on a 24-hour basis during sunspot maximum, an average of about 100 bursts per day is recorded by equipment with the sensitivity of the Harvard spectrograph (5×10^{-22} MKS units at 100 Mc/s). Second harmonic radiation may be recognized on the spec-

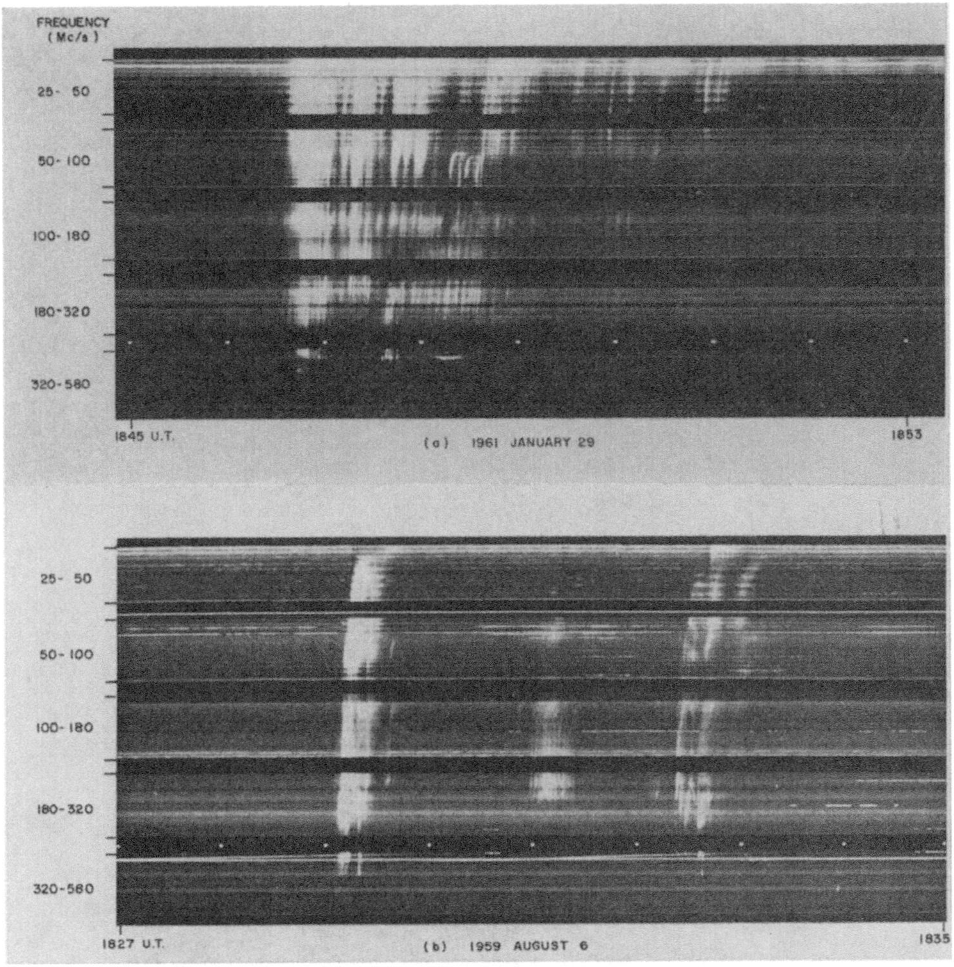

Fig. 14. Groups of Type-III (fast-drift) bursts.

tral records (Figure 15a, b) but is generally difficult to distinguish. Virtually no bursts are recorded with slopes between the two extremes of the slow-drift and fast-drift bursts.

8.2 *Polarization ; Brightness Distribution*

KOMESAROFF (1958) from an analysis of 500 bursts, suggests that about half of them have quasi-circular polarization. AKABANE and COHEN (1961) examined the polarization of the bursts at 200 Mc/s with a receiver of very narrow bandwidth, 10 kc/s; the narrow band receiver minimizes smearing by differential polarization which occurs with wider receiver bandwidths. Some Type-III bursts were observed which were weakly linearly, or highly elliptically, polariz-

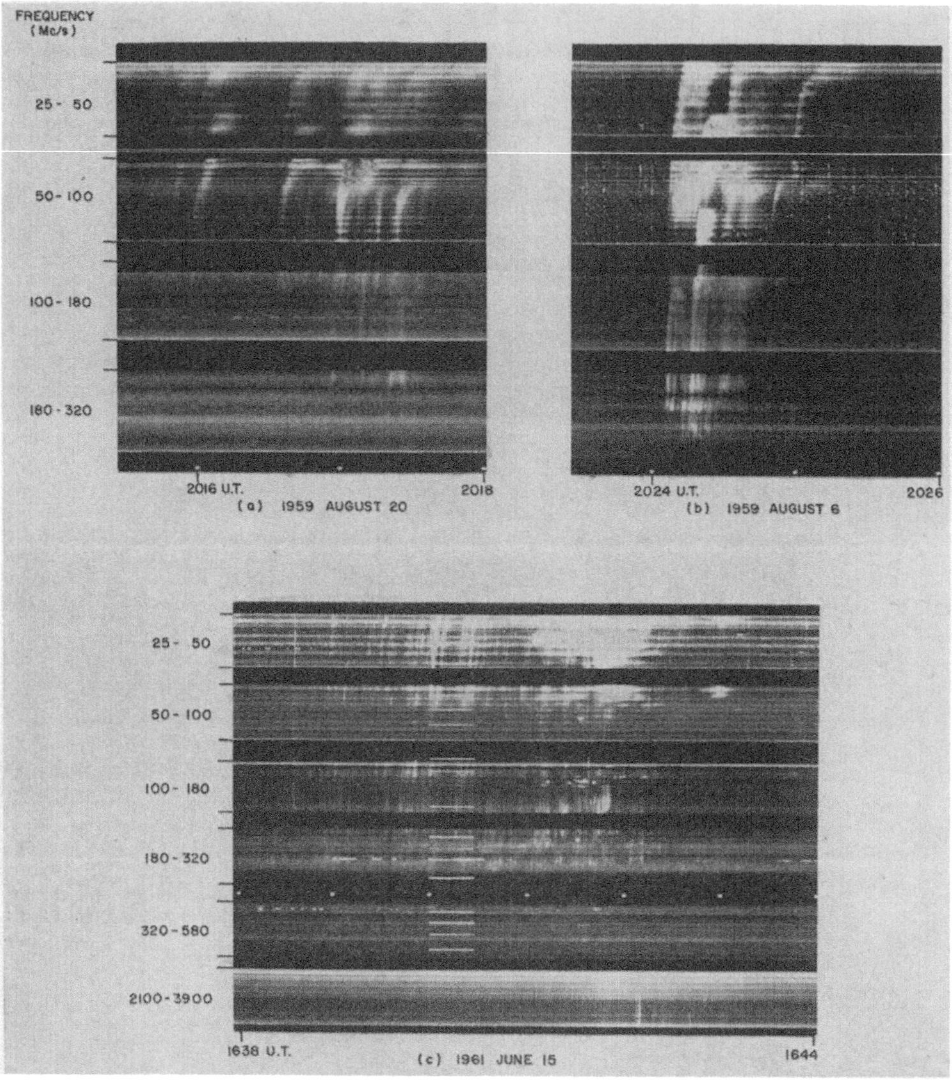

Fig. 15. (a) and (b) Second harmonics in the spectra of Type-III bursts. (c) Group of Type-III bursts showing extension into the microwave band. An enlargement of the microwave bursts is given in Figure 3.

ed. The observations indicated that the Faraday rotation in the corona is often of the order of 10^4 radians.

Measurements by WEISS and SHERIDAN (1962) of the brightness distribution across emitting centers of Type III bursts, at frequencies of 40 and 60 Mc/s, suggest a core and halo structure of dimensions similar to that for Type II bursts: the core half-width is approximately 4 arc min and the halo approximately 40 arc min. Similar measurements at 26.3 Mc/s (ERICKSON, 1963) confirm the core and halo model and indicate a central emitting core of < 10 arc min in diameter with a halo about 40 arc min.

8.3 *Exciter Dispersion*

HUGHES and HARKNESS (1963) estimated the dispersion of the Type III exciter as it moves through the solar corona by two different methods: (i) from the duration of the bursts, after correction for theoretical collision damping time; and (ii) from the instantaneous bandwidth of the bursts, defined as the instantaneous frequency extent at any given time. Their results are plotted in Figure 16, which shows that the radial extent of the exciter is about 1/3 its

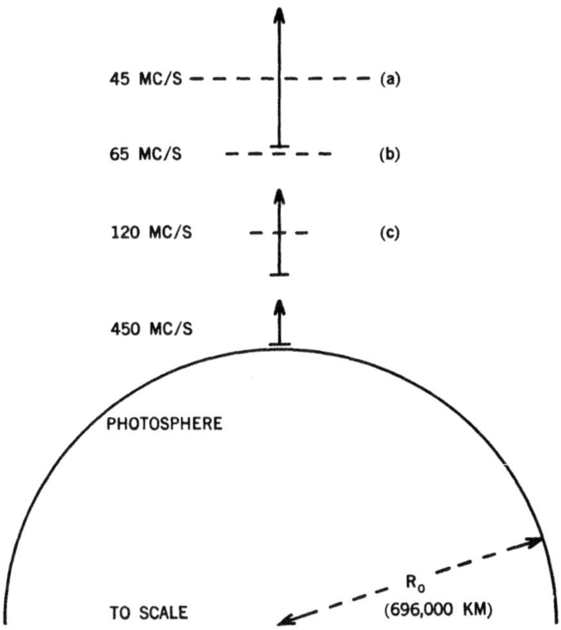

Fig. 16. Radial and transverse dimensions of the exciters of Type-III bursts are shown by the solid and dashed lines respectively. The radial lengths were estimated by HUGHES and HARKNESS (1963). The transverse dimensions (a) and (b) were measured by WILD and SHERIDAN (1958), and (c) by GOLDSTEIN (1959).

height above the photosphere. MORIMOTO (1961) finds that at a frequency of 200 Mc/s the emitting regions have approximately the same extension in both horizontal and vertical directions: about 1.5 arc min, corresponding to 70000 km. This result is in fair accord with the estimate of Hughes and Harkness.

Measurements of the angular diameter of the bursts, also plotted in Figure 16, show that the transverse dimensions of the exciter increase as the exciter moves outward from the chromosphere.

8.4 *Flare Association*

Type-III bursts, particularly those of strong intensity occurring in groups of 10 or more, are generally associated with solar flares. The probability of association is increased if the flare is (i) of increased area or height, or (ii) accompanied by an optical surge or spray (SWARUP, STONE, and MAXWELL, 1960). In general the bursts occur within a few minutes of the start of the accompanying flare. MALVILLE (1962) finds a high correlation between Type-III bursts and the flash phases of flares. HUGHES and HARKNESS (1963) have further shown that there is a tendency for the radio bursts to start at the same time as, or just precede, the flash phase.

8.5 *Velocity and Nature of Primary Disturbance*

As in the case of Type-II bursts, the existence of harmonics in the Type-III bursts suggests that they are generated by electrons oscillating at the appropriate plasma frequency. If we again assume that bursts are generated by a disturbance traveling approximately radially outwards along a coronal streamer in which the electron density is some ten times that of the surrounding corona, we find that the velocity of the primary disturbance, as determined from the slope of the fast-drift bursts (Figure 11b) and an appropriate frequency-height transformation, is of the order of 0.3–0.5 c (Figure 12b). The slope of the curves in Figure 12b indicates either that the exciters decelerate as they move outward, or that they follow curved trajectories in the high corona, or that the gradient of electron density may be less steep than was assumed for the model calculations.

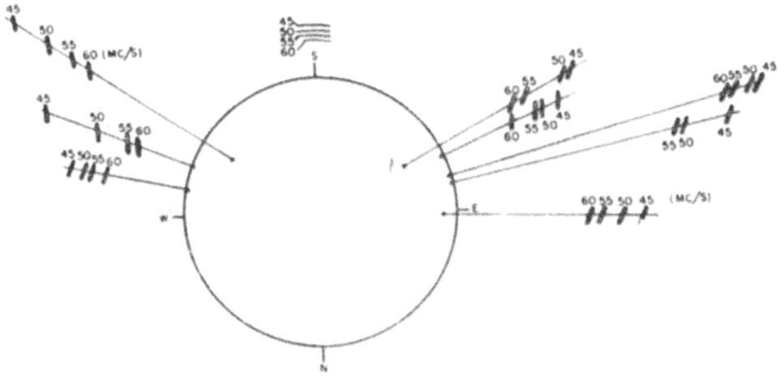

Fig. 17. The emission positions for 8 Type-III bursts associated with limb flares. The optical positions of the flares are shown as open circles. The lines on which the radio sources are located by the sweep-frequency interferometer are shown in part only, viz., that part which is close to the radial line drawn through the optical flare. The circular arcs show the levels of the fundamentals of the plasma frequencies for Baumbach-Allen electron densities in the corona. (WILD, SHERIDAN, and NEYLAN, 1959).

Direct measurements with a sweep-frequency interferometer over the frequency band 40–70 Mc/s (WILD, SHERIDAN, and NEYLAN, 1959) confirmed that the emission height for the bursts increases with decreasing frequency (Figure 17). The results also experimentally confirmed the hypothesis that the frequency drift is caused by a fast-traveling disturbance moving outward through regions of decreasing electron density, and exciting these regions to radiate at their natural plasma frequency. From the differences in the times of arrival of different frequencies, and their relative positions on the solar disk, it was deduced that the velocity of the disturbance causing the fast-drift bursts was of the order of 0.45 c. As shown in Figure 17, the direct observations also support the hypothesis that the primary disturbance is propagated along a coronal streamer in which the density is much greater than the Baumbach-Allen values.

The nature of the primary disturbance is uncertain. It is unlikely that the Type-III bursts could be generated by magnetohydrodynamic shock waves (WESTFOLD, 1957), since, for such waves to reach velocities appropriate to Type-III bursts, magnetic fields of some hundreds of gauss would have to exist high in the corona. It has been suggested that Type-III bursts are caused by the ejection of streams of electrons of near-relativistic velocities (WILD, SHERIDAN, and NEYLAN, 1959; DE JAGER, 1962; UCHIDA, 1962). Alternatively, the disturbance may have the nature of fast-traveling large-amplitude *transverse* shock waves (electron plasma shocks); such waves would have velocities intermediate between the thermal velocity of the electrons and the velocity of light (KROOK, unpublished).

WILD (1963) suggests that the quasi-periodic spacing in groups of Type-III bursts, shown in Figure 14a, may be attributed to a repeated pinch effect, such as might occur in a discharge process in a flare region.

8.6 *Related Bursts*

Although about 95 per cent of the bursts on the spectral records fall into the spectral categories I, II, III, and IV, there are a few bursts which cannot be so classified. Some of these bursts have distinct affinities to Type-III bursts. Examples are shown in Figure 18 and we now briefly describe them in turn.

Figure 18a shows a group of *inverted U* bursts. These bursts are characterized by a smooth change from negative to positive frequency drift in the frequency type time profile. They are fairly common on the Harvard spectral records, about 50 well-defined bursts being recorded each year. They occur mainly at frequencies below 300 Mc/s. The exciters of these bursts are believed similar to those producing Type-III bursts, but in the case of the inverted U bursts are believed to be guided along closed magnetic field lines joining photospheric regions of opposite polarity (MAXWELL and SWARUP, 1958).

Figure 18b shows a group of fast-drift bursts, of limited frequency range, that appear to be accompanied by echoes. The bursts may have a negative or a positive frequency-time profile, and they are described as *forward-drift* and *reverse-drift pairs* respectively. A detailed analysis of these bursts has been published by ROBERTS (1958). They are only rarely seen on the Harvard spectral records.

About 5–10 minutes after some Type-II bursts, there may occur a series of

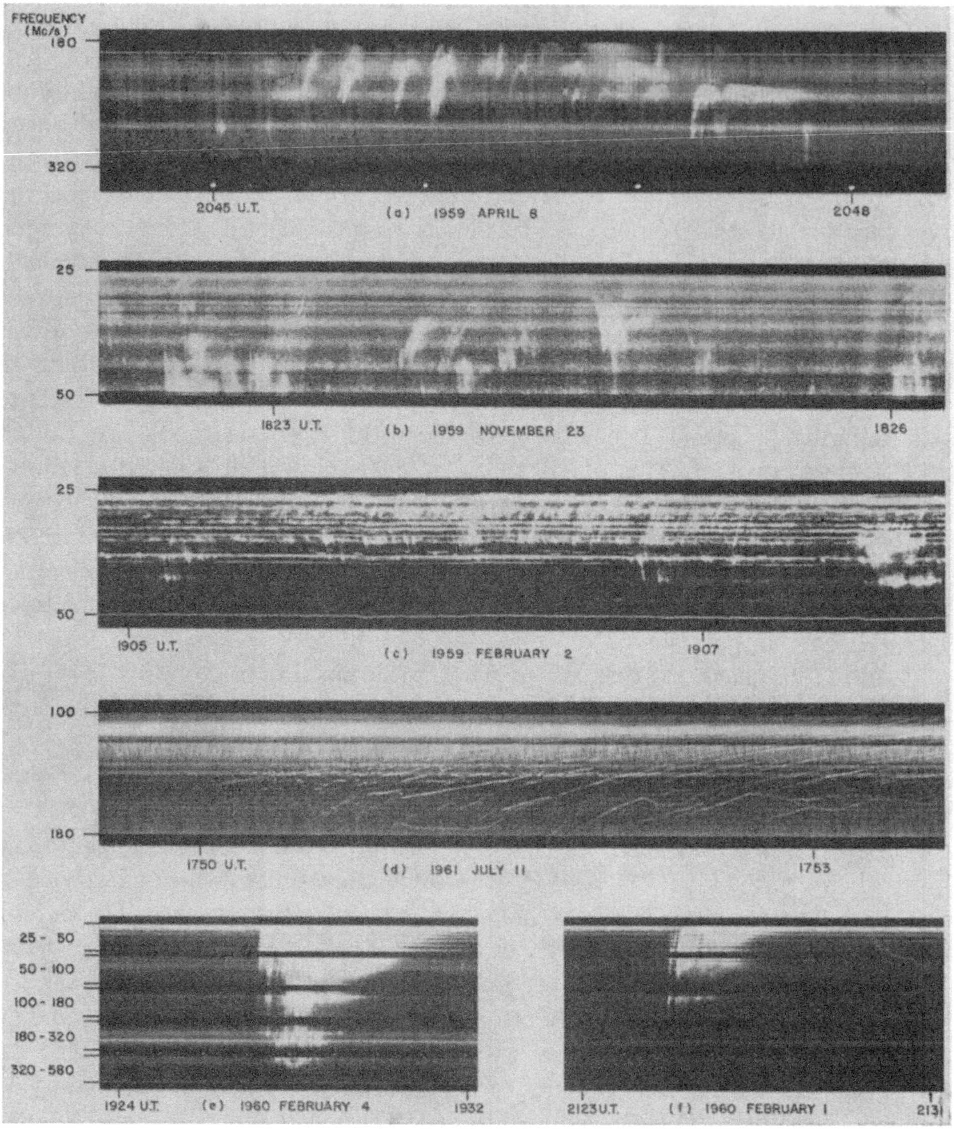

Fig. 18. Spectra of bursts related to Type-III bursts. (a) A group of inverted U bursts. (b) Forward and reverse drift pairs. (c) Herringbone structure following a Type-II burst. (d) Intermediate drift bursts occurring during a Type-IV burst. (e) Continuum Type-V burst. (f) Continuum Type-V burst.

short bursts at low frequencies, which have a *herringbone* structure (Figure 18c). This structure has also been discussed by ROBERTS (1959).

On rare occasions, bursts are recorded with drift rates intermediate between those of Type-II bursts and Type-III bursts. An example of these *intermediate-drift* bursts, in the band 25–50 Mc/s, is shown in Figure 18d.

Examples of short bursts of continuum radiation following groups of Type-III bursts are shown in Figure 18e, f. These bursts were distinguished on the Harvard spectrograph records in 1956 and were subsequently named Type-V bursts by WILD, SHERIDAN, and NEYLAN (1959). They occur with about 2 per cent of groups of Type-III bursts. The broadband emission suggests that the generating mechanism may be the synchrotron process, but this is not certain.

9. Continuum Burst (Spectral Type IV)

9.1 *General Characteristics*

In this section we shall be concerned with the Type-IV continuum bursts that accompany solar flares. The main spectral features of these bursts are their broad frequency bandwidth and the absence of sharp changes of intensity with frequency or time (in contrast to the sharp structure in Types I, II, and III). Examples from the Harvard records are shown in Figures 19 and 20. The emission usually covers many octaves of the radio spectrum, and the duration of the bursts may vary from a few minutes to many hours. For the largest flares, the radio emission may cover the whole observable radio band from < 10 Mc/s to > 50 Gc/s. Reckoned on a continuous observing basis, about 40 bursts were recorded each year at sunspot maximum. In about 70 per cent of cases they were preceded or accompanied by Type-II emission.

The Type-IV bursts were first recognized by BOISCHOT (1957), from observations with a grating array and a fixed-frequency receiver operating at 169 Mc/s. The spectral characteristics of the bursts have been described by MAXWELL, SWARUP, and THOMPSON (1958) and by THOMPSON and MAXWELL (1962). A group of papers describing recent observations of the bursts has been published in the *Proceedings of the International Conference on Cosmic Rays and the Earth Storm* (*J. Phys. Soc. Japan* **17** (1962), A-2, 195–249). The bursts are of considerable importance in studies of geophysical phenomena, since they are associated with the emission of both high-unity and low-energy cosmic ray particles from the sun.

The physical characteristics of Type-IV bursts, as observed up to the present, are summarized in Table 4.

9.2 *Flare Association*

Type-IV bursts of duration < 15 min are generally associated with flares of importance $\leqslant 1$, whereas the longest lived bursts, > 4 hours, are generally associated with flares of importance 3. The onset of the radio bursts follows the start of associated flare with a mean delay of about 12 min. The flares are concentrated toward the central meridian of the sun and 85 per cent lie between longitudes $\pm 60°$, suggesting that the radio emission is concentrated in a radial beam of corresponding angular width (THOMPSON and MAXWELL, 1962).

ELLISON, MCKENNA, and REID (1961) have observed a dark absorbing halo,

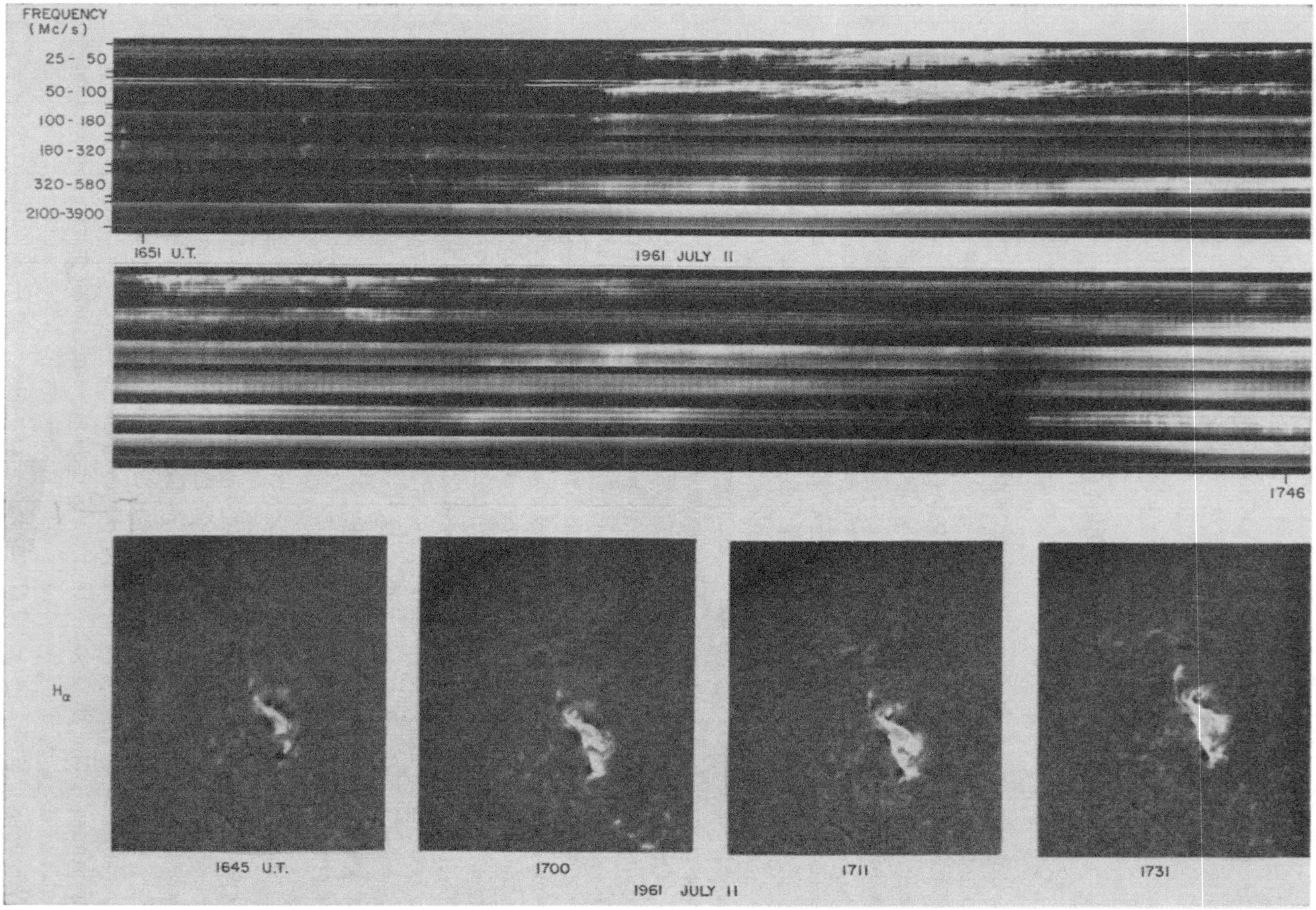

Fig. 19. Type-IV (continuum) burst accompanied by a Type-II burst. Notice the gradual extension of the Type-IV radiation from the centimeter band into the decimeter band, and then into the meter band. The burst continued until approximately 1845 UT. The development of the associated class 3 flare is shown in Hα photographs from the Lockheed Solar Observatory (courtesy G. E. Moreton).

TABLE 4

CHARACTERISTICS OF TYPE-IV BURSTS

	Wavelength			*Reference*
	Centimeter	*Decimeter*	*Meter*	
Diameter of source (arc min)	1.5 expanding to 4	Several arc min	—	TANAKA and KAKINUMA (1962)
	—	2–8 (70 per cent < 4 at 340 Mc/s)	> 10 (87 Mc/s)	KUNDU and FIROR (1961)
	—		7–12 (169 Mc/s)	BOISCHOT (1958)
	—	—	2–6 (169 Mc/s)	PICK-GUTMANN (1961)
	—	—	6 (core: 40–70 Mc/s)	WEISS (1963b)
Minimum height above photosphere (units of R_0: $R_0 = 696000$ km)	0.05	0.1–0.4	—	TANAKA and KAKINUMA (1962)
		< 0.06 (340 Mc/s)	> 0.3 (87 Mc/s)	KUNDU and FIROR (1961)
		—	0.3 (200 Mc/s)	MORIMOTO (1961)
		—	1.1 (40–70 Mc/s)	WEISS (1963b)
		—	0.2–0.3 (169 Mc/s)	PICK-GUTMANN (1961)
Motion (projected transverse component)		< 1′ (340 Mc/s) -	1000 km/sec (up to 6 R_0)	KUNDU and FIROR (1961)
		< 2′ (1420 Mc/s)	1100 km/sec (up to 2 R_0)	KRISHNAN and MULLALY (1961)
		—	1000 km/sec (up to 5 R_0)	BOISCHOT (1958)
		—	2200–4400 km/sec (up to 4 R_0; 4 cases)	WILD, SHERIDAN, and TRENT (1959); WEISS (1963b)
		—	Stationary (8 cases)	
Polarization (percentage circular)	10–40 per cent E-wave	100 per cent O-wave	Weak E-wave	TANAKA and KAKINUMA (1962)
			10–20 per cent E-wave for first 10 min (moving); 50–100 per cent O-wave later (stationary)	WILD, SHERIDAN, and TRENT (1959); FOKKER (1960); PICK-GUTMANN (1961); WEISS (1963b)
Intensity ($\times 10^{-22}$ MKS units)	Usually 100–2000 Sometimes up to 5000	Usually 100–2000 Sometimes up to 50000	Usually 100–2000 Sometimes up to 10000	TAKAKURA and KAI (1961); THOMPSON and MAXWELL (1962)
Duration (minutes)	5–40	5–120	5– < 120	THOMPSON and MAXWELL (1962)

Fig. 20. Type-IV (continuum) burst preceded by Type-II and Type-III bursts.

which they call the flare nimbus, which sometimes surrounds class 3 or 3+ flares and persists during the lifetime of a Type-IV burst. They presume that the nimbus is the optical counterpart of the cloud of relativistic electrons generating the Type-IV burst.

9.3 *Pulsations*

Although Type-IV bursts are generally characterized on spectral records by a smooth featureless appearance, on occasions a pulsating structure is observed in the decimeter band (YOUNG *et al.*, 1961; THOMPSON and MAXWELL, 1962). As may be seen in Figure 21a, b, the pulses are of wide frequency bandwidth, generally more than an octave in extent, and occur in rapid succession. They often resemble a storm of Type-III bursts. The cause of the pulsations is uncertain but, if Type-IV radiation is generated by the synchrotron mechanism, the pulses may result from the injection of high-energy electrons into the magnetic field in discrete jets.

9.4 *Polarization*

Observations of the polarization of Type-IV bursts are summarized in Table 4. In the centimeter band the bursts are generally circularly polarized to the extent of 10–40 per cent, and the sense of polarization corresponds to that of the extraordinary (E) wave of magneto-ionic theory. TANAKA and KAKINUMA (1962) have established the important fact that the sense of polarization of the

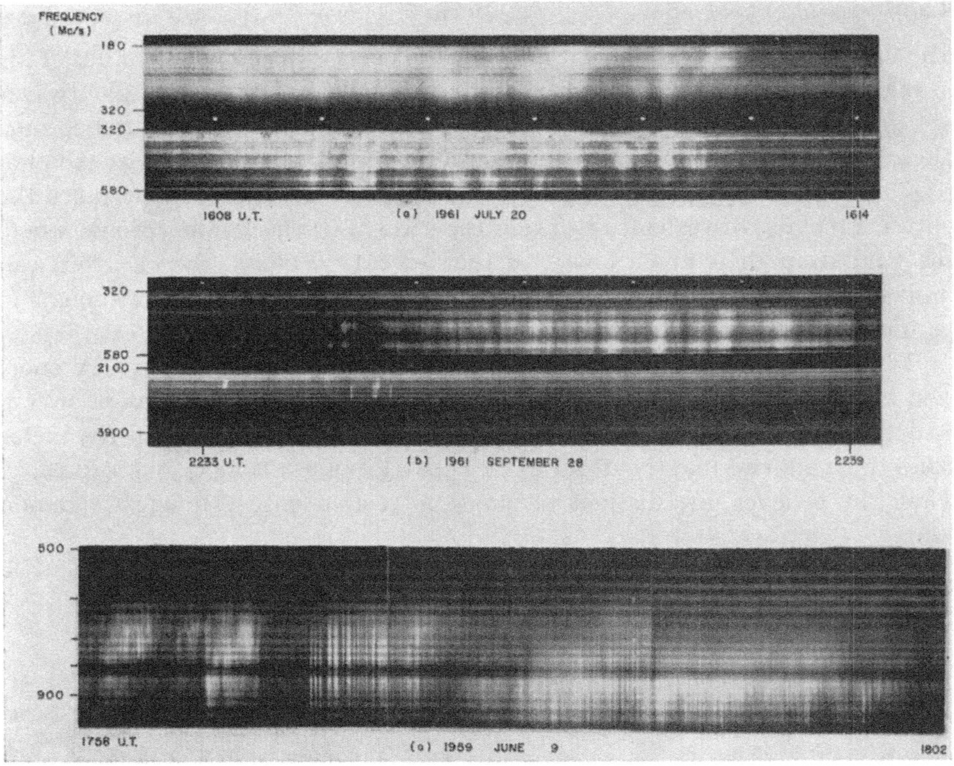

Fig. 21. Pulsating structure in Type-IV bursts. (a) This record is a sectional enlargement of the radio burst shown in Figure 2e. (b) Pulsating structure with bursts in microwave band. (c) Pulsating structure observed with a spectrograph covering range 500–950 Mc/s. (Figure (c) from YOUNG *et al.*, 1961.)

radiation changes from the extraordinary component in the centimeter band to the ordinary component in the decimeter band. Observations in the decimeter band are relatively meager but suggest that the polarization is approximately 50–100 per cent circularly polarized in the ordinary (O) mode. In the meter band the observations indicate that Type-IV emission is initially fairly weakly polarized and the sense of polarization corresponds to that of the extraordinary wave. This polarization probably corresponds to that of the *moving* component of Type-IV emission. Subsequently the polarization may increase in intensity and changes its sense of polarization to the ordinary mode.

The latter polarization is typical not only of *stationary* Type-IV bursts (WEISS, 1963b) but also of storm-continuum (PICK-GUTMANN, 1961) and of noise storms (PAYNE-SCOTT and LITTLE, 1951). The similarity in polarization, burstiness, height, and directivity of stationary Type-IV bursts, storm-continuum, and noise storms strongly suggest that all meter wavelength phenomena are closely related, if not essentially the same.

9.5 *Motion of Source*

The data of Table 4 show that the over-all Type-IV phenomenon is not static in the solar atmosphere, but that a primary disturbance often moves outwards at velocities of the order of 1 000 km/sec, at least in the middle and outer corona. Microwave Type-IV emission generated in the strong field region near the photosphere appears first, and is nearly completed in the time that it takes the outward-moving disturbance to reach the coronal levels. In the corona, where the field strength is much lower, of the order 1–10 gauss, the electrons can radiate at lower frequencies and the critical frequency is also low enough to permit the escape of meter and decameter wavelength radiation.

Weiss (1963b) has examined the properties of 12 sources of Type IV emission at meter wavelengths, from results obtained with the Australian sweep-frequency interferometer in the range 40–70 Mc/s. He finds that in the meter band there are two major categories of Type-IV bursts: *moving* and *stationary*. These, he believes, are distinct phenomena, related only through a common

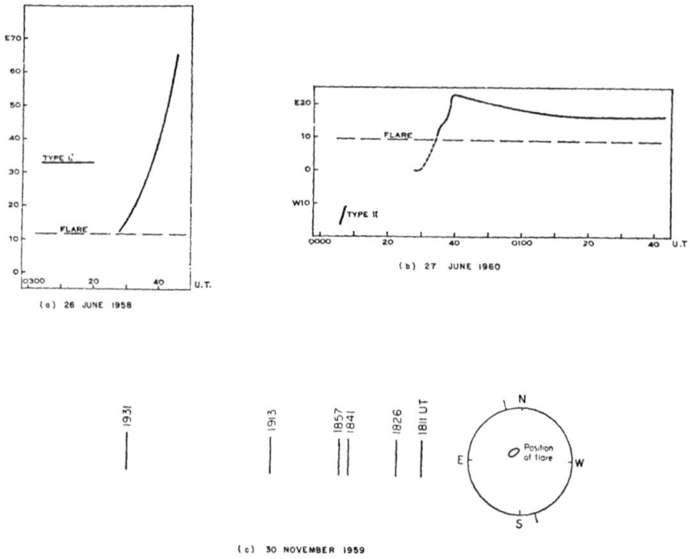

Fig. 22. Transverse velocity components for *moving* Type-IV bursts. (a, b) Measurements at 40–70 Mc/s (Weiss, 1963b). (c) Measurements at 87 Mc/s (Kundu and Firor, 1961).

initiating disturbance low in the solar atmosphere. Weiss suggests that the stationary Type-IV burst may be a separate event, the initial stages of which may be obscured by the more intense but short-lived moving burst. Of the observed Type-IV bursts, 8 out of 12 were stationary. Weiss also finds that the initial position of the meter wavelength Type-IV source, whether moving or stationary, is often widely separated from the position of the associated Type-II burst. Hence the sources of Type-II and Type-IV bursts, although presumably both generated in the flare region, do not always follow the same trajectories through the solar atmosphere.

The *moving* Type-IV burst is characterized by rapid outward movement through the corona (Figure 22), fairly short duration, ill-defined spectral features, a broad cone of emission, and polarization in the extraordinary mode (if it is assumed that the magnetic field of the leading spot predominates). Often the outward movement is followed by a slower inward movement to a final stable position. The *stationary* Type-IV burst is characterized by little or no source movement (Figure 23), long duration, broad continuous spectrum, relatively small source diameter, height close to the local plasma level, and

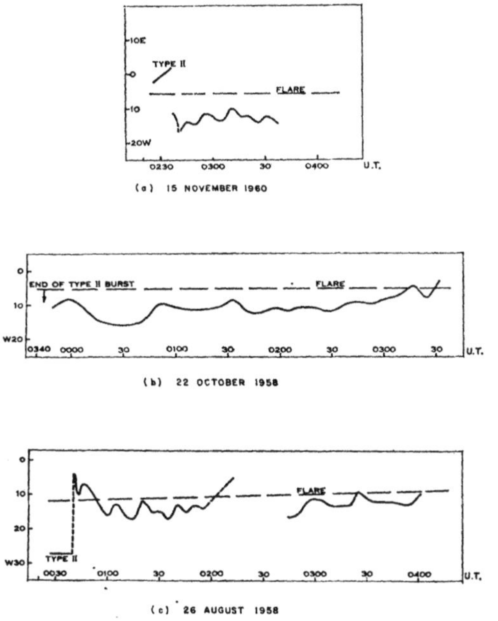

Fig. 23. Position measurements on *stationary* Type-IV bursts at 40–70 Mc/s (Weiss, 1963b).

strong polarization in the ordinary mode (on the leading spot hypothesis). These bursts have a characteristic initial development in intensity and polarization, and an initial spectral drift from high to low frequencies.

9.6 *Generalized Model Spectra*

Generalized model spectra for Type-IV bursts have been suggested by various authors. Four of these models are reproduced in Figure 24. It will immediately be seen that the models differ significantly among themselves and only approximate the observed spectral characteristics of Type-IV bursts illustrated in Figures 19 and 20. The reasons for this disagreement are fairly clear. In the first place, the bursts do not run to stereotyped form: they are complex and their characteristics vary widely from case to case. This complexity is understandable, since the magnetic field configuration above sunspot regions undoubtedly varies tremendously and the density and energy range of electrons generating the bursts will also vary extensively. Second, the data on which the models are based are still incomplete, the lack of information in the decimeter band being particularly marked. To establish detailed models, one would require spectral data over the range 10 Mc/s–50 Gc/s and information, in at least every octave of this band, about the polarization, position, motion and diameter of the radiating source. It would also be necessary to have all these data available simultaneously for the largest bursts. At the present time, most such data are unavailable, and many important questions will therefore not be settled until the next sunspot maximum, when it is hoped that much more complete observations will be made. In the interim, it seems expedient to discuss the

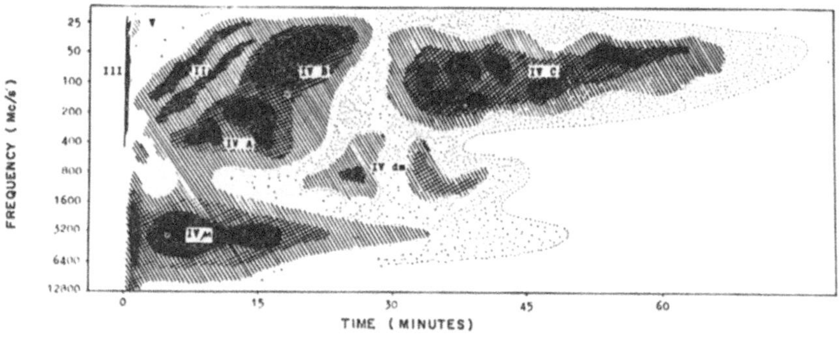

(a)

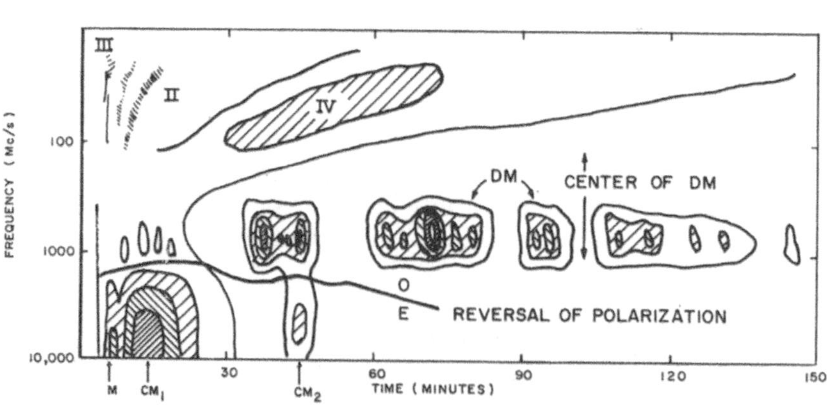

(b)

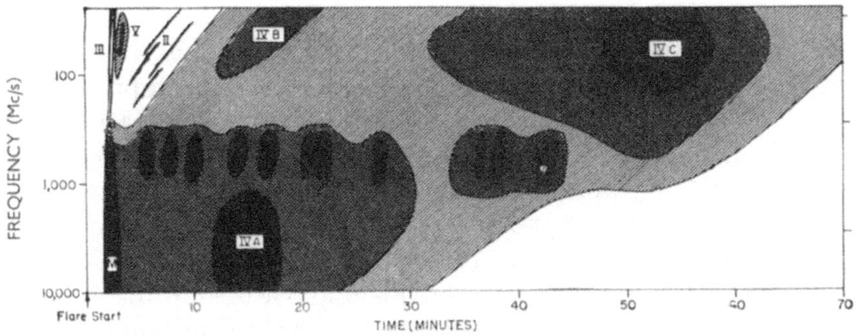

(c)

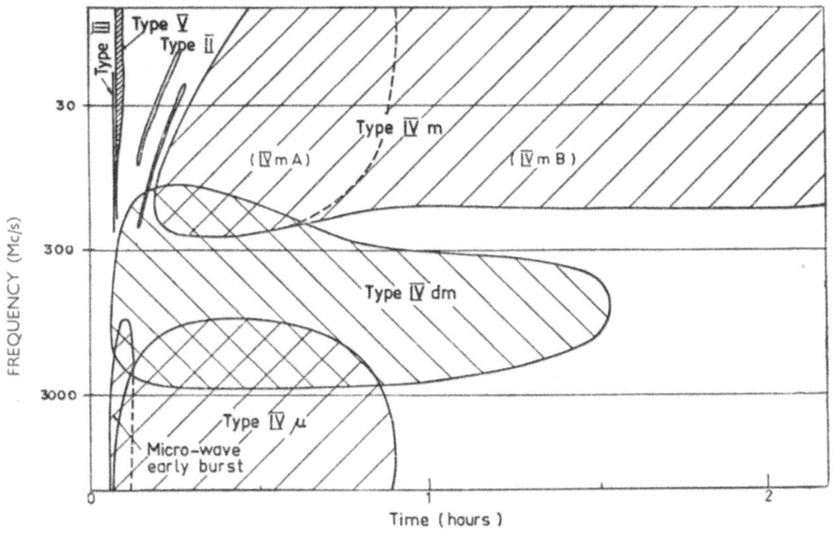

(d)

Figure 24. Hypothetical generalized model spectra for Type-IV bursts. (a) FOKKER (1961). (b) TANAKA and KAKINUMA (1962). (c) KUNDU and SMERD (1962). (d) WILD (1962). The diagrams show that there is at present no agreement on the precise characteristics of the Type-IV components, nor on nomenclature.

characteristics of Type-IV bursts in terms of their observed properties in the centimeter, decimeter, and meter bands (literally, 1–10 cm, 10 cm–1 m, 1–10 m) and the data of Table 4 have been subdivided in this manner.

9.7 *Emission Mechanisms*

The intensity of Type-IV radiation, which is frequently $> 10^{-19}$ MKS units at 100 Mc/s, equivalent to temperatures $> 5 \times 10^9$ °K over an emitting source 10 arc min in width, indicates that it is non-thermal in origin. It is now generally believed that emission in the initial stages of Type-IV bursts is generated by synchrotron radiation from electrons which have been accelerated to relativistic energies at the time of a flare, and which have been trapped in magnetic fields. This mechanism was first suggested by BOISCHOT and DENISSE (1957). Measurements on Type-IV bursts taken with the Australian sweep-frequency interferometer (WILD, SHERIDAN, and TRENT, 1959) show that, in the band 40–70 Mc/s, all frequencies apparently originate at the same source position (in contradistinction to observations of Type-II and Type-III bursts). The synchrotron mechanism explains the broad-band nature of the radiation, its stability over long periods of time, and the fact that the instantaneous position of the emitting source is largely independent of frequency.

We can reasonably account for the observed radio emission spectrum by the synchrotron mechanism in terms of known magnetic field strengths, 250–500 gauss in the chromosphere and 1–10 gauss in the corona, for electron energies in the range 0.5–5 Mev. The densities of relativistic electrons in the corona would require to be about 100 cm^{-3}. (It is possible that the synchrotron emission may extend to optical frequencies, although for fields of 1 000 gauss near the photosphere, the optical emission would require electrons of energy at least 100 Mev. It is not yet certain whether electrons are accelerated to such high energies by flares.)

Radiation in the centimeter band is generated in the upper chromosphere. Unless there is a continuous supply of high-energy electrons, the radiation in this band will be relatively short-lived, since the collisional damping time for electrons of relativistic energies in the chromosphere, where the density is of the order of 10^9 particles cm^{-3}, is approximately 20 min (TAKAKURA, 1960). In point of fact, this figure accords with the observed duration of the bursts in the microwave band, indicating that the electron acceleration processes in flares are probably rapid and of only a few minutes' duration. Radiation in the meter band is generated in the corona, where the density is of the order of 10^7 particles cm^{-3} and the theoretical collisional damping time for relativistic electrons is about 100 hours for $\beta = 0.7$, and 1 000 hours for $\beta = 0.9$. The radiation damping time in fields of 10 gauss is about 300 hours for $\beta = 0.7$–0.9 (TAKAKURA, 1960). Type-IV bursts in the meter band wave have typical durations of several hours and we therefore deduce that the duration of the bursts is probably limited by the dispersion of the magnetic fields.

DENISSE (1960) has suggested that the "storm continuum" following a Type-IV burst, may be generated by a Čerenkov plasma wave process. He believes that, at the time of a flare, high-energy particles are ejected high into the corona where they may remain trapped in coronal magnetic fields. These particles may subsequently diffuse downwards, possibly along radial magnetic field lines in a coronal streamer extending above a sunspot. Provided the velocity of the particles is greater than the thermal velocity of electrons in the plasma, Čerenkov plasma waves may be excited. Energy may then be transferred from the plasma waves into electromagnetic modes by scattering on density fluctuations or by "combination" scattering. The energy transfer into electromagnetic radiation will occur preferentially at the plasma frequency or harmonics thereof. At the present time, however, it does not appear possible, from observations of emission height, polarization, or directivity, to distinguish between a synchrotron emission or a Čerenkov plasma wave to account for the later components of Type-IV bursts, storm continuum, or the enhanced background radiation accompanying noise storms.

10. The Relation of Radio Bursts to X-Ray and High-Energy Nucleon Emission from Solar Flares

Let us now briefly examine the acceleration and emission processes in a large flare that give rise to radio emission, X rays, and high-energy nucleons. Some of these problems have already been considered by DE JAGER (1962) and by WILD (1962). The relation of X rays to radio bursts in the centimeter band has been considered by KUNDU (1961).

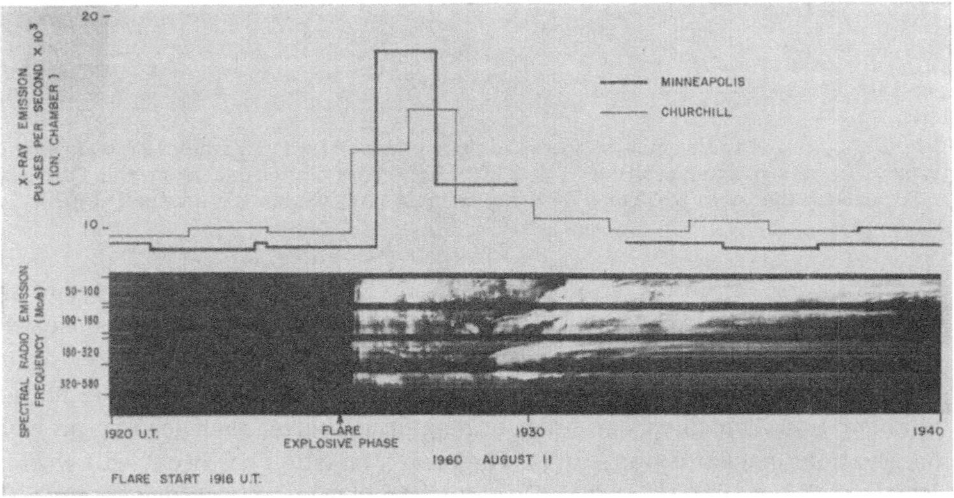

Fig. 25. X-ray and radio emission from the class 3 flare of 1960 August 11, commencing 1916 UT, heliographic position N23 E27. High-energy nucleons first reached the earth at approximately 2400 UT (X-ray data: WINCKLER, MAY and MASLEY, 1961.)

Two typical sets of data, showing the relation of the radio bursts to X rays and accompanying optical phenomena, are shown in Figures 25 and 26 respectively. In this discussion, we shall refer to one of these cases, namely that shown in Figure 26, to illustrate the sequence of events that occurs at the time of a flare, and to summarize current theories invoked to account for the various phenomena. The flare under consideration was that of 1961 September 28 commencing 2202 UT, optical importance 3, and heliographic coordinates N15 E29. It was accompanied by radio emission of spectral Types II, III, and

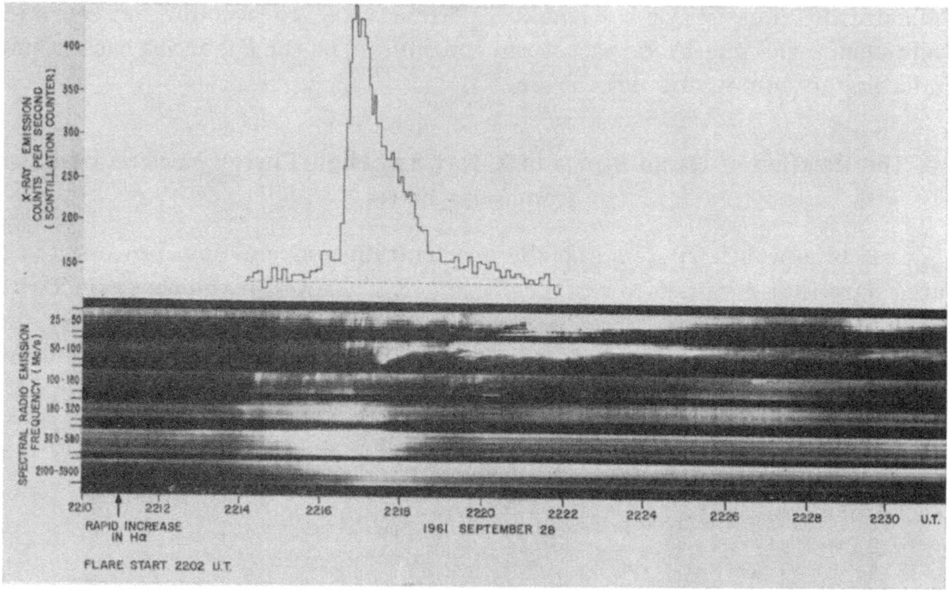

Fig. 26. X-ray and radio emission from the class 3 flare of 1961 September 28, commencing 2202 UT, heliographic position N15 E29. High-energy nucleons from the flare first reached the earth at 2245 UT (X-ray data: ANDERSON and WINCKLER, 1962.)

IV, by a burst of X rays, and by the emission of protons with energies of the order of a few hundred Mev. The X-ray emission from the flare has been described in some detail by ANDERSON and WINCKLER (1962) from observations made with balloon-borne scintillation counters; their data had a time resolution of about 1 sec, which was an order of magnitude better than for any previous observations of solar X-ray bursts. The proton emission has been dealt with by BRYANT, CLINE, DESAI, and MCDONALD (1962) from observations made with scintillation and geiger counters carried aboard the satellite Explorer XII. The solar radio spectral data have been discussed by MAXWELL (1963).

The source of energy in the flare is believed to be the magnetic field but the precise manner in which magnetic energy is transformed into kinetic and other forms of energy is not known. It has been suggested that the energy may be released by partial annihilation of the magnetic field, in the manner proposed by SWEET (1958) and GOLD and HOYLE (1959). The flare itself generally occurs in a weak-field region, near, but not in, a sunspot, and is first seen in the optical spectrum. About ten minutes later there is sometimes an explosive or flash phase, in which jets of matter and large-scale shock waves, or what are interpreted as shock waves on flare patrol films, move out from the flare area.

At the time of the explosive phase we speculatively assume that electrons and protons are accelerated, possibly by a Fermi mechanism (FERMI, 1949, 1954), from collisions with oppositely moving shock fronts of magnetohydrodynamic waves (PARKER, 1957, 1958). The efficiency of the acceleration process depends on the number of effective collisions of the particles with irregularities in the magnetic field. However, the density of field fluctuations with linear dimensions capable of reversing the direction of motion of an electron may be much greater than for a proton, as the Larmor radius of an electron is much lower than for a proton. Hence, rather than protons and electrons being accelerated to equal velocities in the flare plasma to give protons of energy 200 Mev, say, and electrons of 0.1 Mev, it is possible that the electrons may attain much higher energies.

Turning to the flare of 1961 September 28, we note that the time sequence of the various events was as follows. The flare commenced at 2202 UT and, although there was no explosive or flash phase, a rapid rise in the Hα emission was seen at 2211–2213 (MORETON, 1962). Radio bursts of spectral Types III and IV commenced at 2214, the main X-ray burst commenced at 2216.5, a Type-II burst commenced at 2217.5, and solar protons of energy > 100 Mev reached the earth at 2245. The acceleration period for the 1961 September 28 flare is therefore presumed to be the interval between the rapid optical growth phase at 2211–2213, and the initiation of the radio and X-ray bursts at 2214–2216. This interval of approximately 3 minutes agrees fairly well with PARKER's (1957) estimate of 2 minutes, as the period required to accelerate ions from thermal to relativistic energies in a flare region, by Fermi's mechanism.

We assume that electrons of energy up to about 100 Kev were then injected downwards into regions of high density in the lower chromosphere, where they gave rise to X-ray emission by bremsstrahlung or by the inverse Compton effect. The 3-minute duration of the X-ray burst, and its extremely rapid rise time, suggest that the acceleration processes were short lived and that the high energy electrons were released in a relatively short burst. The flux of X rays of energy > 20 keV incident on the earth's atmosphere was estimated by Anderson and Winckler at about 110 photons cm^{-2} sec^{-1} and the observed energy spectrum was typical of bremsstrahlung emission. No gross changes of the X-ray energy spectrum occurred during the event.

Simultaneously: (i) Type-IV radiation commenced in the centimeter band of the radio spectrum, presumably generated by relativistic electrons spiraling in fields of about 250 gauss in the upper chromosphere. If it is assumed that the acceleration process finished at approximately 2217 and that the duration of the Type-IV emission in the centimeter band was governed thereafter by the collisional damping time of the relativistic electrons ($\beta = 0.7$–0.9) in the chromospheric plasma (density $= 10^9$ cm^{-3}), the calculated decay time is about

20 min. This is in accord with the measured duration of the microwave burst.

(ii) Groups of Type-III bursts were generated by exciters moving outwards through the solar corona at velocities of about 0.4 c. These exciters might be jets of electrons of energy about 200 Kev, or might be fast-traveling electron plasma shock waves. Such Type-III bursts can be observed in Figure 26 from 2214–20.

(iii) A magnetohydrodynamic shock front, carrying field and material, is presumed to have moved outward from the flare region at a velocity of about 1 500 km/sec. When the shock front reached the outer corona about 5 minutes later it gave rise to a Type-II burst. As may be seen in Figure 25b, the Type-II burst commenced at 2217.4 and consisted of emission at fundamental and second harmonic frequencies. The highest frequency of the fundamental was in this case 50 Mc/s. Assuming that the radiation results from plasma oscillations in a coronal streamer of density 10 times the Baumbach-Allen model, we deduce the radiating height at 50 Mc/s to be 600000 km.

For large flares, Type-II disturbances are often followed closely by an outward moving cloud of electrons of relativistic energies, 0.5–5 Mev, constrained by magnetic fields to generate Type-IV emission by the synchrotron mechanism. This radiation gradually spreads to lower and lower frequencies as the cloud migrates outwards. In the case of the 1961 September 28 flare, however, there was little Type-IV radiation in the meter band.

Nearly all observed cases of very high-energy solar nucleon emission are associated with Type IV bursts and it is generally assumed that the flare mechanism which accelerates electrons to relativistic velocities also accelerates the heavier nucleons to sub-relativistic or relativistic energies. The relativistic nucleons escape more or less directly from the solar plasma. In the case of the 1961 September 28 flare, high-energy protons from the flare were first observed near the earth at approximately 2245, that is about 40 min after the flare was first sighted optically. For the following 1.4 hours, the direction of arrival of flare protons of energy $>$ 100 Mev was anisotropic. Thereafter, protons in the solar beam continued to arrive at gradually decreasing intensity for 30 hours.

Forty-six hours after the flare there was a geomagnetic storm with a sudden commencement, and a Forbush decrease of galactic cosmic rays. GOLD (1959) has suggested that the sudden commencement preceding a geomagnetic storm results from the arrival at the earth of a shock front that has propagated through the interplanetary plasma, over the normal solar wind, at a velocity of about 1000 km/sec. It is tempting to identify the passage of such a shock front through the corona with the generation of a Type-II burst. Behind the shock front one would expect to find storm plasma moving with a velocity slightly smaller than that of the front itself, but of the same order of magnitude.

Acknowledgements

The author gratefully acknowledges the help of Mrs. Ann Capri, who prepared the diagrams, and of Mrs. Carol Giles, who typed the text and tables. The research programs in solar radio astronomy at the Harvard Station are supported financially by the U. S. Air Force Cambridge Research Laboratories and the Sacramento Peak Observatory, under contract AF19(628)–2370.

References

AKABANE, K. and COHEN, M. H.: 1961, *Astrophys. J.* **133**, 258.
ALLEN, C. W.: 1947, *Mon. Not. Roy. Astron. Soc.* **107**, 426.
ALLEN, C. W.: 1963, *Astrophysical Quantities* (2nd Ed., Univ. of London, Athlone Press).
ANDERSON, K. A. and WINCKLER, J. R.: 1962, *J. Geophys. Res.* **67**, 4103.
BILLINGS, D. E.: 1963, see page 280 of this volume.
BOHM, D. and GROSS, E. P.: 1949, *Phys. Rev.* **75**, 1851, 1864.
BOHM, D. and GROSS, E. P.: 1950, *Phys. Rev.* **79**, 992.
BOISCHOT, A.: 1957, *Compt. Rend.* **244**, 1326.
BOISCHOT, A.: 1958, *Ann. Astrophys.* **21**, 273.
BOISCHOT, A. and DENISSE, J. F.: 1957, *Compt. Rend.* **245**, 2194.
BOISCHOT, A. and SIMON, P.: 1959, *Paris Symposium on Radio Astronomy* (ed. by R. N. BRACEWELL), Stanford Univ. Press, p. 140.
BRYANT, D. A., CLINE, T. L., DESAI, U. D., and MCDONALD, F. B.: 1962, *J. Geophys. Res.* **67**, 4983.
CHRISTIANSEN, W. N. and MATHEWSON, D. S.: 1958, *Proc. Inst. Radio Engr.* **46** ,127.
CHRISTIANSEN, W. N., MATHEWSON, D. S., PAWSEY, J. L., SMERD, S. F., BOISCHOT, A., DENISSE, J. F., SIMON, P., KAKINUMA, T., DODSON-PRINCE, H., and FIROR, J.: 1960, *Ann. Astrophys.* **23**, 75.
CHRISTIANSEN, W. N. and WARBURTON, J. A.: 1953, *Australian J. Phys.* **6**, 190, 262.
COHEN, M. H.: 1961, *Astrophys. J.* **133**, 978.
COVINGTON, A. E.: 1948, *Proc. IRE* **36**, 454.
COVINGTON, A. E.: 1959, *Paris Symposium on Radio Astronomy* (ed. by R. N. BRACEWELL), Stanford Univ. Press, p. 159.
COVINGTON, A. E. and DODSON, H. W.: 1953, *J. Roy. Astron. Soc. Canada* **47**, 207.
COVINGTON, A. E. and HARVEY, G. A.: 1958, *J. Roy. Astron. Soc. Canada* **52**, 161.
COVINGTON, A. E. and HARVEY, G. A.: 1961, *Nature (London)* **192**, 152.
DENISSE, J. F.: 1960, URSI XIII General Assembly, London; and *Information Bulletin Solar Radio Observatories, Europe* **4**, 3.
DENISSE, J. F. and DELCROIX, J. L.: 1961, *Theorie des Ondes dans les Plasmas*, Dunod, Paris.
DODSON, H. W.: 1958, *Proc. IRE* **46**, 149.
ELGAROY, O.: 1961, *Astrophys. Norvegica* **7**, 123.
ELLISON, M. A., MCKENNA, S. M. P., and REID, J. H.: 1961, *Mon. Not. Roy. Astron. Soc.*, **122**, 491.
ERICKSON, W. C.: 1963, *General Dynamics-Astronautics Space Science Laboratory Report* No. ERR-AN-233.
FERMI, E.: 1949, *Phys. Rev.* **75**, 1169.
FERMI, E.: 1954, *Astrophys. J.* **119**, 1.
FIROR, J. W.: 1959, *Paris Symposium on Radio Astronomy* (ed. by R. N. BRACEWELL), Stanford Univ. Press, p. 136.
FOKKER, A. D.: 1960, Doctoral Thesis, Univ. of Leiden.
FOKKER, A. D.: 1961, *Information Bulletin Solar Radio Observatories, Europe* **8**, p. 7.
GINZBURG, V. L. and ŽELEZNIAKOV, V. V.: 1958, *Soviet Astron.-A.J.* **2**, 653.
GINZBURG, V. L. and ŽELEZNIAKOV, V. V.: 1959, *Soviet Astron.-A.J.* **3**, 235.

GINZBURG, V. L. and ŽELEZNIAKOV, V. V.: 1961, *Soviet Astron.-A.J.* **5**. 1,
GIOVANELLI, R. G.: 1948, *Nature (London)* **161**, 133.
GIOVANELLI, R. G. and ROBERTS, J. A.: 1958, *Australian J. Phys.* **11**, 353.
GOLD, T.: 1959, *J. Geophys. Res.* **64**, 1665.
GOLD, T. and HOYLE, F.: 1959, *Mon. Not. Roy. Astron. Soc.* **120**, 89.
GOLDSTEIN, S. J.: 1959, *Astrophys. J.* **130**, 393.
GROOT, T. DE: 1959, *Paris Symposium on Radio Astronomy* (ed. by R. N. BRACEWELL), Stanford Univ. Press p. 245.
HACHENBERG, O. and WALLIS, G. Z.: 1961, *Z. Astrophys.* **52**, 42.
HUGHES, M. P. and HARKNESS, R. L.: 1963, *Astrophys. J.* **138**, 239.
IHSANOVA, V.: 1959, *Paris Symposium on Radio Astronomy* (ed. by R. N. BRACEWELL), Stanford Univ. Press, p. 171.
JAGER, C. DE: 1962, *Space Sci. Rev.* **1**, 487.
KAI, K.: 1962, *Publ. Astron. Soc. Japan* **14**, 1.
KAKINUMA, T.: 1956, *Proc. Res. Inst. Atm. Nagoya Univ.* **4**, 78.
KAKINUMA, T.: 1958, *Proc. Res. Inst. Atm. Nagoya Univ.* **5**, 71.
KAKINUMA, T. and SWARUP, G.: 1962, *Astrophys. J.* **136**, 975.
KOMESAROFF, M.: 1958, *Australian J. Phys.* **11**, 201.
KRISHNAN, T. and MULLALY, R. F.: 1961, *Nature (London)* **192**, 58.
KUNDU, M. R.: 1959, *Paris Symposium on Radio Astronomy* (ed. by R. N. BRACEWELL), Stanford Univ. Press, p. 222.
KUNDU, M. R.: 1961, *J. Geophys. Res.* **66**, 4308.
KUNDU, M. R. and FIROR, J. W.: 1961, *Astrophys. J.* **134**, 389.
KUNDU, M. R. and HADDOCK, F. T.: 1961, *Trans. Inst. Rad. Eng. AP* **9**, 82.
KUNDU, M. R. and SMERD, S. F.: 1962, *Information Bulletin Solar Radio Observatories, Europe* **11**, 4.
LEHANY, F. J. and YABSLEY, D. E.: 1949, *Australian J. Sci. Res.* **A2**, 48.
LITTLE, C. G. and PAYNE-SCOTT, R.: 1951, *Australian J. Sci. Res.* **A4**, 489.
MALVILLE, J. M.: 1962, *Astrophys. J.* **135**, 834.
MAXWELL, A.: 1963, *Plan. Space Sci.* **11**, 897.
MAXWELL, A., HOWARD, W. E., and GARMIRE, G.: 1959, *Scientific Report No. 14 under USAF Contract AF19(604)-1394* with Harvard University.
MAXWELL, A., HOWARD, W. E., and GARMIRE, G.: 1960, *J. Geophys. Res.* **65**, 3581.
MAXWELL, A. and SWARUP, G.: 1958, *Nature (London)* **181**, 36.
MAXWELL, A., SWARUP, G., and THOMPSON, A. R.: 1958, *Proc. IRE* **46**, 142.
MAXWELL, A. and THOMPSON, A. R.: 1962, *Astrophys. J.* **135**, 138.
MCCREADY, L. L., PAWSEY, J. L., and PAYNE-SCOTT, R.: 1947, *Proc. Roy. Soc.* **A190**, 357.
MORETON, G. E.: 1962, *J. Geophys. Res.* **67**, 1648.
MORIMOTO, M.: 1961, *Publ. Astron. Soc. Japan* **13**, 285.
NEWKIRK, G.: 1961, *Astrophys. J.* **133**, 983.
O'BRIEN, P. A.: 1953, *Mon. Not. Roy. Astron. Soc.* **113**, 597.
PARKER, E. N.: 1957, *Phys. Rev.* **107**, 830.
PARKER, E. N.: 1958, *Phys. Rev.* **109**, 1328.
PAYNE-SCOTT, R. and LITTLE, A. G.: 1951, *Australian J. Sci. Res.* **A4**, 508.
PICK-GUTMANN, M.: 1961, *Ann. Astrophys.* **24**, 183.
ROBERTS, J. A.: 1958, *Australian J. Phys.* **11**, 215.
ROBERTS, J. A.: 1959, *Australian J. Phys.* **12**, 327.
SCHATZMAN, E. L.: 1963, see p. 313 of this volume.
SCHOTT, G. A.: 1912, *Electromagnetic Radiation and the Mechanical Reactions Arising from It*, Cambridge University Press.
SCHWINGER, J.: 1949, *Phys. Rev.* **75**, 1912.
SEATON, M. J.: 1963, see p. 273 of this volume.
SHAIN, C. A. and HIGGINS, C. S.: 1959, *Austral. J. Phys.* **12**, 357.
SHKLOVSKY, I. S.: 1946, *Astron. J. U.R.S.S.* **23**, 333.
SMERD, S. F.: 1963, see p. 398 of this volume.
SMERD, S. F., WILD, J. P., and SHERIDAN, K. V.: 1962, *Australian J. Phys.* **15**, 180.
SMITH, E. v. P. and MCINTOSH, P. S.: 1962, *J. Geophys. Res.* **67**, 1013, 3797.

STANIER, H. M.: 1950, *Nature (London)* **165**, 354.
STURROCK, P. A.: 1961, *Nature (London)* **192**, 58.
SUZUKI, S.: 1961, *Annals Tokyo Astron. Obs.* **7**, 75.
SWARUP, G., KAKINUMA, T., COVINGTON, A. E., HARVEY, G. A., MULLALY, R. F., and ROME J.: 1963, *Astrophys. J.* **137**, 1251.
SWARUP, G. and PARTHASARATHY, R.: 1958, *Australian J. Phys.* **11**, 338.
SWARUP, G., STONE, P. H., and MAXWELL, A.: 1960, *Astrophys. J.* **131**, 725.
SWEET, P. A.: 1958, *IAU Symposium No.* 6 *(Stockholm,* 1956*)* p. 123.
TAKAKURA, T.: 1956, *Publ. Astron. Soc. Japan* **8**, 182.
TAKAKURA, T.: 1959, *Publ. Astron. Soc. Japan* **11**, 55.
TAKAKURA, T.: 1960, *Publ. Astron. Soc. Japan* **12**, 325, 352.
TAKAKURA, T.: 1961, *Publ. Astron. Soc. Japan* **13**, 312.
TAKAKURA, T. and KAI, K.: 1961, *Publ. Astron. Soc. Japan* **13**, 94.
TANAKA, H. and KAKINUMA, T.: 1959, *Paris Symposium on Radio Astronomy* (ed. by R. N. BRACEWELL), Stanford Univ. Press, p. 215.
TANAKA, H. and KAKINUMA, T.: 1960, *Proc. Res. Inst. Atm. Nagoya Univ.* **7**, 72.
TANAKA, H. and KAKINUMA, T.: 1962, *J. Phys. Soc. Japan* **17**, A-2, 211.
THOMPSON, A. R.: 1961, *Astrophys. J.* **133**, 643.
THOMPSON, A. R. and MAXWELL, A.: 1962, *Astrophys. J.* **136**, 546.
TWISS, R. Q. and ROBERTS, J. A.: 1958, *Australian J. Phys.* **11**, 424.
UCHIDA, Y.: 1960, *Publ. Astron. Soc. Japan* **12**, 376.
UCHIDA, Y.: 1962, *J. Phys. Soc. Japan* **17**, A-2, 234.
VITKEVIČ, V. V. and GORELOVA, M. V.: 1960, *Soviet Astron.-AJ* **4**, 595.
VLADIMIRSKY, V. V.: 1948, *Ž. Eksperim. i Teor. Fiz.* **18**, 392.
WALDMEIER, M.: 1956, *Z. Astrophys.* **40**, 221.
WALDMEIER, M. and MULLER, H.: 1950, *Z. Astrophys.* **27**, 58.
WEISS, A. A.: 1963a, *Australian J. Phys.* **16**, 240.
WEISS, A. A.: 1963b, *Australian J. Phys.* **16**, 526.
WEISS, A. A. and SHERIDAN, K. V.: 1962, *J. Phys. Soc. Japan.* **17**, A-2, 223.
WESTFOLD, K. C.: 1957, *Phil. Mag.* **2**, 1287.
WILD, J. P.: 1962, *J. Phys. Soc. Japan* **17**, A-2, 249.
WILD, J. P.: 1963, *The Solar Corona* (ed. by J. W. EVANS), Academic Press, p. 115.
WILD, J. P. and MCCREADY, L. L.: 1950, *Australian J. Sci. Res.* **A3**, 387.
WILD, J. P., MURRAY, J. D., and ROWE, W. C.: 1954, *Australian J. Phys.* **7**, 439.
WILD, J. P. and SHERIDAN, K. V.: 1958, *Proc. IRE* **46**, 160.
WILD, J. P., SHERIDAN, K. V., and NEYLAN, A. A.: 1959, *Australian J. Phys.* **12**, 369.
WILD, J. P., SHERIDAN, K. V., and TRENT, G. H.: 1959, *Paris Symposium on Radio Astronomy* (ed. by R. N. BRACEWELL), Stanford Univ. Press, p. 176.
WINCKLER, J. R., MAY, T. C., and MASLEY, A. J.: 1961, *J. Geophys. Res.* **66**, 316.
YOUNG, C. W., SPENCER, C. L., MORETON, G. E., and ROBERTS, J. A.: 1961, *Astrophys. J.* **133**, 243.
ŽELEZNIAKOV, V. V.: 1962, *Soviet Astron.-A.J.* **6**, 3.

NON-THERMAL SOURCES AND AMPLIFIED EMISSION IN SOLAR RADIO BURSTS

S. F. SMERD

Radiophysics Laboratory, CSIRO, Sydney, Australia

Abstract. The possibility of amplifying radio waves in a plasma is explored along the lines pioneered by TWISS (1958); amplification depends on the macroscopic absorption coefficient assuming a negative value. The conditions for negative absorption are specified in terms of the electron-energy distribution, the statistical weight of the energy levels of the radiating electrons and their mean emissivity. The latter is characteristic of particular microscopic radiation mechanisms.

One necessary condition for negative absorption is a positive gradient in the electron-energy distribution; this condition is probably met by the streams of fast electrons which have been identified as the sources of solar radio bursts.

A second condition – a negative gradient in the product of statistical weight and electron emissivity – has to be investigated for the various microscopic radiation mechanisms.

We find that bremsstrahlung and synchrotron radiation from highly relativistic electrons cannot lead to amplification; gyro-radiation from mildly relativistic electrons and Čerenkov plasma waves can. In the latter case BOHM and GROSS (1949) demonstrated that amplification was due to coherent emission from bunched electrons in a stream.

These conclusions on amplified and temperature-limited radiations from non-thermal electron streams are applied to different types of solar radio bursts. It is suggested that the drifting bursts of Types II and III originate in radiation from amplified longitudinal plasma waves. Microwave bursts and several phases of Type-IV emission are probably non-thermal, but temperature limited, gyro-synchrotron radiation. It is tempting to invoke amplified gyro radiation to explain Type-I and other bursts superimposed on long continuum radiation.

1. Introduction

The speed and amplitude with which the intensity from the whole Sun varies in solar bursts, particularly at the metre waves, make it clear that we are here dealing with non-thermal radiating systems. The burst sources have been identified with fast and turbulent streams of electrons.

The absolute brightness temperatures attained in such bursts – up to 10^{12} °K – suggest very high energies up to 100 MeV for the radiating electrons.

In an interpretation of solar bursts we have to account therefore for the observed spectra in terms of the radiations emitted when high-energy electron streams permeate the background plasma of the Sun's atmosphere.

The possibility of wave-amplification in these circumstances was first propounded by BAILEY[1] on the basis of complex propagation constants and frequencies in the appropriate dispersion equations. At certain frequencies and with suitable magnetic fields and stream velocities the signs of the imaginary parts suggested instability – i.e. growth with time – and amplification – i.e. growth with distance. It was later pointed out[2] that such general conclusions from the dispersion equations could be misleading. For instance, without specifying a particular situation by initial and boundary conditions it may not

 All rights reserved.

be possible to distinguish apparent wave growth in one direction from an actual damping of the reflected wave in the opposite direction.

TWISS[3] starting with MARTYN's[4] quantum formulation for the macroscopic radio-absorption coefficient derived some general conditions under which this absorption coefficient becomes negative, that is conditions for wave amplification. These conditions were obtained in terms of the energy distribution of the radiating electrons and their mean emissivity. The latter is characteristic of a particular microscopic radiating mechanism.

The possibility of wave amplification is clearly of interest in our search for the origin of the high intensities observed in solar bursts.

2. Positive and Negative Absorption

The radio frequencies are much lower than the optical frequencies and the inequality $hf/kT \ll 1$ invariably applies in radio-astronomical sources. As a consequence stimulated emission is enormously more likely and important in the radio spectrum. The macroscopic absorption coefficient – that is the difference between all stimulated absorptions and all stimulated emissions – is therefore not, as it is in the optical case, merely the sum of all stimulated absorptions.

The correspondence between classical and quantum formulation is introduced by identifying the mean electron emissivity $Q_f(\varepsilon)$ with the spontaneous emission. Thus

$$Q_f(\varepsilon) = A_{\varepsilon \to \varepsilon - hf} hf .$$

The volume emissivity is then

$$\eta_f = \int_0^\infty Q_f(\varepsilon)\, n(\varepsilon)\, \mathrm{d}\varepsilon \qquad (1)$$

where $n(\varepsilon)\,\mathrm{d}\varepsilon$ is the number density of electrons with energy in the range ε to $\varepsilon + \mathrm{d}\varepsilon$.

In deriving the corresponding expression for the absorption co-efficient we assume that the Principle of Detailed Balancing applies for all processes and that the refractive index does not depend on energy. We can then relate the Einstein transition co-efficients in the usual manner, namely

$$g(\varepsilon)\, B_{\varepsilon \to \varepsilon - hf} = g(\varepsilon - hf)\, B_{\varepsilon - hf \to \varepsilon}$$

and

$$A_{\varepsilon \to \varepsilon - hf} / B_{\varepsilon \to \varepsilon - hf} = \mu^2(f)\, hf^3/c^2$$

Here $g(\varepsilon)\,\mathrm{d}\varepsilon$ is the statistical weight of the (continuous) energy levels and is defined by

$$NF(\varepsilon)\, g(\varepsilon) = n(\varepsilon) \qquad (2)$$

where N is the number density of electrons and $F(\varepsilon)$ is their energy distribution in momentum space. From classical statistics (Liouville's theorem) the statistical weight is that volume in momentum space which corresponds to a given energy interval ε, $\varepsilon + \mathrm{d}\varepsilon$; thus for an *isotropic* distribution of momentum, p,

$$g(\varepsilon)\,\mathrm{d}\varepsilon = 4\pi p^2 \mathrm{d}p \tag{3}$$

In the non-relativistic case ($\varepsilon \ll m_0 c^2$)

$$g(\varepsilon) = 4\pi(2m_0^3)^{\frac{1}{2}}\varepsilon^{\frac{1}{2}}\ ; \tag{4}$$

in the highly relativistic case ($\varepsilon \gg m_0 c^2$)

$$g(\varepsilon) = (4\pi/c^3)\varepsilon^2\ . \tag{5}$$

If we assume further that $hf \ll \varepsilon$ which applies well for radio emission from energetic electrons we obtain for the absorption co-efficient

$$K = -\frac{c^2 N}{\mu^2 f^2}\int_0^\infty \frac{\mathrm{d}F}{\mathrm{d}\varepsilon} Q_f(\varepsilon)\, g(\varepsilon)\,\mathrm{d}\varepsilon\ . \tag{6}$$

If Q and F are well-behaved functions and provided $Q(0) = 0$, integration by parts leads to an alternative form

$$K = \frac{c^2 N}{\mu^2 f^2}\int_0^\infty F(\varepsilon)\frac{\mathrm{d}}{\mathrm{d}\varepsilon}\{Q_f(\varepsilon)\, g(\varepsilon)\}\,\mathrm{d}\varepsilon\ . \tag{7}$$

It is clear from Equation (7) that whatever the electron energy distribution: (A) *the absorption coefficient is always positive if the product of statistical weight and electron emissivity increase continuously with energy*. Similarly, it is clear from Equation (6) that whatever the statistical weight and the electron emissivity: (B) *the absorption coefficient is always positive if the energy distribution decreases continuously with increasing energy*. Such is the case in most natural circumstances. It is only then that we can specify the radiation yield by the radiation temperature T_r; the latter is, from Equations (1), (2), (6) and (7) and the Rayleigh-Jeans law, given by

$$kT_r = -\frac{\int_0^\infty F(\varepsilon)\, Q_f(\varepsilon)\, g(\varepsilon)\,\mathrm{d}\varepsilon}{\int_0^\infty \frac{\mathrm{d}F}{\mathrm{d}\varepsilon} Q_f(\varepsilon)\, g(\varepsilon)\,\mathrm{d}\varepsilon} = \frac{\int_0^\infty F(\varepsilon)\, Q_f(\varepsilon)\, g(\varepsilon)\,\mathrm{d}\varepsilon}{\int_0^\infty F(\varepsilon)\frac{\mathrm{d}}{\mathrm{d}\varepsilon}\{Q_f(\varepsilon)\, g(\varepsilon)\}\,\mathrm{d}\varepsilon} \tag{8}$$

where k is Boltzmann's constant.

In the case of a thermal source of Maxwellian energy distribution – when $F(\varepsilon) = (2\pi m_0 kT)^{-3/2} \exp(-\varepsilon/kT)$ – the radiation temperature is identical with the electron temperature. For non-thermal sources, but still under conditions of positive absorption, Equation (8) essentially gives kT_r as a weighted mean energy of the radiating electrons. We refer to such radiation from non-thermal sources as temperature limited.

Twiss[3] has pointed out that there are circumstances when the absorption becomes negative. It is clear from Equation (6) that: (C) *a necessary condition for negative absorption is a positive gradient in the electron energy distribution.* This is the classical, or continuous-energy-distribution, equivalent of the quantum requirement for over population of the high-energy states.

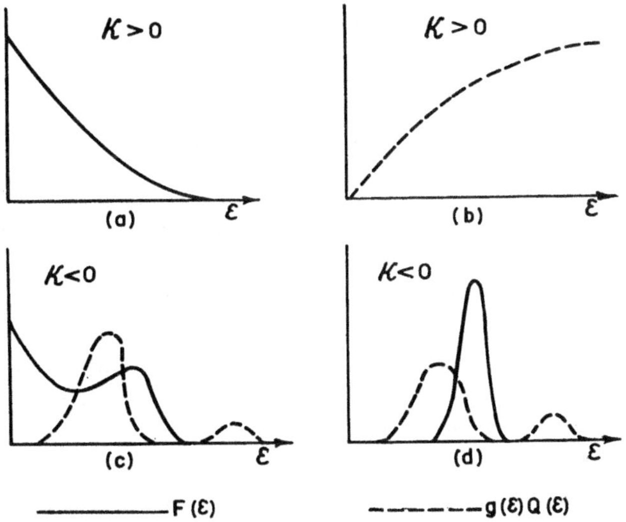

Fig. 1. Examples of electron energy distributions, $F(\varepsilon)$, and electron emissivity $Q_f(\varepsilon)$ multiplied by the statistical weight $g(\varepsilon)$ which lead to positive and negative absorption in accordance with the criteria (A), (B), (C) and (D). (a): Positive absorption when dF/d_ε is negative as in condition (B); a thermal radiation source corresponds to this case. (b): Positive absorption when $(d/d_\varepsilon)\{g(\varepsilon)Q_f(\varepsilon)\}$ is positive as in condition (A); bremsstrahlung is an example. (c) and (d): Two possible situations leading to negative absorption when dF/d_ε is positive and $(d/d_\varepsilon)\{g(\varepsilon)Q_f(\varepsilon)\}$ is negative as in conditions (C) and (D). This can apply to gyro-radiation under favourable conditions.

It is also clear from Equation (7) that even then: (D) *the absorption coefficient will be negative only if in addition to condition* (C) *the product of statistical weight and electron emissivity has a negative gradient in the appropriate energy range.*

We illustrate the above conditions for positive and negative absorption in Figure 1 by the following four examples:

(a) the gradient $dF/d\varepsilon$ is negative for all ε (as it is, for instance, in a thermal plasma); then K must be positive whatever the form of $g(\varepsilon)$ and $Q(\varepsilon)$. (Condition B.)

(b) the gradient $(d/d\varepsilon)\{g(\varepsilon)Q_f(\varepsilon)\}$ is positive for all ε; then K must be positive whatever the form of $F(\varepsilon)$. (Condition A.)

(c) the electron emissivity $Q_f(\varepsilon)$ has one or several "resonance" peaks in the ε-range; then K will be negative if $dF/d\varepsilon$ is positive across one such peak and $F(\varepsilon)$ is insignificant across the others. (Conditions C and D.)

(d) the electron energy $F(\varepsilon)$ has one or more "resonance" peaks in the ε-range; then K will be negative if $(d/d\varepsilon)\{g(\varepsilon)Q_f(\varepsilon)\}$ is negative across one such peak and $Q_f(\varepsilon)$ is insignificant across the others. (Conditions C and D.)

One necessary requirement for negative absorption, namely a positive gradient in the electron energy distribution, is probably met by most electron streams which we have identified as sources of solar bursts. The subsidiary requirements to make the absorption coefficient negative have to be explored for a particular emission mechanism and with due allowance for possible anisotropy in the radiating electrons and the radiating process.

3. The Effects of Anisotropy

So far we have assumed isotropic emission and electron momenta; both these properties can, however, depend on direction. For instance if the source is permeated by a magnetic field the emission at the fundamental gyro-frequency is beamed very broadly along the magnetic field direction, the emission of the high harmonics by highly relativistic electrons is practically confined to the direction of the instantaneous electron velocity. We can allow for the anisotropy by defining the mean electron emissivity $Q_f(\varepsilon, \theta - \phi)$ as a quantity per unit conical angle (here, ϕ and θ are the angles which electron velocity and emission vectors make with the field direction).

The volume emissivity and the absorption coefficient then become

$$\eta_f(\theta) = N \int_0^\infty \int_0^{2\pi} F(\varepsilon, \phi) Q_f(\varepsilon, \theta - \phi) g(\varepsilon, \phi)\, d\varepsilon\, d\phi \tag{9}$$

and

$$K(\theta) = -\frac{c^2 N}{\mu^2 f^2} \int_0^\infty \int_0^{2\pi} \left\{\left(\frac{\partial}{\partial\varepsilon} + \frac{\Delta\phi}{hf}\frac{\partial}{\partial\phi}\right) F(\varepsilon, \phi)\right\}$$
$$\times \{Q_f(\varepsilon, \theta - \phi) g(\varepsilon, \phi)\} d\varepsilon\, d\phi \tag{10}$$

respectively, where $\Delta\phi$ is the change in the direction of the electron motion due to the emission or absorption of a photon. The statistical weight is now that volume in phase space which comprises the electron energies between ε and $\varepsilon + d\varepsilon$ and the electron velocities in the angular interval ϕ to $\phi + d\phi$; thus $g(\varepsilon, \phi)\, d\varepsilon\, d\phi = 2\pi p^2 \sin\phi\, dp\, d\phi = \frac{1}{2} g(\varepsilon) \sin\phi\, d\varepsilon\, d\phi$, where $g(\varepsilon)$ is defined by

Equation (3) and is again given by Equations (4) and (5) for non-relativistic and highly-relativistic electrons respectively. The arguments on positive and negative absorption remain unaltered as long as we replace (d/dε) by

$$\left(\frac{\partial}{\partial\varepsilon} + \frac{\Delta\phi}{hf}\frac{\partial}{\partial\phi}\right).$$

4. The Emission Processes

A. BREMSSTRAHLUNG

This is the term applied to the electromagnetic radiation from charged particles deflected by other charged particles. For non-relativistic electrons we find[5]

$$Q_f(\varepsilon) \propto \xi(\varepsilon)/\varepsilon^{\frac{1}{2}}$$

with

$$\xi(\varepsilon) = l\,n\,(2\varepsilon)^{3/2} + \text{const.}$$

From this and Equation (4) we find that $(\mathrm{d}/\mathrm{d}/\varepsilon\{Q_f(\varepsilon)g(\varepsilon)\}$ is positive for all values of ε. We conclude therefore that *non-relativistic bremsstrahlung cannot lead to amplification.* The temperature-limited bremsstrahlung from a non-thermal source bears close analogy to the thermal case. The brightness temperature cannot exceed the electron temperature (more generally the radiation temperature given by $kT_r \approx \bar{\varepsilon}$); it will reach that value only if the source is optically thick and the opacity depends on the temperature (or mean energy) characteristically as $T^{-3/2}$. Thus, an increase in the electron energy *will decrease the brightness of an optically thin source and decrease the opacity of an originally optically thick source.*

The optical depth of the undisturbed corona at the metre waves is typically about unity; local heating beyond several 10^6 °K will reduce the brightness temperature. Bremsstrahlung bursts are not likely to be observed at the metre waves.

The transition region and chromosphere on the other hand are optically thick to microwaves. The brightness temperature is likely to reach about 10^8 °K before a further increase in the mean electron energy ($>$ 10 keV) no longer leads to brighter microwave emissions. Non-thermal bremsstrahlung bursts so caused, while not of spectacular brightness, need not be confined to gradual bursts, since the bremsstrahlung will follow the rapid fluctuations characteristic of non-thermal electrons. We conclude therefore that impulsive and rapidly fluctuating microwave bursts of not more than $\sim 10^8$ °K brightness temperature could be bremsstrahlung bursts due to non-thermal, fast electrons.

B. GYRO-SYNCHROTRON RADIATION

In general an electron gyrating at an angle ϕ with the magnetic field will be observed to emit harmonics $s = 1, 2, 3 \ldots$ of the Doppler shifted gyro frequency.

For observation at an angle θ to the magnetic field

$$f_H(\beta, \phi, \theta) = f_H(1 - \beta^2)^{\frac{1}{2}}/(1 - \mu\beta \cos \phi \cos \theta) ,$$

where βc is the electron velocity. We denote by

$$f_H(0, \phi, \theta) \equiv f_H = eH/2\pi m_0 c$$

the non-relativistic gyro-frequency.

The absorption co-efficient in the general case has not been evaluated analytically. Numerical solutions have been found by TAKAKURA and KAI[6,7,8] who have interpreted the observed properties of the microwave impulsive burst and of several phases of Type-IV emission in terms of gyro-synchrotron radiation from intermediate-energy electrons ($\varepsilon \approx m_0 c^2$) in magnetic fields of $10 - 10^3$ gauss. The assumed electron energy distribution, though non-thermal, decreased with increasing energy and according to condition (B) could not lead to negative absorption.

The two limiting cases $\varepsilon \lesseqgtr m_0 c^2$ can be treated analytically with the approximations given by Schwinger[9] for these conditions.

Gyro-radiation ($\varepsilon \ll m_0 c^2$)

The emission from such mildly-relativistic electrons is concentrated in the first few harmonics of the gyro frequency. In the case of continuously decreasing $F(\varepsilon)$, as exemplified by the Maxwellian distribution, the absorption coefficient–unlike that of bremsstrahlung–increases with temperature as $T^{s-3/2}$ $(s > 2)$. GINZBURG and ŽELEZNIAKOV[10] noted this property and used it to explain broad continua at the metre-waves as well as the peak emission in the slowly-varying components at the microwaves.

Next we consider the possibility of amplification and allow for $\mathrm{d}F/\mathrm{d}\varepsilon$ to be positive somewhere in the energy range. TWISS[3] argues that, similarly to the discrete set of harmonic frequencies for a given energy, there exist discrete values of energy responsible for emission and absorption at a given frequency. For the latter we obtain[11]

$$\varepsilon_s = m_0 c^2 \{s (f'_H/f) - 1\}$$

where f'_H is the Doppler-shifted value of the non-relativistic f_H. Thus, neglecting Doppler broadening and field variations, $Q_f(\varepsilon)$ consists of lines separated by

$$\varepsilon_s - \varepsilon_{s-1} = m_0 c^2 (f'_H/f) .$$

It follows from Equation (6) and the general conditions discussed in Section 2 that negative absorption and amplification at the sth harmonic are possible if $(\mathrm{d}F/\mathrm{d}\varepsilon)_{\varepsilon_s} > 0$ and $F(\varepsilon)$ does not cover more than one line.

With Doppler and magnetic line broadening, amplification is still possible if the line widths do not exceed their spacing. This restricts[11] such "resonance" amplification to the lower harmonics where

$$s < c/\Delta v \text{ and } s < H/\Delta H .$$

Situations which could lead to amplified gyro radiation are illustrated in Figures 1c and 1d. It is tempting to invoke such radiation to explain Type-I bursts and other bursts superimposed on intense continua.

Synchrotron radiation ($\varepsilon \gg m_0 c^2$)

In this case emission occurs over a large number of high harmonics of the relativistic gyro-frequency; the radiation can be treated as a continuum. BOISCHOT and DENISSE[12] suggested synchrotron radiation from 3 Mev electrons gyrating in a magnetic field of one gauss to account for the moving Type-IV burst at the metre waves.

The emission is highly anisotropic and we have to determine the sign of

$$\left(\frac{\partial}{\partial \varepsilon} + \frac{\Delta \phi}{hf} \frac{\partial}{\partial \phi}\right) g(\varepsilon, \phi) Q_f(\varepsilon, \phi, \theta)$$

in order to find whether amplification is possible. WILD, SMERD and WEISS[11] obtain the gradient as proportional to

$$\int_Z^\infty K_{5/3}(z)\, \mathrm{d}z + 2Z K_{5/3}(Z)$$

where $Z = f/f_c$ and the critical frequency

$$f_c = (3/2)\,(\varepsilon/m_0 c^2)^2 f_H \sin \phi \ ;$$

$K_{5/3}(z)$ is the modified Hankel function. The above gradient is always positive and, contrary to TWISS's[3] result, we conclude that *synchrotron radiation cannot be amplified.*

C. PLASMA WAVES

The third and last known microscopic mechanism which can result in the emission of radio waves from the corona has the added complication that the primary wave is an electron plasma wave; it is only after partial conversion by mode coupling or scattering[10] that observable electromagnetic radiation can escape from the Sun. However, the metre-wave drifting bursts of Types II and III provide strong evidence that in these cases we are dealing with radiation from plasma waves.

These longitudinal waves exist in the region $\mu > 1$ and can therefore be excited analogously to Čerenkov radiation in a neutral medium by the passage through the plasma of electrons at speeds greater than the phase velocity in the medium. Such waves can be set up incoherently by a single electron or a group of electrons which are sufficiently widely spaced to act independently. GINZBURG and ŽELEZNIAKOV[15] have pointed out that this phenomenon cannot happen in streams of fast electrons on the Sun; the smallest number of electrons needed to account for the observed burst intensities is far too large for incoherent emission.

The properties of *coherent* plasma waves have been derived by BOHM and GROSS[13,14]; these authors have introduced the concept of "trapped" electrons. The latter have velocities sufficiently close to the phase velocity of the plasma wave to remain trapped for a while in a potential trough; in this way the electrons can interact strongly with he wave. Those electrons which are slightly faster than the wave on the whole give up energy and the wave grows. The slower electrons lead to damping. If we denote the velocity distribution in the electron stream by $F_s(v)$ the condition for growth[14] is

$$\left(\frac{\partial F_s}{\partial v}\right) > 0 \text{ when } v = v_{ph} .$$

This is equivalent to the condition for wave amplification illustrated in Figure 1c. In this case of a growing plasma wave, as also in all previous cases of wave growth, the final amplitude of the wave–and hence, with the Rayleigh-Jeans radiation law, the effective brightness temperature–has to be obtained by balancing the energy supply to and the energy dissipation from the wave. BOHM and GROSS[14] calculated the resultant amplitude for a coherent plasma wave which dissipates by collision damping. The wave exists mainly near the plasma frequency $f_0 = (e^2 N/\pi m)^{\frac{1}{2}}$ and becomes evanescent at not much more than twice that frequency.

GINZBURG and ŽELEZNIAKOV[15] have used Bohm and Gross's value for the amplified plasma wave and expressions for the efficiency of conversion to electromagnetic radiation to estimate the stream velocities and densities which could account for the observed features of Type-III bursts. Similar calculations by WILD, SMERD and WEISS[11] on the basis of more recent observational results suggest that we can explain a brightness temperature of 10^{11}°K from a $c/4$-stream with $N_s/N \sim 0.02$ and a $c/2$-stream with $N_s/N \sim 0.006$. The extreme instability of such streams is reflected in the high powered functional dependence of the calculated brightness temperature[11] of the fundamental radiation after mode conversion through scattering, viz.

$$T'_b \sim 10^{16} (N_s/N)^8 (v_0/\Delta v)^9 \text{ °K} ,$$

where v_0 is the mean velocity of the stream and Δv its random velocity spread. The similar intensities of the first and second harmonics can also be satisfactorily explained if the conversion is due to scattering of the large plasma waves on thermal fluctuations in the plasma.

A similar interpretation of the slow-drifting type-II burst is possible only if we assume the Type-II source to harbour and at times eject electrons of Type-III velocities. This same assumption was needed to explain[16] the relative positions of the fundamental and harmonic sources in Type-II bursts. On present evidence it seems likely that the slowly-moving Type-II source does indeed carry electrons whose speed is many times the mean thermal velocity of the coronal plasma.

We conclude therefore that both kinds of metre-wave drifting burst are probably electromagnetic radiation emitted when amplified plasma waves scatter on small fluctuations in the coronal plasma.

Acknowledgment

The extension of Twiss's work reported here was done jointly with Dr. J. P. Wild whom the author would also like to thank for this critical reading of the manuscript.

References

1. Bailey, V. A.: 1948, *J. and Proc. Roy. Soc. New South Wales* **82**, 107.
2. Twiss, R. Q.: 1952, *Phys. Rev.* **88**, 1392.
3. Twiss, R. Q.: 1958, *Australian J. Phys.* **11**, 564.
4. Martyn, D. F.: 1948, *Proc. Roy. Soc.* **A193**, 44.
5. Scheuer, P. A. G.: 1960, *Mon. Not. Roy. Astron. Soc.* **120**, 231.
6. Takakura, T.: 1960, *Publ. Astron. Soc. Japan* **12**, 325.
7. Takakura, T.: 1960, *Publ. Astron. Soc. Japan* **12**, 352.
8. Takakura, T. and Kai, K.: 1961, *Publ. Astron. Soc. Japan* **13**, 94.
9. Schwinger, J.: 1949, *Phys. Rev.* **75**, 1912.
10. Ginzburg, V. L. and Železniakov, V. V.: 1961, *Astron. Ž.* **38**, 3; *Soviet Astron.-AJ* **5**, 1.
11. Wild, J. P., Smerd, S. F., and Weiss, A. A.: 1963, *Ann. Rev. Astron. Astrophys.* **1**, 291.
12. Boischot, A. and Denisse, J. F.: 1957, *Comptes Rend.* **245**, 2194.
13. Bohm, D. and Gross, E. P.: 1949, *Phys. Rev.* **75**, 1851.
14. Bohm, D. and Gross, E. P.: 1949, *Phys. Rev.* **75**, 1864.
15. Ginzburg, V. L. and Železniakov, V. V.: 1958, *Astron. Ž.* **35**, 694; *Soviet Astron.-AJ* **2**, 653.
16. Smerd, S. F., Wild, J. P., and Sheridan, K. V.: 1962, *Australian J. Phys.* **15**, 180.

SOME ASPECTS OF TYPE-IV CONTINUUM RADIATION ON DECIMETER WAVELENGTHS*

M. R. KUNDU

Astronomy Department, Cornell University, Ithaca, New York, USA

The flare associated continuum bursts of Type IV in the 500–1 000 Mc/s range in general consist of a smooth continuum with different variabilities superimposed upon it (KUNDU and SPENCER, 1963). These variabilities can take various forms. They are often short-duration fast-drift bursts which appear to be different in nature from the high frequency components of meter-wave bursts of Type III. The intense fast-drift elements which are the most dominant radiation in some dm-λ continuum events are sometimes so blended together as to make the continuum radiation appear to be composed of a succession of fast-drift bursts of fine structure. Even in such events, the smooth background continuum appears clearly whenever the superimposing fast-drift elements are weak or absent (Figure 1a). Occasionally the fast-drift elements occur in the form of "sprays" (Figure 1b). Sometimes drifting patches occur over short

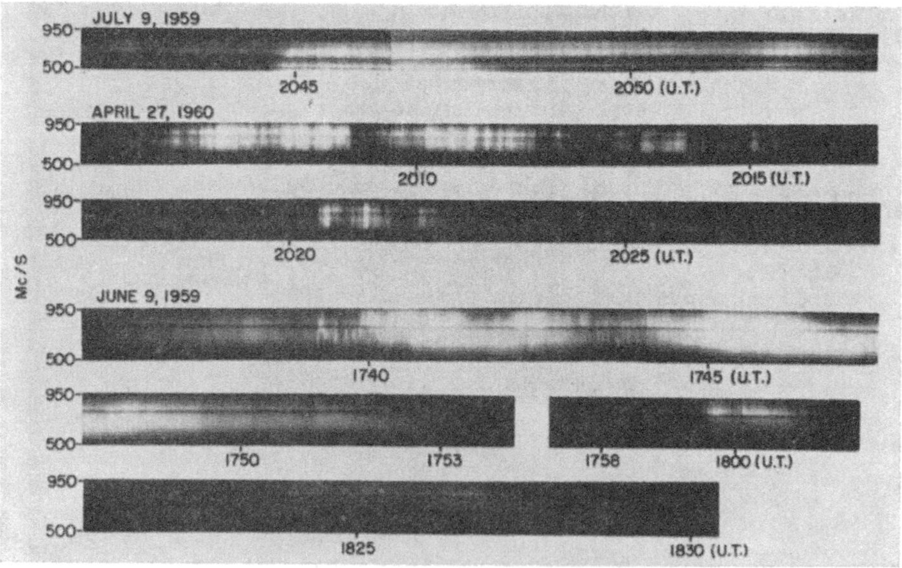

(a)

* The observations described in this note were taken at the Owens Valley Radio Observatory and the Radio Astronomy Observatory of the University of Michigan.

C. de Jager (ed.), The Solar Spectrum. All rights reserved.

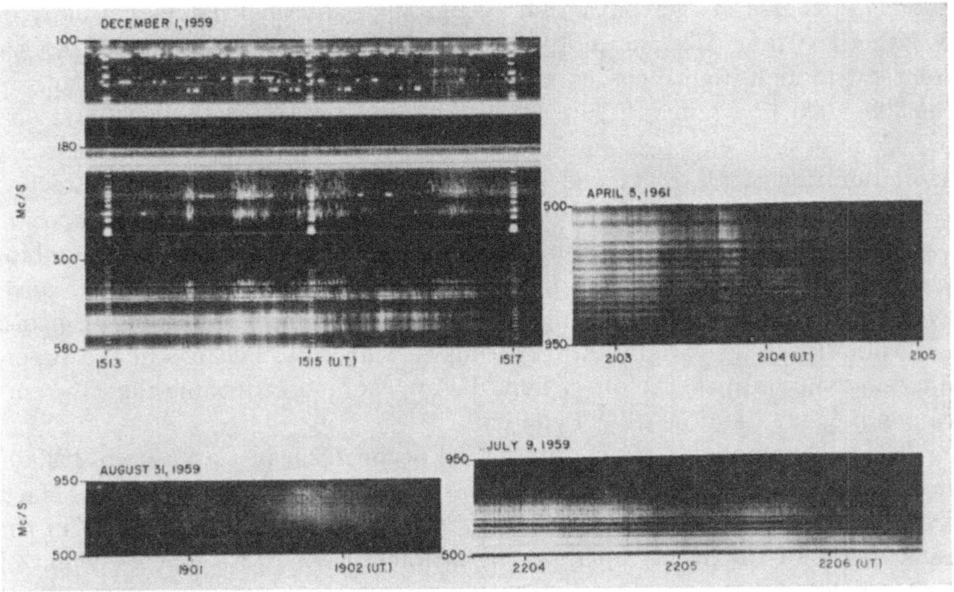

(b)

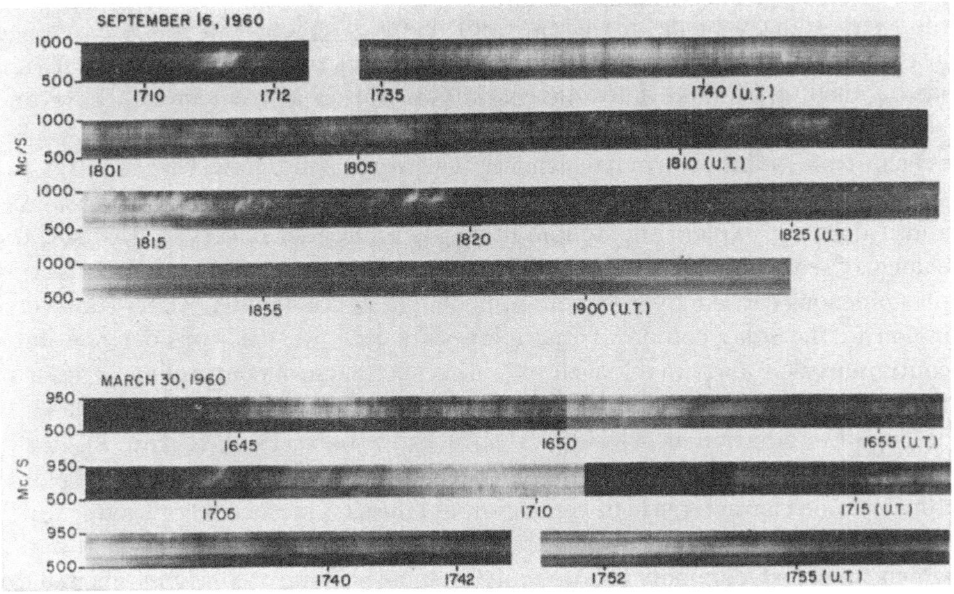

(c)

Fig. 1. Illustrating the variabilities superimposed upon smooth continuum radiation in the 500–1000 Mc/s range. (a) Fine-structure fast-drift bursts, (b) fast-drift bursts in the form of sprays, and (c) blobs of radiation drifting slowly from low to high frequencies. Note that the superimposing variabilities are almost completely absent in the late stage of the events.

periods in course of the continuum event. The individual patches, usually of duration less than 1/2 minute, often drift at the rate of several Mc/s per sec from low to high frequencies over bandwidths of the order of 200–400 Mc/s in the 500–1 000 Mc/s range (Figure 1c). In addition, intermediate-drift bursts are occasionally superimposed upon the continuum (YOUNG *et al.* 1961); these are short duration (< 1 sec at one frequency) bursts drifting from high to low frequencies at rates of about 30 Mc/s per second. Often the superimposing variabilities are so mixed–drifting patches, fine-structure bursts, or a combination of both–as to make the continuum event appear rather complex. However, in the later stages of such events, the continuum becomes very smooth with practically no variabilities superimposed upon it. Because of the superimposing variabilities, the dm-λ Type IV appears more variable than the corresponding cm-λ and meter-λ Type IV.

The background smooth continuum on decimeter waves appears to behave in the same manner as the associated cm-λ continuum. Thus, its time of start, and of maximum coincide with those of the cm-λ continuum, at least in the early stages of the event. Further, the characteristics of its source–height of origin $\not>$ 40000 km above the photosphere, angular size < 4′ arc, no significant motion–appear to be the same in the cm- and dm-λ region. The reversal of the sense of polarization in the dm-λ region–that is, ordinary mode as compared to the extraordinary mode in the cm- and meter-λ region–can be be explained in the same manner as that used for impulsive microwave burst. Thus, TAKAKURA (1960) attributed it to differential absorption at frequencies above and below the gyro-frequency of the extraordinary and ordinary components of the synchrotron radiation from relativistic electrons. This, however, requires too strong a magnetic field (> 1300 gauss for a reversal at 3750 Mc/s) at the source, and it does not explain the double reversals sometimes observed. Instead, the change of sense of polarization can be explained (COHEN, 1961) as a propagation phenomenon, caused by magneto-ionic mode coupling in a quasi-transverse region in the solar corona. Thus, it appears that we can consider the dm-λ continuum as a long wavelength extension of the cm-λ continuum, at least in the early stages of the event and therefore, a part of the first phase (Type IVA) of Type-IV radiation as a whole (KUNDU and SMERD, 1962) (Figure 2).

The different variabilities superimposed on the dm-λ background continuum appear to be characteristic of the region of emission of decimeter waves, which are possibly the least understood at present. The fine-structure fast-drift bursts which are most common are possibly due to plasma oscillations excited by electron streams moving outward or inward through the lowest coronal regions, which will explain their forward and reverse drifts. The reverse drift can also occur if the outward moving streams encounter hills of electron density. The blobs of patchiness drifting slowly from low to high frequencies appear to indicate inward motion of small clouds of gas. They are possibly due to Type-II-like sources exciting plasma oscillations by shock fronts or by turbulent fast

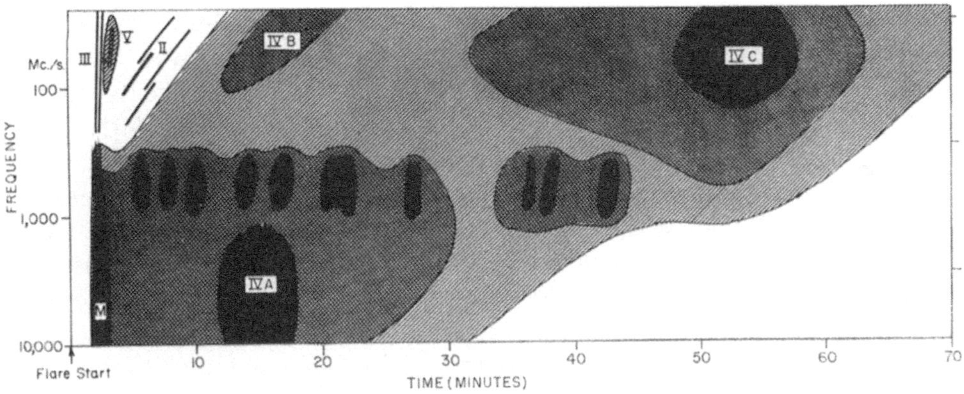

Fig. 2. Composite dynamic spectrum of Type-IV radiation, proposed by KUNDU and SMERD (1962).

electrons in the shock front. The intermediate-drift bursts are much less easy to understand. Assuming that they are due to plasma oscillations excited in the corona by outward moving disturbances and assuming a standard coronal electron density distribution, the inferred speeds of the exciting disturbances are found to lie between approximately 50 and 2000 km sec^{-1}. If the disturbances travel along coronal streamers, the speeds will be about twice these values. These low velocities of the bursts are less than the mean thermal electron velocity of about 7000 km/s in a 10^6 °K corona and hence the possibility of excitation of Čerenkov electron plasma waves can be excluded. The simplicity of these bursts indicates that they are probably not generated in the same manner as Type-II bursts, that is, by turbulent fast electrons in an outward moving shock front. It should be noted, that a small density gradient could result in a considerable decrease in the drift rate for a disturbance of a given velocity. However, the existence of fast-drift bursts on decimeter wavelengths indicates a real difference in velocities between the fast-drift and intermediate-drift bursts.

The fast-drift bursts or other variabilities occur for a considerable period (several tens of minutes) after the start of the associated flare. They are not associated with any visible eruptions or new brightenings of the flare region. On the plasma hypothesis, the continued occurrence of dm-λ fast-drift bursts for a considerable period after the flare start indicates the occurrence of similarly repeated electron jets. These electron jets must be related to some kind of plasma instability, whose effects can be traced through a limited range of plasma levels in the solar atmosphere. Present observations do not permit us to decide whether the repeated instabilities occur in the region of the original explosion or in the region of acceleration where the trapped electrons produce Type-IVA radiation through the synchrotron mechanism. As mentioned earlier, in the late stage of the event, the continuum becomes very smooth with practically no variabilities superimposed upon it, which indicates that the plasma instabilities cease before the synchrotron radiation is over.

The simultaneous presence of different variabilities in some events indicates that the region of emission of dm-waves can be excited very irregularly. Whatever the origin of these different variabilities, it appears that the dm-λ continuum, although a long wavelength extension of the cm-λ continuum, is complicated by other emissions which peculiarly belong to the decimeter-wave region. These peculiar aspects of the decimeter-wave emission may, in part, arise from the fact that the catastrophic explosion of the flare probably occurs in the region of emission of decimeter waves. The streams of electrons which are ejected from this explosion in different directions into the corona and the chromosphere obviously behave in an irregular and complex fashion in the immediate vicinity of the flare explosion. This could account for the irregular senses of drift – forward and reverse – of the fast-drift bursts. High resolution directional observations will certainly contribute considerably towards a better understanding of the fast phenomena in the decimeter-wave region. It appears that an angular resolution of several seconds of arc will be necessary to determine any possible source movement in the 500–1000 Mc/s range. Such a determination seems extremely important to substantiate the plasma hypothesis and, if possible, to determine the structure of the decimeter-wave emissive regions, which appear to be closely related to the seat of the flare explosion.

References

Cohen, M. H.: 1961, *Astrophys. J.* **133**, 978.

Kundu, M. R. and Spencer, C. L.: 1963, *Astrophys. J.* **137**, 572.

Kundu, M. R. and Smerd, S. F.: 1962, *Information Bull. Solar Rad. Obs., Europe* **11**, 4.

Takakura, T.: 1960, *Publ. Astron. Soc. Japan* **12**, 325.

Young, C. W., Spencer, C. L., Moreton, G. E., and Roberts, J. A.: 1961, *Astrophys. J.* **133**, 243.

DISCUSSION ON RADIO AND PARTICLE EMISSION FROM THE SUN

M. MINNAERT
Chairman
Sterrewacht 'Sonnenborgh', Utrecht, The Netherlands

A. FOKKER
Secretary
Sterrewacht 'Sonnenborgh', Utrecht, The Netherlands

L. BIERMANN*

Remarks on the Acceleration Mechanisms

Professor Schatzman fixed our attention to two kinds of acceleration mechanisms, considered by a number of workers in the past, that is the betatron-like acceleration, which was first dealt with by Swann already 30 years ago, and the statistical acceleration proposed by Fermi in 1948. Let me start with a few remarks on the first kind of processes.

The essential idea is to consider the electric field induced by a magnetic field changing with time (say by a displacement of the spots producing the field). If we take quantities which seem appropriate for an active sunspot group, we easily get for the line integral of induced field a value of the order of 10^9 eV or so. To the energy there is, in certain cases, no limit which would depend on the initial energy. A model, discussed by Bagge and myself in 1948 is characterized by an electric field parallel to a neutral line. All this, together with the severe limitations to the applicability of this type of theory, has been discussed in a review paper (*Ann. Rev. of Nucl. Sci.* **2** (1953) 335) ten years ago (for important improvements see: P. A. SWEET, *Stockholm Symposium on Electromagnetic Phenomena in Cosmical Physics*, 1958). A perhaps more appropriate way of looking at the acceleration problem was motivated a few years ago by the observations of the radio-bursts, particularly those of Type III and their interpretation, which Dr. Maxwell just described for us. The starting point is the recognition, that acceleration processes leading to energies of the order of 100 keV or some 100 keV are extremely frequent on the active sun, with frequencies up to 10^{-2} sec^{-1}. Hence we should look for the interpretation of these events by causes not depending on any special circumstances.

Furthermore experience gained in plasma experiments connected with the problem of controlled fusion has shown, that non-thermal emissions and the acceleration of a fraction of the charged particles to highly superthermal velocities are very common in plasmas confined by magnetic fields, and that the ideal conditions ordinarily assumed in magneto-hydrodynamics are more

* Max Planck Institut für Physik und Astrophysik, München, Germany.

C. de Jager (ed.), The Solar Spectrum. All rights reserved.

often violated than fulfilled. New types of instabilities have been discovered, such as the so-called two-stream instability, the universal instabilities depending on temperature and pressure gradients, in addition to the magnetohydrodynamical instabilities known for a somewhat longer time.

I looked a little bit into the special case of a shock wave propagating perpendicular to a magnetic field, such as might be expected to be produced often in an active region. If the shock front becomes thinner, the current density required by the momentum balance must increase, and it seems that current densities leading e.g. to some kind of a two-stream instability or universal instability might well be attained. Under these circumstances non-thermal phenomena should set in of the general kind I described. This suggestion has been qualitatively described some time ago (see the volume dedicated to W. Heisenberg on the occasion of his 60th birthday, published by Friedr. Vieweg & Sohn, Braunschweig, 1961).

Since much more information bearing on this model should soon become available from current experimental and theoretical work on plasmas, I did not make an attempt to fill in the details which have still to be worked out.

Finally, I would like just to draw your attention to some recent work done jointly with R. Lüst.

The point pertaining to the discussions at this conference is the following: in a theory of the magnetic solar cycle proposed by Babcock in 1961, a salient point is that above the solar surface there should continuously be a severing and reconnection of magnetic lines of force. Disregarding its possible cause, we might use the law of induction to estimate the induced electric fields related to the concomitant change of magnetic flux through the solar plasma. Using Babcock's figures, it may be seen that these fields are quite large. Since this is more a subject for the forthcoming Tegernsee Conference, I will content myself to say that the details may be found in a note to be published as a contribution to the volume dedicated to S. Roseland on the occasion of his 70th birthday.

L. BIERMANN. The mean free path given by Schatzman for $T \simeq 10^6$, $n \simeq 10^8$ is probably on the high side. The correct figure (see e.g. SPITZER's *Physics of Fully Ionized Gases*) should be $\simeq 10^8$ cm instead of $> 10^{10}$ cm.

R. MICHARD. You have described a fine theory of the acceleration of particles by a shock wave. What are your ideas on the origin of these shocks?

E. SCHATZMAN. The preferential region for acceleration is where the magnetic lines of force are parallel to the shock front, which is likely to be parallel to the solar surface.

If we consider a filament hanging in the magnetic field, the filament is bordered by two regions where the magnetic field is parallel to the solar surface; such regions are favorably placed for acceleration.

C. DE JAGER. (1) Prof. Schatzman referred to an earlier paper by me, saying that I had predicted that one beam of electrons, accelerated during a flare, should produce X-ray emission and Type-III emission *at the same time*. However, what I predicted was that this should be done by *different* electron beams (*Space Research* **1** (1960) 628; *Liège Colloquium* **20** (1962) 635).

The *observation* was that Type-III jets ($v \simeq c/2$) and solar X-ray bursts (~ 0.1 Å) could both be produced by equally fast electron streams, accelerated in a field of 10^5 eV. At that time I still thought that flares were rather irregular structures, and I assumed that it might easily occur that part of a flare would accelerate electrons *downward* (these, then, would be braked in the photosphere and produce X-ray bremsstrahlung near 0.1 Å), whereas other parts would accelerate electrons *upward*. The latter electrons have a very long mean free path and would therefore shoot through the corona and outward, thereby exciting the plasma oscillations observed as Type-III bursts.

Subsequently Kundu showed that the correlation between X-ray bursts and Type-III bursts is a rather loose one ($\simeq 10\%$), so that we have to change our picture: a flare might be a more regular structure than I thought before, sometimes accelerating electrons upward, sometimes downward but seldom down- and upward at the same time.

(2) An essential problem, to be solved by the observers, is that of the fine-structure in time of the non-thermal X-ray bursts. If these bursts would prove to consist of shorter bursts with a duration of–say–one second, this would perhaps indicate that these bursts are due to accelerated electron jets similar to those producing Type-III radio bursts. Could X-ray bursts of $\leqslant 1$ sec duration still be reconciled with the assumption of trapping of fast electrons?

E. SCHATZMAN. If the emission of X rays by trapped electrons takes 1 second, it means that the density of the emitting region is of the order of 10^{10} cm^{-3}.

A. UNSÖLD. We can observe the acceleration of large amounts of matter to several keV going on in eruptive prominences, while optically almost nothing particular is seen. How can this observation be explained with the presented theoretical considerations?

E. SCHATZMAN. The present considerations do not apply to the phenomena observed in an eruptive prominence.

R. MICHARD. In the process for conversion of plasma waves into electromagnetic radiation by scattering on thermal fluctuations, are there any special requirements on the geometrical size and amplitude of such fluctuations?

S. F. SMERD. If plasma waves are converted to electromagnetic radiation through mode coupling, then the inhomogeneity (in N, T or H) must be large enough for the dispersion curves of the two modes to touch. If conversion is due to scattering on thermal fluctuations, it suffices for the plasma to be hot.

We invoke scattering on thermal fluctuations to account for the escape of electromagnetic radiation at the plasma frequency from anywhere on the sun's disc, as observed.

R. MICHARD. I have doubts on the figure of 50000 km (later corrected to 10^4)

for the height of the initial explosion giving rise to Type-II and -III bursts. In all accelerating mechanisms we need strong magnetic fields to make them efficient, and I expect these strong fields are more likely to be present at low heights.

S. F. SMERD. The radio evidence of outward and inward-moving disturbances, as well as the extrapolation of Type-II and Type-III bursts to a common origin, suggest that the flare explosion occurs at a height of the order of 10^4 km.

A. UNSÖLD. The height of the "flash explosion" in a flare might be obtained from observations of the metallic Fraunhofer lines blended with he strong Hα emission wings.

H. ZIRIN. The 1960 Nov. 20 flare occurred above a sunspot beyond the limb. It had large decimeter and X-ray bursts; these must have originated some 15000 km above the solar surface.

G. MORETON. There have been comments indicating that a two-minute time interval is too long for particle acceleration (by the Fermi mechanism as applied by Parker), leading to subsequent production of Bremsstrahlung X rays. Yet, as I mentioned earlier, there certainly seems to exist observational evidence for such a lag.

E. SCHATZMAN. (1) The puzzling problem is the increase in density by which the flare becomes visible. I would suggest that the increase in density takes place first; only then the acceleration process becomes possible and this takes place in a few seconds only.

(2) The emission on centimeter wavelengths can be explained by the same number of electrons that is required to produce the X rays.

H. E. HINTEREGGER. Can perhaps the 2-minute delay in the appearance of energetic X rays after flash explosion have something to do with an instrumental effect – meaning some change in spectral intensity distribution more or less accidentally bringing the X-ray burst into the spectral detectivity range?

A. MAXWELL. Parker (1957) calculates the period required to accelerate ions from thermal to relativistic energies in a flare region, by the Fermi mechanism, as about 2 minutes. This time agrees fairly well with the observational lag of approximately 2–3 minutes between the flash phase of a flare and the emission of X rays and microwave radio bursts (for example, the flare of 1961 September 28, commencing 22.02 UT).

M. R. KUNDU. I think that cm- and dm-waves have been confused a little here. It is only the cm-λ bursts which are closely associated with high-energy ($>$ 20 keV) X-ray bursts. Only when the cm-λ burst intensity is very high, do we observe occasionally dm-λ bursts in association with $>$ 20 keV X-ray bursts. So the acceleration process should occur near the levels of cm-λ burst emission ($\sim$ 10000 km). Similarly, from the very high correlation of solar cosmic rays with cm-λ Type-IV bursts we conclude that the protons are accelerated in about the same region.

H. ZIRIN. Lindsay and White have measured with the OSO the X-rays in the 1–10 Å range with very sensitive apparatus. They find that even the smallest

2800 Mc/s events give increases in the soft X-ray flux at the same time.

E. SCHATZMAN. Is it possible that the first cosmic rays are produced at the same time as the electrons producing X rays and Type-III bursts?

What is their path from the sun to the earth?

M. R. KUNDU. From the time delay of solar protons observed at the earth with respect to cm-λ or X-ray bursts, as well as the very close association between solar cosmic rays and intense cm-λ bursts (Type IVA), it would seem quite possible that the acceleration of fast particles starts at the time of the initial intense cm-λ burst of Type-IV or the high-energy X-ray burst.

Summary of main points

At the end of the session the main points of the discussions were summarized by professor Minnaert as follows:

Ejection from flare:	Electrons, protons, shock waves.
Origin of instability:	In the photosphere?
Acceleration mechanism:	Swan mechanism: Parker, Biermann.
Fermi mechanism:	Contracting magnetic bottle Schatzman (shock front starting in the chromosphere, motion $\perp$ magnetic field, through clumps). Micro-instabilities (Biermann, Kiepenheuer).
Flash explosion:	At a height of about 10000 km?
X-rays:	Due to Bremsstrahlung.
Distribution of ejected particles:	Electrons, protons, with magnetic fields, partly kept, partly ejected.
Type III:	Produced by plasma oscillations; coherent effect of many electrons.
Type II:	Plasma shock-waves, magnetohydrodynamic waves.
Plasma oscillations $\rightarrow$ radio waves:	Small scale fluctuations, scattering of plasma waves.
Amplification of radio waves:	Fluctuations in N_e, velocity near phase velocity, thermal fluctuations.
Scattering of radio waves:	Core, surrounded by halo of scattered radiation.
Type IV:	(a) synchrotron radiation: moving type–meter waves; stable type–cm/dm waves. Fine-structure in dm waves. (b) Plasma radiation (stable type–meter waves).

GPSR Compliance

The European Union's (EU) General Product Safety Regulation (GPSR) is a set of rules that requires consumer products to be safe and our obligations to ensure this.

If you have any concerns about our products, you can contact us on ProductSafety@springernature.com

In case Publisher is established outside the EU, the EU authorized representative is:

Springer Nature Customer Service Center GmbH
Europaplatz 3
69115 Heidelberg, Germany

Batch number: 10824476

Printed by Printforce, the Netherlands